Methods in Molecular Biology

For further volumes:
http://www.springer.com/series/7651

For over 35 years, biological scientists have come to rely on the research protocols and methodologies in the critically acclaimed *Methods in Molecular Biology* series. The series was the first to introduce the step-by-step protocols approach that has become the standard in all biomedical protocol publishing. Each protocol is provided in readily-reproducible step-by-step fashion, opening with an introductory overview, a list of the materials and reagents needed to complete the experiment, and followed by a detailed procedure that is supported with a helpful notes section offering tips and tricks of the trade as well as troubleshooting advice. These hallmark features were introduced by series editor Dr. John Walker and constitute the key ingredient in each and every volume of the *Methods in Molecular Biology* series. Tested and trusted, comprehensive and reliable, all protocols from the series are indexed in PubMed.

Cytoskeleton

Methods and Protocols

Fourth Edition

Edited by

Ray H. Gavin

Professor Emeritus, Brooklyn College – CUNY, Brooklyn, NY, USA

Humana Press

Editor
Ray H. Gavin
Professor Emeritus
Brooklyn College – CUNY
Brooklyn, NY, USA

ISSN 1064-3745 ISSN 1940-6029 (electronic)
ISBN 978-1-0716-1660-4 ISBN 978-1-0716-1661-1 (eBook)
https://doi.org/10.1007/978-1-0716-1661-1

This Humana imprint is published by the registered company Springer Science+Business Media, LLC, part of Springer Nature.
The registered company address is: 1 New York Plaza, New York, NY 10004, U.S.A.

Preface

In 1998, I received an email from John Walker, series editor of *Methods in Molecular Biology.* Professor Walker's email described plans for a new volume on the cytoskeleton, and he invited me to organize and edit the new book. I thought that I was an unlikely candidate for the role of editor, but I considered that the opportunity to engage in this type of scholarship might never be repeated. My aim for the new book was to develop a volume that broadly interpreted the concept of the cytoskeleton and emphasized diverse experimental protocols and models. The preface to the first edition of *Cytoskeleton: Methods and Protocols* (CMP) pointed out that: "Rather than a network of disparate fibers, [cytoskeletal] polymers are often interconnected and display synergy..." The first edition's book cover, an image of a cytoskeletal network of microtubules and actin filaments, reflected the emphasis on the interconnectedness of cytoskeletal fibers. In the years since the publication of the first edition, technological advances in microscopy and in vitro studies continue to advance our understanding of the cytoskeleton, and CMP has focused on many of these advances in numerous chapters.

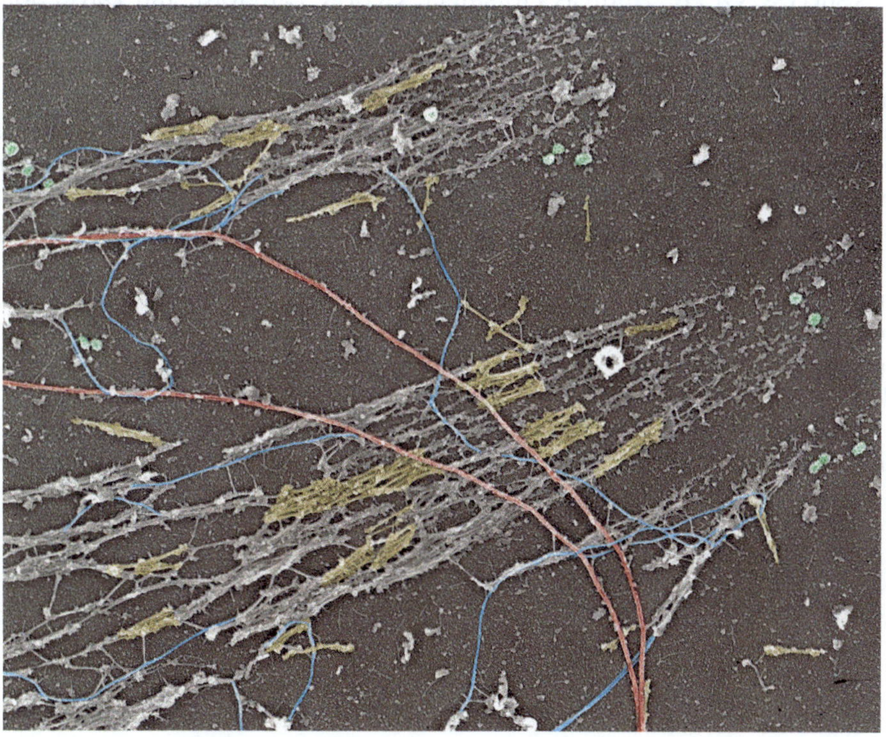

Fig. 1 PREM of the cytoskeleton of a cultured cell after depletion of actin filaments by gelsolin treatment. The image shows focal adhesions, myosin II filaments (yellow), microtubules (red), intermediate filaments (blue), and clathrin-coated vesicles (green). Additional details are in Chapter 2.

As in previous editions, the cover image for the fourth edition of CMP shows an array of cytoskeletal fibers (Fig. 1). The organization of the fourth edition follows the pattern established in the earlier editions, and each chapter adheres to the easy-to-follow instructional format and includes a helpful Notes section, undoubtedly an aid to troubleshooting. The fourth edition continues the emphasis on diverse model systems and experimental approaches for studies of the cytoskeleton. Experimental models include animal and plant cultured cells, flatworms, roundworms, yeast, plants, ciliates, flagellates, and fish. The fourth edition maintains a clear focus on imaging techniques in chapters devoted to platinum replica electron microscopy (PREM), total internal reflection fluorescence (TIRF)-structured illumination microscopy (SIM), and in vitro and in vivo reagents for enhancing visualization of actin filaments. Other chapters feature analyses of cell and organelle motility, dynamics of cytoskeleton components, and genetic and in silico analysis of cytoskeleton proteins.

I have edited CMP for 20 years with four editions. I am hopeful that Springer Press will continue publication of this volume, and I look forward to reading the fifth edition with a new editor at the helm. I have enjoyed my time as an editor; I have had a good run. Editing CMP has been one of the highlights of my career. Some 200 colleagues from 12 countries have lent their expertise to contribute chapters, and I thank them. They are responsible for the book's success. Qi He, Andreas Holzinger, George Langford, and David Soll contributed to all four editions of CMP. And of course, I thank John Walker for inviting me to organize this volume on the cytoskeleton and for continually instructing me in the stylistic requirements of Humana Press and later Springer Press.

I am often mindful of The Gambler, an American song written by Don Schlitz and popularized by Kenny Rogers in the 1970s. It has been reported that Schlitz's inspiration for the song derived from advice (apparently unrelated to gambling) given to him by his father. Some of the lyrics offer an admonition to card players—and anyone contemplating a decision.

"You've got to know when to hold 'em
Know when to fold 'em,
Know when to walk away
And know when to run"

Brooklyn, NY, USA ***Ray H. Gavin***

Contents

Contributors

DEREK A. APPLEWHITE • *Department of Biology, Reed College, Portland, OR, USA*

CARLOS BARRETO • *Department of Biology, Brooklyn College of the City University of New York, Brooklyn, NY, USA*

NICOLAS BIAIS • *Department of Biology, Brooklyn College of the City University of New York, Brooklyn, NY, USA; Graduate Center of the City University of New York, New York, NY, USA; Laboratoire Jean Perrin, UMR 8237 Sorbonne Université/CNRS, 4 Place Jussieu, Paris, France*

ELISON B. BLANCAFLOR • *Noble Research Institute LLC, Ardmore, OK, USA*

ISTVAN R. BOLDOGH • *Department of Pathology and Cell Biology, Columbia University, College of Physicians and Surgeons, New York, NY, USA*

DERRICK BRAZILL • *Department of Biological Sciences, Hunter College, New York, NY, USA; The Graduate Center, City University of New York, New York, NY, USA*

CHENGLIN CHAI • *Noble Research Institute LLC, Ardmore, OK, USA*

SABRINA CHIN • *Noble Research Institute LLC, Ardmore, OK, USA*

TRACY CHONG • *Howard Hughes Medical Institute, University of Wisconsin-Madison, Madison, WI, USA; Regenerative Biology, Morgridge Institute for Research, Madison, WI, USA*

GILBERTO GARCIA • *Department of Molecular & Cellular Biology, Howard Hughes Medical Institute, The Glenn Center for Aging Research, University of California, Berkeley, Berkeley, CA, USA*

MICHAEL GOTESMAN • *Department of Biology, New York City College of Technology of the City University of New York, Brooklyn, NY, USA*

ERIC R. GRIFFIS • *Nikon Imaging Center, University of California San Diego, San Diego County, CA, USA*

QI HE • *Department of Biology, Brooklyn College of the City University of New York, Brooklyn, NY, USA*

RYO HIGUCHI-SANABRIA • *Department of Molecular & Cellular Biology, Howard Hughes Medical Institute, The Glenn Center for Aging Research, University of California, Berkeley, Berkeley, CA, USA*

ANDREAS HOLZINGER • *Functional Plant Biology, Institute of Botany, University of Innsbruck, Innsbruck, Austria*

STEFAN HOMENTCOVSCHI • *Department of Molecular & Cellular Biology, Howard Hughes Medical Institute, The Glenn Center for Aging Research, University of California, Berkeley, Berkeley, CA, USA*

ROLAND E. HOSEIN • *Department of Biology, Brooklyn College of The City University of New York, Brooklyn, NY, USA*

LINDA A. HUFNAGEL • *Department of Cell and Molecular Biology, University of Rhode Island, Kingston, RI, USA*

NAAME KELET • *Department of Molecular & Cellular Biology, Howard Hughes Medical Institute, The Glenn Center for Aging Research, University of California, Berkeley, Berkeley, CA, USA*

CHRISTINA KING-SMITH • *Department of Biology, Saint Joseph's University, Philadelphia, PA, USA*

DAVID A. KNECHT • *Department of Molecular and Cell Biology, University of Connecticut, Storrs, CT, USA*
CHRISTINE A. LACY • *Department of Biology, University of Richmond, Richmond, VA, USA*
GEORGE M. LANGFORD • *Biology Department, Syracuse University, Syracuse, NY, USA*
ANTONIO LOPEZ • *Department of Biology, Brooklyn College of the City University of New York, Brooklyn, NY, USA*
DANIEL F. LUSCHE • *Developmental Studies Hybridoma Bank and W.M. Keck Dynamic Image Analysis Facility, Department of Biology, The University of Iowa, Iowa City, IA, USA*
SHARATH MADASU • *Protein Division, Ibex Biosciences, LLC, Cumberland, MD, USA*
CARTER A. MITCHELL • *Protein Purification, Kemp Proteins, Frederick, MD, USA*
EMILIA LAURA MUNTEANU • *Department of Biology, Brooklyn College of the City University of New York, Brooklyn, NY, USA; Columbia University Irving medical Center, New York, NY, USA*
PHILLIP A. NEWMARK • *Regenerative Biology, Morgridge Institute for Research, Madison, WI, USA; Howard Hughes Medical Institute, University of Wisconsin-Madison, Madison, WI, USA; Department of Integrative Biology, University of Wisconsin-Madison, Madison, WI, USA*
SHOICHIRO ONO • *Department of Pathology, and Winship Cancer Institute, Emory University School of Medicine, Atlanta, GA, USA; Department of Cell Biology, and Winship Cancer Institute, Emory University School of Medicine, Atlanta, GA, USA*
LIZA A. PON • *Confocal and Specialized Microscopy Shared Resource, Herbert Irving Comprehensive Cancer Center, Columbia University, New York, NY, USA; Department of Pathology and Cell Biology, Columbia University, New York, NY, USA*
OMAR A. QUINTERO • *Department of Biology, University of Richmond, Richmond, VA, USA*
CHRISTOPHER ROBLODOWSKI • *Department of Biological Sciences and Geology, Queensborough Community College of the City University of New York, Bayside, NY, USA*
LABIB ROUHANA • *Department of Biological Sciences, Wright State University, Dayton, OH, USA*
AVDAR SAN • *Department of Biology, Brooklyn College of the City University of New York, Brooklyn, NY, USA*
LUIS CARLOS SANTOS • *Department of Biology, Brooklyn College of the City University of New York, Brooklyn, NY, USA; Icahn School of Medicine at Mount Sinai, New York, NY, USA*
ANDRIELE SILVA • *Department of Biology, Brooklyn College of the City University of New York, Brooklyn, NY, USA*
CIERRA N. SING • *Department of Pathology and Cell Biology, Columbia University, College of Physicians and Surgeons, New York, NY, USA*
SHANEEN SINGH • *Department of Biology, Brooklyn College of the City University of New York, Brooklyn, NY, USA*
DAVID R. SOLL • *Developmental Studies Hybridoma Bank and W.M. Keck Dynamic Image Analysis Facility, Department of Biology, The University of Iowa, Iowa City, IA, USA*
TATYANA SVITKINA • *Department of Biology, University of Pennsylvania, Philadelphia, PA, USA*
THERESA C. SWAYNE • *Confocal and Specialized Microscopy Shared Resource, Herbert Irving Comprehensive Cancer Center, Columbia University, New York, NY, USA*
CATHERINE A. TSANG • *Department of Pathology and Cell Biology, Columbia University, College of Physicians and Surgeons, New York, NY, USA*

EDWARD VOSS • *Developmental Studies Hybridoma Bank and W.M. Keck Dynamic Image Analysis Facility, Department of Biology, The University of Iowa, Iowa City, IA, USA*
YUEYANG WANG • *Genetics Division, Ibex Biosciences, LLC, Rockville, MD, USA*
DEBORAH WESSELS • *Developmental Studies Hybridoma Bank and W.M. Keck Dynamic Image Analysis Facility, Department of Biology, The University of Iowa, Iowa City, IA, USA*
ELIZA WIECH • *Department of Biology, Brooklyn College of the City University of New York, Brooklyn, NY, USA*
SELWYN A. WILLIAMS • *Department of Biology, Miami Dade College, Miami, FL, USA*
TORSTEN WÖLLERT • *Biology Department, Syracuse University, Syracuse, NY, USA*
EMILY J. YANG • *Department of Pathology and Cell Biology, Columbia University, College of Physicians and Surgeons, New York, NY, USA*

Part I

Imaging the Cytoskeleton

Chapter 1

Super-Resolution Imaging of the Actin Cytoskeleton in Living Cells Using TIRF-SIM

Torsten Wöllert and George M. Langford

Abstract

Super-resolution (SR) imaging techniques have advanced rapidly in recent years, but only a subset of these techniques is gentle enough to be used by cell biologists to study living cells with minimal photodamage. Our research is focused on studies of the dynamic remodeling of the actin cytoskeleton in living pancreatic beta cells during insulin secretion. These studies require super-resolution light microscopic techniques that are gentle enough to record rapid changes of the actin cytoskeleton in real time. In this chapter, we describe an SR technique that breaks the diffraction limit of the conventional light microscope called TIRF-SIM. Using this SR techniques, we have been able to show that (1) microvilli on pancreatic beta cells translocate in the plane of the plasma membrane and (2) the cortical actin network reorganizes when cells are stimulated to secrete insulin. We describe the FIJI plugins that were used to process and analyze the TIRF-SIM images to obtain quantitative data.

Key words Pancreatic beta cells, Cortical actin cytoskeleton, Microvilli, Live cell imaging, Super-resolution microscopy, TIRF-SIM, CLSM, FIJI, Image processing and analysis, Airyscan

1 Introduction

1.1 Super-Resolution Imaging of the Actin Cytoskeleton

Recent advances in imaging technology have enhanced the resolution of fluorescence microscopes by incorporating the newest developments in optical engineering, labeling technologies, AI, and computational capacity [1]. Fluorescence microscopy is now an indispensable tool in the life sciences. With the development of advanced fluorescence imaging, it is possible to acquire high-resolution images of the substructure of cells as well as detailed maps of the macromolecular assemblies within living cells and how these complexes respond to changing physiological conditions [2].

In our research, we use a super-resolution (SR) imaging technique called TIRF-SIM (total internal reflection fluorescence (TIRF)-structured illumination microscopy (SIM)) (Fig. 1) to study the dynamic properties of the actin cytoskeleton in living

Ray H. Gavin (ed.), *Cytoskeleton: Methods and Protocols*, Methods in Molecular Biology, vol. 2364,
https://doi.org/10.1007/978-1-0716-1661-1_1,

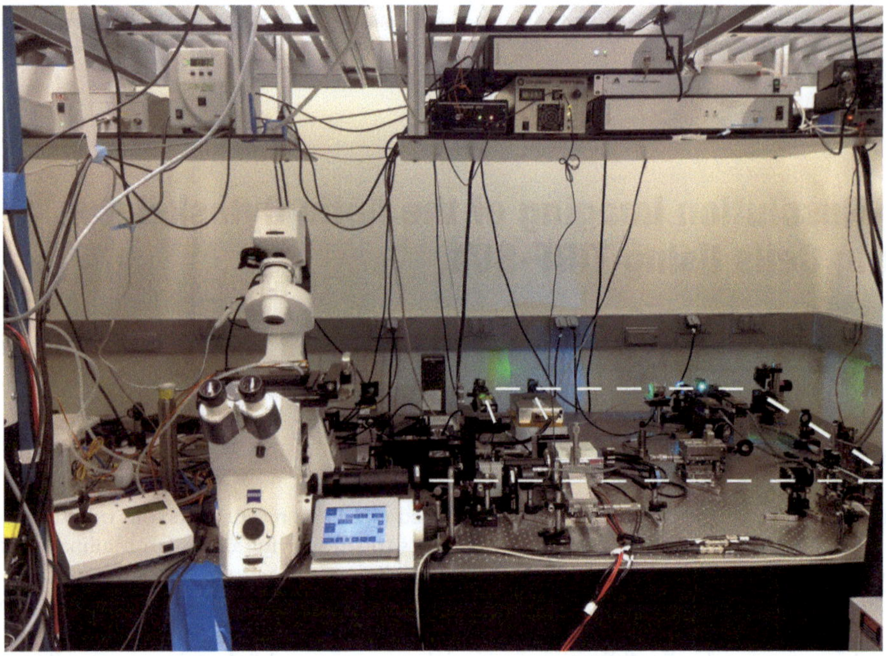

Fig. 1 TIRF-SIM imaging system at the Advanced Imaging Center (AIC), Janelia Research Campus, Howard Hughes Medical Institute, Ashburn, VA. This system was designed and built by Matts Gustafsson and colleagues [16] and was available to users of the AIC until August 2018. White dashed line illustrates the light path

pancreatic beta cells (Fig. 2). The actin cytoskeleton is one of the most important and structurally complex components of living cells. It is involved in many critically important functions including maintenance of cell shape; cell division; cell migration; cargo transport; formation of membrane protrusions such as lamellipodia, filopodia, and microvilli; and secretion. In all of these functions, the actin cytoskeleton exists in a state of constant flux, capable of rapidly remodeling upon changing physiological conditions.

The actin filaments comprising the actin cytoskeleton nucleate at specific sites; grow, shrink, and slide relative to each other; cross-link into small bundles; and form networks of differing porosity. Furthermore, the curvature and stiffness of actin filament bundles change under different physiological conditions. Specific actin-binding proteins interact with actin filaments to confer these different dynamic states [3, 4]. The ability of the actin filaments to assume these different molecular configurations allows them to serve as *tracks* for molecular motors carrying cargo in the cell, as *force generators* producing pushing forces by subunit addition at growing tips, as force generators due to relative sliding produced by molecular motors, as *tethering sites* for membrane bound organelles, or as *physical barriers* between secretory vesicles and the plasma membrane [3, 4].

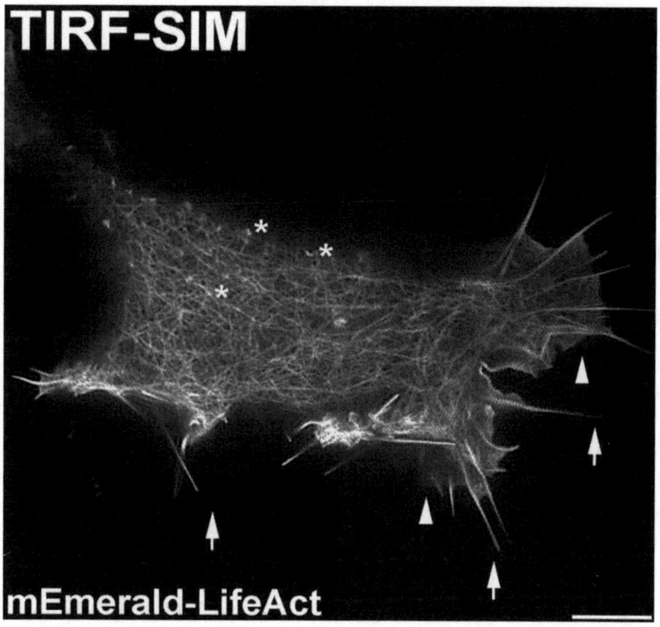

Fig. 2 Pancreatic beta cell (INS-1 832/13) imaged with super-resolution TIRF-SIM microscopy. TIRF-SIM image shows the cortical actin network in a living INS-1 cell transiently transfected with mEmerald-LifeAct and grown for 24 h. Super-resolution TIRF-SIM image shows that the cortical actin network is a dense, meshwork of actin bundles. Arrows and arrow heads indicate filopodia and lamellipodia, respectively. Asterisks indicate the actin-containing microvilli on the cortical actin network. Scale bar: 5μm

The individual actin filaments that form the actin cytoskeleton are 6 nm in diameter and therefore well below the resolution of the conventional light microscope, which is approximately 200 nm. Individual actin filaments align to form bundles of ten or more filaments, but the diameter of these bundles is also smaller than the diffraction limit of the light microscope. Therefore, to image actin filament bundles and to record the rapid changes resulting from the dynamic nature of these bundles, we require a super-resolution imaging technique that allows us to acquire images at framing rates close to real time in living cells.

In this chapter, we describe a custom-built TIRF-SIM microscope (Fig. 1), a form of SR imaging that breaks the diffraction limit of the conventional light microscope and is gentle enough to meet our requirement for acquisition of rapid time-domain information. TIRF-SIM improves spatial resolution by twofold, allowing us to detect objects that are smaller than the theoretical diffraction limit. Localization techniques such as PALM and STORM [1] that have resolution in the 15–20 nm range work best with fixed samples and therefore not suitable for our studies (Table 1).

Table 1
Super-resolution techniques for live cell imaging

Methods	Spatial resolution	Temporal resolution	Imaging of fast cytoskeletal dynamics	Manufacturer
SIM	100 nm	ms	✓	Zeiss, Nikon, Deltavision
STED	30 nm	>s	x	Leica, PicoQuant, Abberior Instruments
PALM	15–20 nm	>s	x	Zeiss, Nikon, abbelight instruments, ONI
spt-PALM	15–20 nm	ms	x	Nikon, abbelight instruments
STORM	15–20 nm	>s	x	Zeiss, Nikon, Leica, abbelight instruments, ONI
MINFLUX	<10 nm	>s	x	Abberior Instruments

In addition to TIRF-SIM, we describe confocal laser scanning microscopy (CLSM) with Airyscan technology [5], another imaging technique gentle enough for live cell imaging. Each technique has advantages and disadvantages, often requiring the use of both techniques to capture the full range of rapid structural changes exhibited by the actin cytoskeleton. These techniques are gentle enough to allow short-term recordings without significant bleaching of fluorophores and photodamage to living cells. TIRF-SIM microscopy is the best method to use for live cell imaging although CLSM with Airyscan is an excellent alternative (Fig. 3). The resolution achieved with Airyscan, however, is less than twofold, and therefore it is not defined as a super-resolution technique. It is referred to as an enhanced resolution technique.

Using these two imaging techniques, each with a high signal-to-noise ratio and fast acquisition speed, we determined that (1) microvilli on pancreatic beta cells translocate in the plan of the plasma membrane due to filament assembly/disassembly and (2) the cortical actin filament network remains intact but changes porosity and stiffness when beta cells are stimulated to secrete insulin. FIJI, an open-source software platform, was used to process and analyze the images to obtain quantitative data. Motion analysis of microvilli motility was performed using the plugin TrackMate [6], and curvature and stiffness of the filaments of the cortical actin cytoskeleton were determined using the software plugin Kappa [7].

1.2 The Actin Cytoskeleton and Glucose-Stimulated Insulin Secretion (GSIS) in Pancreatic Beta Cells

We used rat insulinoma cell line (INS-1 832/13) [8] grown in culture to study the role of the actin cytoskeleton in glucose-stimulated insulin secretion (GSIS). GSIS is the process by which insulin is released from dense-core granules in pancreatic beta cells when the level of blood glucose increases, such as after a meal. The rapid uptake of glucose by pancreatic beta cells is followed by the

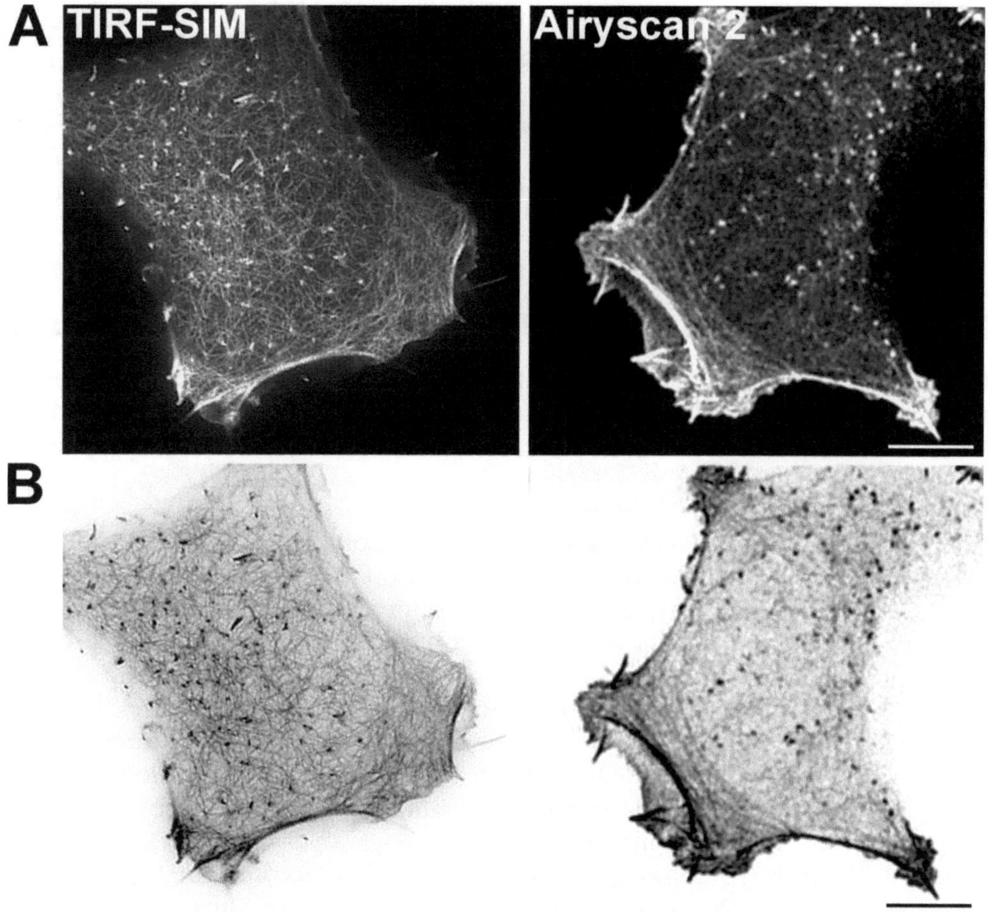

Fig. 3 Side-by-side comparison of the difference in spatial resolution between TIRF-SIM microscopy (~100 nm) and confocal laser scanning microscopy (CLSM) with Airyscan 2 technology (~120 nm). (**a**) Pancreatic beta cells were transiently transfected with mEmerald-LifeAct and grown for 24 h. The cortical actin network was imaged with TIRF-SIM (left panel) or the Zeiss LSM 980 with Airyscan 2. The TIRF-SIM image shows finer detail of the filaments comprising the cortical actin cytoskeleton. (**b**) The contrast of both images was inverted to show individual actin bundles more clearly. Scale bar: 5 μm

metabolic breakdown of glucose to produce ATP, which binds to K_{ATP} ion channels causing the channels to close. As a result, beta cells depolarize, which allows Ca^{2+} channels to open. The entry of Ca^{2+} activates a signaling cascade that triggers exocytosis of insulin-containing dense core granules [9].

Pancreatic beta cells are structurally complex with a dense cortical actin filament network and prominent microvilli projecting from their surfaces [10]. These actin-rich finger-like projections (Fig. 4) play an essential role in glucose sensing, the first step in glucose-stimulated insulin secretion. Microvilli, therefore, are key structural components in glucose uptake, while the cortical actin filament network plays a major role in glucose-stimulated insulin secretion.

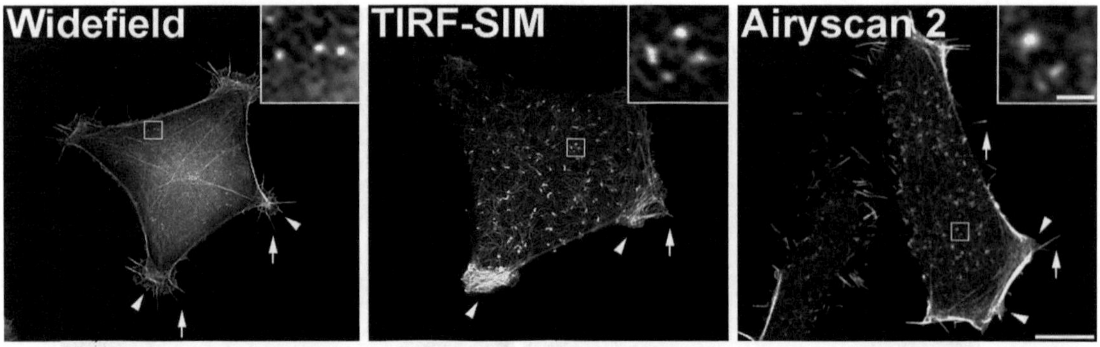

Fig. 4 Microvilli on INS-1 cells imaged using widefield, TIRF-SIM microscopy, and confocal laser scanning microscopy with Airyscan 2 technology. Microvilli (insets) are detectable by all three imaging modalities, but the actin filament network comprising the cortical actin cytoskeleton is not resolvable with widefield microscopy. Scale bar: 5μm

Current models of the role of the actin cytoskeleton in pancreatic beta cells focus primarily on the cortical actin filament network and its potential role in insulin secretion [9]. These models view the cortical actin cytoskeleton as a physical barrier that prevents the dense core granules from gaining access to docking sites on the plasma membrane, thereby preventing fusion and release of insulin. These models propose that insulin secretion requires depolymerization of the cortical actin filaments, which allows the insulin-containing dense core granules to dock and fuse with the plasma membrane for exocytosis.

These models do not take into account the key role of microvilli, which serve as the primary sites of glucose sensing. Glucose uptake in rat pancreatic beta cells occurs via GLUT2, the liver-type facilitated glucose transporter. GLUT2 transporters are located primarily on microvilli where the density of GLUT2 is six times higher than on flat regions of the plasma membrane [11]. Therefore, microvilli are the principal sites of glucose sensing in these cells. We focused our studies on both the role of the actin cytoskeleton in microvilli during glucose sensing and the role of the cortical actin filament network in GSIS.

Our ability to image the actin cytoskeleton in living cells allowed us to discover that microvilli translocate in the plane of the membrane (Fig. 5). Microvilli motility was dependent on the dynamic properties of the actin filaments. One important question to answer is why microvilli translocate. One obvious answer is to provide a search strategy for ligands or binding partners that exist at low concentrations [12]. To definitively demonstrate that microvilli motility represents an efficient search strategy for ligands, we need imaging data acquired in real time in living cells at sufficient spatial and temporal resolution and with multicolor fluorescent markers of membrane and cytoskeletal components to record the membrane and cytoskeletal dynamics.

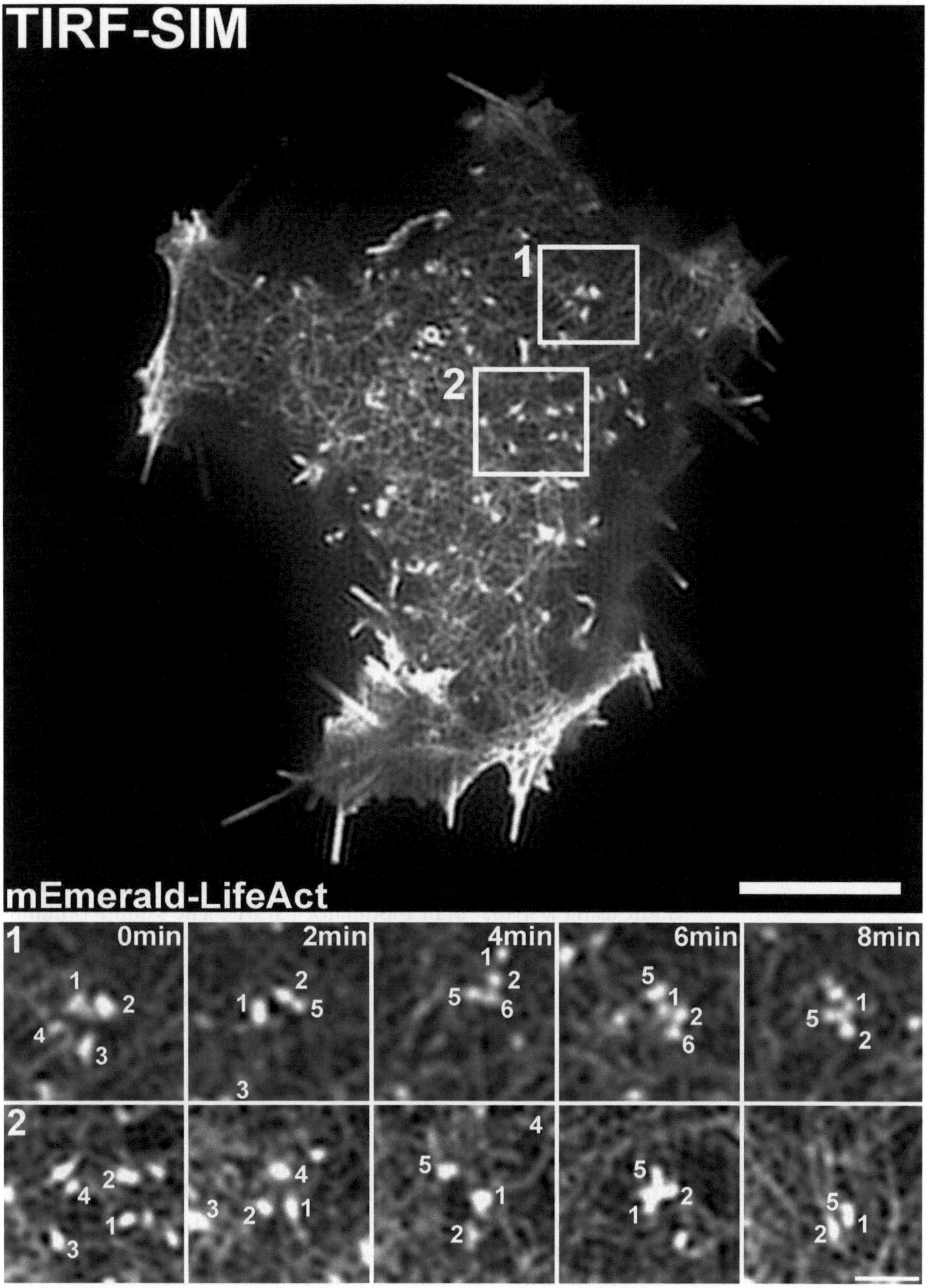

Fig. 5 Microvilli are motile actin-containing structures that translocate on the surface of pancreatic beta cells. Single frame from a TIRF-SIM time-lapse movie of a living INS-1 cell expressing mEmerald-LifeAct (top panel) and sequential micrographs (0–8 min) showing motile microvilli translocating on INS-1 cells. Individual microvilli are numbered. Bars: 5μm and 300 nm

Our data show that microvilli translocate in the fluid lipid bilayer as a structural unit. Membrane fluidity at the junction between the microvillus and the plasma membrane determines the ability of microvilli to translocate as integral units. These observations have profound implications for how we understand the sensory function of structures like microvilli and how we design studies to understand complex diseases like type 2 diabetes.

1.3 TIRF-SIM Microscopy

To image the actin cytoskeleton in microvilli of INS-1 832/13 cells, we used TIRF-SIM (total internal reflection fluorescence (TIRF)-structured illumination microscopy (SIM)) microscopy. TIRF-SIM is an SR imaging technique that has the advantage of imaging live cells far faster and with orders of magnitude less light than required for other forms of SR fluorescence microscopy.

TIRF, also known as evanescent field microscopy [13, 14], breaks the diffraction limit in the axial or *z*-dimension by illuminating cells with a subwavelength (~100 nm deep) field of excitation light. This evanescent field illuminates only the membrane region, producing a high-contrast, low background image of the cortical actin network-plasma membrane complex. With ultrahigh numerical aperture objectives, the depth of illumination can be as shallow as 50 nm, the thickness of the cortical actin cytoskeleton.

When TIRF is combined with SIM to form a TIRF-SIM imaging system, the diffraction limit is broken in both the axial and lateral dimensions, and the lateral resolution is twofold better than a conventional microscope [15, 16]. As a result, objects like actin filament bundles and vesicles that are smaller than the diffraction limit can be tracked and studied over time in multiple colors. Thus, both cytoskeletal structure and dynamics can be observed.

SIM achieves super-resolution by modulating the excitation light using a spatial light modulator (SLM). The sample is illuminated with a grating pattern of light [15]. When this excitation pattern and the spatial pattern of the sample superpose, an interference pattern called a Moiré fringe is produced. The excitation pattern is translated and rotated to generate a series of images with different Moiré fringes. As the spacing of the illumination grating is known, the grating pattern can be mathematically removed from the resulting Moiré fringes to gain access to the normally irresolvable higher-resolution information in the sample.

To generate the sinusoidal excitation patterns required for TIRF-SIM, a ferroelectric spatial light modulator is used. The binary ferroelectric SLM is capable of sub-millisecond pattern switching. Binary grating patterns are displayed on the SLM, and the resulting ±1 diffraction orders are filtered, relayed, and focused into the back aperture of the objective lens. The necessary phase shifts and rotations of the gratings are applied by changing the displayed SLM image. Reconstruction of TIRF-SIM images requires the acquisition of a minimum of three phase shifts per pattern rotation; therefore the SLM pattern period must be divisible by 3 [16].

The twofold improvement in resolution achieved with TIRF-SIM is sufficient to detect filamentous actin in microvilli that averaged 150 nm in diameter, a size that is difficult to detect without SR capability (Fig. 4). Even though the spatiotemporal resolution of TIRF-SIM is not sufficient to capture the dynamic changes in the sub-millisecond timeframe or in the sub-100-nm size domain, it is able to capture super-resolution images at a rate that allowed us to quantitatively measure the velocity of movement of microvilli on pancreatic beta cells (Figs. 5 and 6) as well as the structural changes in the cortical actin filament network (Figs. 7 and 8) under basal and elevated glucose conditions.

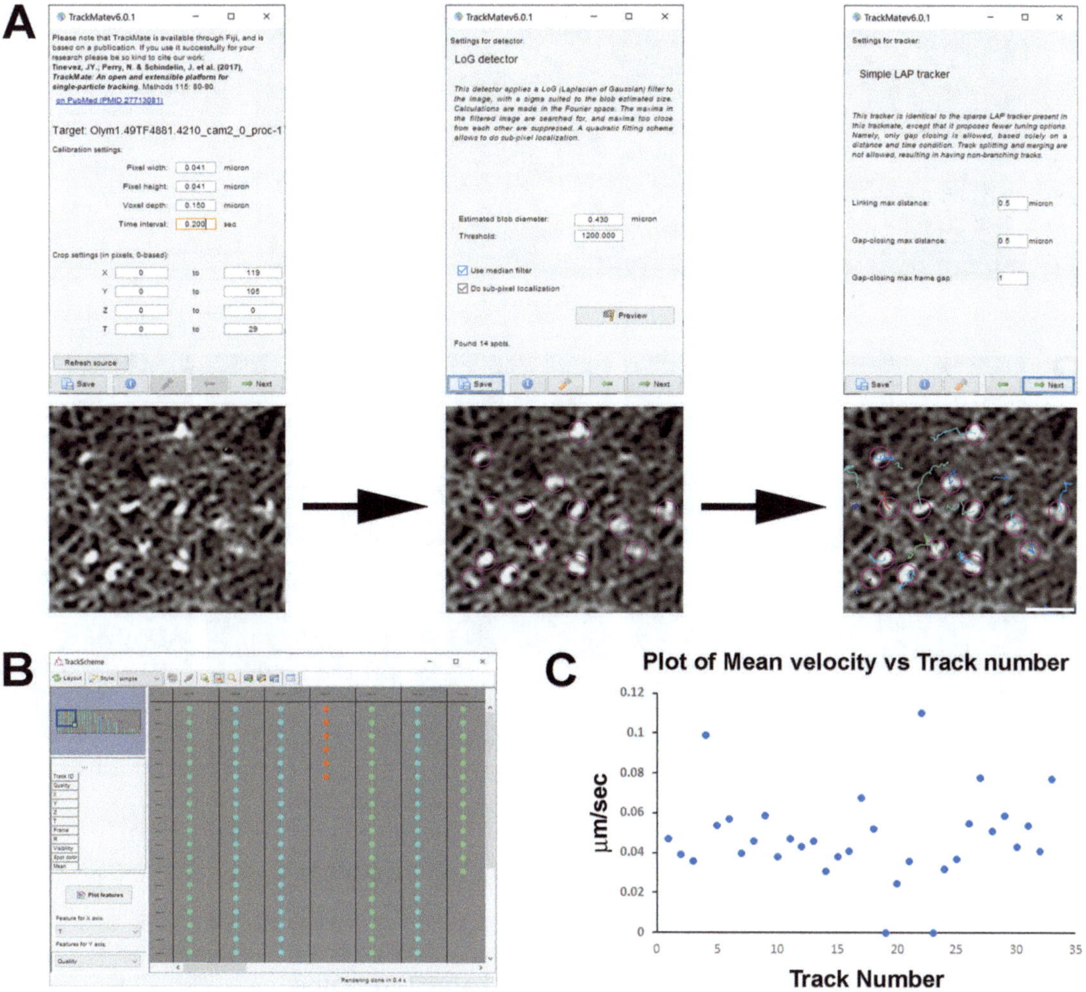

Fig. 6 Motion analysis of motile microvilli on surfaces of pancreatic beta cells. (**a**) Workflow for automatic tracking of microvilli by the FIJI plugin TrackMate [6]. (**b**) Display of the time sequence of each microvilli tracked. (**c**) Plot of the mean velocities of the microvilli (track number). Bars: 5 µm and 500 nm

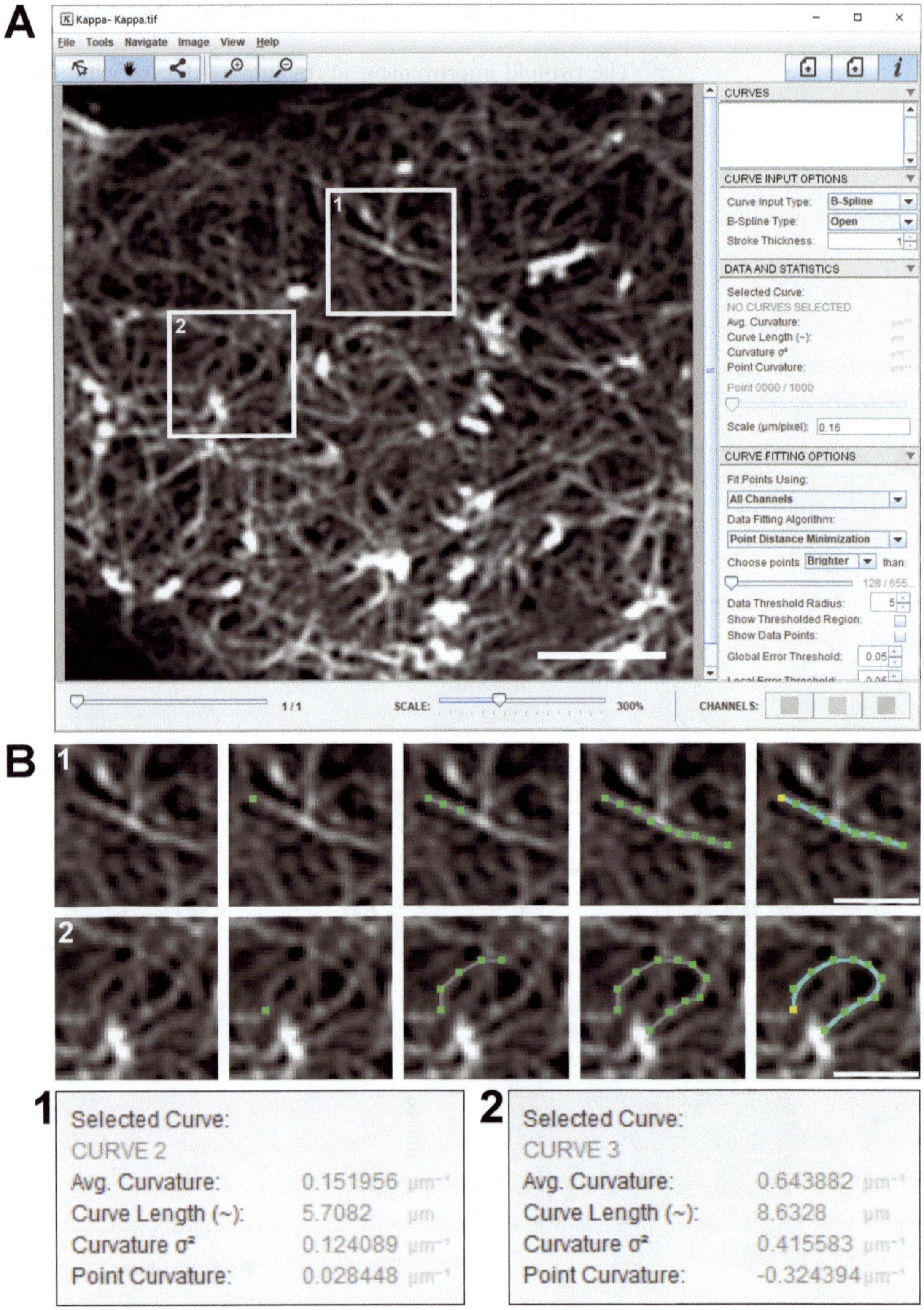

Fig. 7 Analysis of actin filament curvature in the cortex of pancreatic beta cells. (**a**) Screenshot of the Kappa curvature analysis FIJI plugin [7] during an online measurement of INS-1 832/13 cells expressing mEmerald-LifeAct. Images were taken with the TIRF-SIM imaging system. Boxed regions show enlargement of two small actin bundles used for B-spline analysis. (**b**) Micrographs 1 and 2 show individual steps to enter a curve using the Control Points and Curve Fitting tool. Scale bar: 1 µm

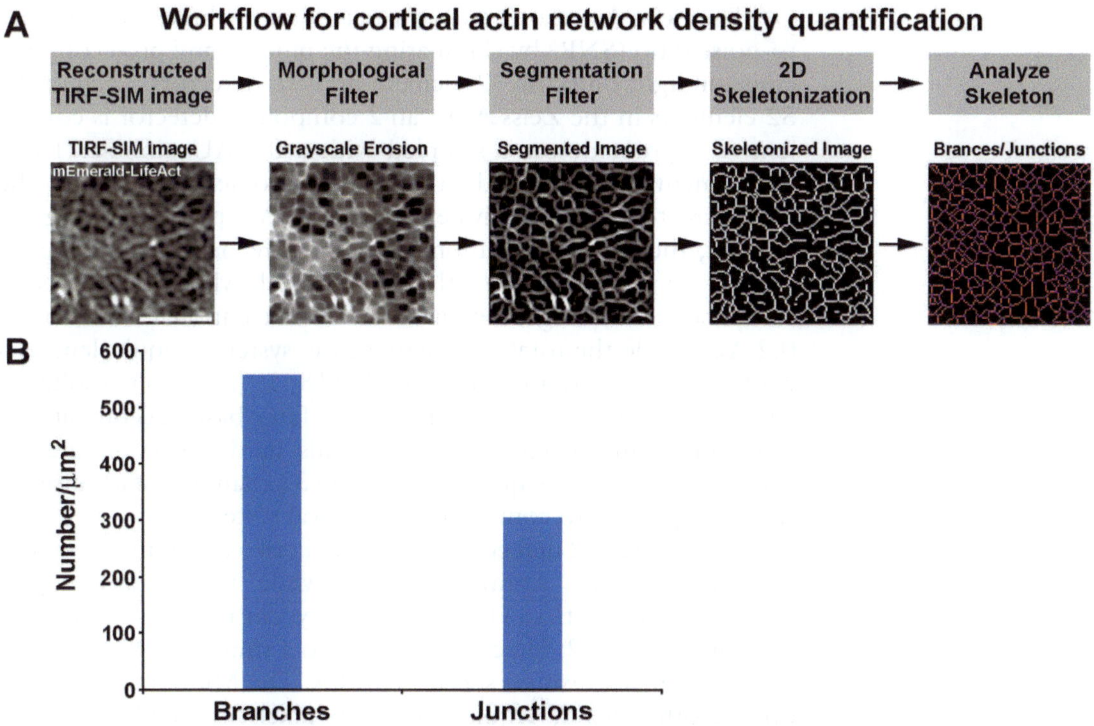

Fig. 8 Workflow for actin network density quantification using multiple FIJI plugins. (**a**) The TIRF-SIM image was processed using the following steps: grayscale erosion followed by segmentation and then skeletonization. (**b**) Plot of the resultant quantitative analysis of the cortical actin network density. Shown are the number of branches and junctions in a selected region of an INS-1 832/13 cell cultured under basal glucose conditions. Scale bar: 2μm

1.4 Confocal Laser Scanning Microscopy with Airyscan 2 Technology

We used confocal laser scanning microscopy (CLSM) with Airyscan technology [17], an enhanced resolution technique, as a second imaging method to study microvilli motility and cortical actin filament dynamics in pancreatic beta cells. We used the Zeiss LSM 980 with Airyscan 2 microscope, which has a 32-element GaAsP-PMT array detector, that replaces the physical confocal pinhole aperture and unitary detector assembly found in traditional confocal systems. Each of the 32 detector elements acts as its own small pinhole with positional information that allows for increased contrast of high spatial frequency information previously unavailable in traditional confocal systems. The increase in spatial frequency information enables Airyscan to produce images with substantially increased signal-to-noise ratio and resolution without having to increase laser power or sampling. Because the Airyscan 2 detector effectively reduces out-of-focus light, the imaging system is able to work on optically dense samples. The resolution of the Airyscan system is approximately 120 nm, slightly less than a twofold increase in resolution required to be defined as super-resolution imaging. Therefore, Airyscan is referred to as an enhanced resolution imaging technique.

The Zeiss Airyscan 2 detector technology achieves high signal-to-noise ratio (SNR) by eliminating the pinhole and projecting the emission light onto the 32-element array of detectors. Each of the 32 elements in the Zeiss Airyscan 2 compound detector is equivalent to a point detector sampled with a 0.2-AU pinhole. These 32 elements are arranged in a circular geometry, such that the total detector area is equivalent to a 1.25-AU pinhole setting. As a result, the spatial resolution of the final reconstructed image is defined by the sampling of the central pixel, which in this case is comparable to images acquired with conventional confocal at 0.2 AU, while the total sensitivity of the system is equivalent to a confocal image acquired at 1.25 AU [18, 19]. Further resolution gains are achieved by the use of a Wiener filter-based deconvolution step during image reconstruction. This hardware configuration allows simultaneous acquisition using all 32 channels, at acquisition rates comparable to conventional confocal systems.

The Airyscan 2 detector can operate in two different fast modes that increase the scanning speeds while maintaining super-resolution capabilities. The standard Airyscan mode utilizes information from all 32 detectors assigned to a single pixel in the final image for an acquisition speed of 15 fps. The Multiplex 4Y mode, on the other hand, acquires 4 image pixels in parallel with the Airyscan 2 detector, thereby increasing the image acquisition speed from 15 fps to 25 fps. In Multiplex SR-8Y mode, the acquisition speed increases to 47 fps at 512 × 512 resolution for a single channel through acquisition of 8 image pixels in parallel and additional beam-shaping stretching of the excitation spot in the *y*-dimension.

The enhanced resolution achieved with Airyscan 2 is coupled with an increase in SNR compared to conventional confocal. The enhanced sensitivity eliminates the need for high photon budgets (energetic flux on to the sample) to achieve high signal-to-noise ratio, and thus Airyscan is a gentle imaging technique that does not degrade sample integrity. Airyscan 2 is different from other enhanced resolution CLSM light microscopic acquisition methods, in that it is a pinhole-plane, laser point-scanning detector-based hardware method and not a software deconvolution and processing method typically combined with a closed pinhole. The latter limits photon collection efficiency compared to 1 Airy unit or 1.25 Airy unit pinhole sizes.

In summary, the Airyscan 2 technology improves the spatial resolution and SNR by exploiting a combination of the equivalent of confocal imaging with a 0.2-AU pinhole setting, Wiener filter-based deconvolution, and the pixel reassignment principle [5, 20, 21]. Airyscan technology benefits from the collection of all of the emitted light, which otherwise would be rejected at the pinhole in a conventional confocal architecture.

1.5 Image Processing, Analysis, and Workflow

We used image processing and analysis software to transform our morphological data into quantitative information that we then used to rigorously test our hypotheses. Importantly, quantitative image analysis allowed us to develop a more hypothesis-driven approach to sample preparation, image processing, and data management. In other words, quantitative information improved our ability to identify the most appropriate fluorescent probes needed to label cells and the most appropriate type of microscopic technique needed to obtain the requisite data.

We used FIJI, an open-source image processing and analysis software developed at the National Institutes of Health and the University of Wisconsin, for our studies. Motion analysis of microvilli motility was performed using the plugin TrackMate [6] (Fig. 6), and the curvature and stiffness of the filaments of the cortical actin cytoskeleton were determined using the software plugin Kappa [7] (Fig. 7). Zen Blue was used for image acquisition and analysis on the CLSM Zeiss LSM980 with Airyscan 2 microscope. Zen Blue software controls the CLSM microscope system and was used to process and analyze 3D image stacks. Zen Blue was used to view metadata offline, to manage images/video sequences, and to perform basic image processing.

Standard image analysis techniques and algorithms were used for image processing and image analysis (Fig. 8). Images were first processed to remove microscope or sample "artifacts" using background and/or shading correction. Then, image processing steps were applied to enhance or suppress level of detail using deconvolution or digital image filters such as averaging/smoothing with a Gaussian kernel or linear filters. Other options for image processing include edge detection for finding boundaries or contrast enhancement filters such as unsharp or Laplacian filters when bright or dim objects are present in digital images. To automatically identify objects from background, the commonly used technique of thresholding was applied utilizing pixel intensities for foreground and background separation. This method was also used to generate binary images which are suitable for further image processing using erosion/dilation to detect or identify objects.

1.6 The Future of Super-Resolution Microscopy

Imaging techniques are improving at a rapid rate, but a significant gap exists between the super-resolution (SR) techniques that rely on fixed specimen and those that have the ability to image living cells in real time. TIRF-SIM is one SR technique that has the advantage of imaging live cells far faster and with orders of magnitude less light than required for other forms of SR fluorescence microscopy. However, the resolution achieved with TIRF-SIM is limited to a twofold gain beyond conventional fluorescence microscope, or ~100 nm with visible light.

Betzig and colleagues [22] developed a way to extend TIRF-SIM to the sub-100 nm spatial domain by exploiting the spatially patterned activation of a reversibly photoswitchable fluorescent protein to reach 45–62 nm resolution at sub-second acquisition called PS NL-SIM. Consequently, they were able to acquire substantially more frames at an improved signal-to-noise ratio by this technique.

The improvement in spatiotemporal resolution achieved by PA NL-SIM is of great benefit to cell biologists, but this imaging modality is not currently available commercially. The best commercially available systems for live cell imaging fail to achieve the level of resolution required to capture dynamic changes at sub-second resolution with high spatial resolution. Continuing improvements in SR techniques are clearly needed.

Super-resolution microscopy remains the method of choice for nanoscale imaging of protein dynamics in living cells. As new, enhanced resolution systems become available commercially, studies requiring long-term imaging of living cells will benefit greatly and advance research more rapidly in these important fields of cell biology.

2 Materials

2.1 Cell Culture

1. Rat insulinoma cells (INS-1 832/13) ([8] *see* **Note 1**).
2. RPMI 1640 medium with L-Glutamine (commercially available) supplemented with 10% fetal bovine serum (FBS).
3. Penicillin-streptomycin (optional).
4. Sterile phosphate-buffered saline (PBS) (1×) (commercially available).
5. Solutions of 0.25% trypsin and ethylenediamine tetraacetic acid (EDTA).
6. Freezing media: 10% DMSO in 1 mL RPMI 1640 medium.
7. 10 cm plastic cell culture dishes.
8. 6-well culture plates.
9. 5 mL disposable serological pipets.
10. 10 mL disposable serological pipets.
11. 15 mL centrifuge tubes.
12. 45 mm bottle top filter for rapid sterile filtration.

2.2 Transient Cell Transfection

1. Optifect transfection reagent from Invitrogen (Thermo Fisher Scientific, Waltham, MA) (*see* **Note 2**).
2. Opti-MEM reduced serum medium supplemented with HEPES (4-(2-hydroxyethyl)-1-piperazineethanesulfonic acid), 2.4 g/L sodium bicarbonate and L-glutamine from Gibco (Thermo Fisher Scientific, Waltham, MA).

3. mEmerald-LifeAct-7 vector (RRID:Addgene_54148, Watertown, MA).
4. 25 mm round coverslips and 35 mm glass bottom dishes.

2.3 TIRF-SIM Live Cell Imaging

1. 25 mm round uncoated glass coverslips.
2. Immersion oil type LDF (Cargille Laboratories, Cedar Grove, NJ).
3. Zeiss Axio Observer Z1 motorized microscope equipped with differential interference optics (DIC), an Olympus 100×/1.49 NA oil objective lens (Olympus America Inc., Center Valley, PA) and filter sets for FITC, TRITC, and DAPI fluorescence (Carl Zeiss Microscopy, LLC, White Plains, NY).
4. X-Cite 120 light source for epifluorescence microscopy (Excelitas Technologies Corp., Waltham, MA).
5. 488 nm (Sapphire, Coherent, Santa Clara, CA) and 565 nm laser (2RU-VFL-P-1000-560-B1R, MPB Communications Inc., Quebec, Canada).
6. Acousto-optic tunable filter (AOTFnC-400.650-TN, Quanta Tech, New York, NY).
7. Phase-only modulator (BVO AHWP3, Bolder Vision Optik, Inc., Boulder, CO).
8. Ferroelectric spatial light modulator (SXGA-3M, Forth Dimension Displays Ltd., Dalgety Bay, Scotland, Great Britain).
9. Polarization rotator consisting of a liquid crystal variable retarder (SWIFT, Meadowlark Optics Inc., Frederick, CO) and an achromatic quarter-wave plate (BVO AQWP3, Bolder Vision Optik, Inc., Boulder, CO).
10. 2× Orca-Flash 4.0 V2 C11440-22CU digital sCMOS cameras (Hamamatsu Photonics, Bridgewater, NJ).
11. *X*–*Y* non-flat top stage with controller (MS-2000, Applied Scientific Instrumentation, Eugene, OR).
12. Stage top environmental control chamber (H301, Okolab, Naples, Italy) with chamber inserts for 25 mm round coverslips.
13. Late model PC computer with fast processor and large random-access memory (RAM).
14. *See* **Notes 3–4**.

2.4 CLSM (Zeiss LSM 980 with Airyscan 2) Live Cell Imaging

1. 35 mm glass bottom dishes.
2. Immersion oil, Immersol 518N (Carl Zeiss Microscopy, LLC, White Plains, NY).
3. LSM980 confocal microscope with Airyscan 2 detector with Multiplex 4Y and 8Y (beam shaping) modules (Carl Zeiss

Microscopy, LLC, White Plains, NY) equipped with Definite Focus2, Z-Piezo, a dry 10×/0.3 NA objective lens, and a 40×/1.2 NA and 63×/1.4 NA oil objective lenses.

4. 405 nm, 488 nm, 561 nm, and 639 nm diode lasers.
5. Colibri 5 Type RGB-UV light source for epifluorescence microscopy.
6. Environmental chamber for temperature, CO_2, and humidity control.
7. Late model PC computer with fast processor, NVIDIA graphics card, and large random-access memory (RAM).
8. Image acquisition and environmental control software (Zen Blue 3.2).

2.5 Image Processing and Analysis

FIJI open-source image processing and analysis software (NIH, https://imagej.net/Fiji).

3 Methods

3.1 Cell Culture

1. Combine 50 mL FBS with 450 mL RPMI 1640 medium, filter sterilized with a 45 mm bottle top filter, and store at 4 °C.
2. Maintain the pancreatic beta cell line, INS-1 832/13 in 10 mL sterile RPMI 1640 medium at 37 °C in a 5% CO_2 in air atmosphere. Grow the cells in 10 cm plastic dishes, and change the medium every 2 days before microscopic observation.
3. Passage INS-1 832/13 cells when they reached ~70% confluence with 0.25% trypsin/EDTA. Remove the RPMI 1640 medium from the cell culture dish and wash cells with 10 mL PBS. Add 2 mL 0.25% trypsin/EDTA solution to the cells and incubate them for 2–10 min at 37 °C until the cells are rounded up and detached spontaneously from the cell culture dish. Collect cells with 8 mL RPMI 1640 medium to inactivate the trypsin, and transfer them to a 15 mL centrifugation tube. Centrifuge the cells at 900 rpm for 5 min in a clinical centrifuge. Aspirate supernatant to remove trypsin before adding fresh medium and seeding in culture dishes. Split the INS-1 832/13 cells at a 1:3 ratio, which provides experimental cell cultures that will reach ~70% confluence after 3–5 days. Cryopreserve cells in freezing medium containing 10% DMSO (v/v) followed by long-term storage in liquid nitrogen.

3.2 Transfection of INS-1 (832/13) Cells with Actin Expression Vector

1. For transient transfections using the mEmerald-LifeAct-7 vector, plate cells on 35 mm round coverslips placed individually in 6-well culture plates or 35 mm glass bottom dishes 2 days before transfection. Importantly, grow the cells in RPMI-1640 without antibiotics 1 day before transfection to prevent cell death.

2. The following protocol used to transfect pancreatic beta cells was optimized in our laboratory.
3. To prepare the DNA-Optifect complexes, add 4μL (~4μg) of mEmerald-LifeAct DNA to 246μL of Opti-MEM reduced serum medium, and gently mix the solution. Dilute 17μL of Optifect transfection reagent with 233μL Opti-MEM, mix gently, and incubate for 5 min at room temperature. Combine the diluted DNA with the diluted Optifect solution, and incubate the mixture for 20 min at room temperature.
4. Add the DNA-Optifect complexes (500μL total volume) to 2 mL of culture medium containing the INS-1 832/13 cells while rocking the 6-well culture plate or 35 mm glass bottom dish. Replace the cell culture medium after 4 h, and incubate the cells for a period of 24 h at 37 °C prior to live cell imaging.

3.3 Super-Resolution Live Cell Imaging

3.3.1 TIRF-SIM Microscopy

1. Grow and transfect rat pancreatic beta cells on 25 mm sterilized round coverslips.
2. Turn on the temperature controller, CO_2 regulator, and air pump at least 30 min before starting live cell imaging.
3. Turn on the 488 nm and 560 nm lasers. Set the "relative power" to about 60% for the 488 nm laser, and set the laser power to 250 mW for the 560 nm laser.
4. Apply a drop of immersion oil to the 100×/1.49 NA objective lens, and place sample onto microscope stage. Use epifluorescence microscopy to find the focus and to detect positively transfected cells within the cell population.
5. Switch the light path from the ocular to the camera, select camera mode in the image acquisition software to ensure emissions go to camera, and choose proper sequence mode for the objective to acquire a single color image.
6. Use Multi-Scan mode for multi-channel sequential image acquisition.
7. Set the time lapse as desired and run experiment.
8. Image cells over periods of 30 min at 37 °C for defined time intervals (e.g., 0, 2, 5, 10, 15, 20, 25, and 30 min) to monitor actin filament dynamics and microvilli motility in pancreatic beta cells.
9. Reconstruct images and choose a wiener constant between 0.005 and 0.01 (*see* **Note 5**).

3.3.2 Enhanced Resolution Confocal Laser Scanning Microscopy (CLSM)

1. Grow and transfect rat pancreatic beta cells in 35 mm glass bottom dishes.
2. Turn on the temperature control at least 1 h before the beginning of imaging. This will allow all components of the microscope and the stage incubator to reach 37 °C. Set the desired CO_2 concentration (usually 5%).

3. Apply a drop of immersion oil to the 63×/1.4 NA objective lens, and place the 35 mm glass bottom dish in the suitable attachment of the microscope stage.
4. Identify transfected cells within the cell population in the 35 mm glass bottom dish, and find the focus using epifluorescence microscopy. Choose Fluorescence in the Locate tab for opening the dye finder to select a dye for sample localization.
5. Load an existing configuration or use the Smart Setup wizard to set dyes and wavelength(s) using the Acquisition tab. Select the detectors for imaging by choosing between LSM for standard confocal imaging and Airyscan for imaging using the Airyscan 2 detector. Adjust excitation power, diameter of the pinhole in LSM mode, and detector settings in the Channels box.
6. Set the scan area, image zoom, and scan speed in the Acquisition Mode box, and set the Z-stack boundaries in the Z-stack menu box.
7. For imaging of the actin cytoskeleton within the region of interest (ROI), select the size of the scanning area by positioning a box around the object or ROI using the Crop dialog box.
8. Set the time-lapse conditions such as interval and total number of scans for the live cell imaging using the Time Series dialog box and start the experiment.

3.3.3 Image Processing and Analysis

We used the FIJI plugins TrackMate [6] and Kappa [7] for quantitative analysis of the cortical actin cytoskeleton and movement of microvilli. We developed a workflow for actin network density quantification to regularize the processing and analysis steps (*see* Subheading "Quantification of Actin Filament Network Density").

Automated Tracking of Motile Microvilli Using TrackMate

1. Open a time-lapse sequence in FIJI and choose TrackMate from the Plugins menu.
2. Adjust brightness or contrast of the image using the Adjust Brightness/Contrast dialog box or Auto enhance function from the Image menu.
3. Set the Time interval in the Calibration settings menu (Fig. 6a, left panel).
4. Select the LoG (Laplacian of Gaussian) detector from the drop-down list.
5. Define the Estimated blob diameter and set Threshold in the LoG detector settings. Check the Use Median Filter box. Preview the current settings (Fig. 6a, middle panel).
6. Set the Initial Thresholding on the Quality feature to restrict the number of spots.

7. Select the HyperStack Displayer.
8. Set Filters on spots, e.g., color, intensity, signal/noise, diameter.
9. Choose Simple Linear Assignment Problem (LAP) Tracker from the Select a Tracker menu and set the parameters for the LAP Tracker (Fig. 6a, right panel).
10. Set Filters on tracks, e.g., velocity, track displacement, distance traveled.
11. Select TrackScheme and Analysis from the Display Options menu to display all tracks (Fig. 6b) and to obtain data on track statistics including displacement, distance traveled, and velocity (Fig. 6c).
12. Select Capture Overlay and Execute to create a new window displaying all microvilli and their corresponding tracks (Fig. 6a, right panel).
13. *See* **Note 6** for download and installation instructions.

Curvature Analysis of Cortical Actin Bundles Using B-Splines

1. Open an image from the hard drive or an active image from FIJI using the File menu in the Kappa plugin (Fig. 7a).
2. Adjust brightness or contrast of the image using the Adjust Brightness/Contrast dialog box or Auto enhance function from the Image menu.
3. Use the Zoom tool icons to zoom-in or zoom-out the image.
4. Choose B-Spline from the Curve Input Type box in the Curve Input Options menu (Fig. 7a, right menu).
5. Select Open from the B-Spline Type box in the Curve Input Options menu (Fig. 7a, right menu).
6. Select the Control Point tool from the Tools menu to enter control points on the entire length of individual actin bundles (Fig. 7b, panels 1 and 2).
7. Press Enter to create a curve along individual actin bundles (Fig. 7b, panels 1 and 2).
8. Adjust curves by using the Fit Curve tool from the Tools menu. Choose the Curve Fitting Options dialog box to select between individual RGB channels, the Thresholded Radius, and Region and Data Points.
9. Use the Selection tool from the Tool menu to switch between individual B-spline curves.
10. Save curves using the Save Curve Data option from the File menu.
11. Export curve data as .csv files using the Export panel.
12. *See* **Note 7** for download and installation instructions.

Quantification of Actin Filament Network Density

1. Open a TIRF-SIM image in FIJI using the File menu.
2. Adjust brightness or contrast of the image using the Adjust Brightness/Contrast dialog box or Auto enhance function from the Image menu (Fig. 8a).
3. Choose a ROI and apply the morphological filter GreyscaleErode from the Morphology plugin in the Plugins menu. Check the White Foreground box (Fig. 8a) (*see* **Note 8**).
4. Apply the segmentation filter Tubeness [23] from the Analyze menu in the Plugins menu. Adjust brightness and contrast with the Auto enhance function (*see* **step 2**) (Fig. 8a).
5. Create an 8-bit image by changing the image Type from the Image menu.
6. Skeletonize the 8-bit image using the Skeletonize2D/3D plugin (Fig. 8a).
7. Analyze the skeletonized image using the Analyze Skeleton2D/3D from the Analyze menu (Fig. 8a).
8. For quantitative data of the actin filament network density, the number of branches and junctions per unit area was calculated (Fig. 8b).

4 Notes

1. Rat insulinoma cell line (INS-1 832/13) is widely used as beta cell model for diabetes research and to study insulin secretion. This cell line can be obtained from Sigma-Aldrich.
2. The Optifect transfection reagent from Invitrogen is designed for lipid-based transfections and should be used for low-density cell cultures (10–70% confluence) or cells that are sensitive to other commercially available transfection reagents because of its reduced toxicity.
3. HHMI Janelia offers the design for constructing a TIRF-SIM imaging system for free for Non-Profit Research by downloading designs at Flintbox (see link below) and rights and designs available for Commercial License https://hhmi.flintbox.com/technologies/ab07753a-d138-4946-a180-5dac7aa21cf3.
4. An article published in JOVE [24] provides another guide to build a TIRF-SIM microscope that is capable of imaging up to 10 Hz in three colors at a 100 nm resolution.
3. A Wiener constant of 0.01 should be used for TIRF-SIM image reconstruction. This constant should be increased to reduce the noise or lower the contrast of acquired TIRF-SIM images.
5. The TrackMate source code for download is available at https://github.com/fiji/TrackMate/releases/tag/

TrackMate_-4.0.0. The TrackMate manual for download can be found at https://imagej.net/File:TrackMate-manual.pdf.

6. The Kappa source code for download and installation instructions are available at https://github.com/brouhardlab/Kappa.
7. The download for morphological filtering such as erosion and dilation, morphological reconstruction, and watershed segmentation for 2D/3D images can be found at https://github.com/ijpb/MorphoLibJ/.

Acknowledgments

The authors would like to thank Dr. Teng-Leong Chew, Director of the Advanced Imaging Center, and his team at Howard Hughes Medical Institute (https://www.aicjanelia.org/) for technical support with the TIRF-SIM imaging system. We would also like to thank Dr. Tanay Desay from Carl Zeiss Microscopy, LLC, for technical support with the LSM980 with Airyscan 2 and Drs. George G. Holz and Oleg Chepurny from SUNY Upstate Medical University for advice with the INS-1 832/13 cell culture.

References

1. Galbraith JA, Galbraith CG (2011) Super-resolution microscopy at a glance. J Cell Sci 3:247–255
2. Taraska JW (2019) A primer on resolving the nanoscale structure of the plasma membrane with light and electron microscopy. J Gen Physiol 151:974–985
3. Blanchion L, Boujeema-Paterski R, Sykes C, Plastino J (2014) Actin dynamics architecture and mechanics in cell motility. Physiol Rev 94:235–263
4. Svitkina T (2018) The actin cytoskeleton and actin-based motility. Cold Spring Harb Perspect Biol 10:a018627
5. Sheppard CJ, Mehta SB, Heintzmann R (2013) Superresolution by image scanning microscopy using pixel reassignment. Opt Lett 38:2889–2892
6. Tinevez JY, Perry N, Schindelin J, Hoopes GM, Reynolds GD, Laplantine E, Bednarek SY, Shorte SL, Eliceiri KW (2017) TrackMate: an open and extensible platform for single-particle tracking. Methods 115:80–90
7. Mary H, Brouhard GJ (2019) Kappa (*K*): analysis of curvature in biological image data using B-splines. bioRxiv https://doi.org/10.1101/852772
8. Hohmeier HE, Mulder H, Chen G, Henkel-Rieger R, Prentki M, Newgard CB (2000) Isolation of INS-1-derived cell lines with robust ATP-sensitive K+ channel-dependent and -independent glucose-stimulated insulin secretion. Diabetes 49:424–430
9. Trexler AJ, Taraska JW (2017) Regulation of insulin exocytosis by calcium-dependent protein kinase C in beta cells. Cell Calcium 67:1–10
10. Orci L, Gabbay KH, Malaisse WJ (1972) Pancreatic beta cell web: its possible role in insulin secretion. Science 175:1128–1130
11. Orci L, Thorens B, Ravazzola M, Lodish HF (1989) Localization of pancreatic beta cell glucose transporter to specific plasma membrane domains. Science 245:295–297
12. Bénichou O, Moreau M, Suet PH, Voituriez R (2007) Intermittent search process and teleportation. J Chem Phys 126:234109
13. Steyer JA, Almers W (2001) A real-time view of life within 100 nm of the plasma membrane. Nat Rev Mol Cell Biol 2:268–275
14. Axelrod D (2013) Evanescent excitation and emission in fluorescence microscopy. Biophys J 104:1401–1409

15. Gustafsson MG (2000) Surpassing the lateral resolution limit by a factor of two using structured illumination microscopy. J Microsc 198:82–87
16. Kner P, Chhun BB, Griffis ER, Winoto L, Gustafsson MG (2009) Super-resolution video microscopy of live cells by structured illumination. Nat Methods 6:339–342
17. Korobchevskaya K, Lagerholm CB, Colin-York H, Fritzsche M (2017) Exploring the potential of Airyscan microscopy for live cell imaging. Photo-Dermatology 4:41
18. Sarder P, Nohorai A (2006) Deconvolution methods for 3-D fluorescence microscopy images. IEEE Signal Process Mag 23:32–45
19. Huff J (2015) The Airyscan detector from Zeiss: confocal imaging with improved signal-to-noise ratio and super-resolution. Nat Methods:12, i-ii
20. Sheppard CJ (1988) Superresolution in confocal imaging. Optik 80:53–54
21. Bertero M, Brianzi P, Pike ER (1999) Super-resolution in confocal scanning microscopy. Inverse Probl 3:195–212
22. Li D, Shao L, Chen BC, Zhang X, Zhang M, Moses B, Milkie DE, Beach JR, Hammer JA III, Pasham M, Kirchhausen T, Baird MA, Davidson MW, Xu P, Betzig E (2015) Advanced imaging extended-resolution structured illumination imaging of endocytic and cytoskeletal dynamics. Science 349: aab3500
23. Sato Y, Nakajima S, Shiraga N, Atsumi H, Yoshida S, Koller T, Gerig G, Kikinis R (1998) Three-dimensional multi-scale filter for segmentation and visualization of curvilinear structures in medical images. Med Image Anal 2:143–168
24. Young LJ, Ströhl F, Kaminski CF (2016) A guide to structured illumination TIRF microscopy at high speed with multiple colors. J Vis Exp 111:e53988

Chapter 2

Imaging Cytoskeleton Components by Electron Microscopy

Tatyana Svitkina

Abstract

The cytoskeleton is a complex of detergent-insoluble components of the cytoplasm playing critical roles in cell motility, shape generation, and mechanical properties of a cell. Fibrillar polymers—actin filaments, microtubules, and intermediate filaments—are major constituents of the cytoskeleton, which constantly change their organization during cellular activities. The actin cytoskeleton is especially polymorphic, as actin filaments can form multiple higher-order assemblies performing different functions. Structural information about cytoskeleton organization is critical for understanding its functions and mechanisms underlying various forms of cellular activity. Because of the nanometer-scale thickness of cytoskeletal fibers, electron microscopy (EM) is a key tool to determine the structure of the cytoskeleton.

This article describes application of rotary shadowing (or platinum replica) EM (PREM) for visualization of the cytoskeleton. The procedure is applicable to thin cultured cells growing on glass coverslips and consists of detergent extraction (or mechanical "unroofing") of cells to expose their cytoskeleton, chemical fixation to provide stability, ethanol dehydration and critical point drying to preserve three-dimensionality, rotary shadowing with platinum to create contrast, and carbon coating to stabilize replicas. This technique provides easily interpretable three-dimensional images, in which individual cytoskeletal fibers are clearly resolved and individual proteins can be identified by immunogold labeling. More importantly, PREM is easily compatible with live cell imaging, so that one can correlate the dynamics of a cell or its components, e.g., expressed fluorescent proteins, with high-resolution structural organization of the cytoskeleton in the same cell.

Key words Electron microscopy, Cytoskeleton, Critical point drying, Rotary shadowing, Actin, Microtubules, Immunogold, Correlative microscopy

1 Introduction

Electron microscopy (EM) has been instrumental in discovering the cytoskeleton in the first place and also in investigating its structural organization in different cells and conditions. The initial progress in the cytoskeletal studies closely paralleled the development of EM techniques. Thus, the introduction of heavy metal fixation led to the discovery of actin filaments in non-muscle cells [1], while the discovery of microtubules [2] was made possible after the introduction of aldehyde fixation [3].

Ray H. Gavin (ed.), *Cytoskeleton: Methods and Protocols*, Methods in Molecular Biology, vol. 2364, https://doi.org/10.1007/978-1-0716-1661-1_2,

A great value of EM is its ability to obtain structural information at a high-resolution level, which for biological samples is limited by a sample preparation procedure rather than by the power of a transmission electron microscope (TEM). Vacuum in the TEM column and electron beam irradiation impose strict restrictions on how samples should be prepared, which in turn greatly affect the quality of images and the rate of success. A large number of different EM protocols have been developed over the years to improve the quality of samples and the amount of collected information and to avoid artifacts. Each technique has its pluses and minuses, making it more suitable for some applications than for others.

The thin sectioning technique was initially a dominant way to visualize the cytoskeleton [4, 5]. It involves the embedding of chemically fixed specimens into a resin followed by thin sectioning to allow for beam penetration. Contrast is generated by positive staining of the sections with heavy metal salts, and the limited ability of stains to bind bioorganic material reduces the resolution of this technique. Thin sections provide a 2D view of the sample at a single plane, and a series of sections is required to retrieve the 3D information. Such reconstruction works well with relatively large and simple objects but is not efficient in revealing the details of complex and delicately organized cytoskeletal structures. Among cytoskeleton components, actin filaments are notoriously difficult to preserve and visualize by thin section EM. Although actin filament bundles can be reasonably well visualized in plastic-embedded samples after appropriate fixation, delicate actin filament networks, such as those in lamellipodia, in the bulk cytoplasm, or at the surface of membrane organelles, remain largely uninterpretable even in best examples [6–9].

Different versions of whole mount EM have been used to investigate the structural organization of the cytoskeleton in its entirety. Thus, the structural arrangement of actin filaments in lamellipodia was first visualized by the negative staining EM of cultured cells [10]. In this technique, partially permeabilized cells growing on EM grids are immersed into a heavy metal stain solution, which is blotted off shortly after, and the samples are dried in open air. The dried stain generates a dark amorphous background on which the structures appear as translucent shapes. Negative staining EM provides high resolution and allows one to see the thin regions of a cell all the way through. In combination with electron tomography, this approach can successfully reveal structural organization of branched actin filament networks [11–13]—the cytoskeleton structures that are particularly challenging for high-resolution structural studies. Weaknesses of the negative staining EM include partial sample flattening during air drying, relatively low contrast, and low stability of the samples. In cryo EM technique, the samples are quickly frozen (to prevent ice crystal formation) and viewed while still embedded in amorphous ice, either as whole mounts or after cryosectioning, so that they remain hydrated and the proteins retain their natural conformation [14–16]. To view frozen samples, a TEM should be equipped with a chilled sample holder, and electron beam power and the observation time should be minimized to keep the specimen frozen. No contrasting procedures are used in this technique except for the specimen's own contrast. Because of low intrinsic contrast and sensitivity of samples to irradiation, cryo EM has been most useful for analyses of samples suitable for averaging, such as purified macromolecules [16, 17]. Application of cryo EM to intact cells is challenging due to limited opportunity of applying averaging approaches, although subtomogram averaging has been successfully used for imaging of a few suitable subcellular structures [14]. Cryo EM, especially in combination with electron tomography, can image only small regions of the cell due to lengthy image acquisition and sample damage during imaging, so that a larger context of the imaged region is often unavailable. Another limitation of cryoEM is its incompatibility with immunochemistry [18]. Because of higher intrinsic contrast of lipids, membrane organelles are more permissive for cryo EM analyses in intact cells [19–24], as compared to purely proteinaceous structures, such as the cytoskeleton. Among cytoskeletal components, microtubules are more detectable by cryo EM [25–27] than actin filaments, while actin filament bundles are easier to visualize and analyze [28, 29] than branched actin networks [30, 31].

In metal replica EM, heavy metals are evaporated onto a 3D sample at an angle, which reveals its surface topography [32, 33]. The quality of the samples is greatly enhanced if rotary, and not unilateral, coating is used, as it helps to avoid deep featureless shadows. As metal coating is not cohesive, it is subsequently stabilized by a layer of carbon, which keeps metal grains together and is fairly transparent for the electron beam. The coated sample, or just a metal-carbon replica, is subsequently removed from its original support and placed onto EM grids. The resolution of replica EM is quite high, but it depends on the metal grain size, the thickness of the coating, and the angle of shadowing. Platinum is the most popular metal, as it provides a good compromise between the grain size and an ease of evaporation. The replica technique was initially introduced to study freeze-fractured samples [34], but it is applicable for a large range of samples, such as single molecules [35–37], cells [33, 38–48], and tissues [49–51]. This approach can reveal the 3D structure in great detail, but it is limited by the depth of shadowing penetration.

Given that PREM (platinum replica electron microscopy) reveals only the sample surface, analyses of the cytoskeleton can be accomplished only after the cytoskeleton is exposed to metal shadowing. This is typically achieved by detergent extraction, which reproducibly exposes the entire cytoskeleton of the cell.

For best results, the extraction conditions need to be carefully adjusted to properly preserve the cytoskeleton architecture. Nonionic detergent Triton X-100 is most commonly used to dissolve cell membranes, whereas preservation of the cytoskeleton relies on the composition of the buffer and is improved by various stabilizing supplements, such as phalloidin for actin filaments, taxol for microtubules, and nonspecific stabilizers polyethelete glycol or glutaraldehyde [33, 46, 52–54]. Since detergent extraction dissolves all cellular membranes, analyses of the membrane-associated cytoskeleton are typically accomplished using mechanical rupture of the cell (unroofing), which can expose cell interior without dissolving membranes. In two main unroofing approaches employed for cytoskeleton studies, the mechanical impact is delivered either through an ultrasonic burst [44, 47, 55–57] or by applying an adhesive material to the cell surface and subsequently peeling it off [48, 57–61]. Mechanical rupture methods are not highly reproducible, but successful samples excellently reveal interactions between the cytoskeleton and membrane organelles at high resolution.

In most cells, the cytoskeleton is heavily dominated by actin filaments, which largely conceal other cytoskeletal components, such as microtubules, intermediate filaments, myosin II bipolar filaments, etc. Such cytoskeleton components can be uncovered by treating the chemically or mechanically exposed cytoskeletons with the actin-severing protein gelsolin, which dissolves actin filaments without perturbing non-actin structures [53, 62–65]. For our studies of the cytoskeleton in cultured cells, we chose the PREM approach, in which either detergent extraction or unroofing is used to expose the cytoskeleton; chemical fixation helps to preserve the sample structure; ethanol dehydration followed by critical point drying (CPD) preserves the cell's 3D organization; and rotary shadowing with platinum creates contrast. Over the years, we have found a good combination of individual steps to develop a reliable and relatively simple protocol that consistently produces highly informative images with excellent yield that can be combined with immunochemistry [33, 38, 39, 42, 48, 53, 54, 62, 66, 67]. However, this approach is not universal but is limited to relatively thin samples attached to glass surfaces. Also, because of extensive fixation and dehydration, it can achieve the molecular level of resolution only for very large molecules, such as myosin II [62], plectin [63], or spectrin [68], but is optimal for analyses of the fine cytoskeletal architecture with a single filament resolution at the scale of a whole cell.

As EM, in general, cannot work with live samples, investigators can only guess the kind of activity the cell was involved in at the moment of fixation and what it would do next. A partial solution for this problem is provided by correlative light and EM (CLEM), in which the dynamics of a living cell is followed by time-lapse optical imaging, and the same sample is subsequently analyzed by EM. Our PREM protocol made it possible to perform CLEM routinely, as it allowed us to obtain high-quality structural information for a cell of interest with high probability [38, 46, 48, 53, 54, 66]. Other labs also successfully used CLEM in combination with PREM using similar or slightly modified protocols [43, 47, 69, 70]. Several other EM techniques have also been used in a correlative approach [15], including resin-embedded samples [21, 71–75], cryoEM [28, 29, 31], and negatively contrasted cells [12, 76].

In this article, I present detailed protocols for PREM of the cytoskeleton and its extensions, such as immunogold labeling, correlative PREM, and actin depletion by gelsolin. Given that the key steps ensuring high quality of PREM samples are the same for the cytoskeletal preparations prepared by detergent extraction and by unroofing, only detergent-extracted cells are used here as an example. However, mechanically ruptured cells prepared by various unroofing protocols are also compatible with these procedures [48, 56, 57].

2 Materials

2.1 Cell Culture and Extraction

1. Small (6–12 mm) coverslips that can be made by cutting regular coverslips with a diamond pencil. Trapezoidal (square with one oblique side) coverslips lacking mirror symmetry are helpful to easily determine the cell-containing side. Commercially available 12 mm round coverslips are also acceptable (*see* **Note 1**).
2. Phosphate-buffered saline (PBS) with Ca^{2+} and Mg^{2+}(-commercially available).
3. PEM buffer: 100 mM PIPES (free acid), pH 6.9 (adjust with KOH), 1 mM $MgCl_2$, and 1 mM EGTA. Make working buffer from 2× stock solution, which can be stored up to 1 month at 4 °C (*see* **Note 2**).
4. Extraction solution: 1% Triton X-100 in PEM buffer supplemented (optionally) with 1–4% polyethylene glycol (PEG) (MW 20,000–40,000), 2–10 μM phalloidin, and/or 2–10 μM taxol (paclitaxel) (*see* **Note 3**). Use a stirrer and allow 15–20 min to dissolve PEG. Extraction solutions can be stored for up to 3 days at 4 °C, but phalloidin and taxol should be added before use. Stock solutions (1000×) of phalloidin and taxol are made in dimethylsulfoxide (DMSO) and stored at −20 °C in aliquots.

2.2 Fixation

1. Glutaraldehyde solution: 2% glutaraldehyde (EM grade, Sigma) in 0.1 M sodium cacodylate, pH 7.3. The working

solution can be stored at 4 °C for up to 1 week, 2× stock solution of sodium cacodylate is stable at 4 °C. *Caution*: Glutaraldehyde is toxic and volatile, so a fume hood should be used when working; sodium cacodylate is toxic.

2. Tannic acid solution: 0.1% tannic acid (Fisher) in distilled water (*see* **Note 4**). Use within a day.
3. Aqueous uranyl acetate solution: 0.2% uranyl acetate in distilled water. Use a stirrer to dissolve. Store at room temperature. *Caution*: Uranyl acetate is toxic.

2.3 Dehydration and Critical Point Drying (CPD)

1. Graded ethanol solutions: 10%, 20%, 40%, 60%, 80%, and 100% ethanol in distilled water.
2. Alcohol uranyl acetate solution: 0.2% uranyl acetate in 100% ethanol. Use a stirrer to dissolve. Use within a day. *Caution*: Uranyl acetate is toxic.
3. Dehydrated ethanol: Wash molecular sieves (4 Å, 8–12 mesh) (Fisher) with several changes of water to remove dust. Dry in the air, bake at 160 °C overnight, cool down, and add to a bottle of 100% ethanol (~50 g per 500 mL). Seal with Parafilm and store at room temperature. Do not shake, as the beads are fragile and easily generate dust.
4. CPD sample holder and scaffolds: A holder with a lid and two scaffolds are homemade with a stainless-steel wire mesh. The holder should fit into the chamber of the CPD apparatus. A scaffold is required to maintain the holder above the stirring bar during stirring. It should fit a beaker in which dehydration will be processed, e.g., a 50 mL glass beaker.
5. CPD device: We use Samdri PVT-3D (Tousimis) CPD with manual operation, but other devices are also appropriate.
6. Carbon dioxide: Use liquid dehydrated CO_2 (bone dry grade) in a tank with a siphon (deep tube) and a water and oil-absorbing filter (Tousimis). Siphonized tanks make it possible to take the liquid phase of CO_2 from the bottom of the tank.

2.4 Platinum and Carbon Coating

1. Vacuum evaporator: We use Auto 306 coater (Boc Edwards) equipped with a water-cooled diffusion pumping system, carbon and metal evaporation sources, a rotary stage, and a thickness monitor.
2. Metals for evaporation (Ted Pella): Tungsten wire (0.76 mm), platinum wire (0.2 mm), and carbon rods (3 mm).

2.5 Preparation of Replicas

1. Hydrofluoric acid (HF): 5–10% HF in distilled water is prepared from the concentrated acid (49%). Do not use glassware to handle acid-containing solutions. *Caution*: It is

extremely volatile and toxic; it causes severe skin burns. Use a fume hood and wear gloves.

2. Platinum loop: The optimal loop diameters are 3–5 mm.
3. EM grids: Formvar-coated EM grids with low mesh size (e.g., 50) to provide a large viewing area. Other options are acceptable.

2.6 Immunogold PREM

1. Immunogold buffer: 20 mM Tris–HCl, pH 8.0, 0.5 M NaCl, and 0.05% Tween 20. For dilution of antibodies, the buffer is supplemented with 1% bovine serum albumin (BSA); for washing, 0.1% BSA is added. Stock solutions (5×) are stable at 4 °C for several months. Sodium azide can be added to the stock solution to prevent microbial contamination. *Caution*: Sodium azide is toxic.
2. Quenching solution: 2 mg/mL sodium borohydrate ($NaBH_4$) in PBS. Use immediately.
3. Blocking solution: 1 mg/mL glycine (or lysine) in PBS. Stable at 4 °C.

2.7 Correlative PREM

1. Marked coverslips: Homemade coverslips with reference marks are prepared by evaporating gold through a finder grid placed in the middle of a 22 × 22 square or 25 mm round coverslip. A variety of finder grids are available commercially. Baking (160 °C overnight) of gold-coated coverslips is necessary for the firm adhesion of gold to glass (*see* **Note 5**). For light microscopy, choose cells on the clear footprint of the finder grid.
2. Light microscopy: The light microscopic system should be equipped with an environmental chamber to maintain normal cell behavior; it should allow for fast exchange of culture medium to extraction solution to quickly stop cellular activity during imaging. A simple option is to use open dishes on a heated stage with the cells growing in a bicarbonate-free medium. We use UNO Stage Top Incubator from Okolab, which has an easily removable chamber lid.

2.8 Actin Depletion

1. Gelsolin buffer: 50 mM MES-KOH, pH 6.3, 0.1 mM $CaCl_2$, 2 mM $MgCl_2$, and 0.5 mM DTT.
2. Gelsolin: dialyze 0.1–0.2 mg/ml of gelsolin against gelsolin buffer. Both full-length gelsolin [62, 77, 78] and recombinant Ca^{2+}-independent gelsolin fragment [46, 63, 65] can be used for actin depletion.

3 Methods

In a basic form, PREM can be used to study the cytoskeleton architecture in a cell population. In an advanced form, it can be combined with immunogold staining to detect specific proteins in the cytoskeleton (Fig. 1), with light microscopy to correlate the cytoskeleton organization with cell behavior or with the distribution and dynamics of fluorescent probes (Figs. 1 and 2) and with actin depletion to fully reveal the organization of microtubules, intermediate filaments, and myosin II filaments (Fig. 3). The basic PREM procedure consisting of detergent extraction, fixation, dehydration, CPD, metal shadowing, and preparation of replicas is described first, and it is followed by the description of the advanced applications.

A major source of artifacts in PREM is the failure to perform a genuine CPD, which may occur if wet samples are transiently exposed to air, or water is not fully exchanged to ethanol or ethanol to liquid CO_2, or if the dried samples absorb ambient humidity. In order to get a high-quality preparation, it is critical not to allow a liquid-gas interface to touch the samples at any point during the procedure. Practically, it means keeping the cells away from air while they are wet and away from water while they are dry. Changes of solutions need to be done quickly, with a layer of liquid always being retained above the cells. After drying, the cells should be kept at low humidity until they are coated with carbon.

3.1 Cell Culture and Extraction

Detergent extraction is used to expose the internal cytoskeletal structures, while the carrier buffer is designed to maximally preserve them until fixation. Additional preservation may be achieved using specific and nonspecific stabilizers.

1. Put small glass coverslips into a culture dish and plate the cells. Cell culture conditions are specific for each system and are not discussed here. If several coverslips are placed into the same dish, make sure that they do not overlap and that there are no bubbles underneath.
2. When the cells are ready, gently remove the culture medium from the dish either (a) using a pipette, or (b) by pouring it out, or (c) by taking a coverslip out with forceps and placing it into the next solution; quickly rinse with the pre-warmed to 37 °C PBS (*see* **Note 6**).
3. Immediately, but gently, add extraction solution equilibrated to room temperature to the dish with coverslips (or place the coverslip into another dish with the extraction solution if the coverslip was taken out); gently stir the dish with the extraction solution to ensure that the detergent instantly reaches all the cells (*see* **Note 7**). Incubate for 3–5 min at room temperature.

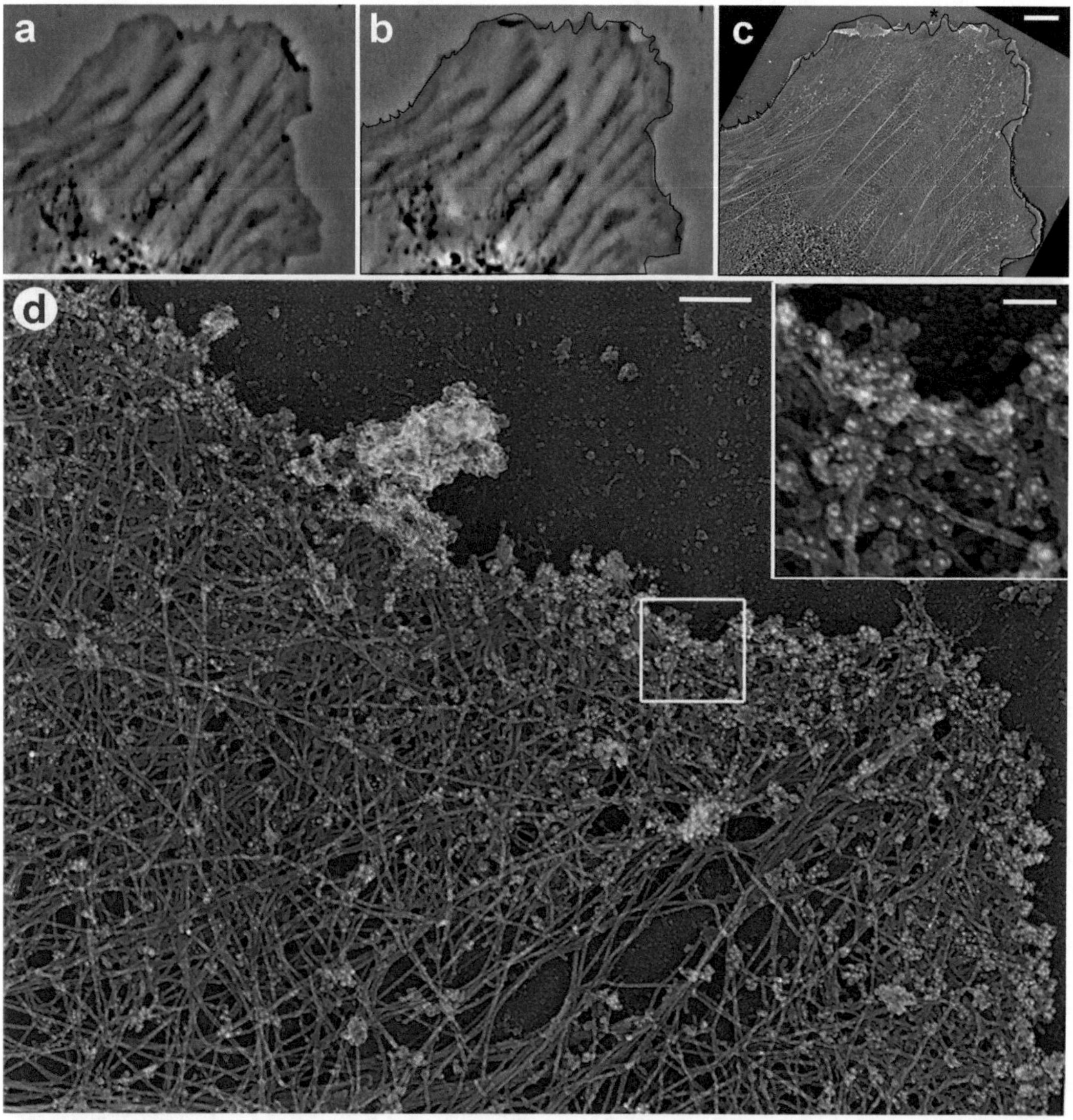

Fig. 1 Correlative phase-contrast and PREM of cultured Rat-2 fibroblast combined with immunogold staining of ADF/cofilin. (**a**, **b**) Frames from time-lapse sequence showing the last live cell image (**b**) and an image 12 s earlier (**a**). *Black line* in (**b**) shows the cell edge outline from (**a**). (**c**) Low magnification PREM image of the cell overlaid with the cell outline, as in (**b**). (**d**) PREM of the protruding edge (*asterisk* in **c**) comprising a lamellipodium filled with dense actin network. Before the PREM processing, the sample was immunogold labeled with cofilin antibody; inset in (**d**) shows gold particles as white dots. Scale bars: 5μm (**a**–**c**), 200 nm (**d**), 50 nm (inset in **d**)

4. Rinse cells with PEM buffer two to three times, for a few seconds each time, at room temperature (*see* **Note 8**).

3.2 Fixation

Chemical fixation provides cytoskeletal structures with physical resistance against damage by subsequent procedures, especially

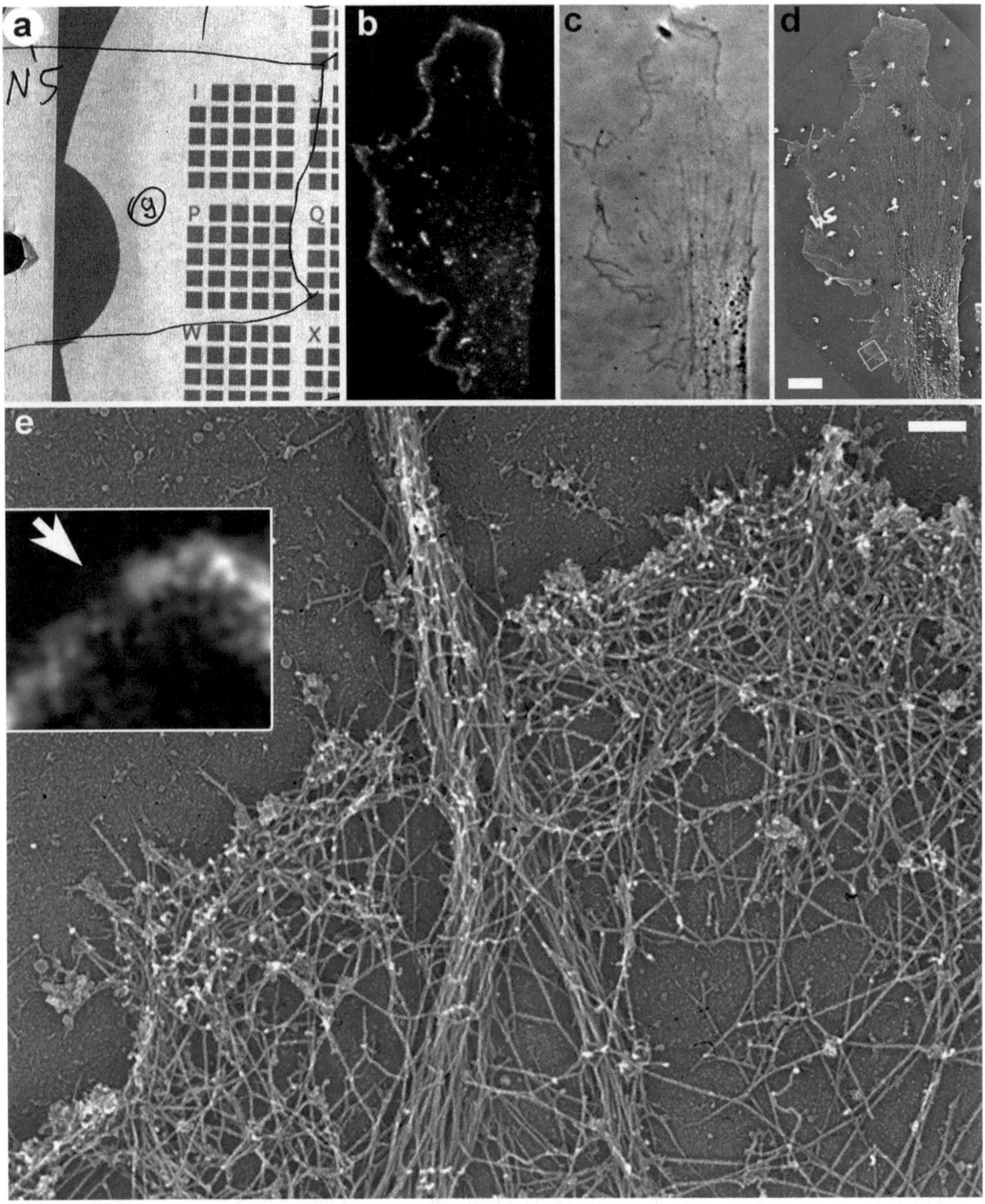

Fig. 2 Correlative fluorescence microscopy and PREM of cultured B16F1 mouse melanoma cell expressing EGFP-capping protein. (**a**) Map showing position of the cell (number 9 in a circle) relative to the reference marks on the coverslip. (**b**) Fluorescence image of the cell showing localization of EGFP-capping protein to the edge of lamellipodia and puncta in lamella. (**c**) Phase contrast image of the same cell. (**d**) Low magnification PREM image of the same cell. Box indicates a region enlarged in (**e**). (**e**) High magnification PREM of the boxed region from (**d**) showing actin filament bundle in a filopodium in the center and dense branched network of actin filaments in lamellipodia. Inset shows the same region by fluorescence microscopy. Bright fluorescence corresponds to lamellipodia, while the dim region (*arrow*) corresponds to the filopodium. Scale bars: 5μm (**b**–**d**), 200 nm (**e**)

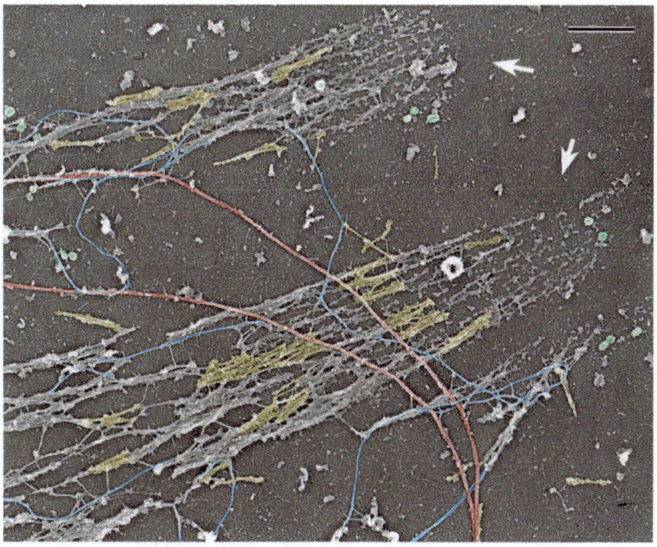

Fig. 3 PREM of the cytoskeleton of a cultured PtK1 potoroo epithelial cell after depletion of actin filaments by gelsolin treatment. Image shows a peripheral cell region containing two large focal adhesions (arrows). Abundant myosin II filaments, a subset of which is highlighted in yellow, accumulate along the actin-depleted stress fibers extending away from focal adhesions, while some individual myosin II filaments can be seen elsewhere in the cytoplasm. Exposed microtubules (red), intermediate filaments (blue), and clathrin-coated vesicles (green) are also visible. Scale bar: 500 nm

dehydration and CPD. It is a three-step procedure using different fixatives: glutaraldehyde, tannic acid, and uranyl acetate.

1. After the last PEM rinse, add glutaraldehyde solution and incubate for at least 20 min at room temperature. If necessary, the samples can be stored at this stage at 4 °C for up to 3 weeks. Take care to prevent evaporation during storage.
2. Transfer coverslips to another container with tannic acid solution, or change solutions in the same dish (*see* **Note 9**). Make sure that the cells remain covered with liquid during transfer. No washing is necessary before this step. Incubate for 20 min at room temperature.
3. Take the coverslips out of the tannic acid solution one by one, rinse by dipping sequentially into two water-filled beakers, and place in a new plate with distilled water. Do not keep the coverslips out of the solution longer than necessary. Incubate for 5 min (*see* **Note 10**).
4. Take the coverslips out of the water one by one, rinse twice again by dipping into water, and place in a new plate with aqueous uranyl acetate solution. Avoid drying during transfer. Incubate for 20 min at room temperature.

5. Wash off uranyl acetate solution with distilled water by transferring the samples or exchanging the solutions.

3.3 Dehydration and CPD

Drying of the samples is necessary to expose the surfaces for metal coating in a vacuum. However, plain drying in the open air generates major structural distortions. When the liquid-gas interface passes through the samples, the forces of the surface tension that are enormous at the cellular scale flatten the samples. During CPD, the temperature and pressure of a liquid are raised above its critical point, at which the phase boundary and surface tension do not exist. In this state, the liquid can well be considered as compressed gas. When the pressure is released, the samples remain dry with the 3D organization intact because they never experienced the surface tension (*see* **Note 11**). Carbon dioxide has reasonably low values of critical point pressure and temperature that can be tolerated by biological samples. However, a direct transfer of the samples from water to CO_2 is not possible, and ethanol, which is freely miscible with both water and CO_2, is used as an intermediate. For dehydration and CPD, the coverslips are stacked in the sample holder with pieces of lens tissue as spacers and processed simultaneously.

1. Cut lens tissue with loosely arranged fibers into pieces fitting the size of the CPD sample holder or a little larger (*see* **Note 12**).
2. Place the CPD sample holder into a wide container with distilled water. Put a piece of lens tissue at the bottom of the holder, place a coverslip with the cell side up on the lens tissue, and cover with another piece of lens tissue. Continue loading the coverslips into the holder, alternating them with pieces of lens tissue. All loading should be done under water. Make sure that the coverslips are minimally exposed to air during loading. Do not overload the holder, as it will interfere with the exchange of solutions (*see* **Note 13**). Keep track of sample identity based on its position in the holder.
3. Put a stirrer bar into a 50 mL glass beaker, place a scaffold over it, and add 10% ethanol in the amount sufficient to cover the CPD holder. Quickly transfer the CPD holder with samples from the water-filled container into the beaker. Place the beaker on a magnetic stirrer, and stir for at least 5 min.
4. Prepare another beaker with a stirrer bar and a scaffold and add 20% ethanol. Transfer the CPD holder from the first beaker into the second, and stir for at least 5 min.
5. Continue dehydration by transferring the CPD holder sequentially through the remaining graded ethanols, at least 5 min in each: 40%, 60%, 80%, and 100% (twice). Alternate the two beakers with their scaffolds and stirrer bars (*see* **Note 14**).

6. Place the CPD holder into a beaker with 0.2% uranyl acetate in ethanol to fully cover it. Incubate for 20 min. Stirring is not necessary.
7. Transfer the CPD holder through beakers with 100% ethanol (twice) and dehydrated 100% ethanol dried over a molecular sieve (twice), as in **steps 3–5**. Stir for at least 5 min in each.
8. Fill the chamber of the CPD apparatus with dehydrated ethanol, just sufficient to cover the CPD holder. Place the holder in the ethanol and close the lid. Operate the CPD according to the manufacturer's instructions or following the procedure below (*see* **Note 15**).
9. Open the CO_2 tank. Cool down the CPD chamber to 10–15 °C to keep the CO_2 in a liquid state. Maintain this temperature until the heating step (*see* **Note 16**).
10. Open the inlet valve on the CPD to allow the CO_2 to fill the chamber. If the device is equipped with a magnetic stirrer, turn it on and keep it running till the heating step. Wait for 5 min. This is a mixing step, during which the ethanol in the chamber, the sample holder, and the samples are equilibrated with the liquid CO_2 from the tank.
11. With the inlet valve still open, open the outlet valve slightly until you clearly see that the liquid exchange is happening in the chamber, but do not allow the level of the liquid to go below the top of the sample holder. Wait for 30 s. This is a washing step, when the ethanol–CO_2 mixture is released from the chamber and replaced with pure CO_2 from the tank. If the CPD is not equipped with a magnetic stirrer, shake the CPD manually during the washing to mix the chamber contents better (*see* **Note 17**).
12. Close the outlet valve. Wait for 5 min.
13. Repeat **steps 11** and **12** nine more times (total of ten washes). After the last wash, wait until the chamber is completely filled with CO_2; that may take less than 5 min.
14. Close both the inlet and outlet valves and turn on the heat. As the chamber is isolated, heating will raise both the temperature and the pressure. Wait until both parameters exceed the critical values for CO_2 (critical pressure = 1072 psi or 73 atm, critical temperature = 31 °C) and reach values of ~1250 psi (85 atm) and ~40 °C. If one value is reached sooner than the other, maintain the former at a steady-state level by turning the heater or the outlet valve on and off until both the values are reached.
15. Open the outlet valve slightly to slowly release the pressure until it reaches the atmospheric pressure. It should take about 10 min (*see* **Note 18**).
16. Open the CPD chamber, remove the sample holder and immediately put it into a desiccator.

3.4 Platinum and Carbon Coating

Platinum shadowing generates the contrast of the samples. The angle and the thickness of the coating are critical parameters influencing the quality of the image. Lower angles provide higher contrast but do not penetrate deep into the sample. Thinner coats provide higher resolution, but lower contrast. For cellular studies, we shadow platinum at a ~45° angle with rotation to achieve a ~2 nm thickness of the coat, which is controlled by the thickness monitor. Carbon is applied from the top of the samples with a thickness of 3.5–5 nm. The basic steps of coating are listed below. Use the equipment manual for detailed operation.

1. Open the coating chamber of the vacuum evaporator. Load the evaporation materials, platinum and carbon. Adjust the angles of coating.
2. Mount the samples onto the rotary stage of the vacuum evaporator using a double-sided tape. This will prevent the dislodging of the samples during rotation (*see* **Note 19**). To prevent damage of the samples by ambient humidity, perform mounting as quickly as possible (*see* **Note 20**).
3. Pump down the coating chamber till ~5×10^{-6} mbar.
4. Turn on the stage rotation, shadow with platinum (2 nm) and then with carbon (3.5–5 nm) (*see* **Note 21**).
5. Vent the coating chamber. Remove the samples together with the mounting tape and place in a Petri dish. The samples can be safely stored at ambient conditions at this stage.

3.5 Preparation of Replicas

The release of the replicas from the coverslips is achieved by floating the coverslips onto the surface of hydrofluoric acid solution, which dissolves glass. After that, the replicas are washed and mounted on EM grids.

1. While the coverslips are still attached to the sticky tape, scratch the coated surface of each coverslip with a needle or a razor blade to make regions fitting the size of an EM grid.
2. Fill the wells of a 12-well plate with ~10% HF almost to the top, which makes replica handling easier.
3. Detach a coverslip from the sticky tape and place it on the surface of the HF solution, so that the coverslip remains floating. Wait until the glass sinks, leaving the platinum replica floating. The replica will fall apart along the scratches made in **step 1** (*see* **Note 22**).
4. Fill the wells of another 12-well plate with distilled water, and add a trace amount of a detergent to decrease the surface tension of water. Stock solution of the detergent is prepared by dissolving ~1 drop of detergent in ~20 mL of water. To make a working solution, dip a platinum loop into the stock

solution, take it out (this will create a film on the loop), and dip the loop into a water-filled well. The final concentration of detergent is ~10^{-6}% (*see* **Note 23**).

5. Using a platinum loop, transfer the replica pieces onto the surface of the detergent-containing water. Wait for 1 min or more.
6. Fill the wells of another 12-well plate with distilled water without detergent. Transfer the replica pieces onto the surface of the water. Wait for 1 min or more.
7. Pick up the replica pieces onto Formvar-coated EM grids with the lower side of the replica attached to the Formvar film. The technique of replica mounting is similar to the way thin sections are picked up onto grids. Fasten a grid in a pair of forceps, partially submerge the grid into water at a ~45° angle, bring the grid close to a piece of replica and allow them to make contact; then, gently pull the grid out of the water making sure that the replica piece remains attached to, and spreads over, the grid (*see* **Note 24**). Dry grids.
8. Examine the samples in TEM. Present the images in inverse contrast (as negatives) because it gives a more natural view of the structure, as if illuminated with scattered light.

3.6 Immunogold PREM

Structural information has much greater value if the identity of the structures is known. Immunostaining is a conventional way to identify cellular components. For EM purposes, the antibodies are labeled with electron-dense markers. A popular marker, colloidal gold, has a higher electron density than platinum and thus is appropriate for PREM. For successful immunogold PREM, a primary antibody should work after glutaraldehyde fixation, which optimally preserves the structure (*see* **Note 25**).

1. After glutaraldehyde fixation (Subheading 3.2, **step 1**), wash the samples with three changes of PBS. Incubate for at least 5 min in the last change.
2. Quench samples with $NaBH_4$ in PBS for 10 min at room temperature. Shake off the bubbles occasionally. Rinse with PBS as in **step 1**.
3. Incubate with blocking solution for 20 min at room temperature.
4. Apply primary antibody at concentration giving bright staining by light microscopy. Incubate for 30–45 min at room temperature. Wash with PBS as in **step 1**.
5. Rinse once in immunogold buffer with 0.1% BSA. Apply gold-conjugated secondary antibody diluted ~1:10 in immunogold buffer with 1% BSA. Incubate overnight at room temperature in a sealed container in moist conditions (*see* **Note 26**).

6. Rinse in immunogold buffer with 0.1% BSA as in **step 1**, and perform the remaining steps starting from Subheading 3.2, **step 1** (*see* **Note 27**).

3.7 Correlative Light Microscopy and PREM (CLEM)

CLEM combines the advantages of both microscopic techniques, namely, the high spatial resolution of EM and the high temporal resolution of optical live imaging. In this procedure, the cell dynamics is recorded by light microscopy, and then the same cell is analyzed by EM. The correlative analysis is extremely important from at least two points of view: to control for potential artifacts and to establish functional connections between the cytoskeletal organization and the cell's motile behavior or the dynamics of cytoskeletal components [79, 80]. Modifications of the basic procedure as required for correlative PREM are described below.

1. Cell Culture and Light Microscopy.
   (a) Grow cells in dishes with marked coverslips mounted over a hole in the bottom of the dish. To make a dish, drill a 18 mm hole in the bottom of a 35 mm dish; polish the edges to remove any burrs; apply a thin layer of vacuum grease just outside the edges of the hole; place a marked coverslip symmetrically over the hole with the coated side facing up; and press firmly to spread the grease until it forms a clear circle and all the air bubbles are gone (*see* **Note 28**).
   (b) While imaging a region of interest, mark its position on a map with a pattern of reference marks.
   (c) Because of the large difference in the resolution of light microscopy and PREM, perform light microscopy at the highest possible resolution.
   (d) PREM is able to reveal even minor photodamage, not recognizable at the light microscopic level; therefore, keep the illumination of the samples to a minimum.
2. *Extraction*: Change the culture medium to extraction solution as soon as possible after the acquisition of the last live image, and take another image after extraction. It will serve as a reference to correlate the light and PREM images.
3. Fixation: Perform all the fixation and washing steps with the marked coverslips still attached to the dishes, by exchanging solutions.
4. Dehydration: Before loading the coverslips into the CPD sample holder, excise the central marked area of the coverslip as following:
   (a) Wipe away the immersion oil from the bottom of the dish using dry cotton swabs first, followed by ethanol-soaked swabs.

(b) Detach the marked coverslip from the bottom of the dish, and immediately place it into a water-filled 100 mm Petri dish.

(c) Lightly press down the coverslip to allow the vacuum grease on the underside to secure the coverslip in the dish. Make sure that grease does not contaminate the region of interest.

(d) Use a diamond pencil as a cutter and a razor blade as a ruler to cut off the greasy margins of the coverslip. Do not press the pencil hard, but instead make several light cuts along the same line, guided by the razor blade, until the margin detaches. Move it out of the way and cut off another margin. This procedure should be done with the coverslip completely submerged in the water (*see* **Note 29**). When done, transfer the excised central region to another water-filled container. Use a new dish for the next coverslip, as the remaining glass crumbs may cause shattering of the coverslip.

All subsequent processing, including coating, is performed as in the basic procedure.

5. Preparation of replicas: After the samples are coated, the region of interest needs to be specifically recovered for PREM analysis as follows:

(a) Immobilize a coated coverslip in the middle of a 100 mm Petri dish, with two pieces of double-sided tape positioned under opposite corners of the coverslip, so that the region of interest is not obstructed and the coverslip is not attached too strongly.

(b) Under a dissection microscope, localize the cells of interest using locator marks. Dried and shadowed cells have sufficient contrast for their shape to be seen even at low magnification. If necessary, use a regular light microscope with a low power lens.

(c) Using a razor blade (or a needle), make cuts in the platinum–carbon layer around the region of interest (*see* **Note 30**). To facilitate the separation of this region from the rest of the replica after HF treatment, make additional cuts connecting the region of interest with the edges of the coverslip.

(d) Make a drawing on the map to depict the exact shape of the outlined region of interest, which will help to identify it from among the other replica pieces during replica preparation.

(e) Perform the washing and mounting on grids, as described in the basic protocol, handling only the replica piece with the region of interest. While mounting on a grid, use a dissection microscope to make sure that the cell of interest does not go to a grid bar; or use single-hole grids.

3.8 Actin Depletion

Depletion of abundant actin filaments from the cytoskeletons by gelsolin treatment exposes other normally obscured cytoskeletal structures, thus permitting detailed analyses of their organization in cells (*see* **Note 31**)

1. After detergent extraction (Subheading 3.1, **step 4**), rinse coverslips once with gelsolin buffer supplemented with 10 μM taxol to preserve microtubules (*see* **Note 32**).
2. Make small drops of gelsolin solution on a piece of Parafilm placed on the bottom of a Petri dish, one drop per coverslip at sufficient distances from one another (*see* **Note 33**).
3. Place a coverslip onto the gelsolin drop with cell side up and press the coverslip down to allow the gelsolin solution to completely cover the coverslip surface. Close the dish and incubate 1 h at room temperature in moist conditions (*see* **Note 34**).
4. Rinse three times with PEM buffer and proceed to sample fixation (Subheading 3.2, **step 1**).

4 Notes

1. Small coverslips allow for better exchange of solutions during dehydration and CPD and thus for better quality of samples at the end.
2. Stock solutions with a concentration of more than 2× relative to the working concentration would change pH significantly after dilution to the working concentration. Free acid PIPES is not soluble in water and forms a milky suspension but becomes soluble upon neutralization. KOH granules can be used for neutralization initially, until the solution almost clears. However, remember to allow enough time for the granules to dissolve before adding more. Finish the pH adjustment with 1 N KOH. KOH is preferable over NaOH because K^+-containing buffers more faithfully imitate the cytoplasm composition.
3. PEG is a nonspecific stabilizer of the cytoskeleton; phalloidin and taxol are specific stabilizers of actin filaments and microtubules, respectively.
4. Low molecular weight tannic acid (MW ~322), also called digallic acid, which was originally manufactured by Mallinckrodt Inc. under the product number 1764, has been shown to yield best results in preserving cells structures, including

fibrillar components of the cytoplasm [81] and the actin cytoskeleton, in particular [82], as compared with high molecular weight tannins (≥1400). We are still using the Mallinckrodt brand of low molecular weight tannic acid that was previously offered by Fisher but currently discontinued. Digallic acid is now offered by other companies, e.g., Santa Cruz Biotechnology, and appears to be a valid substitute, although we did not test it.

5. Commercially available etched coverslips are not suitable for PREM, as the etched marks are not visible in TEM.
6. Rinsing with PBS is optional, but if omitted, the extraction solution at the next step should be added in sufficient quantity to overcome the potentially harmful effects from the remaining medium and serum.
7. High viscosity of the extraction solution makes it poorly miscible with PBS or medium, in which cells were submerged in the preceding step. Without stirring, slow diffusion of the detergent would lead to gradual extraction and result in severe cell distortion and cytoskeleton disruption. On the other hand, vigorous stirring would disrupt the cytoskeleton mechanically. Thus, a fine compromise should be found. The choice of the extraction solution depends on a cell type and a goal. For a new experimental system, try different options in preliminary experiments. A basic extraction solution (Triton X-100 in PEM) produces better clarity of the cytoskeleton but makes it easier to damage the cells during extraction. If using this protocol, handle the samples extremely gently, and use phalloidin and taxol to better preserve the actin filaments and microtubules, respectively. The addition of PEG to the extraction solution allows for better preservation of the structure, but it also retains many cytoskeleton-associated components, which may partially obscure the filament arrangement. Such an effect is increased with increasing PEG concentration and molecular weight, but PEGs in the range of 20,000–40,000 act similarly. We typically use 2% PEG (35,000). For extremely fragile and poorly attached cells, low concentrations of glutaraldehyde can be added to the extraction solution as stabilizing supplements. In this case, the detergent and fixative compete with each other, and the results depend on their ratio. The extraction solution containing 0.5% Triton X-100 and 0.25% glutaraldehyde in PEM buffer worked well in our experiments.
8. For PEG-containing extraction solutions, use a longer washing time, at least 1 min in each change. If the stabilizing drugs are used during extraction, add them also to the rinsing buffer in a fourfold to fivefold lower concentration.

9. It is convenient to use a multiwell plate with numbered wells (24-well for 6–8 mm coverslips or 12-well for 9–12 mm coverslips) to transfer the samples. This makes it possible to combine samples from different experiments for PREM processing while keeping parallel samples in the original container as a backup.
10. Uranyl acetate and tannic acid react with each other and form a precipitate. Extensive washing is important to avoid the formation of debris on the samples.
11. Two alternative procedures have been used to dry samples before metal coating: freeze-drying [41, 49, 57, 83–87] and air-drying from an organic solvent hexamethyldisilazane (HMDS) [47, 88–90]. Freeze-drying, when successful, results in excellent sample quality but is limited by a low success rate. In some published images of freeze-dried samples, neighboring filaments appear fused suggesting inadequate drying. This problem might arise from insufficient rate of ice sublimation or partial sample melting. Air-drying from HMDS is becoming popular due to its simplicity. As compared with water, drying from HMDS causes less sample distortion because HMDS has low, but not zero, surface tension. Although CPD and HMDS drying are believed to be equally good, direct comparisons of these two procedures were conducted only using microscopy techniques that were insufficient to resolve individual cytoskeletal filaments [91–93]. We directly compared sample quality by PREM of cytoskeletal preparations prepared using either CPD or air-drying from HMDS. We found that the samples were comparable at low and medium magnifications, but at high magnifications required to resolve individual actin filaments, the quality of HDMS-dried samples was clearly inferior. Specifically, in CPD samples, actin filaments had uniform diameters and were clearly distinguishable from one another even in tight bundles. In contrast, in the HDMS-dried samples, actin filaments often fused with each other and with an underlying surface. Similar artifacts can also be seen in published PREM images obtained after HMDS drying.
12. Pieces of lens tissue slightly larger, than the holder's bottom area, will make minor wrinkles which promote looser packing of the coverslips in the holder and facilitate the liquid exchange.
13. The acceptable number of samples for a load depends on the sizes of both the holder and the coverslips. For an 18 × 12 mm holder and ~7 × 7 mm coverslips, the maximum load is 12. For larger coverslips, the load should be decreased. Larger holders may accept more samples, especially if the coverslips are staggered.

14. It is not necessary to dry the beakers before the next incubation, as the ethanol concentration may not be exact, except for 100% ethanol, when it is better to dry the beakers and scaffolds. Incubation for 5 min is minimal. For larger coverslips or greater loads, increase the incubation time.
15. The process of CPD is most commonly used for scanning EM and production of microelectronics. Consequently, the protocols suggested by the manufacturers or incorporated into automated procedures of CPDs are designed for those applications. PREM, however, is more demanding in terms of sample quality. We adjusted the CPD processing to fully remove all traces of ethanol from the samples before bringing the CO_2 to the critical point; this helps to eliminate minor artifacts that appear as a fusion of closely positioned filaments in the cytoskeleton. The CPD operation described here is applicable to manual CPDs, such as Samdri PVD-3D (Tousimis), which we use in the lab, or to semi-automatic CPDs switched to a manual mode of operation, e.g., Samdri-795 (Tousimis).
16. Lower temperatures are acceptable, but the diffusion of ethanol from the samples will be slower, so longer washing time is needed. Warming up the chamber till the ambient temperature is allowed if the outlet valve of the CPD is closed, and the CO_2 remains pressurized and in liquid form. However, it is important to cool down the chamber back to 15 °C before opening the outlet valve for purging out the ethanol–CO_2 mixture.
17. Letting the liquid level go below the samples will irreversibly damage them. On the other hand, a too low rate of liquid exchange is also a mistake. Adjust the outlet valve to get a steady-state liquid level, about halfway from the top of the holder to the top of the chamber. This will also make the liquid mixing more efficient. Although the shaking step sounds a bit amusing, it does make difference by helping to remove the ethanol from the samples.
18. Fast release of pressure may cause condensation of CO_2 back to liquid state and ruin the dried samples.
19. Conventional Scotch double-sided tape becomes too sticky in a vacuum, preventing the safe detachment of the samples after coating. To avoid this problem, sandwich the double-sided tape between the glued parts of two Post-It notes, with the sticky sides exposed.
20. Humidity in the room should be below 35%; the 35–50% humidity level may be acceptable, but much caution and the speedy mounting of the samples is required; humidity >50% is not acceptable. Try to run a powerful dehumidifier in the latter case.

21. If the evaporator is not equipped with a thickness monitor, the thickness of the coating may be adjusted in the preliminary experiments based on the amount of coating material loaded (for platinum) or used (carbon) for evaporation.

22. To safely float a coverslip, grab it with the forceps from the top for parallel edges, lift the coverslip, and carefully place it onto the liquid surface, keeping it in a horizontal position. Practice first by placing a coverslip onto a clean solid surface. Alternatively, a coverslip can be placed first on a platinum loop bent at 90° angle and then loaded horizontally onto the liquid surface. If the replica does not fall apart along the scratches, use the platinum loop to reach the replica from below, lightly touch it, and pull or shake it to detach it from other pieces. Extreme care should be used not to ruin the replicas with these manipulations.

23. Water has much greater surface tension than HF, which may cause severe replica breakage, if detergent is not added. Test the detergent concentration before applying it to the samples. An overdose of detergent causes shrinkage and even drowning of the replicas. Stock solutions should be changed at least every 2 weeks. Old detergents leave contamination on the samples, looking like semi-transparent films between filaments. Household non-colored detergent, such as Ivory, works fine. Triton X-100 can also be used, but it should be prepared fresh every time.

24. Sometimes, replicas appear to be repelled by the grid, making it difficult to establish the initial contact between a replica piece and a grid. Try to gently guide a piece of replica to the wall of the well to restrict its motility, and then pick it up. However, there is a danger of smashing the replica against the wall with this approach. Treating grids with glow discharge prior to capturing replicas also helps.

25. The efficiency of staining may be improved if the cells are fixed with a lower (e.g., 0.2%) glutaraldehyde concentration before staining. For some antibodies that do not work after glutaraldehyde fixation, it may be possible to stain unfixed samples by incubating them with primary antibodies diluted in PEM with phalloidin and taxol for 10–15 min, then fixing with glutaraldehyde, quenching, washing and staining with a secondary antibody.

26. Gold size of 10–20 nm is optimal for this technique, as smaller particles are poorly visible and larger particles are too disruptive for an image.

27. Second glutaraldehyde fixation helps retaining gold-conjugated secondary antibody during subsequent procedures.

28. The coverslips can be mounted either inside or outside the dish, but inside mounting is more convenient at later stages, when the centerpiece of the coverslip needs to be cut out. For mounting, use a minimal amount of grease, just sufficient to seal the dish; excessive grease causes complications at later stages. Commercially available glass-bottom dishes can also be used, but it may be difficult to remove the coverslip from the bottom for PREM processing.
29. Cutting under water is more difficult than in the air; therefore, use a sharp diamond pencil and avoid glass crumbs.
30. To reduce the effect of shaky hands, hold a razor blade with one hand with the sharp blade corner pointing down; stabilize the blade by putting the index finger of the other hand onto the blunt blade corner pointing up; rest the forearms on the table and the other fingers of both hands on the microscope stage and/or dish edges; keep the blade above the sample and find its unfocused image in the microscope; slowly bring down the sharp corner of the blade until it almost comes to focus; bring the blade corner to a region where a cut is to be made; and under microscope control, bring it down to the sample and make a scratch.
31. It appears that a similar goal of depleting actin filaments could be achieved using actin-depolymerizing drugs, such as latrunculins or cytochalasins. However, these drugs need to be applied to living cells, which severely affects cell physiology and causes multiple structural rearrangements of the cytoskeleton. In contrast, gelsolin is applied to cytoskeletal preparations, not to living cells, and non-actin structures retain their endogenous organization in the course of treatment [46, 64, 65].
32. Taxol should be added to all solutions for preservation of microtubules, because otherwise all microtubules will be destroyed by the Ca^{2+}-containing gelsolin buffer. If using calcium-independent N-terminal fragment of gelsolin [94], rather than the full length gelsolin, it is possible to use PEM buffer instead of gelsolin buffer as vehicle. Yet, addition of taxol greatly helps in proper preservation of microtubules in either case.
33. Hydrophobic surface of Parafilm prevents gelsolin solution from spreading. The volume of drops depends on the size of coverslips and should be sufficient to cover the coverslip surface.
34. One option to generate moist conditions is to attach a water-soaked filter paper to the lid to the Petri dish, shake off extra water, and close the lid. Another option is to put wet filter paper around the dish periphery.

Acknowledgments

The author acknowledges the current support from NIH grant R01 GM 095977.

References

1. Wohlfarth-Bottermann KE (1962) Weitreichende fibrillare Protoplasmadifferenzierungen und ihre Bedeutung fur die Protoplasmastromung. I. Elektronenmikroskopischer Nachweis und Feinstruktur. Protoplasma 54:514–539
2. Ledbetter MC, Porter KR (1963) A "microtubule" in plant cell fine structure. J Cell Biol 19 (1):239–250
3. Sabatini DD, Bensch K, Barrnett RJ (1963) Cytochemistry and electron microscopy. The preservation of cellular ultrastructure and enzymatic activity by aldehyde fixation. J Cell Biol 17:19–58
4. Abercrombie M, Heaysman JE, Pegrum SM (1971) The locomotion of fibroblasts in culture. IV. Electron microscopy of the leading lamella. Exp Cell Res 67(2):359–367
5. Wohlfarth-Bottermann KE (1964) Differentiations of the ground cytoplasm and their significance for the generation of the motive force of ameboid movement. In: Allen RD, Kamiya N (eds) Primitive motile systems in cell biology. Academic, pp 79–109
6. Kukulski W, Schorb M, Kaksonen M, Briggs JA (2012) Plasma membrane reshaping during endocytosis is revealed by time-resolved electron tomography. Cell 150(3):508–520. https://doi.org/10.1016/j.cell.2012.05.046
7. Luduena MA, Wessells NK (1973) Cell locomotion, nerve elongation, and microfilaments. Dev Biol 30(2):427–440
8. Tilney LG, Derosier DJ, Mulroy MJ (1980) The organization of actin filaments in the stereocilia of cochlear hair cells. J Cell Biol 86 (1):244–259
9. Wu M, Huang B, Graham M, Raimondi A, Heuser JE, Zhuang X, De Camilli P (2010) Coupling between clathrin-dependent endocytic budding and F-BAR-dependent tubulation in a cell-free system. Nat Cell Biol 12 (9):902–908
10. Small JV, Isenberg G, Celis JE (1978) Polarity of actin at the leading edge of cultured cells. Nature 272(5654):638–639
11. Damiano-Guercio J, Kurzawa L, Mueller J, Dimchev G, Schaks M, Nemethova M, Pokrant T, Bruhmann S, Linkner J, Blanchoin L, Sixt M, Rottner K, Faix J (2020) Loss of Ena/VASP interferes with lamellipodium architecture, motility and integrin-dependent adhesion. elife 9:e55351. https://doi.org/10.7554/eLife.55351
12. Mueller J, Szep G, Nemethova M, de Vries I, Lieber AD, Winkler C, Kruse K, Small JV, Schmeiser C, Keren K, Hauschild R, Sixt M (2017) Load adaptation of lamellipodial actin networks. Cell 171(1):188–200.e116. https://doi.org/10.1016/j.cell.2017.07.051
13. Sturner T, Tatarnikova A, Mueller J, Schaffran B, Cuntz H, Zhang Y, Nemethova M, Bogdan S, Small V, Tavosanis G (2019) Transient localization of the Arp2/3 complex initiates neuronal dendrite branching in vivo. Development 146(7):dev171397. https://doi.org/10.1242/dev.171397
14. Chakraborty S, Jasnin M, Baumeister W (2020) Three-dimensional organization of the cytoskeleton: a cryo-electron tomography perspective. Protein Sci 29(6):1302–1320. https://doi.org/10.1002/pro.3858
15. Kopek BG, Paez-Segala MG, Shtengel G, Sochacki KA, Sun MG, Wang Y, Xu CS, van Engelenburg SB, Taraska JW, Looger LL, Hess HF (2017) Diverse protocols for correlative super-resolution fluorescence imaging and electron microscopy of chemically fixed samples. Nat Protoc 12(5):916–946. https://doi.org/10.1038/nprot.2017.017
16. Murata K, Wolf M (2018) Cryo-electron microscopy for structural analysis of dynamic biological macromolecules. Biochim Biophys Acta 1862(2):324–334. https://doi.org/10.1016/j.bbagen.2017.07.020
17. Stark H, Chari A (2016) Sample preparation of biological macromolecular assemblies for the determination of high-resolution structures by cryo-electron microscopy. Microscopy 65 (1):23–34. https://doi.org/10.1093/jmicro/dfv367
18. Lucic V, Rigort A, Baumeister W (2013) Cryo-electron tomography: the challenge of doing structural biology in situ. J Cell Biol 202 (3):407–419. https://doi.org/10.1083/jcb.201304193

19. Engel BD, Schaffer M, Albert S, Asano S, Plitzko JM, Baumeister W (2015) In situ structural analysis of Golgi intracisternal protein arrays. Proc Natl Acad Sci U S A 112 (36):11264–11269. https://doi.org/10.1073/pnas.1515337112
20. Fischer TD, Dash PK, Liu J, Waxham MN (2018) Morphology of mitochondria in spatially restricted axons revealed by cryo-electron tomography. PLoS Biol 16(9):e2006169. https://doi.org/10.1371/journal.pbio.2006169
21. Hoffman DP, Shtengel G, Xu CS, Campbell KR, Freeman M, Wang L, Milkie DE, Pasolli HA, Iyer N, Bogovic JA, Stabley DR, Shirinifard A, Pang S, Peale D, Schaefer K, Pomp W, Chang CL, Lippincott-Schwartz J, Kirchhausen T, Solecki DJ, Betzig E, Hess HF (2020) Correlative three-dimensional super-resolution and block-face electron microscopy of whole vitreously frozen cells. Science 367(6475):eaaz5357. https://doi.org/10.1126/science.aaz5357
22. Hurley JH, Nogales E (2016) Next-generation electron microscopy in autophagy research. Curr Opin Struct Biol 41:211–216. https://doi.org/10.1016/j.sbi.2016.08.006
23. Jorgens DM, Inman JL, Wojcik M, Robertson C, Palsdottir H, Tsai WT, Huang H, Bruni-Cardoso A, Lopez CS, Bissell MJ, Xu K, Auer M (2017) Deep nuclear invaginations are linked to cytoskeletal filaments—integrated bioimaging of epithelial cells in 3D culture. J Cell Sci 130(1):177–189. https://doi.org/10.1242/jcs.190967
24. Mahamid J, Pfeffer S, Schaffer M, Villa E, Danev R, Cuellar LK, Forster F, Hyman AA, Plitzko JM, Baumeister W (2016) Visualizing the molecular sociology at the HeLa cell nuclear periphery. Science 351 (6276):969–972. https://doi.org/10.1126/science.aad8857
25. Chakraborty S, Mahamid J, Baumeister W (2020) Cryoelectron tomography reveals nanoscale organization of the cytoskeleton and its relation to microtubule curvature inside cells. Structure 28(9):991–1003.e1004. https://doi.org/10.1016/j.str.2020.05.013
26. Grange M, Vasishtan D, Grunewald K (2016) Cellular electron cryo tomography and in situ sub-volume averaging reveal the context of microtubule-based processes. J Struct Biol 197(2):181–190. https://doi.org/10.1016/j.jsb.2016.06.024
27. Paul DM, Mantell J, Borucu U, Coombs J, Surridge KJ, Squire JM, Verkade P, Dodding MP (2020) In situ cryo-electron tomography reveals filamentous actin within the microtubule lumen. J Cell Biol 219(9): e201911154. https://doi.org/10.1083/jcb.201911154
28. Anderson KL, Page C, Swift MF, Suraneni P, Janssen ME, Pollard TD, Li R, Volkmann N, Hanein D (2017) Nano-scale actin-network characterization of fibroblast cells lacking functional Arp2/3 complex. J Struct Biol 197 (3):312–321. https://doi.org/10.1016/j.jsb.2016.12.010
29. Marston DJ, Anderson KL, Swift MF, Rougie M, Page C, Hahn KM, Volkmann N, Hanein D (2019) High Rac1 activity is functionally translated into cytosolic structures with unique nanoscale cytoskeletal architecture. Proc Natl Acad Sci U S A 116(4):1267–1272. https://doi.org/10.1073/pnas.1808830116
30. Jasnin M, Crevenna AH (2016) Quantitative analysis of filament branch orientation in Listeria actin comet tails. Biophys J 110 (4):817–826. https://doi.org/10.1016/j.bpj.2015.07.053
31. Jasnin M, Beck F, Ecke M, Fukuda Y, Martinez-Sanchez A, Baumeister W, Gerisch G (2019) The architecture of traveling actin waves revealed by cryo-electron tomography. Structure 27(8):1211–1223.e1215. https://doi.org/10.1016/j.str.2019.05.009
32. Heuser J (1981) Preparing biological samples for stereomicroscopy by the quick-freeze, deep-etch, rotary-replication technique. Methods Cell Biol 22:97–122
33. Svitkina TM (2017) Platinum replica electron microscopy: imaging the cytoskeleton globally and locally. Int J Biochem Cell Biol 86:37–41. https://doi.org/10.1016/j.biocel.2017.03.009
34. Steere RL (1957) Electron microscopy of structural detail in frozen biological specimens. J Cell Biol 3(1):45–60
35. Friend JE, Sayyad WA, Arasada R, McCormick CD, Heuser JE, Pollard TD (2018) Fission yeast Myo2: molecular organization and diffusion in the cytoplasm. Cytoskeleton (Hoboken) 75(4):164–173. https://doi.org/10.1002/cm.21425
36. Heuser J (1989) Protocol for 3-D visualization of molecules on mica via the quick-freeze, deep-etch technique. J Electron Microsc Tech 13(3):244–263
37. Loesser KE, Franzini-Armstrong C (1990) A simple method for freeze-drying of macromolecules and macromolecular complexes. J Struct Biol 103(1):48–56
38. Chikina AS, Svitkina TM, Alexandrova AY (2019) Time-resolved ultrastructure of the cortical actin cytoskeleton in dynamic

membrane blebs. J Cell Biol 218(2):445–454. https://doi.org/10.1083/jcb.201806075
39. Efimova N, Svitkina TM (2018) Branched actin networks push against each other at adherens junctions to maintain cell-cell adhesion. J Cell Biol 217(5):1827–1845. https://doi.org/10.1083/jcb.201708103
40. Henson JH, Ditzler CE, Germain A, Irwin PM, Vogt ET, Yang S, Wu X, Shuster CB (2017) The ultrastructural organization of actin and myosin II filaments in the contractile ring: new support for an old model of cytokinesis. Mol Biol Cell 28(5):613–623. https://doi.org/10.1091/mbc.E16-06-0466
41. Heuser JE, Kirschner MW (1980) Filament organization revealed in platinum replicas of freeze-dried cytoskeletons. J Cell Biol 86 (1):212–234
42. Ong K, Svitkina T, Bi E (2016) Visualization of in vivo septin ultrastructures by platinum replica electron microscopy. Methods Cell Biol 136:73–97. https://doi.org/10.1016/bs.mcb.2016.03.011
43. Scott BL, Sochacki KA, Low-Nam ST, Bailey EM, Luu Q, Hor A, Dickey AM, Smith S, Kerkvliet JG, Taraska JW, Hoppe AD (2018) Membrane bending occurs at all stages of clathrin-coat assembly and defines endocytic dynamics. Nat Commun 9(1):419. https://doi.org/10.1038/s41467-018-02818-8
44. Sochacki KA, Shtengel G, van Engelenburg SB, Hess HF, Taraska JW (2014) Correlative super-resolution fluorescence and metal-replica transmission electron microscopy. Nat Methods 11(3):305–308. https://doi.org/10.1038/nmeth.2816
45. Svitkina TM, Shevelev AA, Bershadsky AD, Gelfand VI (1984) Cytoskeleton of mouse embryo fibroblasts. Electron microscopy of platinum replicas. Eur J Cell Biol 34(1):64–74
46. Svitkina TM, Verkhovsky AB, Borisy GG (1995) Improved procedures for electron microscopic visualization of the cytoskeleton of cultured cells. J Struct Biol 115(3):290–303
47. Vassilopoulos S, Gibaud S, Jimenez A, Caillol G, Leterrier C (2019) Ultrastructure of the axonal periodic scaffold reveals a braid-like organization of actin rings. Nat Commun 10(1):5803. https://doi.org/10.1038/s41467-019-13835-6
48. Yang C, Svitkina TM (2019) Ultrastructure and dynamics of the actin-myosin II cytoskeleton during mitochondrial fission. Nat Cell Biol 21(5):603–613. https://doi.org/10.1038/s41556-019-0313-6
49. Hirokawa N (1989) Quick-freeze, deep-etch electron microscopy. J Electron Microsc (Tokyo) 38 Suppl:S123–S128
50. Meyer HW, Richter W (2001) Freeze-fracture studies on lipids and membranes. Micron 32 (6):615–644
51. Suleiman H, Zhang L, Roth R, Heuser JE, Miner JH, Shaw AS, Dani A (2013) Nanoscale protein architecture of the kidney glomerular basement membrane. elife 2:e01149. https://doi.org/10.7554/eLife.01149
52. Small JV (1981) Organization of actin in the leading edge of cultured cells: influence of osmium tetroxide and dehydration on the ultrastructure of actin meshworks. J Cell Biol 91(3 Pt 1):695–705
53. Svitkina TM, Borisy GG (1998) Correlative light and electron microscopy of the cytoskeleton of cultured cells. Methods Enzymol 298:570–592. https://doi.org/10.1016/S0076-6879(98)98045-4
54. Svitkina T (2007) Electron microscopic analysis of the leading edge in migrating cells. Methods Cell Biol 79:295–319. https://doi.org/10.1016/S0091-679X(06)79012-4
55. Collins A, Warrington A, Taylor KA, Svitkina T (2011) Structural organization of the actin cytoskeleton at sites of clathrin-mediated endocytosis. Curr Biol 21(14):1167–1175. https://doi.org/10.1016/j.cub.2011.05.048
56. Heuser J (2000) The production of 'cell cortices' for light and electron microscopy. Traffic 1 (7):545–552
57. Morone N, Usukura E, Narita A, Usukura J (2020) Improved unroofing protocols for cryo-electron microscopy, atomic force microscopy and freeze-etching electron microscopy and the associated mechanisms. Microscopy (Oxf) 69(6):350–359. https://doi.org/10.1093/jmicro/dfaa028
58. Brands R, Feltkamp CA (1988) Wet cleaving of cells: a method to introduce macromolecules into the cytoplasm. Application for immunolocalization of cytosol-exposed antigens. Exp Cell Res 176(2):309–318
59. Massaad MJ, Oyoshi MK, Kane J, Koduru S, Alcaide P, Nakamura F, Ramesh N, Luscinskas FW, Hartwig J, Geha RS (2014) Binding of WIP to actin is essential for T Cell actin cytoskeleton integrity and tissue homing. Mol Cell Biol 34(23):4343–4354. https://doi.org/10.1128/MCB.00533-14
60. Peitsch CF, Beckmann S, Zuber B (2016) iMEM: isolation of plasma membrane for cryoelectron microscopy. Structure 24 (12):2198–2206. https://doi.org/10.1016/j.str.2016.09.016
61. Sato F, Asakawa H, Fukuma T, Terada S (2016) Semi-in situ atomic force microscopy imaging of intracellular neurofilaments under physiological conditions through the

'sandwich' method. Microscopy 65 (4):316–324. https://doi.org/10.1093/jmicro/dfw006

62. Shutova MS, Spessott WA, Giraudo CG, Svitkina T (2014) Endogenous species of mammalian nonmuscle myosin IIA and IIB include activated monomers and heteropolymers. Curr Biol 24(17):1958–1968. https://doi.org/10.1016/j.cub.2014.07.070
63. Svitkina TM, Verkhovsky AB, Borisy GG (1996) Plectin sidearms mediate interaction of intermediate filaments with microtubules and other components of the cytoskeleton. J Cell Biol 135(4):991–1007
64. Verkhovsky AB, Borisy GG (1993) Non-sarcomeric mode of myosin II organization in the fibroblast lamellum. J Cell Biol 123 (3):637–652
65. Verkhovsky AB, Svitkina TM, Borisy GG (1995) Myosin II filament assemblies in the active lamella of fibroblasts: their morphogenesis and role in the formation of actin filament bundles. J Cell Biol 131(4):989–1002
66. Svitkina TM, Borisy GG (2006) Correlative light and electron microscopy studies of cytoskeletal dynamics. In: Celis J (ed) Cell biology: a laboratory handbook, vol 3. vol 28, 3rd edn. Elsevier, pp 277–285
67. Efimova N, Yang C, Chia JX, Li N, Lengner CJ, Neufeld KL, Svitkina TM (2020) Branched actin networks are assembled on microtubules by adenomatous polyposis coli for targeted membrane protrusion. J Cell Biol 219(9): e202003091. https://doi.org/10.1083/jcb.202003091
68. Efimova N, Korobova F, Stankewich MC, Moberly AH, Stolz DB, Wang J, Kashina A, Ma M, Svitkina T (2017) BetaIII spectrin is necessary for formation of the constricted neck of dendritic spines and regulation of synaptic activity in neurons. J Neurosci 37 (27):6442–6459. https://doi.org/10.1523/JNEUROSCI.3520-16.2017
69. Ripoll L, Heiligenstein X, Hurbain I, Domingues L, Figon F, Petersen KJ, Dennis MK, Houdusse A, Marks MS, Raposo G, Delevoye C (2018) Myosin VI and branched actin filaments mediate membrane constriction and fission of melanosomal tubule carriers. J Cell Biol 217(8):2709–2726. https://doi.org/10.1083/jcb.201709055
70. Sochacki KA, Dickey AM, Strub MP, Taraska JW (2017) Endocytic proteins are partitioned at the edge of the clathrin lattice in mammalian cells. Nat Cell Biol 19(4):352–361. https://doi.org/10.1038/ncb3498
71. Avinoam O, Schorb M, Beese CJ, Briggs JA, Kaksonen M (2015) Endocytic sites mature by continuous bending and remodeling of the clathrin coat. Science 348(6241):1369–1372. https://doi.org/10.1126/science.aaa9555
72. Harterink M, Vocking K, Pan X, Soriano Jerez EM, Slenders L, Freal A, Tas RP, van de Wetering WJ, Timmer K, Motshagen J, van Beuningen SFB, Kapitein LC, Geerts WJC, Post JA, Hoogenraad CC (2019) TRIM46 organizes microtubule fasciculation in the axon initial segment. J Neurosci 39(25):4864–4873. https://doi.org/10.1523/JNEUROSCI.3105-18.2019
73. Kukulski W, Picco A, Specht T, Briggs JA, Kaksonen M (2016) Clathrin modulates vesicle scission, but not invagination shape, in yeast endocytosis. elife 5:e16036. https://doi.org/10.7554/eLife.16036
74. Weinhard L, di Bartolomei G, Bolasco G, Machado P, Schieber NL, Neniskyte U, Exiga M, Vadisiute A, Raggioli A, Schertel A, Schwab Y, Gross CT (2018) Microglia remodel synapses by presynaptic trogocytosis and spine head filopodia induction. Nat Commun 9 (1):1228. https://doi.org/10.1038/s41467-018-03566-5
75. Wesolowska N, Avilov I, Machado P, Geiss C, Kondo H, Mori M, Lenart P (2020) Actin assembly ruptures the nuclear envelope by prying the lamina away from nuclear pores and nuclear membranes in starfish oocytes. elife 9:e49774. https://doi.org/10.7554/eLife.49774
76. Nemethova M, Auinger S, Small JV (2008) Building the actin cytoskeleton: filopodia contribute to the construction of contractile bundles in the lamella. J Cell Biol 180 (6):1233–1244
77. Shutova M, Yang C, Vasiliev JM, Svitkina T (2012) Functions of nonmuscle myosin II in assembly of the cellular contractile system. PLoS One 7(7):e40814. https://doi.org/10.1371/journal.pone.0040814
78. Verkhovsky AB, Surgucheva IG, Svitkina TM, Tint IS, Gelfand VI (1987) Organization of stress fibers in cultured fibroblasts after extraction of actin with bovine brain gelsolin-like protein. Exp Cell Res 173(1):244–255
79. Svitkina TM, Bulanova EA, Chaga OY, Vignjevic DM, Kojima S, Vasiliev JM, Borisy GG (2003) Mechanism of filopodia initiation by reorganization of a dendritic network. J Cell Biol 160(3):409–421
80. Yang C, Czech L, Gerboth S, Kojima S, Scita G, Svitkina T (2007) Novel roles of formin mDia2 in lamellipodia and filopodia formation in motile cells. PLoS Biol 5(11):e317
81. Simionescu N, Simionescu M (1976) Galloylglucoses of low molecular weight as mordant in electron microscopy. I. Procedure, and

evidence for mordanting effect. J Cell Biol 70 (3):608–621

82. Maupin P, Pollard TD (1983) Improved preservation and staining of HeLa cell actin filaments, clathrin-coated membranes, and other cytoplasmic structures by tannic acid-glutaraldehyde-saponin fixation. J Cell Biol 96 (1):51–62
83. Bishai EA, Sidhu GS, Li W, Dhillon J, Bohil AB, Cheney RE, Hartwig JH, Southwick FS (2013) Myosin-X facilitates Shigella-induced membrane protrusions and cell-to-cell spread. Cell Microbiol 15(3):353–367. https://doi.org/10.1111/cmi.12051
84. Bridgman PC, Dailey ME (1989) The organization of myosin and actin in rapid frozen nerve growth cones. J Cell Biol 108(1):95–109
85. Hartwig JH, Shevlin P (1986) The architecture of actin filaments and the ultrastructural location of actin-binding protein in the periphery of lung macrophages. J Cell Biol 103 (3):1007–1020
86. Weng L, Enomoto A, Miyoshi H, Takahashi K, Asai N, Morone N, Jiang P, An J, Kato T, Kuroda K, Watanabe T, Asai M, Ishida-Takagishi M, Murakumo Y, Nakashima H, Kaibuchi K, Takahashi M (2014) Regulation of cargo-selective endocytosis by dynamin 2 GTPase-activating protein girdin. EMBO J 33(18):2098–2112. https://doi.org/10.15252/embj.201488289
87. Wu C, Asokan SB, Berginski ME, Haynes EM, Sharpless NE, Griffith JD, Gomez SM, Bear JE (2012) Arp2/3 is critical for lamellipodia and response to extracellular matrix cues but is dispensable for chemotaxis. Cell 148 (5):973–987. https://doi.org/10.1016/j.cell.2011.12.034
88. Elkhatib N, Bresteau E, Baschieri F, Rioja AL, van Niel G, Vassilopoulos S, Montagnac G (2017) Tubular clathrin/AP-2 lattices pinch collagen fibers to support 3D cell migration. Science 356(6343):aal4713. https://doi.org/10.1126/science.aal4713
89. Ferrari R, Martin G, Tagit O, Guichard A, Cambi A, Voituriez R, Vassilopoulos S, Chavrier P (2019) MT1-MMP directs force-producing proteolytic contacts that drive tumor cell invasion. Nat Commun 10 (1):4886. https://doi.org/10.1038/s41467-019-12930-y
90. Franck A, Laine J, Moulay G, Lemerle E, Trichet M, Gentil C, Benkhelifa-Ziyyat S, Lacene E, Bui MT, Brochier G, Guicheney P, Romero N, Bitoun M, Vassilopoulos S (2019) Clathrin plaques and associated actin anchor intermediate filaments in skeletal muscle. Mol Biol Cell 30(5):579–590. https://doi.org/10.1091/mbc.E18-11-0718
91. Braet F, De Zanger R, Wisse E (1997) Drying cells for SEM, AFM and TEM by hexamethyldisilazane: a study on hepatic endothelial cells. J Microsc 186(Pt 1):84–87. https://doi.org/10.1046/j.1365-2818.1997.1940755.x
92. Bray DF, Bagu J, Koegler P (1993) Comparison of hexamethyldisilazane (HMDS), Peldri II, and critical-point drying methods for scanning electron microscopy of biological specimens. Microsc Res Tech 26(6):489–495. https://doi.org/10.1002/jemt.1070260603
93. Sokolova AI, Pavlova ER, Khramova YV, Klinov DV, Shaitan KV, Bagrov DV (2019) Imaging human keratinocytes grown on electrospun mats by scanning electron microscopy. Microsc Res Tech 82(5):544–549. https://doi.org/10.1002/jemt.23198
94. Kwiatkowski DJ, Janmey PA, Yin HL (1989) Identification of critical functional and regulatory domains in gelsolin. J Cell Biol 108 (5):1717–1726

Chapter 3

Imaging the Actin Cytoskeleton in Live Budding Yeast Cells

Cierra N. Sing, Emily J. Yang, Theresa C. Swayne, Ryo Higuchi-Sanabria, Catherine A. Tsang, Istvan R. Boldogh, and Liza A. Pon

Abstract

Although budding yeast, *Saccharomyces cerevisiae*, is widely used as a model organism in biological research, studying cell biology in yeast was hindered due to its small size, rounded morphology, and cell wall. However, with improved techniques, researchers can acquire high-resolution images and carry out rapid multidimensional analysis of a yeast cell. As a result, imaging in yeast has emerged as an important tool to study cytoskeletal organization, function, and dynamics. This chapter describes techniques and approaches for visualizing the actin cytoskeleton in live yeast cells.

Key words Yeast, Fluorescent proteins, Vital staining, Microscopy, Live cell imaging, Deconvolution, Ratio imaging, Actin, Cytoskeleton

1 Introduction

Thirty-six years ago, the actin cytoskeleton was first visualized in fixed budding yeast cells with fluorescent phalloidin [1]. Subsequently, fluorescence imaging became a useful tool to study actin-containing structures in yeast. Many elements of actin cytoskeletal structure and function are conserved within the eukaryotic lineage. Nevertheless, unlike mammalian cells, budding yeast contains a single actin gene, making it a convenient model organism for studying actin organization and function. The actin cytoskeleton is responsible for similar functions in yeast as in other eukaryotic cells including cell polarity, cell division, endocytosis, and organelle transport.

The actin cytoskeleton of yeast consists of two major structures that are maintained throughout the cell cycle: patches and cables. Actin patches are endocytic vesicles that are coated with actin filaments that are nucleated by the Arp2/3 complex. Other patch constituents include conserved proteins such as clathrin and its

Ray H. Gavin (ed.), *Cytoskeleton: Methods and Protocols*, Methods in Molecular Biology, vol. 2364,
https://doi.org/10.1007/978-1-0716-1661-1_3,

adaptors, endocytic adaptors, nucleation promoting factors (NPFs), kinases, and actin bundling proteins (e.g., Sac6p/fimbrin) [2]. Early in the G1 phase of the cell cycle, actin patches become concentrated at the nascent bud site, reflecting cell polarization toward the small growing daughter cell during bud emergence. Patches remain concentrated in the bud during isotropic growth [3, 4]. Finally, when the bud cell reaches a similar diameter to the mother cell, actin patches redistribute to the bud neck and facilitate cytokinesis [3, 4]. Thus, the distinctive polarity of actin patches can be used as an indicator for the function of the cell polarization machinery.

Actin cables are bundles of actin filaments that extend through the entire length of the yeast cell and reach the mother distal tip. They are essential for yeast cell division because they act as tracks for transport of cellular constituents including mitochondria, Golgi, vacuoles, secretory vesicles, spindle alignment elements, and mRNA from mother to developing daughter cells during cell division [5]. They are also critical for transport of vesicles and other cargos that contribute to cytokinesis.

Actin cables are highly dynamic. They undergo retrograde flow, movement from the bud toward the mother cell tip. In yeast, as in mammalian cells, retrograde actin cable flow (RACF) is driven by the pushing force of actin cable assembly and pulling force of a type II myosin protein, Myo1p [6, 7]. Actin cable assembly occurs in the bud tip and bud neck. During this process, actin filaments are polymerized by formin proteins (Bni1p and Bnr1p) and assembled into bundles by Sac6p (fimbrin), Abp140p, and Bnr1p. Actin cables are further stabilized by two tropomyosin isoforms (Tpm1p and Tpm2p).

Although RACF occurs in the retrograde direction, from daughter to mother cell, actin cable-dependent cargo transport is primarily in the anterograde direction, from mother to daughter cell. Thus, mitochondria and other cargos are effectively "swimming upstream," against the opposing force of RACF as they move into the daughter cells along actin cables. As a result, actin cable dynamics play a role in promoting inheritance of fitter mitochondria by yeast daughter cells, which, in turn, affects daughter cell fitness and life span [8]. Therefore, studying actin cable dynamics is important for understanding yeast cell biology.

Imaging techniques and technical methods have evolved over time to allow for efficient live-cell imaging of actin dynamics in yeast cells. Here, we describe the steps to visualize and characterize actin movement: tagging of yeast genes at their chromosomal loci with fluorescent proteins; determining functionality of tagged proteins; optimizing imaging methods; and analyzing actin dynamics quantitatively.

1.1 Choosing a Fluorescent Protein for Detection of Cellular Structures

Live imaging of the yeast actin cytoskeleton can be carried out by tagging endogenous proteins with fluorescent proteins or expression of tagged heterologous actin-binding peptides. The most commonly used fluorescent protein is GFP and its variants; however, other fluorescent proteins can be used. Here, we describe how to select a fluorescent protein for imaging studies and tag a protein at its chromosomal locus.

The green fluorescent protein (GFP) revolutionized live-cell imaging in cell biology. GFP was first cloned from the jellyfish *Aequorea victoria* and has been used as a foundation for developing variants with a range of colors and molecular properties. Moreover, combining laboratory mutagenesis with novel fluorescent proteins, such as mCherry from *Discosoma* sp. coral and Teal from *Clavularia* sp. coral, has produced fluorescent proteins with new colors, improved brightness, faster folding, and decreased oligomerization [9].

Fluorescent proteins that are used for live-cell imaging must be photostable and able to withstand repeated excitation. In addition, the emission of the selected fluorescent protein must be bright enough to highlight structures of interest above background fluorescence and detector noise while being spectrally distinct from other labels being used. Fluorescent protein brightness is determined by the protein's intrinsic brightness (the product of extinction coefficient and quantum yield) and the behavior of the fluorescent protein upon ectopic expression (rate of folding and stability). The amount of signal that is excited and detected depends on the imaging system's light sources, lenses, filters, and detectors and their compatibility with the fluorescent protein excitation and emission spectra. For more information on fluorescent protein physical and spectral properties, https://www.fpbase.org/ is an invaluable resource. Lastly, the fluorescent protein must not be toxic or disrupt the cellular behavior or protein function to which it is fused.

1.2 Tagging Endogenous Proteins

Uniquely in yeast, homologous recombination methods are commonly used to tag a protein of interest at its chromosomal locus. This involves the use of an insertion cassette, a double-stranded linear DNA that contains the fluorescent tag, and a selection marker, which is inserted into the target site. It is produced by PCR using a tagging vector as a template. Templates for tagging vectors encompass a variety of fluorescent proteins (e.g., GFP, mCherry, mCitrine), epitopes (e.g., HA, myc), or affinity tags (e.g., GST, TAP, His) that are linked with one of many selectable markers that confer drug resistance or auxotrophic rescue. Families of tagging vectors have been constructed to share PCR-priming sequences. Therefore, a single set of primers made from one template family can be used to insert different tags associated with that template for a given target gene. Additionally, template vectors are available to regulate expression levels from endogenous promoters, constitutively active promoters (e.g., *AHD1*, *GPD1*), or regulatable promoters (e.g., *GAL1*), which can be used for N-terminal tagging. Fluorescent protein and epitope tags can also be used for biochemical techniques including affinity purification, immunoprecipitation, Western blot analysis, and immunofluorescence.

Some cassettes, like the pOM family, permit the removal of the selectable marker after tagging [10]. These cassettes are constructed with a marker flanked by LoxP sites, allowing removal of the marker by bacteriophage Cre recombinase that is conditionally expressed from a plasmid [11]. Removing the selectable marker is useful in several situations: (1) to create an N-terminal tag with no separation between the tag and the endogenous promoter of the tagged protein; (2) to insert a tag internally within the coding region of the gene of interest; (3) for multiple rounds of tagging at the same locus; and (4) for use in yeast strains with a limited number of selectable markers. Commonly available tagging vectors are listed in Table 1.

Tagging vector primers should be designed with sequences that hybridize with both the tagging vector and the target site for homologous recombination within the yeast chromosome. The insertion cassette is produced by PCR using the tagging vector as a template. The resulting amplified DNA is transformed into yeast using a standard protocol [12]. Recombinants that carry the inserted tag are selected using the marker from the insertion cassette, followed by PCR-based screening to ensure the correct insertion size, and finally sequencing to confirm accuracy of the coding sequence.

1.3 Visualization of Actin Dynamics with Actin Bundling Protein Abp140p

Live-cell imaging of actin dynamics permits analysis of RACF velocity as well as cargo trafficking, such as the movement of mitochondria along actin cables (Fig. 1) [13, 14]. Here, and in Subheading 1.4, we describe the two methods that we use to visualize actin dynamics in living yeast cells.

Tagging the actin-encoding *ACT1* gene at its C-terminus with GFP compromises actin function: plasmid-borne *ACT1-GFP* does not rescue the loss of the endogenous *ACT1* gene [15]. Generating a functional fusion protein with actin is challenging, because every surface of the actin protein is involved in important protein–protein interactions, including nucleation, polymerization, interaction with motor proteins and other force generators, cable assembly, capping, cross-linking, and severing. As an alternative, actin-binding proteins are often more tolerant of tagging with fluorescent proteins. For example, tagging actin-binding proteins like Abp140p and Abp1p with C-terminal GFP produces a fluorescent signal that localizes to actin cables and actin patches, respectively (Fig. 1) [6, 7]. Other actin proteins that can be tagged with GFP while maintaining normal function are listed in Table 2.

Table 1
Yeast tagging cassette vectors

Plasmid family	Tag position	Promoter	Tags	Markers
pFA6a[a]	C-terminal	Endogenous	GFP(S65T) 3xHA 13xMyc GST GFPEnvy[b] GFPIvy[2] GFPγ[b]	*TRP1* *kanMX6* *HIS3MX6*
pFA6a-PGAL1[a]	N terminal or internal	*GAL1*	GFP(S65T) 3xHA GST	*TRP1* *kanMX6* *HIS3MX6*
pUR[c]	C-terminal	Endogenous	DsRed	*HIS3* *URA3(K.l.)*
pYM[d]	C-terminal	Endogenous	yEGFP EGFP EBFP ECFP EYFP DsRed, DsRedI RedStar, RedStar2 eqFP611 FlAsH 1xHA, 3xHA, 6xHA 3xMyc, 9xMyc 1xMyc+7xHis TAP Protein A	*kanMX4* *hphNT1* *natNT2* *HIS3MX6* *klTRP1*
pKT[e]	C-terminal	Endogenous	yEGFP yECFP yEVenus yECitrine yESapphire yEmCFP[f] yEmCitrine tdimer2[g] yECitrine+3xHA yECitrine+13xMyc yECFP+3xHA yECFP+13xMyc	*KanMX* *SpHIS5* *CaURA3*
pOM[h]	N terminal or internal	Endogenous[i]	yEGFP 6xHA 9xMyc Protein A TEV-ProteinA TEV-GST-6xHis TEV-ProteinA-7xHis	*kanMX6* *URA3(K.l.)* *LEU2(K.l.)*

(continued)

Table 1
(continued)

Plasmid family	Tag position	Promoter	Tags	Markers
pCY[j]	C-terminal	Endogenous	Cerulean yECFP yEmCFP yEGFP Venus yEVenus yECitrine yEmCitrine mCherry mEos2 yEc-MYC yEHA yEFlAsH	*HygromycinB* *Zeocin*

[a][40]
[b][16]
[c][41]
[d][42]
[e][43]
[f]Monomeric version
[g]Tandem dimer of DsRed
[h][10]
[i]After Cre-mediated removal of auxotrophic marker
[j][44]

Even when protein function is preserved, fluorescent protein fusions may require optimization for successful imaging. For example, Abp140p-GFP produces only a low signal in the cells propagated using glucose-based media. However, this can be improved by culturing cells in a non-fermentable carbon source, such as lactate. Cells grown in lactate medium produce cables that are thicker and therefore more visible (*see* **Note 1**). Alternatively, improved signal-to-noise ratio can be achieved by tagging actin-binding proteins with GFPEnvy. GFPEnvy is an improved version of GFP with brighter signal and photostable, which was constructed by combining mutations found in Superfolder GFP and GFPγ [16]. Tagging Abp140p with GFPEnvy improves signal intensity and reduces photobleaching during time-lapse videos. GFP and GFPEnvy do not disrupt actin dynamics when fused to Abp140p; however, other fluorophore proteins such as td-Tomato compromise the function of Abp140p. Moreover, although GFPEnvy was produced from a monomeric form of GFP, recent studies indicate that it can dimerize in vitro [17]. Thus, it is important to test whether actin dynamics are preserved when using fluorescent protein tags.

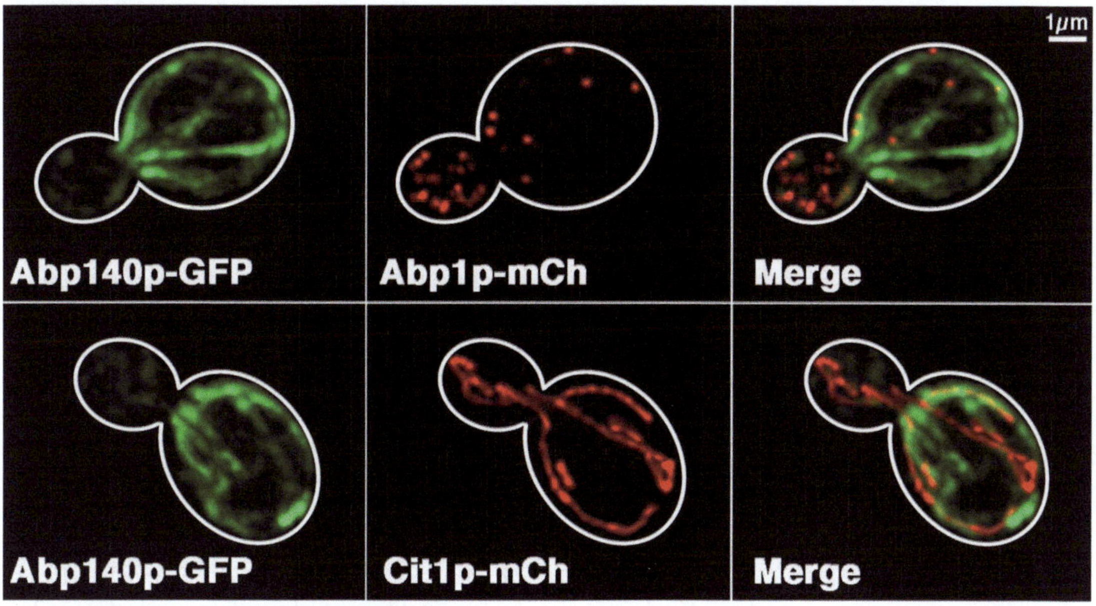

Fig. 1 Colocalization of the actin cytoskeleton with cargo. Yeast cells were grown in YPG and washed in SC medium prior to imaging. **Left panels**: Actin cables visualized with Abp140p-GFP. **Middle panels**: Two cellular structures, actin patches (top) and mitochondria (bottom), visualized by tagging Abp1p and Cit1p, respectively, with mCherry. Images were captured using standard GFP filters and 300 ms exposure time for GFP and standard DsRed filters and 200 ms exposure time for mCherry. **Right panels**: merged images showing partial colocalization of actin patches and mitochondria along actin cables. Scale bar: 1 μm

Table 2
GFP-tagged actin proteins in budding yeast

Actin-containing structure	GFP-tagged proteins
Actin patch (early)	Las17p [45, 46], Sla1p [46, 47], Pan1p [46, 48], End3p [49], Sla2p [46, 50], Bzz1p [51], Vrp1 [52]
Actin patch (mid)	Myo3p [53], Myo5p [54], Bbc1p [51]
Actin patch (late)	Abp1p [15, 46, 55, 56], Arp2p [45], Arp3p [45], Arc15p [46, 57], Sac6p [56, 58], Cap1p [59], Cap2p [60], Scp1p [61], Cof1p [62]
Actin cable (vegetative)	Abp140p [6], Lifeact (Abp140 aa 1-17) [18, 31]

1.4 Visualization of Actin Dynamics with Actin-Binding Peptide Lifeact

An alternative probe for visualizing actin dynamics is Lifeact, an actin-binding peptide, tagged with a fluorescent protein. Lifeact consists of the first 17 amino acids of Abp140p and is sufficient to mediate actin localization of fluorescent proteins without disrupting dynamics [18]. Lifeact has been used across many eukaryotic organisms, including fungi, plants, and invertebrates [19–25]. However, there are some caveats to using an exogenous construct to visualize the actin cytoskeleton. Lifeact fails to label some actin structures, such as stress-induced cofilin-bound actin rods [26]. Moreover, Lifeact overexpression can alter actin dynamics, generating artifacts. For example, overexpressing Lifeact in *Drosophila* germline cells causes loss of cortical actin integrity leading to fertility defects [27]. In fission yeast, highly expressed Lifeact-mCherry or mEGFP-Lifeact affects actin patch formation and cytokinesis [28]. The structure of the Lifeact-F-actin complex indicates that dosage-dependent effects are caused by the reduced binding of cofilin to actin filaments [28–31]. Therefore, expression conditions should be optimized for each application.

Initial reports of Lifeact in budding yeast indicated that it can label actin-containing structures, but its fluorescence intensity was extremely weak, preventing long-term imaging [18, 31]. Our lab has resolved this issue by integrating Lifeact conjugated to GFPEnvy, rather than GFP, into the yeast genome for stable expression.

The stable genomic expression of Lifeact-GFPEnvy (LA-Envy) at the HO locus does not interfere with cell growth (Fig. 2a). Moreover, our previous studies indicate that Abp140p-GFP labels a small number of punctate structures (e.g., Fig. 2c) that either are not actin patches or are a subset of actin patches. On the other hand, we find that LA-Envy localizes to both punctate and filamentous structures, which co-localize with phalloidin-stained actin patches and cables (Fig. 2b). Equally important, we find that the expression of LA-Envy does not result in depolarization of the actin cytoskeleton. The presence of >5 actin patches in the mother cell is a hallmark of cell depolarization, and we found LA-Envy-expressing cells contain <5 punctate structures per mother cell (Fig. 2c).

When we compared labeled cables in cells expressing LA-Envy and Abp140p-GFPEnvy, we observed fewer cables in LA-Envy-expressing cells in both lactate and SC medium, compared to Abp140p-GFPEnvy-expressing cells (Fig. 2c). However, there was no detectable difference in actin cable number in phalloidin-stained cells expressing these two probes (data not shown). Taken together, our results suggest that LA-Envy and Abp140p-GFPEnvy label both patches and cables, but with different affinities; patches are more likely to be labeled with LA-Envy and cables with Abp140p-GFPEnvy. At least in the case of cables, the difference does not appear to be an artifact of fusion protein expression but rather arises from a difference in binding that renders some structures below the threshold of detection. It remains to be determined why this phenomenon occurs, but we speculate that it results from competitive binding between actin cables and patches.

Overall, we find that Lifeact-Envy is the first actin probe that labels both actin cables and patches in live cells. Table 2 lists known functional fluorescent protein-labeled actin-binding proteins that

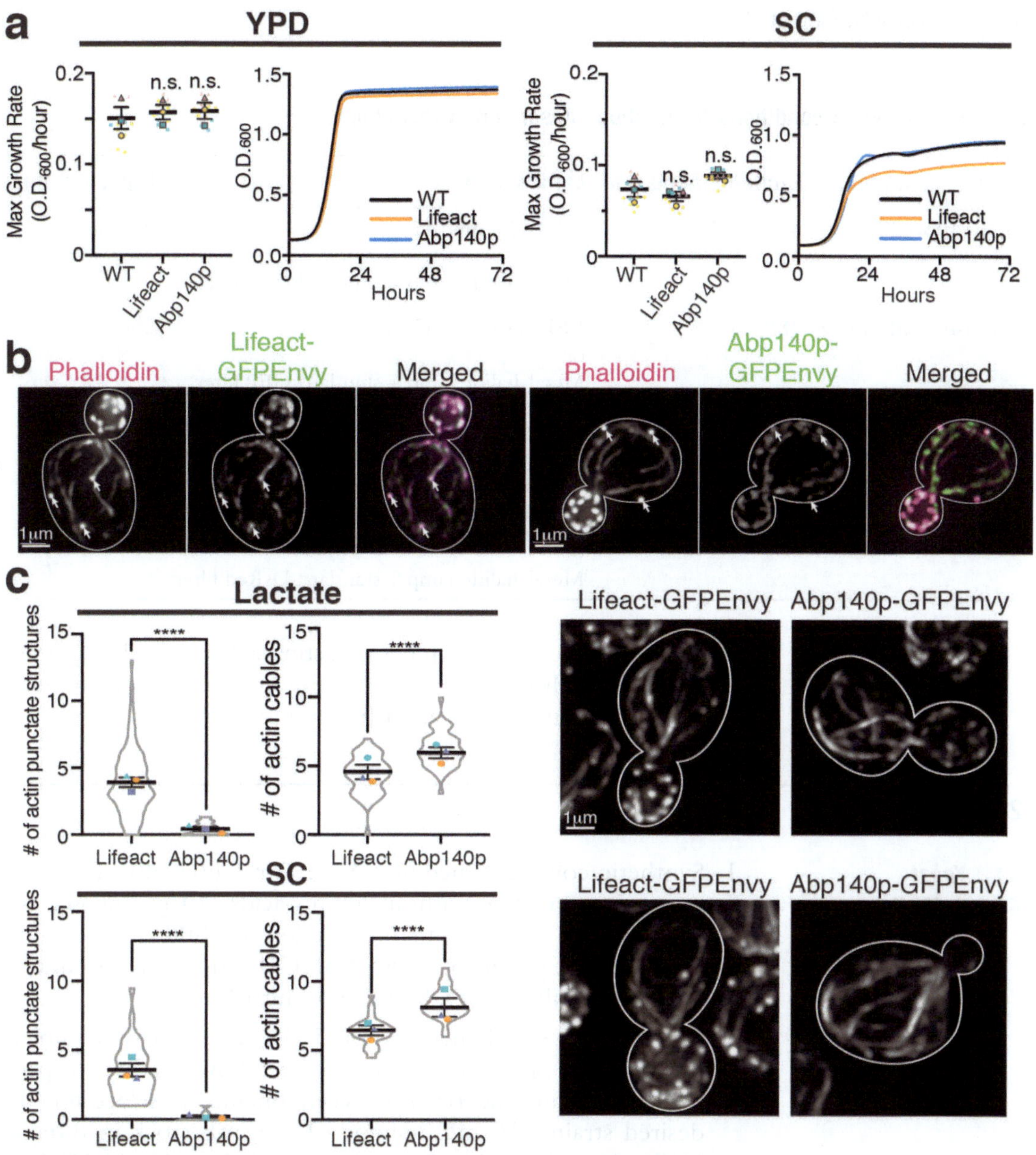

Fig. 2 Visualizing actin cable dynamics with Lifeact-GFPEnvy. (**a**) Maximum growth rates and growth curves of untransformed cells (WT) and cells expressing Lifeact-GFPEvny or Abp140p-GFPEnvy in YPD or SC media at 30 °C. Maximum growth rates were calculated for 2 h intervals ($n = 3$–6 cultures for each independent experiment; statistical significances were calculated by unpaired Student's *t*-test. Error bars represent SEM (standard error of the mean). (**b**) Representative images of fixed mid-log phase yeast cells in lactate medium. **Left panels**: F-Actin structures stained with Alexa568-phalloidin. **Middle panels**: Lifeact or Abp140p tagged with GFPEnvy. **Right panels**: Merged images, showing that actin cables visualized by Lifeact-GFPEnvy or Abp140p-GFPEnvy colocalize with phalloidin. White arrows indicate punctate structures (actin patches) in mother cells. Lifeact-GFPEnvy reveals more punctate structures than Abp140p-GFPEnvy. Scale bars: 1 μm. (**c**) **Left panels**: Quantification of punctate actin structures and actin cables in mid-log phase mother cells expressing Lifeact-GFPEnvy or Abp140p-GFPEnvy in lactate or SC media at 30 °C. $n > 30$ cells/trial. Statistical significances were calculated by unpaired *t*-test (*$p < 0.05$; **$p < 0.01$; ****$p < 0.0001$). Error bars represent SEM. **Right panels**: Representative images of live yeast cells expressing Lifeact-GFPEnvy or Abp140p-GFPEnvy grown in the indicated media. Scale bars: 1μm

Table 3
Suggested imaging conditions for methods described in this chapter

Fluorescent protein	Excitation/emission	Light source	Exposure time
Abp140p-GFP	488/507	LED 470 nm (45%) or Metal-halide lamp + standard GFP filter	250 ms
Abp140p-GFPEnvy	488/507	LED 470 nm (45%) or Metal-halide lamp + standard GFP filter	250 ms
Lifeact-GFPEnvy	488/507	LED 470 nm (45%) or Metal-halide lamp + standard GFP filter	250 ms
Abp1p-mCherry	587/610	LED 550 nm (45%) or Metal-halide lamp + standard DsRed filter	200 ms

serve as probes for actin patches or actin cables in budding yeast. Finally, Table 3 lists suggested imaging conditions for common fluorescent protein-labeled actin-binding proteins.

2 Materials

2.1 Yeast Growth Media

1. Synthetic complete medium (SC): For 1 liter, dissolve 6.7 g yeast nitrogen base without amino acids, 20 g glucose, and amino acid supplements as needed in 800 mL distilled H_2O. Adjust pH to 5.5 with $NaHCO_3$ and bring volume to 1 L with distilled H_2O. Sterilize by autoclaving.
2. Amino acid supplements: For 1 L of synthetic complete medium, supplement with all of the following. For dropout medium, omit one or more components to select for the desired strains: 10 mL adenine (2 mg/mL stock in 0.05 M HCl), 10 mL uracil (2 mg/mL stock in 0.5% $NaHCO_3$), 10 mL arginine (1 mg/mL stock in H_2O), 10 mL histidine (1 mg/mL stock in H_2O), 10 mL leucine (10 mg/mL stock in H_2O), 10 mL lysine (3 mg/mL stock in H_2O), 10 mL methionine (2 mg/mL stock in H_2O), 10 mL phenylalanine (5 mg/mL stock in H_2O), 10 mL tryptophan (2 mg/mL stock in H_2O), and 30 mg tyrosine.
3. Yeast peptone-dextrose medium (YPD): For 1 L, dissolve 10 g yeast extract, 20 g bacto-peptone, and 20 g glucose in distilled H_2O, and bring volume to 1 L with distilled H_2O. Sterilize by autoclaving.
4. Lactate medium: For 1 L, combine 3.0 g yeast extract, 0.5 g glucose, 0.5 g $CaCl_2$, 0.5 g NaCl, 0.5 g $MgCl_2$, 1 g KH_2PO_4,

1 g NH_4Cl, 22 mL 90% lactic acid, and 7.5 g NaOH pellets. Dissolve ingredients in 800 mL distilled H_2O. Adjust pH to 5.5 with NaOH. Bring volume to 1 L with distilled H_2O.

5. Synthetic raffinose medium: For 1 L, combine 6.7 g of yeast nitrogen base without amino acids and amino acid supplements as needed. Dissolve ingredients in 600 mL distilled H_2O. Adjust pH to 5.5 with $NaHCO_3$. Bring volume to 900 mL with distilled H_2O. Supplement after autoclaving with 100 mL filter-sterilized 20% raffinose.
6. Synthetic glycerol medium (SG): For 1 L, combine 6.7 g of yeast nitrogen base without amino acids, 30 mL of glycerol, 0.5 g glucose, and amino acid supplements as needed. Dissolve ingredients in 800 mL distilled H_2O. Adjust pH to 5.5 with $NaHCO_3$.
7. Synthetic galactose medium: For 1 L, combine 6.7 g of yeast nitrogen base without amino acids and amino acid supplements as needed. Dissolve ingredients in 600 mL distilled H_2O. Adjust pH to 5.5 with $NaHCO_3$. Bring volume to 950 mL with distilled H_2O. Supplement after autoclaving with 50 mL filter-sterilized 40% galactose.

2.2 PCR Amplification

1. Betaine solution 5 M, PCR reagent (Sigma, B0300).
2. KAPA HiFi HotStart ReadyMix (2×) (Kapa Biosystems).
3. Sterile distilled H_2O.
4. Forward and reverse primers (diluted to 50μM each).

2.3 Yeast Transformation and Marker Excision

1. 50% w/v polyethylene glycol 3350 in H_2O.
2. 1 M lithium acetate in H_2O.
3. 0.1 M lithium acetate in H_2O.
4. Carrier DNA: 2 mg/mL single-stranded DNA from salmon testes in H_2O. Store at −20 °C.
5. DNA for transformation (1 PCR reaction as described in protocol, or 1μg).
6. Sterile distilled H_2O for resuspending cells before plating.
7. Selective plates for growth of cells containing plasmid or integrated DNA.

2.4 Wide-Field Imaging of Live Yeast

Examples of epifluorescence microscope setups that can be used for wide-field live yeast imaging include the following:

1. Inverted AxioObserver.Z1 microscope (Carl Zeiss Inc., Thornwood, NY) equipped with a metal halide lamp and Colibri LEDs for excitation; 100×/1.3 EC Plan Neofluar oil-immersion objective; and an Orca ER cooled CCD camera

with 1280 × 1024 pixel resolution (Hamamatsu, Bridgewater, NJ). ZEN software (Carl Zeiss) is used to control the camera and microscope hardware.

2. Upright Axioskop 2 microscope (Carl Zeiss Inc., Thornwood, NY) equipped with a pE-4000 LED illumination system (CoolLED, Andover, UK) for excitation; 100×/1.4 Plan-Apochromat objective (Carl Zeiss Inc., Thornwood, NY); and an Orca ER cooled CCD camera. NIS Elements 4.60 Lambda software (Nikon, Melville, NY) is used to control the camera and microscope hardware.

3 Methods

3.1 Modification of Yeast Genes at Their Chromosomal Loci

3.1.1 Primer Design

Generally, primers should be designed with 60 or more bases. The 5′ end of each primer consists of at least 40 bases with perfect homology to the target site. As an example, for C-terminal tagging with a fluorescent protein, the 5′ end of the forward primer should contain the 40–45 bases directly upstream of the stop codon, while the 5′ end of the reverse primer should contain the reverse complement of the 40–45 bases directly downstream of the stop codon. The 3′ ends will consist of 18–25 bases complementary to the sequences that will be inserted from the vector, which include the open reading frame of the fluorescent protein, the transcription termination site, and the selectable marker (*see* **Note 2**).

For further preservation of protein function and folding when tagging with a fluorescent protein, one can insert a short linker between the gene of interest and the fluorescent tag. For instance, GFP fluorescence can often be further optimized by increasing the length of the linker between the target gene and GFP molecule (e.g., by adding five alanines). The linker length and amino acid composition may need to be adjusted for different target proteins. Typically, proteins with processive enzymatic functions, such as polymerases and telomerases, benefit from poly-glycine linkers [32]. When C-terminally tagging, the linker should be incorporated into the forward primer by including codons encoding the appropriate amino acids between the 40 bases of homology to the target gene and the 20 bases corresponding to the template. A similar strategy can be used in the reverse primer for N-terminal tagging.

3.1.2 Amplification of Insertion Cassette from Tagging Vector

1. Prepare the PCR reaction: For 50μL, 10μL Betaine 5 M (Sigma, B0300), 25μL Kapa HiFi HotStart ReadyMix (2×), 12μL sterile distilled H_2O, 1μL each of forward and reverse primers, 1μL template DNA. A single 50μL PCR reaction will provide sufficient DNA for a single yeast transformation (minimally 1μg of DNA) if a high-fidelity, high-yield polymerase is used.

2. Amplify the cassette using the following thermocycler conditions: Initial denaturation cycle at 95 °C for 2–5 min; 35 cycles of 98 °C for 20 s, 60–75 °C (*see* **Note 3**) for 15 s, and 72 °C for 60 s/kb; final extension cycle at 72 °C for 1–5 min.
3. Use the PCR reaction directly in the lithium acetate transformation protocol described below. Alternatively, amplified DNA can be purified using any commercially available PCR purification kit (*see* **Note 4**).

3.2 Lithium Acetate Transformation of Yeast

The lithium acetate transformation protocol is the most commonly used method for yeast transformations [12]. The following protocol is for one transformation reaction. A negative control containing no DNA should always be carried out in parallel.

1. Grow yeast to mid-log phase (0.2–0.5 OD_{600}).
2. Transfer 10^7 cells from the culture to a 1.5 mL microfuge tube.
3. Pellet cells (30 s, 6000 × *g*) and wash with 500μL 0.1 M lithium acetate. Resuspend in 240μL of 50% w/v polyethylene glycol 3350.
4. Add 36μL 1.0 M lithium acetate, 25μL of carrier DNA, and 1μg (50μL) of DNA to be transformed (PCR-amplified insertion cassette or plasmid DNA) or 50μL of H_2O for negative control.
5. Vortex vigorously and incubate in a water bath at 30 °C for 30 min.
6. Heat shock in a water bath at 42 °C for a minimum of 20 min and a maximum of 3 h.
7. Concentrate cells by mild centrifugation (30 s, 6000 × *g*).
8. For transformations using auxotrophic markers, resuspend cell pellet gently in 100μL of sterile distilled H_2O and plate on appropriate selective media.
9. For transformations using drug resistance markers (e.g., KanMX6), resuspend cell pellet gently in 500μL of YPD or synthetic complete medium with dropouts as needed to maintain plasmids. Allow cells to recover at 30 °C for 2–4 h. Concentrate and plate cells as described in **steps** 7 and **8** on appropriate selective medium containing drug.

3.3 Marker Excision by Cre Recombinase

If using a LoxP-containing vector such as the pOM family for tagging a gene of interest, the selectable marker can be removed by the following method.

1. Verify tagging by PCR and sequencing, as described below.
2. Using methods described above, transform the tagged strain with a plasmid that encodes Cre recombinase under control of a galactose-inducible promoter [33]. Select transformants on appropriate medium selecting for the Cre plasmid.

3. Grow cells to mid-log phase in 5 mL liquid medium in a 50 mL conical-bottom tube. Use medium that selects for the Cre plasmid.
4. Pellet cells (5 min, 4000 × *g*) and resuspend in 5 mL galactose medium.
5. Incubate cells in galactose medium in a shaking incubator at 30 °C for 12–16 h.
6. Plate an aliquot of cells on non-selective (YPD or SC) medium.
7. Screen colonies for loss of the tagging cassette marker by replica plating on selective plates.
8. To induce dropout of the Cre plasmid, grow cells in non-selective liquid medium for 24–48 h, plate on non-selective medium, and screen for loss of the plasmid by replica plating on selective plates (*see* **Note 5**).

3.4 Validating and Characterizing Fluorescent Protein-Tagged Cytoskeletal Proteins

3.4.1 PCR Screening and Sequencing

Validating genomic tag insertion at the target locus can be done using a PCR screening method. Use isolated genomic DNA from transformants and untransformed cells as templates for the PCR reaction, and compare fragment sizes amplified by a primer pair that hybridizes up- and downstream of the tagging locus. Sequencing should then be carried out to verify that tagged genes do not carry any mutations.

3.4.2 Verification of Protein Expression and Function

Initial steps to confirm successful expression of tagged proteins in transformed cells can be performed through visual inspection. Prepare cells for short-term imaging as described below, and screen several colonies for consistent fluorescent protein expression.

We recommend using Western blot analysis to study expression of tagged proteins. Benefits of protein expression monitoring include (1) comparing the expression level of tagged protein to that of its native form; (2) troubleshooting protein expression when the levels fall below the threshold of detection by microscopy; and (3) identifying degradation products that may hinder proper localization and function.

It is important to evaluate the localization and functions of tagged constructs and confirm that these are preserved, when knowledge of the gene of interest is available. Tagging proteins with fluorescent proteins may alter their function and/or localizations by disrupting native protein–protein interactions or by inducing oligomerization through the fluorescent protein. Evaluating whether function is preserved in fluorescent protein-tagged strains can be done by examining their cell growth rates and loss of any characterized phenotypes of the gene of interest. Fluorescent protein-tagged strains with compromised protein function will exhibit similar phenotypes to cells bearing mutations in that gene.

For a positive indicator of function, a plasmid-borne version of the fluorescent protein-tagged protein can be used to rescue a wild-type phenotype in yeast cells with a genomic deletion of the gene of interest (*see* **Note 6**). After confirming proper localization and preserved function, strains can be used for live-cell analysis. If, on the other hand, target protein localization and/or function is found to be compromised by a fluorescent protein tag, an alternative approach is to fuse the target protein to a smaller epitope tag such as myc or HA and visualize the protein in fixed cells using immunofluorescence.

3.5 Wide-Field Imaging of Living Cells

Budding yeast are relatively small in comparison to other eukaryotic organisms. The diameter of mother cells is typically 4–6μm. Therefore, a microscope equipped with a high-magnification, high-numerical aperture objective lens and a sensitive camera are required to resolve small intracellular structures with good spatial resolution.

Light throughput and sensitivity of an imaging system are imperative for live-cell imaging because short exposure times and low illumination intensity are needed to reduce photobleaching and phototoxicity. Short exposure times are also important for improving temporal resolution when imaging dynamic structures. Additional time resolution improvement can be achieved by using a direct trigger connection between the acquisition computer and imaging components and maximizing the computer's processor speed, random access memory (RAM), and storage disk speed.

3.6 Multicolor Imaging

Live-cell imaging is not limited to visualizing one organelle at a time. Multiple organelles can be imaged at once with the use of different fluorescent proteins or vital dyes, allowing the analysis of the interaction between organelles or morphological changes within multiple organelles. As described earlier, *S. cerevisiae* is amenable to integration of a wide variety of fluorescent proteins at multiple chromosomal loci. This can be achieved by sequential tagging through multiple homologous recombination events or through crossing different lines.

The limiting factor for multicolor imaging is identifying a compatible collection of distinct fluorescent proteins that are sufficiently bright, spectrally distinct, and photostable and that do not compromise protein function. The most commonly paired fluorescent proteins for two-color imaging are GFP (or GFPEnvy) and red fluorescent protein (RFP), such as mCherry. These fluorescent proteins have distinct spectra and display consistently bright and robust signals. For three-color imaging, the most commonly combined fluorescent proteins are GFP, RFP, and blue fluorescent protein (BFP) [34]. However, BFP requires UV excitation, which can lead to organelle fragmentation, reactive oxygen species production, and cell death [35]. As an alternative, our lab developed a trio of fluorescent proteins that are photostable, fluorescently distinct, and well tolerated by yeast: mTFP1/mCitrine/mCherry [36].

Multiple fluorescent proteins can be detected either simultaneously or sequentially. Simultaneous detection requires either splitting the emitted signal into multiple detectors or dividing the camera area into multiple regions using an image splitter (e.g., Dual-View, Optical Insights, Santa Fe, NM). Although this approach maximizes speed of acquisition, it entails a significant risk of spectral bleedthrough. Therefore, in most cases, sequential imaging offers sufficient speed and more reliable differentiation between labels. There are three strategies to achieve sequential imaging: (1) changing excitation and emission filters, (2) LED excitation switching, and (3) laser excitation switching. Changing excitation and emission filter sets requires a motorized microscope stand that rotates the filter turret or separate filter wheels for excitation and emission. Rotating the filter turret allows the use of specialized cubes for each channel, which will result in the best signal. However, turret rotation may induce vibration in the system and is also quite slow (>1 s) compared to that of separate filter wheels (tens or hundreds of ms).

A faster, more stable approach is switching the excitation wavelength directly at the light source. This is accomplished using either lasers or narrow-spectrum, high-intensity light-emitting diodes (LEDs), such as those listed in the example microscope configurations above. These systems can be configured with multiple wavelengths that match the spectra of commonly used fluorescent proteins and are controlled individually by electrical signals with virtually no lag time. Signal can be collected through a multi-band dichroic/emission filter cube, eliminating any mechanical motion. Thus, these systems are preferred for multicolor imaging of dynamic structures and for excitation ratio imaging.

3.7 Preparing Cells for General Short-Term Imaging

For short-term visualization, concentrated cells can be deposited directly on a slide. However, cells on a microscope slide should not be imaged for more than 10 min. After this point, cells exhibit signs of stress and significant decreases in viability. The following protocol can be used for general imaging of cells expressing fluorescent proteins. Special procedures for imaging actin cable dynamics are detailed in the next section.

1. Grow yeast cells overnight to mid-log phase ($OD_{600} = 0.2–0.5$) in 5 mL of appropriate liquid medium in a 50 mL conical-bottom tube at 30 °C with shaking (*see* **Note 7**).
2. Transfer 1 mL of mid-log phase culture to a 1.5 mL microfuge tube and concentrate cells by centrifugation at ~6000 × *g* for 30 s (*see* **Note 8**).

3. Without disturbing the pellet, remove almost all of the supernatant, leaving a volume of supernatant in the tube approximately twice the volume of the pellet.
4. Resuspend pellet in residual medium and transfer ~1.5μL of the cell suspension to a cleaned microscope slide (*see* **Note 9**).
5. Apply a #1.5 (170-μm-thick) coverslip, taking care to avoid creating bubbles between slide and coverslip (*see* **Note 10**).
6. Acquire images. Prepare a fresh slide from the concentrated cell suspension after 10 min.

3.8 Preparing Cells for Short Time-Lapse Imaging of Actin Cable Dynamics

Short-term imaging can be used to determine the velocity of actin cable movement. Concentrated cells in lactate medium are applied directly to a slide. This protocol can be used to follow actin cable dynamics using fluorescent proteins conjugated to either Abp140p or Lifeact.

1. The morning before the day of the experiment, prepare a pre-culture. Inoculate cells from a colony into 5 mL of appropriate medium in a 50 mL conical-bottom tube and incubate for 4–5 h in a 30 °C shaking incubator.
2. From the pre-culture, inoculate an overnight culture with fresh media to reach mid-log phase (OD_{600} 0.2–0.5) the following day.
3. On the day of the experiment, transfer 1 mL of culture to a 1.5 mL microfuge tube and concentrate cells by centrifugation at ~6000 × *g* for 30 s (*see* **Note 9**).
4. Without disturbing the pellet, remove almost all of the supernatant, leaving a volume of supernatant in the tube approximately twice the volume of the pellet.
5. Resuspend pellet in residual medium and transfer ~1.5μL of the cell suspension to a cleaned microscope slide.
6. Apply a #1.5 (170-μm-thick) coverslip, taking care to avoid creating bubbles between slide and coverslip (*see* **Note 10**).
7. Acquire time-lapse series at a single focal plane, at 500 ms intervals, for 30 s (*see* **Notes 11** and **12**).
8. Analyze cable velocity as described in Subheading 3.13.

3.9 Optimizing Imaging Conditions and Preventing Toxicity

During live-cell imaging, extended illumination of cells can lead to phototoxicity by two primary mechanisms: (1) direct cell damage by photons with wavelengths close to the UV and (2) free radical and reactive oxygen species production by the interaction between photons and cellular molecules.

3.9.1 Checking for Photodamage

The following criteria can be used to check for photodamage to cytoskeletal function.

1. Cytoskeletal structures: intense or long excitation can cause actin cable depolymerization.
2. Dynamics: While behavior of photodamaged cytoskeletal components has not been characterized, photodamage can be suspected if there are alterations in cytoskeletal velocity or other motility parameters in wild-type cells.

3.9.2 Reducing Phototoxicity and Photobleaching

Even if cells remain healthy throughout the imaging process, excessive light exposure still can cause photobleaching of fluorophores. Phototoxicity and photobleaching can be reduced by the following strategies. Resolution, sensitivity, and toxicity must always be balanced. However, in general, longer exposures at low intensity are less phototoxic than shorter exposures at high intensity.

1. Reduce excitation intensity. For LED and/or lasers, adjust intensities through acquisition software. Alternatively, add a neutral density filter.
2. Reduce exposure time. However, be sure to maintain enough dynamic range in pixel intensities to produce an interpretable image.
3. Reduce spatial or time resolution. For example, increase the time interval and/or the z-series interval.

3.9.3 Increasing Signal-to-Noise Ratio

Use the following strategies to increase signal-to-noise ratio without increasing light exposure:

1. Apply binning in the camera. This will result in decreased spatial resolution, but this can be alleviated by adding a projection tube before the camera.
2. Increase camera gain.
3. Reduce camera readout speed, if possible.
4. Use a filter set with a broader spectral window or more efficient coatings to increase throughput.

3.10 Special Considerations for Quantitative Analysis

Two criteria must be met for quantitative fluorescence imaging: (1) emitted fluorescence must be proportional to the number of fluorescent molecules present, and (2) recorded pixel intensities must be proportional to the amount of light emitted by the sample. Under well-controlled conditions, these criteria are met at sufficient levels and can be used to measure many biological factors.

Absolute quantification of fluorescent molecules by imaging is a serious challenge due to many variables that complicate the linearity of fluorescence emission and pixel intensities [37]. However, proper controls and normalization can allow reliable measurements of relative changes in the volume and intensity of fluorescent structures, providing a powerful tool for hypothesis testing.

Minimally, when comparing intensities, the following variables must be controlled in any wide-field fluorescence imaging experiment:

1. Age and concentration of fluorescent probe.
2. Sample age, unless photobleaching is known to be negligible.
3. Objective lens.
4. Illumination spectrum and intensity, including filters, field aperture setting, and light source alignment.
5. Camera model, binning, exposure time, gain, offset, readout speed, and bit depth.

Imaging parameters should always be adjusted to produce pixel values above the background level of the detector and sample, but below saturation if intensities are compared between groups, adjust imaging parameters using the brightest samples, to avoid saturating the detector. In addition, sample size (number of cells imaged) must be sufficient to determine the variability within the population, and fields of view must be selected without conscious or unconscious bias.

3.11 Deconvolution of Wide-Field Fluorescence Images

Typically, the resolution of wide-field fluorescence z-series can be further improved by deconvolution. Deconvolution is a computational method for removing out-of-focus light from wide-field fluorescence images. As a result, images are sharper with a higher signal-to-noise ratio and enhanced three-dimensional information. Various algorithms are available; however, only some allow for quantitation after processing [38]. Deconvolution can be executed by algorithms provided by commercial software such as Volocity (Quorum Technologies) or by free open-source software, such as Fiji/ImageJ. We have found that Volocity and Fiji/ImageJ produce comparable results. For best deconvolution results, use z-series acquired with an interval of 0.2–0.3μm.

3.11.1 Volocity Deconvolution

1. Acquire z-series images using general short-term imaging conditions described above.
2. Load datasets in Volocity and verify that the x, y, and z scale are correctly set.
3. For each channel, generate a new calculated PSF. When prompted, enter the numerical aperture of the objective lens and the maximum wavelength of fluorescence emission (e.g., 507 nm for GFP and 620 nm emission wavelength for mCherry).
4. Using the calculated PSF for each channel, perform iterative deconvolution with 60 iterations, 100% confidence limit (*see* **Note 13**). Number of iterations can be adjusted according to signal intensity and noise level.

3.11.2 Fiji/ImageJ Deconvolution

1. Download and install Diffraction PSF 3D (https://imagej.net/Diffraction_PSF_3D) and Iterative Deconvolve 3D (https://imagej.net/Iterative_Deconvolve_3D) plugins from the Fiji website (https://imagej.net/Fiji/Downloads).
2. Create a calculated PSF by using Diffraction PSF 3D: **Plugins > Diffraction PSF 3D**. Set up PSF parameters by entering immersion medium refractive index, objective lens numerical aperture, emission wavelength, image pixel size, image z step size, image width, image height, and image z slice number. The following are typical settings using our microscopy setup:
   (a) Refractive index: 1.518 (if using immersion oil).
   (b) Numerical aperture: 1.40.
   (c) Emission wavelengths: 507 nm or 610 nm for GFPEnvy and RFP, respectively.
   (d) Longitudinal spherical aberration at maximum aperture: 0.
   (e) Image pixel spacing: 64.50 (substitute your own spacing in nm if applicable).
   (f) Slice spacing: 300 (substitute your own spacing in nm if applicable).
   (g) Image width: 1343 (substitute your own image size in pixels if applicable).
   (h) Image height: 1023 (substitute your own image size in pixels if applicable).
   (i) Image depth: 21 (substitute your own z-stack size if applicable).
   (j) Normalization: sum of pixel values = 1.

   Name the PSF with the corresponding wavelength for easy recognition.
3. Deconvolve selected images: **Plugins > Iterative Deconvolve 3D**. Select the image and corresponding PSF, set Wiener filter gamma to 0, maximum iterations to 60, and delta for termination to 0. The deconvolution will be terminated automatically if iterations do not produce better resolution.
4. Save the deconvolved file.

3.12 Quantification of Fluorescent Signals

To quantify the amount of actin within the whole cell or subregions of the cell, two common measurements of fluorescence signal intensity can be executed: mean and integrated intensity. Mean fluorescence intensity (mean gray value) is the average value of pixels in a cell or region of interest (ROI). Mean intensity can be a proxy for concentration of a fluorophore but is not a good indicator of the total amount of a fluorophore, unless the cells or compartments are of similar size. For example, a small bud and a large bud may have the same number of labeled actin structures, but the small bud will have a higher mean intensity because the structures are crowded into a smaller volume. In scenarios where ROI sizes vary greatly, it is best to approximate the total amount of a fluorescent probe by calculating the integrated intensity (also called integrated density), which is the sum of all of the pixel values in the region. Thus, the integrated intensity takes into consideration not only the mean pixel value but the area of the ROI. Here, we show how to obtain intensity measurements using Volocity and Fiji/ImageJ. For more sensitive and accurate measurements, background should be measured and subtracted from mean intensity values. To correct integrated intensity for background, subtract the product of the background intensity and the area of the ROI.

3.12.1 Using Volocity to Measure Integrated Intensity

1. From the **Measurements** tab, create a protocol by adding the following steps in order: Find Objects Using Intensity, Exclude Objects by Size, Clip Objects to ROIs, Measure Objects.
2. For images with multiple channels, make sure that the correct channel is selected under the **Find Objects Using Intensity** tab.
3. Under **Exclude Objects by Size**, select a size criterion that will exclude background pixels or other artifacts of image capture or deconvolution. This value will depend on imaging conditions and must be determined empirically.
4. Under the **Measure Objects** tab, click the gear icon to select which measurements you wish to use. Check **Intensity and Volume Measurements**.
5. Define the ROI you wish to measure. Volocity allows rectangular, circular, and freehand ROIs. Use the Zoom function to make it easier to outline the desired area.
6. Adjust the threshold under the **Find Objects Using Intensity** tab by manipulating the lower and upper limits. We suggest maximizing the upper limit to include all pixels that have high intensity values. The lower limit should be adjusted to remove enough background to include only signal that is found within the cellular structure of interest.
7. All measurements will appear in a tab below the image. Individual objects will be separated (e.g., if mitochondrial fluorescence intensity is being measured and there are three mitochondrial structures, three measurements will be provided, each color-coded to depict which structure they represent). These measurements can be copied into a spreadsheet application such as Microsoft Excel or saved within Volocity by clicking **Make Measurement Item** under the **Measurement** tab.

8. In the results table, the Sum column reports the integrated intensity of each object. The sum of all of these values is the integrated intensity within the ROI.

3.12.2 Using Volocity to Measure Mean Intensity

1. Define the ROI you wish to measure including background ROIs (outside the cell). Volocity allows rectangular, circular, and freehand ROIs. Use the Zoom function to make it easier to outline the desired area.
2. Click **Measure Object** tab to measure the signal intensity.
3. All measurements will appear in a tab below the image. These measurements can be copied into a spreadsheet application such as Microsoft Excel or saved within Volocity by clicking **Make Measurements** under the **Measurement** tab.
4. Record the mean gray value (total fluorescence of area divided by area pixels).
5. Calculate corrected intensity measurements: subtract the background mean intensity from the mean intensity of each measured ROI.

3.12.3 Using Fiji/ImageJ to Measure Mean or Integrated Intensity

1. From the **Analyze** menu, choose **Set Measurements**. Check the following: Area, **Mean Gray Value**, **Integrated Density**, **Display Label**.
2. Subtract background intensity in the image:
   (a) Generate a maximum projected image: **Image** > **Stacks** > **Z project.** Include all slices by selecting the start and stop slice. Select "Max Intensity" as the projection type.
   (b) Use the **Rectangle selection tool** to select a background region (outside the cells). Measure the mean intensity in this region.
   (c) Subtract the mean background mean intensity from the original image: **Process** > **Math** > **Subtract.** Enter the mean background intensity from the previous step. Process all slices.
3. Using one of the ROI tools, define the ROI you wish to measure.
4. From the **Analyze menu**, choose **Measure** (or press M).
5. Copy or save the measurements that appear in the results window.

3.13 Analysis of Actin Dynamics: Measuring Retrograde Actin Cable Flow Velocity

The methods described here employ Fiji/ImageJ [39] as the tool of choice in our own laboratory. Similar functions are available in most image analysis and processing software packages. Tracking movements involves marking the position of an object at successive timepoints. For structures with elongated, irregular, or dynamic shapes, like actin cables, a point can be defined on the object of interest and can be used as the reference point for velocity measurements. For example, when tracking actin dynamics with Abp140p-GFP-labeled actin, heterogeneity of Abp140p binding produces bright dots along the actin cable, and these areas serve as fiduciary marks to track movement [6]. Tracking can provide several quantitative measurements including velocity, distance traveled, and direction of movement. Here, we show how to calculate velocity.

1. Identify individual cells displaying cable motility by rapidly scrolling through the movie. Configure the **Point Tool** on the Fiji toolbar to **Add to Overlay** and **Label Points**. Then mark and label each cell of interest using the Point Tool.
2. Capture and save a screenshot showing the numbered cells of interest: **Plugin** > **Utilities** > **Capture screen**.
3. For each cell of interest, crop the movie to include only that cell, and save a copy with the cell number.
4. To track actin cable movement, use the Fiji **Manual Tracking** plugin.
5. Manually enter the microscope and image parameters, including **time interval** (e.g., 0.5 s) and **x/y calibration** (pixel size in μm). To generate an overlay showing the track, also check the following boxes: **Tracking** > **Show path**; **Drawing** > **Overlay Dots & Lines**.
6. Configure the Point Tool on the Fiji toolbar to **Auto-measure** and **Auto-Next Slice**.
7. Set the movie to the first timepoint to be analyzed, and click **Add track.**
8. Begin tracking a cable by clicking on a point of interest: either the leading or trailing end or a fiduciary mark.
9. Continue marking the motile structure for the duration of observable movement.
10. Once tracking is finished, click **End track**. This will produce two result windows:
    (a) The tracked path drawn as an overlay on the image. Use this to confirm that the movement of the actin cable matches the manual tracing.
    (b) The Results table, giving position, slice number, distance, and velocity.
11. Use the Manual Tracking results or manually calculate the incremental velocity using the following equation (*see* **Note 14,** Fig. 3):

$$\text{Velocity} = \frac{\text{Distance Traveled}}{\text{Video Time Interval}}$$

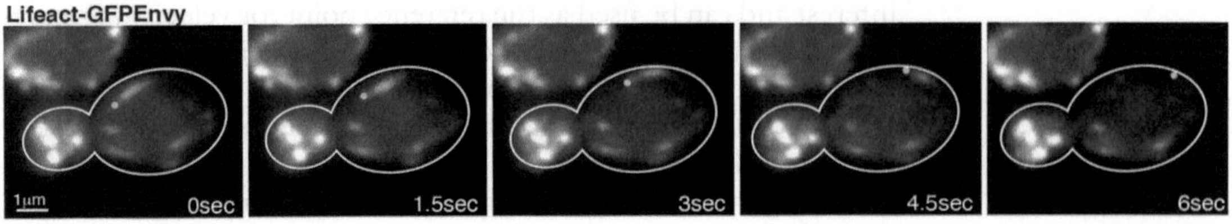

Fig. 3 Tracking actin dynamics: retrograde actin cable flow. Yeast cells expressing Lifeact-GFPEnvy were propagated in lactate medium and imaged every 500 ms for 30 s. Selected frames are shown. Green dots follow a fiduciary mark on an actin cable within a mother cell that undergoes retrograde movement. Scale bar: 1μm

4 Notes

1. Abp140p-GFP signal intensity is higher in yeast propagated using non-fermentable carbon sources. Lactate medium is the preferred choice as it is not autofluorescent and provides more nutrients than synthetic complete medium with glycerol (SG). However, fragmentation of mitochondria may occur in lactate medium. If this is a problem, use glycerol-based media, such as SG.
2. The sequences for the 3′ ends are often provided by the designers of the tagging plasmid. In many cases, the forward primer will include the sequence of the polylinker upstream of the fluorescent protein tag, while the reverse primer will include the reverse complement of the last few nucleotides of the selection marker, or some sequence downstream of the selection marker.
3. The second step of the 35-cycle procedure is the primer annealing phase. The temperature used should match the melting temperature of the DNA. For primers that have secondary structures, or when primer sequences cannot be manipulated to make the melting temperatures comparable, betaine, DMSO, or both can be added to the PCR reaction to a final concentration of 1 M and 5%, respectively (10μL and 5μL, respectively, for a 50μL reaction).
4. For most PCR reactions, we find that product purification is unnecessary. While purification will eliminate some contaminants from the PCR reaction, the accompanying loss of DNA from the process often outweighs the benefits of performing the purification. We specifically find that when using the KAPA HiFi HotStart or Platinum Taq DNA Polymerase High Fidelity PCR reaction kits, purification to eliminate contaminants is unnecessary.
5. The galactose-inducible Cre recombinase plasmids are CEN plasmids, which can often be dropped out by continuous

growth on non-selective medium. Alternatively, pSH47 can be used as this galactose-inducible Cre recombinase expression plasmid contains the *URA3* selectable marker which can be removed from cells by counter-selection using 5-fluoroorotic acid (5-FOA) plates after marker selection.

6. When performing rescue/plasmid complementation assays, it is important to express the fluorescent protein-tagged protein under the endogenous promoter of the gene. This will avoid unexpected artifacts, such as overexpression of a partially functional protein rescuing wild-type phenotypes.
7. For live-cell imaging, avoid using medium containing yeast extract (e.g., YPD, YPG) as the yeast extract is autofluorescent and will require repeated washing prior to imaging. Instead, use synthetic complete or lactate medium.
8. Avoid excessive/extended centrifugation, which causes actin cable depolymerization.
9. Remove as much glass debris and dust from the slide before adding the cell suspension to ensure a single-cell layer is formed when applying the cover slip.
10. Excess volume of cell suspension can cause cells to float and move during image acquisition, while insufficient volume can compress cells or cause uneven spreading. Slight pressure should be applied to the edges of the coverslip to stably trap cells between the slide and coverslip and avoid movement of cells during image acquisition. Take care not to put too much pressure or cells may burst.
11. Actin cable dynamics are quite sensitive; acquire no more than four 30 s datasets per slide. Beyond this point, visible actin motility is greatly reduced due to stress-induced disorganization of the actin cytoskeleton.
12. Videos are captured within a single *z*-plane; therefore, the initial z-slice placement is the focal plane for the entire video. Due to the 3D shape of a yeast cell, actin cables are visually more apparent either just above or just below the center of the cell.
13. Artifacts can occasionally appear in deconvolved images. These may include halo-like structures, amplification of noise pixels, or the disappearance of structures that are visible in the raw data. If artifacts appear, repeat the deconvolution with fewer iterations. However, if artifacts do not appear, but the image is still blurry or faint, repeat with more iterations. If deconvolution fails to exceed the 95% confidence limit, verify that optics are clean, light source is aligned, and cells are mounted under a #1.5 coverslip. Also make sure that images are acquired with pixel values at least in the middle of the detector range and not saturating the detector.

Acknowledgments

We thank the members of the Pon laboratory for support and invaluable advice. This work was supported by awards from the National Institutes of Health (NIH) (GM122589 and AG051047) and Muscular Dystrophy Association (MDA 314107) to LAP; the NIH (AG055326 and DK7647) to CNS; and awards from the National Institute of Aging (NIA) (1K99AG065200-01A1) to RHS. The Confocal and Specialized Microscopy Shared Resource is in the Herbert Irving Comprehensive Cancer Center at Columbia University Medical Center, which is supported in part by an award from the NIH/NCI (5 P30 CA13696).

References

1. Adams AE, Pringle JR (1984) Relationship of actin and tubulin distribution to bud growth in wild-type and morphogenetic-mutant *Saccharomyces cerevisiae*. J Cell Biol 98:934–945
2. Goley ED, Welch MD (2006) The ARP2/3 complex: an actin nucleator comes of age. Nat Rev Mol Cell Biol 7:713–726
3. Engqvist-Goldstein AEY, Drubin DG (2003) Actin assembly and endocytosis: from yeast to mammals. Annu Rev Cell Dev Biol 19:287–332
4. Moseley JB, Goode BL (2006) The yeast actin cytoskeleton: from cellular function to biochemical mechanism. Microbiol Mol Biol Rev 70l:605–645
5. Pruyne D, Legesse-Miller A, Gao L, Dong Y, Bretscher A (2004) Mechanisms of polarized growth and organelle segregation in yeast. Annu Rev Cell Dev Biol 20:559–591
6. Yang H-C, Pon LA (2002) Actin cable dynamics in budding yeast. Proc Natl Acad Sci U S A 99:751–756
7. Huckaba TM, Lipkin T, Pon LA (2006) Roles of type II myosin and a tropomyosin isoform in retrograde actin flow in budding yeast. J Cell Biol 175:957–969
8. Higuchi R, Vevea JD, Swayne TC, Chojnowski R, Hill V, Boldogh IR, Pon LA (2013) Actin dynamics affect mitochondrial quality control and aging in budding yeast. Curr Biol 23:2417–2422
9. Bubnell J, Pfister P, Sapar ML, Rogers ME, Feinstein P (2013) β2 adrenergic receptor fluorescent protein fusions traffic to the plasma membrane and retain functionality. PLoS One 8:e74941
10. Gauss R, Trautwein M, Sommer T, Spang A (2005) New modules for the repeated internal and N-terminal epitope tagging of genes in *Saccharomyces cerevisiae*. Yeast 22:1–12
11. Gueldener U, Heinisch J, Koehler GJ, Voss D, Hegemann JH (2002) A second set of loxP marker cassettes for Cre-mediated multiple gene knockouts in budding yeast. Nucleic Acids Res 30:e23
12. Gietz RD, Schiestl H, Willems AR, Woods RA (1995) Studies on the transformation of intact yeast cells by the LiAc/SS-DNA/PEG procedure. Yeast 11:355–360
13. Fehrenbacher KL, Yang H-C, Gay AC, Huckaba TM, Pon LA (2004) Live cell imaging of mitochondrial movement along actin cables in budding yeast. Curr Biol 14:1996–2004
14. Liao P-C, Higuchi-Sanabria R, Swayne TC, Sing CN, Pon LA (2020) Live-cell imaging of mitochondrial motility and interactions in drosophila neurons and yeast. Methods Cell Biol 155:519–544
15. Doyle T, Botstein D (1996) Movement of yeast cortical actin cytoskeleton visualized in vivo. Proc Natl Acad Sci U S A 93:3886–3891
16. Slubowski CJ, Funk AD, Roesner JM, Paulissen SM, Huang LS (2015) Plasmids for C-terminal tagging in *Saccharomyces cerevisiae* that contain improved GFP proteins, Envy and Ivy. Yeast 32:379–387
17. Bajar BT, Wang ES, Lam AJ, Kim BB, Jacobs CL, Howe ES, Davidson MW, Lin MZ, Chu J (2016) Improving brightness and photostability of green and red fluorescent proteins for live cell imaging and FRET reporting. Sci Rep 6:20889
18. Riedl J, Crevenna AH, Kessenbrock K, Yu JH, Neukirchen D, Bista M, Bradke F, Jenne D, Holak TA, Werb Z, Sixt M, Wedlich-Soldner R,

Riedl J, Crevenna AH, Kessenbrock K, Yu J (2008) Lifeact: a versatile marker to visualize F-actin. Nat Methods 5:605–607

19. Riedl J, Flynn KC, Raducanu A, Gärtner F, Beck G, Bösl M, Bradke F, Massberg S, Aszodi A, Sixt M et al (2010) Lifeact mice for studying F-actin dynamics. Nat Methods 7:168–169
20. Mizoguchi T, Kawakami K, Itoh M (2016) Zebrafish lines expressing UAS-driven red probes for monitoring cytoskeletal dynamics. Genesis 54:483–489
21. van der Honing HS, van Bezouwen LS, Emons AMC, Ketelaar T (2011) High expression of Lifeact in *Arabidopsis thaliana* reduces dynamic reorganization of actin filaments but does not affect plant development. Cytoskeleton (Hoboken) 68:578–587
22. Vidali L, Rounds CM, Hepler PK, Bezanilla M (2009) Lifeact-mEGFP reveals a dynamic apical F-actin network in tip growing plant cells. PLoS One 4:e5744
23. Berepiki A, Lichius A, Shoji J-Y, Tilsner J, Read ND (2010) F-actin dynamics in Neurospora crassa. Eukaryot Cell 9:547–557
24. Peng L, Yu Q, Wei H, Zhu N, Ren T, Liang C, Xu J, Tian L, Li M (2019) The TRP Ca2+ channel Yvc1 regulates hyphal reactive oxygen species gradient for maintenance of polarized growth in *Candida albicans*. Fungal Genet Biol 133:103282
25. Delgado-Alvarez DL, Callejas-Negrete OA, Gómez N, Freitag M, Roberson RW, Smith LG, Mouriño-Pérez RR (2010) Visualization of F-actin localization and dynamics with live cell markers in Neurospora crassa. Fungal Genet Biol 47:573–586
26. Munsie LN, Caron N, Desmond CR, Truant R (2009) Lifeact cannot visualize some forms of stress-induced twisted F-actin. Nat Methods 6:317
27. Spracklen AJ, Fagan TN, Lovander KE, Tootle TL (2014) The pros and cons of common actin labeling tools for visualizing actin dynamics during drosophila oogenesis. Dev Biol 393:209–226
28. Courtemanche N, Pollard TD, Chen Q (2016) Avoiding artefacts when counting polymerized actin in live cells with LifeAct fused to fluorescent proteins. Nat Cell Biol 18:676–683
29. Flores LR, Keeling MC, Zhang X, Sliogeryte K, Gavara N (2019) Lifeact-GFP alters F-actin organization cellular morphology and biophysical behaviour. Sci Rep 9:3241
30. Kumari A, Kesarwani S, Javoor MG, Vinothkumar KR, Sirajuddin M (2020) Structural insights into actin filament recognition by commonly used cellular actin marker. EMBO J 39:e104006
31. Belyy A, Merino F, Sitsel O, Raunser S (2020) Structure of the Lifeact-F-actin complex. PLoS Biol 18:e3000925
32. Sabourin M, Tuzon CT, Fisher TS, Zakian VA (2007) A flexible protein linker improves the function of epitope-tagged proteins in *Saccharomyces cerevisiae*. Yeast 24:39–45
33. Cheng TH, Chang CR, Joy P, Yablok S, Gartenberg MR (2000) Controlling gene expression in yeast by inducible site-specific recombination. Nucleic Acids Res 28:E108
34. Lee S, Lim WA, Thorn KS (2013) Improved blue green and red fluorescent protein tagging vectors for S cerevisiae. PLoS One 8:e67902
35. Godley BF, Shamsi FA, Liang F-Q, Jarrett SG, Davies S, Boulton M (2005) Blue light induces mitochondrial DNA damage and free radical production in epithelial cells. J Biol Chem 280:21061–21066
36. Higuchi-Sanabria R, Garcia EJ, Tomoiaga D, Munteanu EL, Feinstein P, Pon LA (2016) Characterization of fluorescent proteins for three- and four-color live-cell imaging in *S cerevisiae*. PLoS One 11:e0146120
37. Pawley J (2000) The 39 steps: a cautionary tale of quantitative 3-D fluorescence microscopy. BioTechniques 28(884–6):888
38. Wallace W, Schaefer LH, Swedlow JR (2001) A workingperson's guide to deconvolution in light microscopy. BioTechniques 31:1076–1078. 1080 1082 passim
39. Schneider CA, Rasband WS, Eliceiri KW (2012) NIH image to ImageJ: 25 years of image analysis. Nat Methods 9:671–675
40. Longtine MS, McKenzie A, Demarini DJ, Shah NG, Wach A, Brachat A, Philippsen P, Pringle JR (1998) Additional modules for versatile and economical PCR-based gene deletion and modification in *Saccharomyces cerevisiae*. Yeast 14:953–961
41. Rodrigues F, van Hemert M, Steensma HY, Côrte-Real M, Leão C (2001) Red fluorescent protein (DsRed) as a reporter in *Saccharomyces cerevisiae*. J Bacteriol 183:3791–3794
42. Janke C, Magiera MM, Rathfelder N, Taxis C, Reber S, Maekawa H, Moreno-Borchart A, Doenges G, Schwob E, Schiebel E et al (2004) A versatile toolbox for PCR-based tagging of yeast genes: new fluorescent proteins more markers and promoter substitution cassettes. Yeast 21:947–962
43. Sheff MA, Thorn KS (2004) Optimized cassettes for fluorescent protein tagging in *Saccharomyces cerevisiae*. Yeast 21:661–670

44. Young CL, Raden DL, Caplan JL, Czymmek KJ, Robinson AS (2012) Cassette series designed for live-cell imaging of proteins and high-resolution techniques in yeast. Yeast 29:119–136
45. Madania A, Dumoulin P, Grava S, Kitamoto H, Schärer-Brodbeck C, Soulard A, Moreau V, Winsor B (1999) The *Saccharomyces cerevisiae* homologue of human Wiskott-Aldrich syndrome protein Las17p interacts with the Arp2/3 complex. Mol Biol Cell 10:3521–3538
46. Kaksonen M, Sun Y, Drubin DG (2003) A pathway for association of receptors adaptors and actin during endocytic internalization. Cell 115:475–487
47. Warren DT, Andrews PD, Gourlay CW, Ayscough KR (2002) Sla1p couples the yeast endocytic machinery to proteins regulating actin dynamics. J Cell Sci 115:1703–1715
48. Miliaras NB, Park J-H, Wendland B (2004) The function of the endocytic scaffold protein Pan1p depends on multiple domains. Traffic 5:963–978
49. Morishita M, Engebrecht J (2005) End3p-mediated endocytosis is required for spore wall formation in *Saccharomyces cerevisiae*. Genetics 170:1561–1574
50. Sun Y, Carroll S, Kaksonen M, Toshima JY, Drubin DG (2007) PtdIns(4 5)P2 turnover is required for multiple stages during clathrin- and actin-dependent endocytic internalization. J Cell Biol 177:355–367
51. Sun Y, Martin AC, Drubin DG (2006) Endocytic internalization in budding yeast requires coordinated actin nucleation and myosin motor activity. Dev Cell 11:33–46
52. Vaduva G, Martin NC, Hopper AK (1997) Actin-binding verprolin is a polarity development protein required for the morphogenesis and function of the yeast actin cytoskeleton. J Cell Biol 139:1821–1833
53. Evangelista M, Klebl BM, Tong AH, Webb BA, Leeuw T, Leberer E, Whiteway M, Thomas DY, Boone C (2000) A role for myosin-I in actin assembly through interactions with Vrp1p Bee1p and the Arp2/3 complex. J Cell Biol 148:353–362
54. Jonsdottir GA, Li R (2004) Dynamics of yeast myosin I: evidence for a possible role in scission of endocytic vesicles. Curr Biol 14:1604–1609
55. Sekiya-Kawasaki M, Groen AC, Cope MJTV, Kaksonen M, Watson HA, Zhang C, Shokat KM, Wendland B, McDonald KL, McCaffery JM et al (2003) Dynamic phosphoregulation of the cortical actin cytoskeleton and endocytic machinery revealed by real-time chemical genetic analysis. J Cell Biol 162:765–772
56. Huckaba TM, Gay AC, Pantalena LF, Yang H-C, Pon LA (2004) Live cell imaging of the assembly disassembly and actin cable-dependent movement of endosomes and actin patches in the budding yeast *Saccharomyces cerevisiae*. J Cell Biol 167:519–530
57. Boldogh IR, Yang HC, Nowakowski WD, Karmon SL, Hays LG, Yates JR, Pon LA (2001) Arp2/3 complex and actin dynamics are required for actin-based mitochondrial motility in yeast. Proc Natl Acad Sci U S A 98:3162–3167
58. Smith MG, Swamy SR, Pon LA (2001) The life cycle of actin patches in mating yeast. J Cell Sci 114:1505–1513
59. Karpova TS, Reck-Peterson SL, Elkind NB, Mooseker MS, Novick PJ, Cooper JA (2000) Role of actin and Myo2p in polarized secretion and growth of *Saccharomyces cerevisiae*. Mol Biol Cell 11:1727–1737
60. Waddle JA, Karpova TS, Waterston RH, Cooper JA (1996) Movement of cortical actin patches in yeast. J Cell Biol 132:861–870
61. Winder SJ, Jess T, Ayscough KR (2003) SCP1 encodes an actin-bundling protein in yeast. Biochem J 375:287–295
62. Okreglak V, Drubin DG (2007) Cofilin recruitment and function during actin-mediated endocytosis dictated by actin nucleotide state. J Cell Biol 178:1251–1264

Chapter 4

Imaging the Actin Cytoskeleton in Fixed Budding Yeast Cells

Cierra N. Sing, Emily J. Yang, Ryo Higuchi-Sanabria, Liza A. Pon, Istvan R. Boldogh, and Theresa C. Swayne

Abstract

Budding yeast, *Saccharomyces cerevisiae*, is an appealing model organism to study the organization and function of the actin cytoskeleton. With the advent of techniques to perform high-resolution, multidimensional analysis of the yeast cell, imaging of yeast has emerged as an important tool for research on the cytoskeleton. This chapter describes techniques and approaches for visualizing the actin cytoskeleton in fixed yeast cells with wide-field and super-resolution fluorescence microscopy.

Key words Yeast, Fluorescent proteins, Vital staining, Microscopy, Super-resolution structured illumination microscopy (SR-SIM), Actin, Cytoskeleton

1 Introduction

The actin cytoskeleton was first visualized in fixed budding yeast cells with fluorescent phallotoxins, and, since then, fluorescence imaging has emerged as a useful tool to study actin-containing structures [1]. The organization and function of the actin cytoskeleton are conserved within the eukaryotic lineage. However, actin organization in yeast is less complex than that found in other eukaryotes. Moreover, yeast contain a single actin gene. Therefore, yeast is a good model organism to study the actin cytoskeleton.

Yeast contain two actin-based structures that persist throughout the cell cycle: patches and cables. Actin patches are endocytic vesicles that are coated with a network of actin filaments and associated with actin-binding proteins including fimbrin (Sac6p), capping protein, the Arp2/3 complex and its nucleation-promoting factors, as well as clathrin and its adaptor proteins [2]. Actin patches localize to sites of polarized cell surface growth including the incipient bud site at the start of the cell division cycle, the bud tip and periphery though most of the cell cycle, and the

Ray H. Gavin (ed.), *Cytoskeleton: Methods and Protocols*, Methods in Molecular Biology, vol. 2364,
https://doi.org/10.1007/978-1-0716-1661-1_4,

bud neck prior to cytokinesis [3, 4]. The polarized distribution of actin patches can be used as a proxy for the function of the cell polarization machinery.

The other major actin structure is actin cables. Actin cables are bundles of actin filaments that serve as tracks for anterograde (from mother to daughter cell) transport of cellular constituents (e.g., mitochondria, Golgi, vacuoles, secretory vesicles, spindle alignment elements, and mRNA) and are essential for cell division [5]. Late in the cell cycle, actin cables mediate transport of cytokinesis factors to the bud neck. The organization and assembly of actin cables reflect these cell cycle-regulated functions. Actin cables form in the bud tip and neck using formins (Bni1p and Bnr1p), actin bundling proteins (fimbrin, Abp140p and Bnr1p), and tropomyosins (Tpm1p and Tpm2p). During most of the cell cycle, they extend along the mother-bud axis from the bud tip to the mother cell tip. However, toward the end of cell division, actin cables in both the mother cell and bud are oriented with their plus ends toward the bud neck, the site of cytokinesis.

While observations in fixed cells cannot capture the dynamic nature of the actin cytoskeleton, there are cases where fixed-cell imaging is the method of choice. Certain proteins are not amenable to live-cell imaging, either because the tags or methods used for visualization compromise protein function or localization or because the signal produced is below the limit of detection. Ectopic expression of tagged proteins may increase signal but may also compromise the cytoskeleton. Imaging fixed cells can enhance both signal and resolution compared to live-cell imaging: (1) signal can be amplified by tagging or staining methods that are usable only with fixed cells (e.g., adding multiple copies of an epitope tag), and (2) many high- and super-resolution imaging methods (e.g., stochastic optical reconstruction microscopy) are slow and not readily adaptable to live-cell imaging. Finally, stained fixed cell samples are relatively stable and versatile: they can be visualized by more than one imaging method or stored and imaged more than once.

Here, we describe methods for visualizing the actin cytoskeleton by immunofluorescence or phalloidin staining followed by either conventional wide-field fluorescence microscopy or super-resolution structured illumination microscopy (SR-SIM). We also describe how to quantify various aspects of actin cable morphology using freely available image analysis software.

2 Materials

2.1 Yeast Growth Media

Synthetic complete (SC) medium: 6.7 g yeast nitrogen base without amino acids, 20 g glucose. Dissolve ingredients in 800 mL distilled H_2O. Supplement with the following amino acids as needed: 10 mL adenine (2 mg/mL stock in 0.05 M HCl), 10 mL uracil (2 mg/mL stock in 0.5% $NaHCO_3$), 10 mL arginine (1 mg/mL stock in H_2O), 10 mL histidine (1 mg/mL stock in H_2O), 10 mL leucine (10 mg/mL stock in H_2O), 10 mL lysine (3 mg/mL stock in H_2O), 10 mL methionine (2 mg/mL stock in H_2O), 10 mL phenylalanine (5 mg/mL stock in H_2O), 10 mL tryptophan (2 mg/mL stock in H_2O), and 30 mg tyrosine. Adjust pH to 5.5 with $NaHCO_3$ and bring volume to 1 L with distilled H_2O. Sterilize by autoclaving.

2.2 Fixation and Staining of Yeast Cells

1. Wash solution: Mix 12.5 mL 1 M potassium phosphate, pH 7.5, and 400 mL 1 M KCl. Bring volume up to 500 mL with distilled H_2O. Sterilize by autoclaving.
2. Tris/DTT: 1 mL 1 M Tris–SO_4, pH 9.4, 100μL 1 M DTT (final 10 mM). Bring volume to 10 mL with distilled H_2O. Make this solution on the day of use.
3. Zymolyase (Seikagaku Inc., Tokyo, Japan): Dissolve Zymolyase 20T in wash solution to a concentration of 50μg/mL. Make this solution on the day of use.
4. NS: 10 mL 1 M Tris–HCl, pH 7.5, 21.4 g sucrose, 1 mL 0.5 M EDTA, pH 8.0, 0.5 mL 1 M $MgCl_2$, 0.05 mL 1 M $ZnCl_2$, 0.05 mL 0.5 M $CaCl_2$. Adjust volume to 500 mL with distilled H_2O. Filter-sterilize and store in 40 mL aliquots at −20 °C.
5. NS+: On the day of use, supplement 10 mL NS with 40μL 200 mM phenylmethylsulfonyl fluoride (PMSF) dissolved in ethanol (final 0.8 mM), 5μL beta-mercaptoethanol (final 0.05%), and 100μL 10% (w/v) NaN_3 (final 0.1%).
6. 10× PBS stock: 80 g NaCl, 2 g KCl, 14.4 g Na_2HPO_4, 2.4 g KH_2PO_4. Dissolve ingredients in 800 mL distilled H_2O. Adjust pH to 7.4 with HCl/NaOH. Bring volume to 1 L with distilled H_2O.
7. PBS+: 1× PBS, 1% bovine serum albumin, and 0.1% NaN_3 (from a 10% (w/v) stock) prepared in distilled H_2O. Make the solution on the day of use (*see* **Note 1**).
8. PBT: 1× PBS, 1% bovine serum albumin, 0.1% NaN_3 (from a 10% (w/v) stock), and 0.1% Triton X-100 prepared in distilled H_2O. Make the solution on the day of use.
9. Polylysine: Dissolve poly-L-lysine in distilled H_2O to 0.5 mg/mL. Filter-sterilize and store in 1 mL aliquots at −20 °C.
10. ConA (Concanavalin A, Sigma C7275): Dissolve ConA in distilled H_2O to 2 mg/mL. Filter-sterilize and store in aliquots at −20 °C.
11. Fluorescent phalloidin (rhodamine phalloidin or Alexa Fluor phalloidin, Invitrogen, Carlsbad, CA): Dissolve in 100% methanol to 200 units/mL. Store in 50μL aliquots at −20 °C.

12. Mounting solutions: Invitrogen SlowFade Diamond Antifade mounting solution for phalloidin staining imaged by wide-field microscopy. Invitrogen ProLong Diamond Antifade mounting solution for phalloidin staining imaged by super-resolution microscopy.
13. Paraformaldehyde (*see* **Note 2**): 20% solution, EM grade (Electron Microscopy Sciences 15713, Hatfield, PA).

2.3 Antibodies

Several proteins can be used for immunofluorescence visualization of actin in *S. cerevisiae*. For actin, we suggest the C4 mouse monoclonal antibody, which recognizes a conserved region in actin that is common to yeast and many other cell types [6].

2.4 Equipment for Wide-Field Microscopy

Two examples of suitable wide-field microscopy systems:

1. Inverted AxioObserver.Z1 microscope (Carl Zeiss Inc., Thornwood, NY) equipped with a metal halide lamp and Colibri LEDs for excitation; 100×/1.3 EC Plan Neofluar oil immersion objective; and an Orca ER cooled CCD camera with 1280 × 1024 pixel resolution (Hamamatsu, Bridgewater, NJ). ZEN software (Carl Zeiss) is used to control the camera and microscope hardware.
2. Upright Axioskop 2 microscope (Carl Zeiss Inc., Thornwood, NY) equipped with a pE-4000 LED illumination system (CooLED, Andover, UK) for excitation; 100×/1.4 Plan-Apochromat objective (Carl Zeiss Inc., Thornwood, NY); and an Orca ER cooled CCD camera. NIS Elements 4.60 Lambda software (Nikon, Melville, NY) is used to control the camera and microscope hardware.

3 Methods

The paraformaldehyde fixation detailed here (Subheading 3.1) preserves cellular structures by creating covalent hydroxymethylene bridges between spatially adjacent amino acid residues. Other fixatives, such methanol and acetone, are compatible with some immunofluorescence staining, including the actin cytoskeleton. However, these fixative solutions are not recommended when evaluating additional organelles, like mitochondria, because they solubilize many membranes [7]. Fixed cells can be stained by various methods. For immunofluorescence staining, the cell wall must first be removed with zymolyase. The resulting spheroplasts are then exposed to pretreated primary and secondary antibodies in a staining chamber and mounted on a polylysine-coated coverslip (Subheading 3.2). If antibody staining is not required, we recommend leaving the cell wall intact in order to preserve cell integrity and 3D shape. Intact cells can be stained directly with non-antibody agents such as fluorescent phalloidin (Subheading 3.3) and then mounted on a slide. For SR-SIM imaging, intact cells can be further immobilized by binding to a coverslip using the lectin Concanavalin A (Subheading 3.4).

3.1 Fixation of Yeast Cells

1. Grow a 5 mL liquid culture of cells in a 50 mL conical-bottom tube to mid-log phase ($OD_{600} = 0.2–0.5$) (*see* **Note 3**).
2. Add paraformaldehyde to the culture medium to a final concentration of 3.7% (*see* **Note 4**).
3. Incubate cells under normal growth conditions for 50–60 min.
4. Concentrate 1×10^7 cells by centrifugation (30 s at $10{,}000 \times g$).
5. Continue either to non-antibody staining (Subheadings 3.3 or 3.4) or to removal of the cell wall for immunofluorescence (Subheading 3.2).

3.2 Immunofluorescence

3.2.1 Pretreatment of Antibodies with Yeast Cell Walls

Fungi are a common feature of the environment, including that of the animals used to generate polyclonal antisera. As a result, commercial polyclonal antibodies may contain antibodies that recognize the yeast cell wall. Spheroplasts, yeast that have been enzymatically treated to remove the cell wall, can contain residual cell wall components. As a result, staining of spheroplasts with antibodies that recognize cell wall components may result in high background. To remove these nonspecific antibodies, we recommend pretreating polyclonal antibodies with intact yeast cells. Pretreated antibodies can be prepared in bulk and stored for later use.

1. Grow yeast cells to late log phase ($OD_{600} > 1.0$). A 5 mL culture is sufficient to treat 400μL of undiluted antiserum.
2. Dilute antiserum to 1/25 in PBS+.
3. From the yeast cell culture, take a volume equal to ¼ the volume of diluted antibody (250μL per mL of diluted antibody), and transfer to a microcentrifuge tube.
4. Concentrate cells by centrifugation (30 s at $10{,}000 \times g$), and wash three times with 1 mL of PBS+.
5. Resuspend cells in diluted antiserum, and incubate with gentle mixing for 2 h at 4 °C.
6. Pellet cells by centrifugation, and transfer antiserum into a fresh tube.
7. Repeat **steps 3–5** (exposing treated antiserum to a fresh batch of cells).
8. Pellet the cells by centrifugation, and aliquot pretreated antiserum for storage at −20 °C.

3.2.2 Preparation of a Staining Chamber

To protect samples from desiccation and unnecessary light exposure, incubations are performed in a dark, humid chamber.

1. Cut ten 10 × 15 cm sheets of Parafilm, and press them together, crimping slightly at the edges, to make a flat, semi-rigid platform.
2. Place damp rolled paper towels on Parafilm platform.
3. Cover the platform and paper towels with an inverted opaque tray for the duration of the incubation.

3.2.3 Removal of the Cell Wall with Zymolyase

1. Resuspend fixed cell pellet (from Subheading 3.1) in 1 mL Tris/DTT, and incubate for 20 min at 30 °C water bath.
2. Concentrate cells by centrifugation (30 s at 10,000 × *g*).
3. Wash 1× with 500μL wash solution.
4. Concentrate cells by centrifugation, resuspend pellet in 0.5 mL of zymolyase solution, and incubate for 0.5–1.0 h at 30 °C. The actual incubation time varies from strain to strain (*see* **Note 5**).
5. Pellet cells gently (30 s at 6000–8000 × *g*), and wash cell pellet three times with 0.5 mL of NS+ (*see* **Note 6**).
6. Resuspend the final cell pellet with 2–5× pellet volume of NS+.
7. Store fixed spheroplasts at 4 °C for up to 1 week or continue with immunofluorescence staining.

3.2.4 Attachment of Spheroplasts to Coverslip with Polylysine

Polylysine is a polycation that enhances the attachment of the negatively charged spheroplasts to a coverslip.

1. Place 30–40μL of polylysine solution on the Parafilm platform.
2. Lay a 22 × 22 mm coverslip on the polylysine drop, and apply some pressure to allow the entire coverslip to be covered by polylysine (*see* **Note 7**).
3. Incubate for at least 10 s.
4. Remove the coverslip from the polylysine by slowly pipetting 200–300μL distilled water under the edge of the coverslip. This will cause the coverslip to float off the Parafilm for easy removal.
5. Rinse off excess polylysine by pipetting water gently on the coverslip for approximately 0.5 s. Drain excess liquid by touching the edge of the coverslip to filter paper.
6. Place coated coverslip with polylysine-coated side facing up on Parafilm platform.
7. Place 100μL of sterile filtered 1xPBS on polylysine-coated coverslip.

8. Add 10μL of fixed spheroplast cell suspension to the drop of 1× PBS on the coverslip.
9. Mix and spread cells gently over the coverslip with the side of the pipet tip (*see* **Note 8**).
10. Incubate 35 min at room temperature in the staining chamber.
11. Remove unbound spheroplasts by gently pipetting 500μL 1× PBT onto the coverslip and allowing it to run off.
12. Wash coverslip 2× with 500μL 1× PBT and place coverslip with cell side up on filter paper.

3.2.5 Indirect Immunofluorescence Staining of Attached Spheroplasts

For efficient antibody usage, spheroplast-coated coverslips are inverted on a small volume of diluted, pretreated antibody on a Parafilm platform.

1. Place 30μL of diluted primary antibody in 1xPBT on the Parafilm platform. Lay the spheroplast-coated side of the coverslip onto the drop, ensuring the solution is evenly spread over the coverslip.
2. Incubate for 2 h at room temperature in the staining chamber (*see* **Note 9**).
3. Lift the coverslip by slowly pipetting 200–300μL of 1× PBT under the edge of the coverslip. Rinse coverslip by gently pipetting 500μL 1× PBT onto the coverslip and allowing it to run off; repeat for three to four times. Rock coverslip back and forth during washing and wick excess liquid by touching the edge of the coverslip to filter paper.
4. Place 30–40μL of diluted secondary antibody in 1× PBT on the Parafilm platform. Place the coverslip on the antibody as in **step 1**.
5. Incubate 1 h at room temperature in the staining chamber.
6. Lift coverslip, rinse, and drain as in **step 2**.
7. Place 1–2μL of mounting medium on a microscope slide, and gently place the coverslip, sample side down, onto the mounting solution. Dry any residual liquid from the edges of the coverslip. Seal the edges with clear nail polish.
8. After the nail polish is dry, rinse the coverslip surface with distilled water, and dry gently with a Kimwipe. Visualize sample as soon as possible to maximize fluorescence signal. Most samples are stable for several days.

3.3 Visualization of the Actin Cytoskeleton in Fixed Cells with Fluorescent Phalloidin

The following protocol is for staining of cells in suspension. Any fluorophore of choice conjugated to phalloidin can be used (*see* **Note 10**). However, we recommend using Alexa Fluor 488 phalloidin for imaging actin by wide-field microscopy or super-resolution structured illumination microscopy, because of its reproducible, robust, and photostable signal.

1. Grow cells overnight to mid-log phase ($OD_{600} = 0.2–0.5$) in a 50 mL conical bottom tube.
2. Fix 1 mL of cells by adding 20% paraformaldehyde to a final concentration of 3.7%, and incubate in fixative for 50 min to 1 h in a 30 °C shaking incubator (*see* **Note 11**).
3. Wash fixed cell pellet three times with 500μL of either wash solution or 1× PBS, concentrating each time by centrifugation (30 s at 10,000 × *g*).
4. Wash cells with 500μL of 1× PBT and resuspend in 25μL of 1× PBT.
5. Add 20μL of Alexa Fluor 488 phalloidin to a concentration of 2.5μM and incubate for 25 min at room temperature.
6. Wash cells with 100μL of 1× PBS.
7. Mount cells for imaging. For enhanced immobilization of intact cells, resuspend in 30–60μL 1× PBS and mount on a ConA-coated coverslip (Subheading 3.4). For a simpler preparation, continue to the next step.
8. Resuspend cells in 7–10μL of SlowFade mounting solution (*see* **Note 12**). Place ~1.5μL of cell suspension onto a microscope slide. Place a 22 × 22 mm coverslip on top of the cells and seal the edges of the coverslip with nail polish.

3.4 Preparation for Super-Resolution Structured Illumination Microscopy (SR-SIM)

Cells are attached to ConA-coated coverslips either before or after paraformaldehyde fixation (Fig. 1).

1. Wash high-performance coverslips (18–22 mm square) with distilled H_2O. Wick away any excess liquid and place on filter paper to air dry.
2. Place clean coverslip on a Parafilm platform, and add 20μL of 2 mg/mL ConA solution on top of the coverslip, and spread using the side of a pipet tip.
3. Incubate in staining chamber for 30 min at room temperature.
4. Remove unbound ConA by gently adding distilled H_2O to the corner of the coverslip until the coverslip is full of liquid and then rocking the coverslip to allow liquid to run off.

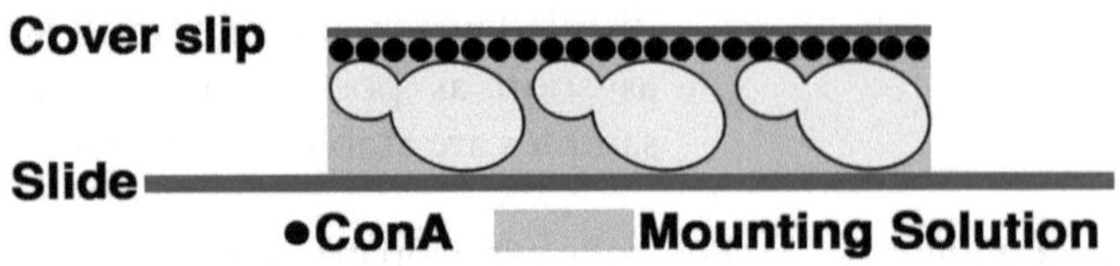

Fig. 1 Schematic of ConA immobilization of cells onto a coverslip. Coverslips are coated with ConA to immobilize cells. ProLong Diamond or another mounting solution is placed onto the slide, and coverslips are placed directly on the mounting medium with cells facing down

5. Allow coverslip to air dry in dark chamber, and store in chamber until needed.
6. Place 30μL of fixed cell suspension on the coverslip, and spread using the side of a pipet tip.
7. Incubate in staining chamber for 15 min.
8. Wash off excess cells by gently applying distilled H_2O as in **step 4**. Repeat wash step three to four times.
9. Remove excess liquid by touching the edge of the coverslip to filter paper.
10. Place 7.5μL of mounting solution (*see* **Note 13**) onto the center of a slide. For optimal signal preservation (which is essential for SR-SIM), we recommend using ProLong Diamond Antifade mounting solution.
11. Place the coverslip cell side down onto the mounting solution. Apply gentle pressure on the coverslip to push out any air pockets that may have formed.
12. If using a hard-set mounting medium such as ProLong Diamond, lay slide flat and allow to cure at room temperature in the dark for 24 h or following the manufacturer's instructions. In this case, there is no need to seal with nail polish.
13. If using a non-curing mounting medium, seal the coverslip edges with nail polish.
14. For best signal intensity, image as soon as possible (within a few days).

3.5 Imaging Fixed Budding Yeast Cells: Wide-Field Microscopy and SR-SIM

Here, we provide two methods for imaging fixed budding yeast cells: wide-field microscopy with deconvolution and super-resolution structured illumination microscopy (SR-SIM). Deconvolution is a computational method that reassigns out-of-focus signal to produce sharper images with enhanced three-dimensional information [8]. Wide-field imaging with deconvolution yields better signal-to-noise ratio and better sensitivity in comparison to point-scanning or spinning-disk confocal imaging. Here, we describe how to deconvolve wide-field images using commercial Volocity software or the free, open-source Fiji/ImageJ software.

Alternatively, even higher-resolution images can be obtained by SR-SIM, which allows resolution to be improved up to twofold over conventional wide-field methods [9]. SR-SIM requires a specialized microscope system (such as Zeiss Elyra or Nikon N-SIM), additional sample preparation, and increased imaging and processing time. Nevertheless, suitable SR-SIM samples yield dramatically improved resolution (Fig. 2). In our experience, SR-SIM is superior for obtaining highly detailed qualitative images of actin structures, assessing actin cable thickness, and counting

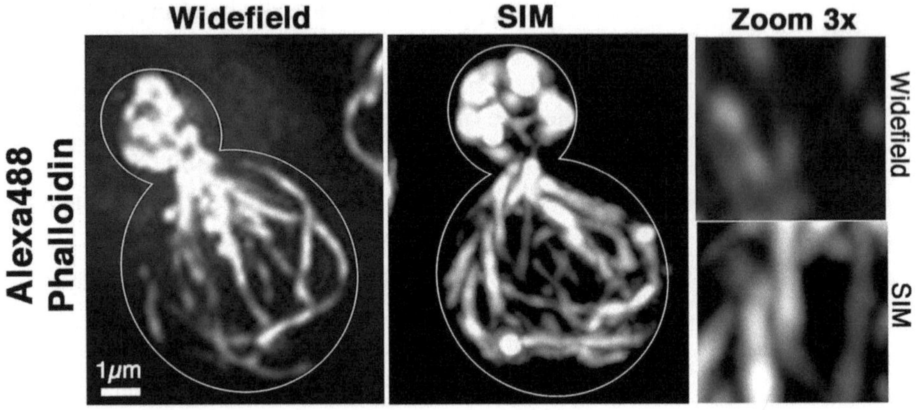

Fig. 2 Comparison of conventional wide-field microscopy with deconvolution and SR-SIM. Wild-type cells fixed in 3.7% paraformaldehyde and stained with 2.5μM Alexa Fluor 488 phalloidin. **Left panel**: Z series were collected at 0.3μm intervals using conventional wide-field microscopy with 1 × 1 binning, 116 gain, 250 ms exposure time, and 470 nm LED excitation with standard GFP filter for Alexa Fluor 488. Images were deconvolved (Volocity) using 60 iterations and 100% confidence interval. **Middle panel**: Z series were collected at 0.125μm intervals using a super-resolution structured illumination microscope (N-SIM, Nikon) with 1x1 binning, 200 gain, 300 ms exposure time, and 488 nm laser at 45% power. Image reconstruction was performed using Nikon Elements software with the following parameters: illumination modulation contrast = 2, high-resolution noise suppression = 1, out-of-focus blur suppression = 0.2. **Right panels**: 3× magnification of the boxed area. Scale bar: 1μm. Quantification of actin cables using the two methods is shown in Table 1

Table 1
Actin cables and patches were counted in mother cells using conventional wide-field microscopy and SR-SIM. Actin cables parallel to the mother-bud axis and greater than half the length of the mother cell were counted

Method	Actin cables/cell (±SEM)	Actin patches/cell (±SEM)	*n* (cells)
Wide-field with deconvolution	8.42 ± 0.21	1.84 ± 0.13	173
SR-SIM	11.37 ± 0.25	2.6 ± 0.20	67

cables. We typically are able to detect more cables in SR-SIM images than in wide-field deconvolved images. However, wide-field imaging recapitulates trends in cable abundance observed by SR-SIM.

3.5.1 Visualization of Fixed Cells by Wide-Field Imaging and Deconvolution

On average, yeast cells are only 4–6μm in diameter, so a high-magnification, high-numerical aperture objective lens and high-resolution camera are essential for successful imaging. It is important for cameras to be able to detect signal under low-intensity illumination and short exposure times to minimize photobleaching. Additionally, it is imperative to use filter sets that maximize light throughput while preventing bleed-through in multicolor imaging experiments.

1. Optimize signal levels by adjusting fixation and staining conditions and using as fresh a sample as possible.
2. Set exposure time and camera gain so that pixel values fill as much as possible of the dynamic range of the camera, without saturation.
3. Acquire z-series of images at a focus interval of 0.2–0.3μm.
4. Deconvolve using either Volocity or Fiji/ImageJ following the steps below.

Deconvolution Using Volocity

1. Load datasets in Volocity and verify that the *x*, *y*, and *z* scale are correctly set.
2. For each channel, generate a new calculated PSF. When prompted, enter the numerical aperture of the objective lens and the maximum wavelength of fluorescence emission (e.g., 507 nm for Alexa Fluor 488 phalloidin).
3. Using the calculated PSF for each channel, perform iterative deconvolution with 60 iterations, 100% confidence limit (*see* **Note 14**).

Deconvolution Using Fiji/Image J

1. Download and install Diffraction PSF 3D (https://imagej.net/Diffraction_PSF_3D) and Iterative Deconvolve 3D (https://imagej.net/Iterative_Deconvolve_3D) plugins from the FIJI website (https://imagej.net/Fiji/Downloads).
2. Create a calculated PSF by using Diffraction PSF 3D: **Plugins** > **Diffraction PSF 3D**. Set up PSF parameters by entering immersion medium refractive index, objective lens numerical aperture, emission wavelength, image pixel size, image z step size, image width, image height, and image Z stack. Typical settings using our microscopy setup:
   (a) Refractive index: 1.518 (if using immersion oil).
   (b) Numerical aperture: 1.40.
   (c) Emission wavelengths: 507 nm or 610 nm for red or green fluorophores, respectively.
   (d) Longitudinal spherical aberration at maximum aperture: 0.
   (e) Image pixel spacing: 64.50 (substitute your own spacing in nm if applicable).
   (f) Slice spacing: 300 (substitute your own spacing in nm if applicable).
   (g) Image width: 1343 (substitute your own image size in pixels if applicable).
   (h) Image height: 1023 (substitute your own image size in pixels if applicable).

(i) Image depth: 21 (substitute your own z-stack size if applicable).

(j) Normalization: sum of pixel values =1.

Name your PSF with the corresponding wavelength for easy recognition.

3. Deconvolve selected images: **Plugins** > **Iterative Deconvolve 3D**. Select the image and corresponding PSF, set Wiener filter gamma to 0, maximum iterations to 60, and delta for termination to 0. The deconvolution will be terminated automatically if iterations do not produce better resolution.
4. Save the deconvolved file.

3.5.2 Visualization of Fixed Cells by Super-Resolution Structured Illumination Microscopy (SR-SIM)

SR-SIM imaging reveals complex details like fine structures of the actin cytoskeleton that are not visible using conventional wide-field imaging (Fig. 2, Table 1). The improved resolution in SR-SIM is generated by modulating the fluorescence excitation pattern with a diffraction grating. The interaction between the illumination pattern and fine sample details produces a Moiré effect. The superimposed grating is then rotated at specific angles to generate multiple sets of Moiré patterns, which can be reconstructed computationally to restore details not captured by standard illumination. Importantly, "super-resolution" or "high-resolution" SIM (SR-SIM) differs from previously marketed structured illumination systems such as Apotome, which only provide optical sectioning. SR-SIM, in contrast, doubles resolution in the *x*, *y*, and *z* dimensions.

Sample Preparation and Imaging

There are several criteria that must be met for optimal SR-SIM imaging:

1. Minimal movement in the sample. We recommend fixing and immobilizing cells and using a hard-set mounting medium. In addition, the room and equipment should be stable in temperature and free from drafts.
2. Maximal signal intensity and photostability. Alexa Fluor 488 phalloidin is preferred for labeling the actin cytoskeleton. Mounting medium should show excellent anti-bleaching performance; ProLong Diamond is recommended.
3. Predictable optical performance. A #1.5 (170μm) coverslip is recommended. Because standard coverslip thickness varies by tens of micrometers, which may induce spherical aberration, high-performance coverslips (manufactured with smaller variation in thickness) should be used. In addition, mounting medium should have a consistent, stable refractive index.
4. Adequate Z sampling. Z-series of images should be acquired at a focus interval of ~0.125μm.

5. Appropriate pixel values in raw data. Exposure time and camera gain should be set so that pixel values fill no more than 1/4 of the dynamic range of the camera. EM gain for EMCCD cameras should not exceed 150.

Reconstruction

Parameters for SR-SIM reconstruction vary with the software used. Here, we describe options for Nikon Elements software (Nikon Instruments, Melville, NY).

The following three parameters should be adjusted for each sample. Recommended values for images of actin are given.

1. Illumination modulation contrast. Higher values reveal more detail but may induce artifacts. Recommended values: 0.5–2.0.
2. High-resolution noise suppression. Higher values suppress artifacts due to pixel noise but also suppress fine details. Recommended values: <1.0.
3. Out-of-focus blur suppression. Higher values suppress out-of-focus signal, essential for densely labeled areas, but also reduce overall signal and resolution. Recommended values: 0.1–0.2.

3.6 Quantification of Fluorescent Signals

Mean fluorescence intensity (mean gray value) is the average value of pixels in a cell or region of interest (ROI). Mean intensity can be a proxy for concentration of a fluorophore but is not a good indicator of the total amount of a fluorophore, unless the cells or compartments are of similar size. For example, a small bud and a large bud may have the same number of labeled actin structures, but the small bud will have a higher mean intensity because the structures are crowded into a smaller volume. In scenarios where ROI sizes vary greatly, it is best to approximate the total amount of a fluorescent probe by calculating the integrated intensity (also referred to as integrated density), which is the sum of all of the pixel values in the region. Thus, the integrated intensity takes into consideration not only the mean pixel value but the area of the ROI. Here, we show how to obtain intensity measurements using Volocity and Fiji/ImageJ. For more sensitive and accurate measurements, background should be measured and subtracted from mean intensity values. To correct integrated intensity for background, subtract the product of the background intensity and the area of the ROI.

3.6.1 Using Volocity to Measure Integrated Intensity

1. From the **Measurements** tab, create a protocol by adding the following steps in order: Find Objects Using Intensity, Exclude Objects by Size, Clip Objects to ROIs, Measure Objects.
2. For images with multiple channels, make sure that the correct channel is selected under the **Find Objects Using Intensity** tab.

3. Under **Exclude Objects by Size**, select a size criterion that will exclude background pixels or other artifacts of image capture or deconvolution. This value will depend on imaging conditions and must be determined empirically.
4. Under the **Measure Objects** tab, click the gear icon to select which measurements you wish to use. Check **Intensity and Volume Measurements**.
5. Define the ROI you wish to measure. Volocity allows rectangular, circular, and freehand ROIs. Use the Zoom function to make it easier to outline the desired area.
6. Adjust the threshold under the **Find Objects Using Intensity** tab by manipulating the lower and upper limits. We suggest maximizing the upper limit to include all pixels that have high intensity values. The lower limit should be adjusted to remove enough background to include only signal that is found within the cellular structure of interest.
7. All measurements will appear in a tab below the image. Individual objects will be separated (e.g., if mitochondrial fluorescence intensity is being measured and there are three mitochondrial structures, three measurements will be provided, each color-coded to depict which structure they represent). These measurements can be copied into a spreadsheet application such as Microsoft Excel or saved within Volocity by clicking **Make Measurement Item** under the **Measurement** tab.
8. In the results table, the Sum column reports the integrated intensity of each object. The sum of all of these values is the integrated intensity within the ROI.

3.6.2 Using Volocity to Measure Mean Intensity

1. Define the ROI you wish to measure including background ROIs (outside the cell). Volocity allows rectangular, circular, and freehand ROIs. Use the Zoom function to make it easier to outline the desired area.
2. Click **Measure Object** tab to measure the signal intensity.
3. All measurements will appear in a tab below the image. These measurements can be copied into a spreadsheet application such as Microsoft Excel or saved within Volocity by clicking **Make Measurements** under the **Measurement** tab.
4. Record the mean gray value (total fluorescence of area divided by area pixels).
5. Calculate corrected intensity measurements: subtract the background mean intensity from the mean intensity of each measured ROI.

3.6.3 Using Fiji/ImageJ to Measure Mean or Integrated Intensity

1. From the **Analyze** menu, choose **Set Measurements**. Check the following: **Area, Mean Gray Value, Integrated Density, Display Label**.
2. Subtract background intensity in the image:
   (a) Generate a maximum projected image: **Image** > **stacks** > **Z project.** Include all slides by set the start and stop slice. Select "Average Intensity" as the projection type.
   (b) Use the **Rectangle selection tool** to select a background region (outside the cells). Measure the mean intensity in this region.
   (c) Subtract the mean background mean intensity from the original image: **Process** > **Math** > **Subtract.** Enter the mean background intensity from the last step. Process all slides.
3. Using the ROI tool, define the ROI you wish to measure.
4. From the **Analyze menu**, choose **Measure** (or press M).
   Copy or save the measurements that appear in the results window.

3.6.4 Quantitative Analysis of Actin Cable Number, Thickness, and Intensity Using Line Profile Measurements

Actin cable number and structure within a cell vary with age, metabolic state, and other factors, and this is controlled by a host of actin-binding and regulatory molecules. To quantify these effects, we use a line profile analysis method. Since actin cables are organized mostly in parallel along the length of the mother cell, a line profile drawn across the center of the mother cell effectively samples the cable population and can be used to determine the number, thickness, and actin content of actin cables (Fig. 3). When the pixel intensity along the line is plotted, cables appear as high-intensity peaks. The number of peaks corresponds to the number of cables, and the width of each peak corresponds to the thickness of the actin cable (Table 2). The height of the peak or the area under the curve can be used to estimate relative actin content. Although this analysis method is relatively objective, it underestimates the abundance of actin cables. Alternatively, cables can be counted manually by viewing the Z stack and marking identified cables with either Volocity or Fiji/ImageJ. The manual method is more sensitive for capturing cable numbers, but it is not as objective as line profile analysis and requires a trained eye to not overestimate cable number. The following protocol describes how to use Fiji/ImageJ to perform these measurements [10].

Creating Line Profiles

1. Collect z-stack images of cells stained with fluorescent phalloidin. It is recommended that images be deconvolved using Volocity or comparable software to remove out-of-focus signal and increase the signal-to-background ratio.

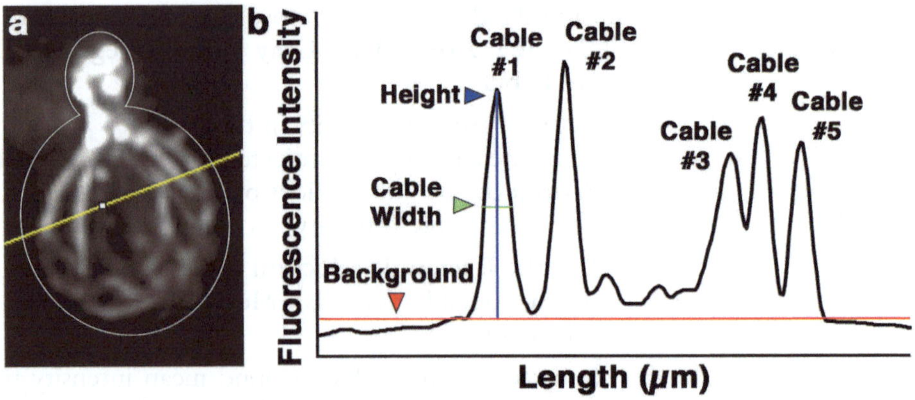

Fig. 3 Schematic for measurements of actin cable number, thickness, and content using line profiles. (**a**) Wild-type cells were stained with Alexa Fluor 488 phalloidin as for Fig. 2 and mounted in ProLong Gold. A line ROI (yellow) is drawn near the center of the mother cell perpendicular to the mother-bud axis. (**b**) Schematic of line profile measurement of actin cable thickness and actin content (actin cable number). The background level, determined by averaging the intensity of regions outside of the cell, is shown in red. A vertical line is then drawn from the maximum point of the peak to the background line as shown in blue. This represents the maximum intensity of the cable. A horizontal line (green) is then drawn across the peak at half of the maximum intensity. This represents the thickness of the actin cable. The number of peaks correlates to the number of actin cables

Table 2
Line profile measurements reveal carbon source-dependent changes in actin cable thickness. Wild-type cells were grown to mid-log phase in glucose or lactate medium. Cells were fixed in 3.7% paraformaldehyde and stained with 1.65μM rhodamine phalloidin for 35 min. Wide-field Z-series were collected throughout the cell (6μm) at 0.3μm intervals using 1 × 1 binning, 200 ms exposure, and a metal halide lamp and standard dsRed filters

Medium	Actin cable thickness, μM (±SEM)	*n* (cells)
Glucose	0.241 ± 0.007	83
Lactate	0.288 ± 0.006	123

2. Make a maximum-intensity projection of the z stack using **Image** > **Stacks** > **Z Project**; choose Projection Type: Max Intensity.
3. If the image has more than one channel, make sure the actin channel is selected.
4. Set up Fiji/ImageJ with the following parameters:
   (a) **Edit** > **Options** > **Input-Output**. Under Results Table Options, check Copy Row Numbers and leave other options unchecked.
   (b) **Edit** > **Options** > **Profile Plot Options**. Set width and height equal to the width and height of your image. Set Minimum Y to 0 and Maximum Y to the image maximum, obtained from Analyze > Histogram.

(c) **Analyze** > **Set Measurements**. Check Area and Display Label.

(d) **Analyze** > **Set Scale**. Make sure that the pixel size is set accurately.

5. Set the Line ROI tool to the Straight Line mode. Double-click on the Line ROI tool and set Line Width to 3. Do not check Spline Fit.
6. Draw a line ROI perpendicular to the mother-bud axis, near the middle of the mother and avoiding actin patches, as shown in Fig. 3a. Both ends of the line should extend into the background outside the cell.
7. **Analyze** > **Plot Profile**. A Plot window should appear showing Gray Value (pixel value) versus Distance.
8. If desired, save the plot image (**File** > **Save**). Data can be saved (click **Save** on the Plot window) or copied and pasted into a spreadsheet application such as Excel (Click **Copy** on the Plot window and then paste into the other application). Proceed to the next section to quantify the profile data using ImageJ.

Measurements of Cable Number, Thickness, and Intensity

1. Count the number of peaks in the profile. This represents the number of actin cables within the mother cell (*see* **Note 15**).
2. Determine a background level by inspection or by using a spreadsheet to average the values from the parts of the profile that are outside the cell. In ImageJ, draw a horizontal line across the plot at that gray value, as shown in red in Fig. 3b.
3. Cable thickness is defined as the width of the peak at half-maximal height above background. To find the height, using the Line ROI tool, draw a vertical line from the maximum height of the peak to the background line, as shown in blue in Fig. 3b. Holding the shift key while drawing a line forces the line to be either vertical or horizontal.
4. Press T to add the vertical line to ROI Manager. **Analyze** > **Measure**. The Length value represents the maximum cable intensity.
5. In the ROI Manager window, check Show All and Labels. A small number will appear in the middle of the vertical line, marking half the peak height.
6. Draw a horizontal line across the peak at half the peak height as shown in green in Fig. 3b. **Analyze** > **Measure.** The length of this horizontal line represents the thickness of the actin cable.
7. Click the Wand tool and select an area under the peak above the background line. **Analyze** > **Measure**. The Area measurement represents the integrated intensity of the cable and can give a relative measurement of actin content for each cable.

4 Notes

1. Filter sterilization can be used to remove any contaminants and/or large aggregates of BSA. Add 0.1 M lysine to 1xPBT to help remove high background.
2. Paraformaldehyde is toxic and carcinogenic. It is readily absorbed through the skin as a liquid and toxic to the respiratory tract as a gas. Gloves and safety goggles are recommended when handling paraformaldehyde, and all handling should be performed in a chemical hood.
3. We recommend fixing cells in mid-log phase because internal structures, including the actin cytoskeleton, are better preserved, compared to cells picked from solid media. The choice of carbon source is also important when imaging actin [11, 12]. Actin cable thickness is increased in cells grown using a non-fermentable carbon source such as lactate.
4. The actin cytoskeleton can be disrupted by centrifugation and rapid changes in temperature [13]. Adding fixative directly to a liquid culture under growth conditions will minimize disruptions. Other important variables are pH and fixative concentration. High paraformaldehyde concentration and low pH increase the number of crosslinks formed and improve structural preservation, but excessive crosslinking can result in artifacts or background fluorescence, owing to crosslinking of fluorescent substances in the media to cellular proteins. Therefore, it is necessary to optimize fixation protocols.
5. To check for cell wall removal, examine 1–2μL of cells under a microscope using transmitted light. Cell edges become less refractile when the wall is removed. Overtreatment with zymolyase will cause cell lysis.
6. Zymolyase treatment changes the integrity of the pellet. When washing, initially add 50μL of buffer and gently aspirate the cells through the pipet tip to resuspend. Then add the remaining wash volume. This will aid in resuspending the cell pellet.
7. Marking one corner of the coverslip with a solution-resistant marker will help differentiate between the coated and uncoated sides.
8. The polylysine-coated coverslips are extremely hydrophobic; try to spread cells to the edges of the coverslip as well as you can.
9. Two antigens can be simultaneously stained by mixing the two primary antibodies in the first incubation and mixing the two secondary antibodies in the second step. Primary antibodies must be raised in different species, and fluorophores used for visualization must have sufficiently separated excitation and emission spectra.

10. This protocol can be adapted for spheroplasts that are attached to coverslips. For adherent spheroplasts, add ¼ volume of fluorescent phalloidin solution (200 units/mL) to the secondary antibody solution and proceed as described in the immunofluorescence protocol.
11. For best results, use 1 mL of cells at $OD_{600} = 0.5$.
12. Cell density should be around 3×10^8 mL but can be adjusted as needed to produce good distribution of cells on the coverslip.
13. Fixation prior to attaching cells to coverslips may result in incomplete immobilization [14]. Moreover, it leads to loss of cells from the coverslip during staining manipulations. We find that using a hard-set mounting medium (such as ProLong Diamond) effectively immobilizes the cells, even if they are fixed before binding. A protocol for fixation of cells after binding to a coverslip can be found in [14].
14. Artifacts occasionally appear in deconvolved images. These may include halo-like structures, amplification of noise pixels, or the disappearance of structures that are visible in the raw data. If artifacts appear, repeat the deconvolution with fewer iterations. If artifacts do not appear, but the image is still blurry or faint, repeat with more iterations. If deconvolution fails to exceed the 95% confidence limit, verify that optics are clean, light source is aligned, and cells are mounted under a #1.5 coverslip. Also make sure that images are acquired with pixel values at least in the middle of the detector range and not saturating the detector.
15. The line plot profile method for counting cables is reliable and more objective than counting by eye. However, it is important to keep in mind that the measurements are only performed in the center of the mother cell. This may result in underrepresentation of the true cable count due to failure to include cables that do not traverse the region where the line is drawn or adjacent cables that are too close for the software to resolve. Alternatively, cables may be counted manually.

Acknowledgments

We thank the members of the Pon laboratory for support and invaluable advice. This work was supported by awards from the National Institutes of Health (NIH) (GM122589 and AG051047) and Muscular Dystrophy Association (MDA 314107) to LAP; the NIH (AG055326 and DK7647) to CNS; and awards from the National Institute of Aging (NIA) (1K99AG065200-01A1) to

RHS. The Confocal and Specialized Microscopy Shared Resource is in the Herbert Irving Comprehensive Cancer Center at Columbia University Medical Center, which is supported in part by an award from the NIH/NCI (5 P30 CA13696).

References

1. Adams AE, Pringle JR (1984) Relationship of actin and tubulin distribution to bud growth in wild-type and morphogenetic-mutant *Saccharomyces cerevisiae*. J Cell Biol 98:934–945
2. Goley ED, Welch MD (2006) The ARP2/3 complex: an actin nucleator comes of age. Nat Rev Mol Cell Biol 7:713–726
3. Engqvist-Goldstein AEY, Drubin DG (2003) Actin assembly and endocytosis: from yeast to mammals. Annu Rev Cell Dev Biol 19:287–332
4. Moseley JB, Goode BL (2006) The yeast actin cytoskeleton: from cellular function to biochemical mechanism. Microbiol Mol Biol Rev 70:605–645
5. Pruyne D, Legesse-Miller A, Gao L, Dong Y, Bretscher A (2004) Mechanisms of polarized growth and organelle segregation in yeas. Annu Rev Cell Dev Biol 20:559–591
6. Lessard JL (1988) Two monoclonal antibodies to actin: one muscle selective and one generally reactive. Cell Motil Cytoskeleton 10:349–362
7. Boldogh IR, Yang H-C, Nowakowski WD, Karmon SL, Hays LG, Yates JR, Pon LA (2001) Arp2/3 complex and actin dynamics are required for actin-based mitochondrial motility in yeast. Proc Natl Acad Sci U S A 98:3162–3167
8. Wallace W, Schaefer LH, Swedlow JR (2001) A workingperson's guide to deconvolution in light microscopy. BioTechniques 31:1076–1078. 1080 1082 passim
9. Gustafsson MG (2000) Surpassing the lateral resolution limit by a factor of two using structured illumination microscopy. J Microsc 198:82–87
10. Schneider CA, Rasband WS, Eliceiri KW (2012) NIH image to ImageJ: 25 years of image analysis. Nat Methods 9:671–675
11. Damsky CH (1976) Environmentally induced changes in mitochondria and endoplasmic reticulum of *Saccharomyces carlsbergensis* yeast. J Cell Biol 71:123–135
12. Visser W, van Spronsen EA, Nanninga N, Pronk JT, Gijs Kuenen J, van Dijken JP (1995) Effects of growth conditions on mitochondrial morphology in *Saccharomyces cerevisiae*. Antonie Van Leeuwenhoek 67:243–253
13. Lillie SH, Brown SS (1994) Immunofluorescence localization of the unconventional myosin Myo2p and the putative kinesin-related protein. Smy1p to the same regions of polarized growth in *Saccharomyces cerevisiae*. J Cell Biol 125:825–842
14. Mund M, Kaplan C, Ries J (2014) Localization microscopy in yeast. Methods Cell Biol 123:253–271

Chapter 5

Imaging of Actin Cytoskeletal Integrity During Aging in *C. elegans*

Gilberto Garcia, Stefan Homentcovschi, Naame Kelet, and Ryo Higuchi-Sanabria

Abstract

The actin cytoskeleton plays a fundamental role in the regulation of multiple cellular pathways, including trafficking and locomotion. The functional integrity of the cytoskeleton is important during aging, as the decline of cytoskeletal integrity contributes to the physiological consequence of aging. Moreover, improving cytoskeletal form and function throughout aging is sufficient to drive life span extension and promote organismal health in multiple model systems. For these reasons, optimized protocols for visualization of the actin cytoskeleton and its downstream consequences on health span and life span are critical for understanding the aging process. In *C. elegans*, the actin cytoskeleton shows diverse morphologies across tissues, potentially due to the significantly different functions of each cell type. This chapter describes an imaging platform utilizing LifeAct to visualize the actin cytoskeleton in live, whole nematodes throughout the aging process and methods to perform follow-up studies on the life span and health span of these organisms.

Key words Actin, Microscopy, Aging

1 Introduction

The actin cytoskeleton is a critical component for the viability of eukaryotic cells, as it contributes to a variety of cellular functions, including maintenance of mechanical support and structure, trafficking of cellular components, and even mediation of cellular stress responses and autophagy [1–5]. The diversity of actin's roles within the cell is mirrored by its complexity: the actin protein is first synthesized as a monomer, which has little to no function on its own. However, by coordinating with an intricate set of regulatory proteins, actin can assemble into complex polymers with multifaceted functions. Thus, maintenance and quality control of the actin cytoskeleton are both significant and exceptionally complex, as it involves regulating not only the actin protein itself but the

Ray H. Gavin (ed.), *Cytoskeleton: Methods and Protocols*, Methods in Molecular Biology, vol. 2364,
https://doi.org/10.1007/978-1-0716-1661-1_5,

hundreds of actin-regulating proteins that exist. Moreover, the importance of maintaining actin is made evident by the numerous diseases in which cytoskeletal dysfunction is observed, including neurodegeneration, muscle myopathies, and the natural aging process [6–8]. Actin is, therefore, a promising area of biological research, with potential discoveries poised to improve our understanding of cellular and organismal health throughout the aging process.

Visualization of the actin cytoskeleton in live organisms proves a unique challenge due to the highly dynamic nature of actin and the protein-protein interactions used by every surface of the monomer to form branches and filaments, making fluorescent tags at the endogenous actin-encoding gene highly impractical [9, 10]. However, a more recent study utilized a split-GFP system, whereby a small fragment of GFP can be used to tag the actin gene at the endogenous locus and the rest of the GFP protein is expressed from a separate construct to allow for GFP assembly in trans. By placing only a very small tag on the actin protein itself, the protein tag does not disrupt the polymerization capacity of actin, even when linked to the rest of the GFP protein in trans [11]. As an alternative, many actin-binding proteins have been tagged with fluorescent proteins, such as Abp140p in *S. cerevisiae* [12, 13]. This approach can be refined by instead tagging and expressing exogenous proteins that bind to actin, such as LifeAct, F-tractin, and Utrophin, all of which have variable benefits and consequences that should be taken into consideration when being utilized [14, 15]. For example, overexpression of many of these constructs can actually alter F-actin stability and/or dynamics [15]. Additionally, some constructs may differentially label various actin structures [16].

1.1 Imaging of the Actin Cytoskeleton in C. elegans

The nematode, *Caenorhabditis elegans*, is an attractive model organism for biological studies as they are easy to grow and have a multicellular anatomy that allows investigation of tissue-specific effects and cell-to-cell communication. Nematodes have been used heavily as an effective model system for large-scale screening due to the ease of genetic perturbations through RNA interference (RNAi), which can be delivered through a bacterial vector. Their quick generation time, isogeny, transparency, genetic tractability, and large evolutionary conservation of cellular pathways make them an ideal system. However, a major disadvantage of *C. elegans* as a model system is that tools for cell biological studies are severely limited, especially for studying cytoskeletal form and function. Moreover, the thick cuticle of the adult animal is difficult to penetrate with traditional large-molecule actin-staining dyes, such as phalloidin. As such, staining animals would generally require the additional step of breaking down the cuticle, making genetic reporters a simpler and more attractive option for visualization of the actin cytoskeleton in live, adult worms.

In *C. elegans*, expression of LifeAct fused to a fluorescent protein provides a simple and efficient way of visualizing the actin cytoskeleton in live worms [17]. LifeAct is the first 17 amino acids of a yeast actin binding protein, Abp104p, which binds to F-actin filaments and can be used to visualize the F-actin network in live cells when fused to a fluorescent protein [14]. While LifeAct is not the only tool available for visualization of the actin cytoskeleton, we have seen that ectopic expression of F-tractin or Utrophin in *C. elegans* caused detrimental effects, and stable lines could not be synthesized. However, further optimization and reduced expression levels could potentially resolve this issue. Alternatively, using the split-GFP system to create direct tags to the actin protein may yield better results, except the genetics of *C. elegans* make this quite challenging. First, *C. elegans* possess five different genes for actin, in which some inherent differences in expression, function, and localization to tissues may exist [18, 19]. This would require complete characterization of all actin genes to determine which gene should be tagged for each functional readout. Moreover, the requirement of creating an additional transgene expressing the remainder of the GFP protein, subsequent crossing of these two lines together, and the necessity to carry two transgene into all future crosses (e.g., to determine cytoskeletal integrity in mutants) make this system less ideal in *C. elegans*. Thus, we present LifeAct here as a robust and reliable approach to actin visualization paired with ease of application.

1.2 Making of C. elegans Transgenic Lines

Transgenic constructs can be introduced into *C. elegans* through two methods: microinjections or biolistic bombardment. Microinjections are accomplished by directly injecting DNA into the nematode's gonads, while biolistic bombardment is achieved through high-speed bombardment of the animal with DNA-coated microparticles. Both methods can produce animals with extra-chromosomal arrays, but only biolistic bombardment can directly result in integration of transgenes into the genome [20–22]. Microinjections require a secondary irradiation step for genome integration or concurrent employment of advanced techniques during the microinjections, such as Mos1-mediated single-copy insertion (MosSCI) [23–26]. Here, we will focus on the latter as our preferred method of strain creation.

DNA introduced into the *C. elegans* germline can form extra-chromosomal arrays consisting of multiple copies of the DNA. These extra-chromosomal arrays are heritable to varying degrees from parent to offspring and from cell to cell within the same animal during development [27]. While the copy number can be coarsely adjusted by altering the concentrations of DNA introduced into animals, these changes are imprecise. Ultimately, the copy-number and heritability of the extra-chromosomal array are stochastic. Even though the creation of transgenic animals with extra-chromosomal arrays is quick and proves uniquely useful for some circumstances, such as mosaic analysis, these animals prove to be less effective for use with cell biology markers [28]. The imprecise expression level and variability from cell to cell can prove to be problematic for constructs that may be toxic at high levels and when consistent expression is required in specific cells. Furthermore, the unstable nature of the arrays requires selection of array-positive animals through markers, such as fluorophores, gene rescues, or dominant mutant alleles, thereby increasing labor requirements for experimental setups and strain maintenance [23]. While integration of these arrays into the genome would solve the variability in expression level and heritability, transgene copy-number and the site of the integration would remain uncontrolled. MosSCI transgenic animals, however, are single-copy insertions at known loci, making them an ideal choice for cell biology marker strains. The single-copy transgenes allow expression levels to have minimal effects on cells, improving the likelihood of successfully creating strains expressing potentially toxic constructs, and the known loci enable coordination of insertion sites with potential future crosses.

1.3 Form and Function of Actin During Aging in Live C. elegans

Transgenic animals were created by fusing LifeAct to mRuby and expressing them under tissue-specific promoters to accomplish robust labeling of the actin cytoskeleton (Fig. 1a, b). These cytoskeletal structures can be manipulated by conditions that alter cytoskeletal dynamics, such as actin knockdown by RNAi, heat-shock, or treatment with drugs that alter actin polymerization, providing evidence that they serve as a reliable tool for interrogating cytoskeletal integrity (Fig. 1c–e). Finally, deterioration in actin cytoskeletal structural integrity can be visualized during the aging process and thus can also be used for aging studies (Fig. 2). The actin cytoskeleton in muscle cells is visualized as linear striations of filamentous actin parallel to muscle fibers, which show disorganization beginning at day 7 of adulthood (Fig. 2a). Actin in the intestine is visualized as two thick cables found along the intestinal lumen, with more complex structures at the pharynx and anus, all of which also begin to show deterioration at day 7 (Fig. 2b). In the hypodermis, there exist two distinct subpopulations of actin structures. The first is observed in the outer region of the syncytium, closer to the cuticle, appearing as large clusters of actin filaments throughout adulthood. A second set of clusters is found in the inner region of the syncytium, and potentially also in the seam cells, but only become visible at approximately days 3–4 of adulthood (Fig. 2c). These structures show loss of organization by day 7 and are completely lost by day 10. All of the constructs in this method have minimal background signal, making them exceptionally easy to image using standard compound microscopes without the need for more refined imaging techniques (e.g., confocal or deconvolution microscopy) to evaluate these actin structures.

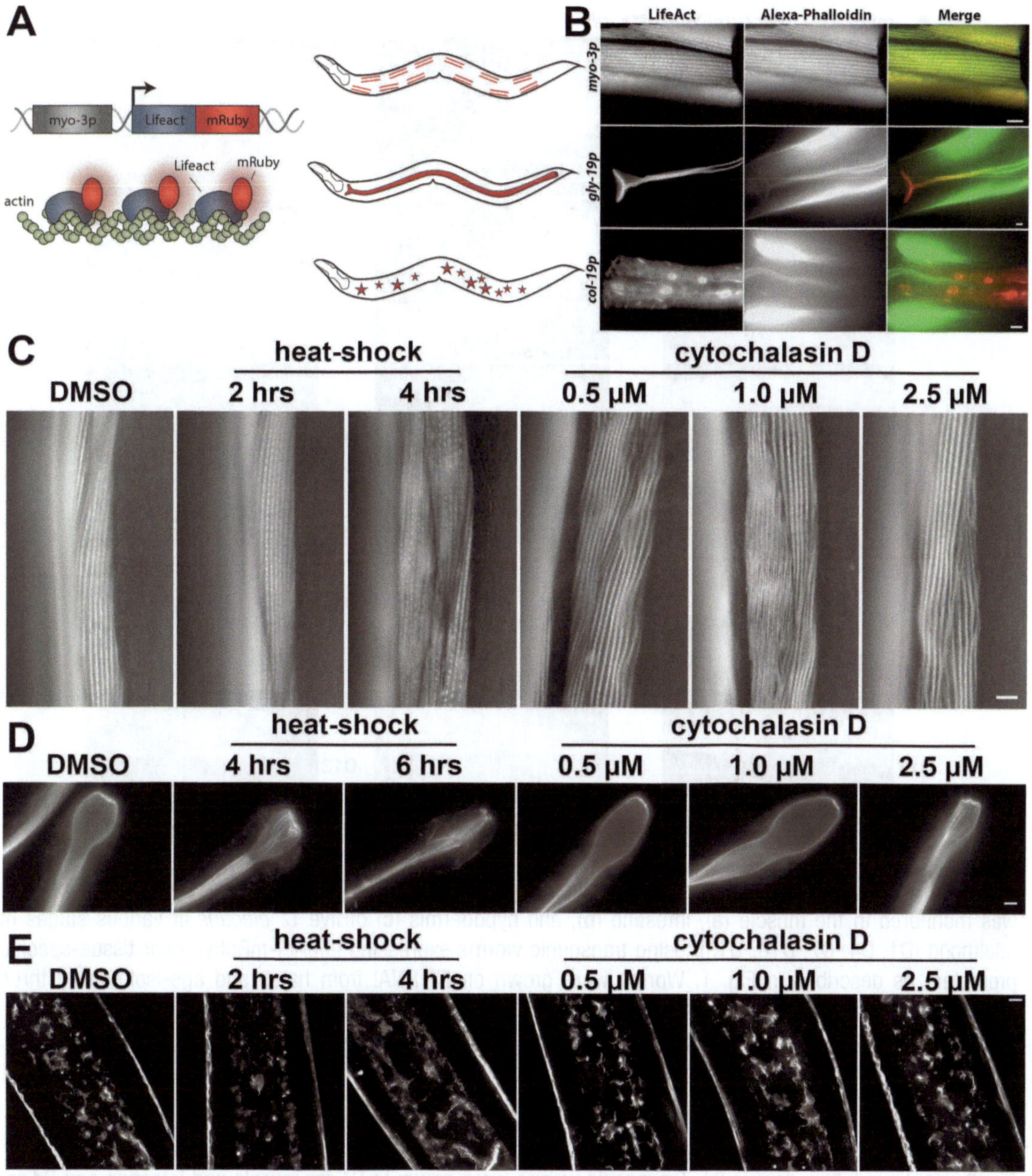

Fig. 1 LifeAct is a robust and reliable reporter for tissue-specific visualization of actin in live animals. (**a**) Left: Schematic for tissue-specific LifeAct expression in *C. elegans*. LifeAct is fused to the red fluorescent protein, mRuby, and expressed in worms using a tissue-specific promoter, such as myo-3p for muscle expression. LifeAct-mRuby allows robust visualization of the actin cytoskeleton by binding to F-actin. Right: schematic of the location where LifeAct-mRuby is expressed and visualized in the muscle, intestine, and hypodermis. (**b**) LifeAct-mRuby-marked structures colocalize with Alexa Fluor 488 phalloidin staining. Transgenic worms expressing LifeAct-mRuby under tissue-specific promoters for the muscle (*myo-3p*), intestine (*gly-19p*), and hypodermis (*col-19p*) were grown to D1 on empty vector (EV) RNAi from hatch, fixed, and stained as described in [17]. All scale bars are 5μm. (**c**–**e**) LifeAct-mRuby-marked actin is susceptible to disruption by actin-destabilizing conditions, such as heat-shock and cytochalasin D treatment in the muscle (**c**), intestine (**d**), and

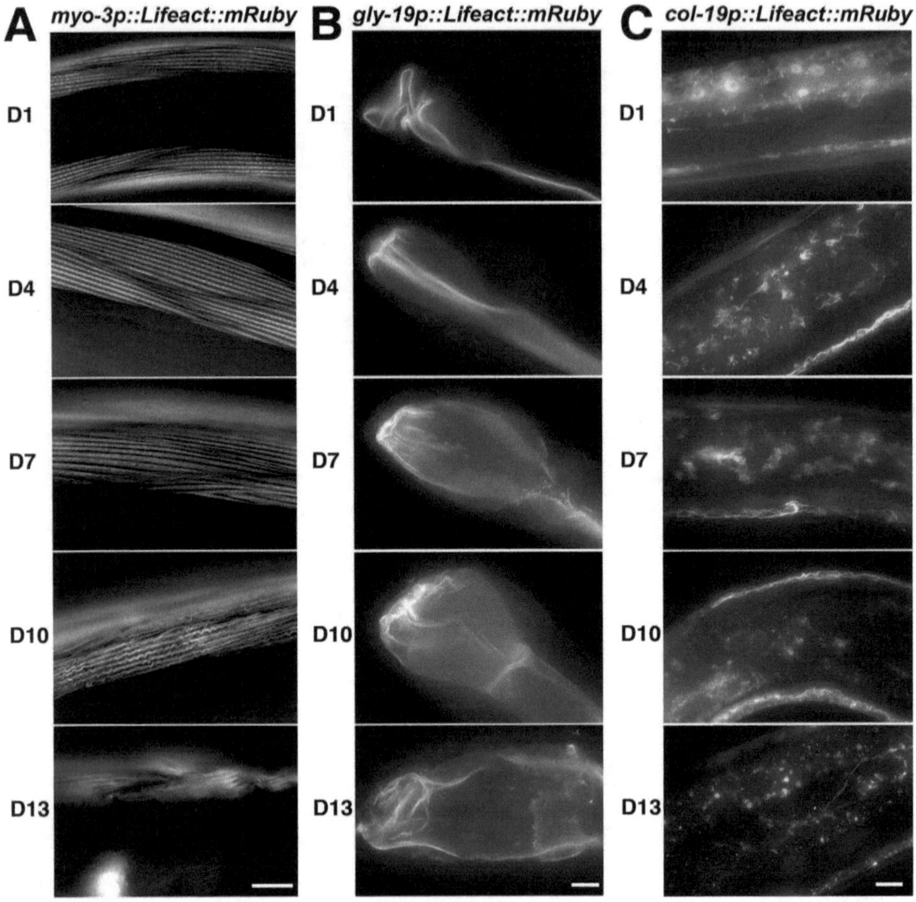

Fig. 2 The actin cytoskeleton shows loss of structure and integrity as a function of age. Cytoskeletal integrity was monitored in the muscle (**a**), intestine (**b**), and hypodermis (**c**) of live *C. elegans* at various stages of adulthood (D1, D4, D7, D10, D13), using transgenic worms expressing LifeAct-mRuby under tissue-specific promoters as described in Fig. 1. Worms were grown on EV RNAi from hatch and age-isolated by three different methods: sedimentation, FUDR treatment, or hand-picking, as described in Subheading 3.6. All three methods show similar trends. Representative data from hand-picking are shown. Scale bars are 5μm. (Original figure from [17])

Fig. 1 (continued) hypodermis (**e**). Transgenic worms were grown to D1 for muscle and intestine and D4 for hypodermis on EV RNAi. Worms were heat-shocked at 34 °C for 2–6 h on solid agar or treated with 0.5–2.5μM cytochalasin D spinning in M9 solution for 4 h at 20 °C and imaged immediately. For actin RNAi, worms were grown from hatch until L4 on EV RNAi and then moved to *act-1* RNAi for 24 h and imaged on D1. Images for *act-1* RNAi were contrast enhanced separately from other images for ease of visualization of phenotypes due to a significant decrease in fluorescent signal under these conditions. Scale bars are 5μm. DMSO treatment of worms did not affect cytoskeletal integrity in any tissue type compared to untreated worms. Therefore, DMSO treatment is shown here as a representative control. (Original figure from [17])

Surprisingly, we did not observe any significant cytoskeletal structures in the neurons or pharynx using LifeAct::mRuby, although these transgenic animals are available.

The primary strength of *C. elegans* as a model system is their genetic tractability. The cytoskeletal reporters described here are amenable for testing the effect of genetic perturbations, including RNAi, introduction of genetic mutations, or overexpression of constructs of interest. As an example, the effect of RNAi knockdown and overexpression of *hsf-1* is described. *hsf-1* encodes the transcription factor, heat-shock factor 1 (HSF-1), which is one of the key transcription factors for regulating the cytosolic heat-shock response, a critical quality control machinery for maintaining cytoplasmic protein homeostasis under stress [29]. HSF-1 also regulates the expression of cytoskeletal-regulatory genes and directly impacts the quality and function of the actin cytoskeleton [30]. RNAi knockdown of *hsf-1* accelerates the destabilization of the actin cytoskeleton during aging, resulting in premature breakdown of actin structures in the muscle, intestine, and hypodermis (Fig. 3a–c). In contrast, overexpression of *hsf-1* is sufficient to protect the cytoskeleton at advanced age (Fig. 3d–f).

1.4 Physiological Assays to Measure Organismal Health

The cytoskeletal network is of critical importance to a diverse array of cellular functions. Accordingly, disruption of the actin cytoskeleton by RNAi knockdown of the actin gene, *act-1*, significantly impacts cellular health and decreases organismal life span (Fig. 4a) [17]. In contrast, preservation of cytoskeletal function through overexpression of *hsf-1* significantly increases life span and resistance to thermal stress in *C. elegans* (Fig. 4b–d) [17, 30]. Moreover, promoting actin dynamics has been shown to contribute to the increased quality of other organelles, such as the mitochondria in yeast, also leading to increased health and life span [3]. Thus, life span and thermotolerance assays can serve as an indirect readout of cytoskeletal health. The short life span and ease of maintenance of *C. elegans* make it an excellent model organism for using these physiological measurements and performing other previously established stress assays to evaluate the impact of actin manipulations on organismal health [31].

2 Materials

2.1 C. elegans Growth Media

1. Lysogeny broth (LB): All standard LB recipes are acceptable, including homemade recipes using bacto-tryptone, yeast extract, and NaCl. For this study, commercial LB from Fisher Scientific was used.
2. Nematode growth media (NGM): 1 mM $CaCl_2$, 5 μg/mL cholesterol, 25 mM KPO_4 pH 6.0, 1 mM $MgSO_4$, 0.25% (w/v) Bacto-Peptone, 51.3 mM NaCl; 2% w/v agar added for plates.

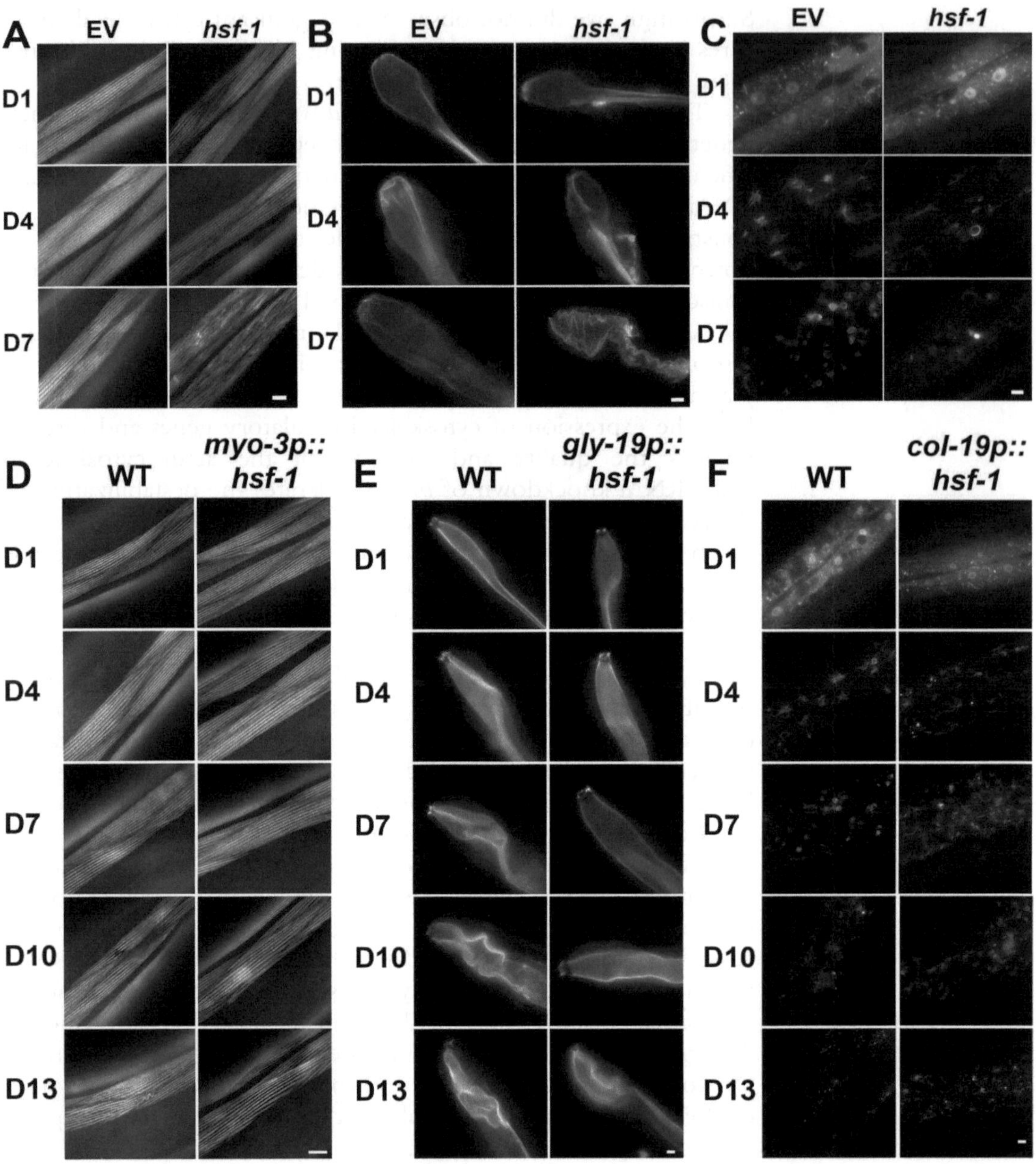

Fig. 3 HSF-1 is required for maintenance of cytoskeletal integrity. Cytoskeletal integrity was monitored in the muscle (**a**), intestine (**b**), and hypodermis (**c**) of transgenic worms expressing LifeAct::mRuby under tissue-specific promoters. For (**a**–**c**) worms were grown on either EV or *hsf-1* RNAi from hatch. Imaging was stopped at D7 due to the large number of dead worms grown on *hsf-1* RNAi beyond D10. (**d**) Cytoskeletal integrity was monitored in muscle of wild-type and *myo-3p::hsf-1* worms at D1, D4, D7, D10, and D13. (**e**) Cytoskeletal integrity was monitored in the hypodermis of wild-type and *col-19p::hsf-1* worms at D1, D4, D7, D10, and D13. (**f**) Cytoskeletal integrity was monitored in the intestine of wild-type and *gly-19p::hsf-1* worms at D1, D4, D7, D10, and D13. For (**d**–**f**) worms were plated on EV RNAi from hatch. All worms were age-isolated by sedimentation, growth on FUDR, or hand-picking. Representative images from hand-picking are shown. Scale bars are 5μm. (Original figure from [17])

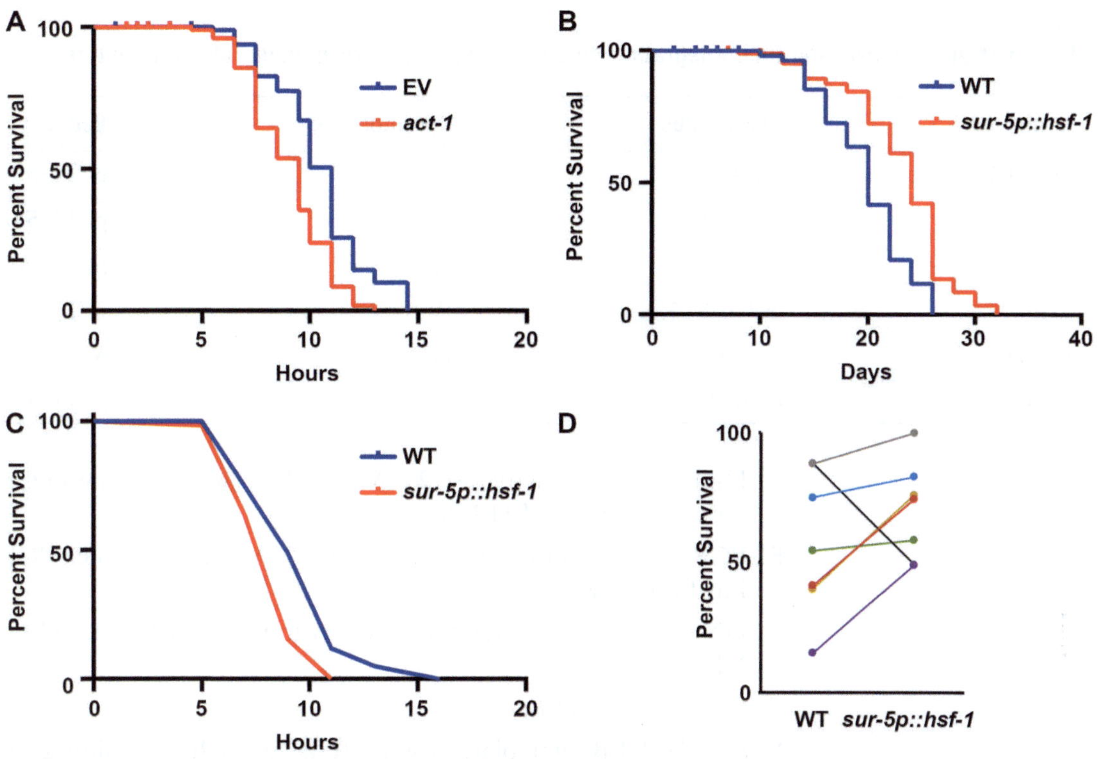

Fig. 4 Overexpression of *hsf-1* promotes organismal health. (**a**) Life span curve of WT animals grown from hatch on EV or *act-1* RNAi (diluted 1:10 with EV) at 20 °C. (**b**) Life span curve of WT and animals overexpressing *hsf-1* in all tissues. WT and *sur-5p::hsf-1* animals were grown from hatch on EV at 20 °C. (**c**) Thermal tolerance survival curve of WT and animals overexpressing *hsf-1* in all tissues. WT and *sur-5p::hsf-1* animals were grown from hatch on EV at 20 °C and transferred to 37 °C at D1. Survival was assayed every 2 h. (**d**) Seven replicates of WT and all-tissue *hsf-1*-overexpressing animals thermal stress survival at 7 h plotted. Each line is a single replicate performed at the same time and is colored for ease of visibility

3. NGM RNAi plates: 1 mM $CaCl_2$, 5 μg/mL cholesterol, 25 mM KPO_4 pH 6.0, 1 mM $MgSO_4$, 2% (w/v) agar, 0.25% (w/v) Bacto-Peptone, 51.3 mM NaCl, 1 mM IPTG, 100 μg/mL carbenicillin/ampicillin. Store at 4 °C in dark for up to 3 months.
4. 5-Fluoro-2′-deoxyuridine (FUDR): 10 mg/mL dissolved in M9 (22 mM KH_2PO_4 monobasic, 42.3 mM KH_2PO_4, 85.6 mM NaCl, 1 mM $MgSO_4$).

2.2 MosSCI Construct Synthesis

1. MosSCI vector (*see* Table 1).
2. Forward and reverse primers with homology overlap for all plasmid components, 10 μM working solution each (*see* Subheading 3.3).
3. Distilled H_2O.

Table 1
MosSCI strains, transposon loci, integration chromosome, and corresponding MosSCI vectors

Strain	Mos1 locus	Chromosome	Vector
EG6701	ttTi4348	I	pCFJ352
EG6702	ttTi4391	I	pCFJ353
EG6699	ttTi5605	II	pCFJ350
EG6703	cxTi10816	IV	pCFJ356
EG6700	cxTi10882	IV	pCFJ351
EG6705	ttTi14024	X	pCFJ355

4. High-Fidelity Polymerase 2× Master Mix (we recommend Q5 Hot Start, NEB M0494S).
5. PCR purification kit: we recommend QIAGEN QIAquick PCR Purification Kit.
6. DNA Assembly Master Mix (we recommend NEBuilder HiFi, NEB E5520S).
7. Competent DH5α cells.
8. Standard LB agar plates for bacterial growth, containing 50 μg/mL carbenicillin.
9. Carbenicillin stock 50 mg/mL.
10. Plasmid mini-prep kit: we recommend QIAGEN QIAprep Spin Miniprep Kit.

2.3 C. elegans *Microinjections*

1. Capillary glass with filament: Sutter Instruments Company, QF100-60-10.
2. Laser-based micropipette puller: Sutter Instruments Company, P-2000.
3. Glass coverslips: #1.5, 22 × 50 mm.
4. Glass coverslips: #1.5, 22 × 22 mm.
5. Injection pad agarose: 2% (w/v) agarose.
6. Injection oil: Halocarbon 700 oil (CAS# 9002-83-9).
7. Dissecting microscope: Leica S7E Stereo Zoom w/ Fiber Optic Light.
8. Fluorescent dissecting microscope: Leica M165 FC, with DSR filter (Leica 10447412).
9. Compound microscope with 10× objective, 40× DIC objective, and Narishige micromanipulator attachments: MN-4, MMO-4, and UT-2.

Table 2
MosSCI injection plasmids and injection mixture concentrations

Plasmid	Transgene	Function	Expression tissue	Working concentration
pCFJ601	*eft-3p::Mos1 transposase*	Transposase	Ubiquitous	50 ng/uL
pGH8	*rab-3p::mCherry::unc-54 UTR*	Co-injection marker	Pan-neuronal	10 ng/uL
pCFJ90	*myo-2p::mCherry::unc-54 UTR*	Co-injection marker	Pharynx	2.5 ng/uL
pCFJ104	*myo-3p::mCherry::unc-54 UTR*	Co-injection marker	Body wall muscle	5 ng/uL
pCFJ35X	*construct, Cb-unc-119(+)*	MosSCI construct	Construct dependent	25 ng/uL

10. FemtoJet 4× and required peripherals: Eppendorf, 5253000025, 5196082001, 5252070054, 5196081005, air compressor: JUN-AIR (1104360).
11. MosSCI strains (*see* Table 1).
12. MosSCI plasmids (*see* Table 2).

2.4 C. elegans *Synchronization*

1. M9 solution: 22 mM KH_2PO_4 monobasic, 42.3 mM KH_2PO_4, 85.6 mM NaCl, 1 mM $MgSO_4$.
2. Bleaching solution: 1.8% (v/v) sodium hypochlorite, 0.375 M KOH.

2.5 Live-Cell Imaging of C. elegans

1. Sodium azide: 1 M (~6.5%) stock solution of sodium azide in water. Store in the dark at 4 °C. This is a 10× solution and is diluted to a 100 mM working stock with M9 for most imaging experiments. A 10 mM working stock can also be used for reduced toxicity over longer-term imaging.
2. Glass slides: we recommend Superfrost Plus Microscope slides.
3. #1.5 cover glass.
4. Nail polish: we recommend clear nail polish.
5. Compound/wide-field microscope: The system used here is a Zeiss AxioObserver.Z1 microscope equipped with a Lumencor Sola light engine, Zeiss Axiocam 506 camera, Zen (blue edition) software, standard GFP filter (Zeiss filter set 46 HE), standard dsRed filter (Zeiss filter set 43), and a 63× Plan-Apochromat oil objective with a 1.4 NA (Zeiss Objective 44 07 60).

3 Methods

3.1 Growth and Maintenance of *C. elegans*

The two bacterial strains most commonly used as a food source for *C. elegans* are OP50, a B strain, and HT115, a K-12 strain that is used almost exclusively for RNAi [32, 33]. There are some major differences between the two bacterial diets, which cause physiological changes in worms, including metabolic profile, mitochondrial quality, and life span [34]. Some of these differences are due to the vitamin B12 deficiency found in OP50 bacteria, which can result in defects in mitochondrial homeostasis and increased sensitivity to pathogens and stress, none of which are observed on an HT115 diet [35]. Therefore, we recommend that all experiments are performed on HT115 if possible, regardless of the necessity of RNAi. However, due to the ease of using OP50, all standard growth (e.g., maintenance and amplification of animals) can be performed on OP50. For all the experiments described in this protocol, we have not seen significant differences in animals maintained on OP50 diets but moved to HT115 post-synchronization.

1. Grow a culture of OP50 in LB for 24–48 h in ambient temperature, ~22–25 °C (*see* **Note 1**).
2. Seed saturated OP50 culture onto NGM plates: ~100–200 μL of volumes on 60 mm NGM plates and approximately 0.8–1 mL of culture onto 100 mm plates. Allow to completely dry before plating worms on bacterial lawn.
3. For standard maintenance, it is recommended to move 10–15 young animals (eggs, L1, or L2 stages) onto a 60 mm plate, although more can be moved if dealing with mutants or transgenic animals with low fecundity (*see* **Note 2**). If kept at 15 °C, animals are moved once per week, and if animals are kept at 20 °C, animals are moved twice per week. A fresh thaw should be performed every 25–30 passages.
4. To expand animals for experimentation, a full 60 mm plate can be chunked onto 100 mm plates. To chunk animals, a small piece of the agar that contains worms of interest is broken off the original plate and placed animal side down onto the bacterial lawn of a fresh NGM plate. As a general frame of reference, animals with wild-type fecundity can have a full 60 mm plate cut into four to six parts and chunked onto individual large 100 mm plates to create full large plates, without starvation of the animals, at 20 °C within 2–3 days.

3.2 Making *C. elegans* *MosSCI* Strains

MosSCI constructs are integrated into the genome through homology-directed repair (HDR) at sites of a double-stranded break created by excision of the Mos1 transposon. More specifically, transgenic lines are created by injecting MosSCI specific strains with a plasmid cocktail composed of the pCFJ601, multiple tissue-specific fluorescent markers, and the transgenic construct within a MosSCI specific vector (refer to "About MosSci," WormBuilder.org Legacy site, 2020) [24, 36]. The MosSCI strains contain the *unc-119(ed3)* mutant allele and the Mos1 transposon at specific loci. pCFJ601 expresses the Mos1 transposase under a germline promoter to induce transposon excision, thereby creating a double-stranded break. Since DNA microinjections can lead to extrachromosomal array formation, the tissue-specific co-injection markers used do not integrate into the genome and serve as indicators of arrays that may form. By including multiple tissue markers, one can reduce the likelihood of extra-chromosomal array false negatives resulting from the mosaic cell to cell heritability of the arrays. Lastly, the MosSCI vectors contain sequence homology for corresponding Mos1 transposon sites in MosSCI strains and the *C. briggsae unc-119(+)* homolog. Alternatively, strains created for the universal MosSCI system have been engineered to have the same sequences surrounding all transposon insertion sites, allowing for use of a single vector with all the universal MosSCI strains [37]. The homology sequences within the vectors allow for HDR-mediated integration of the construct into the genome at single-copy levels, while the *Cb-unc-119(+)* is used as a selection marker for successfully injected worms and potential transgene integration.

3.3 MosSCI Construct Design and Synthesis

To create *C. elegans* strains expressing fluorescent LifeAct, genetic constructs must contain the necessary components to ensure proper expression. All constructs must contain a 5′ UTR (promoter), our gene of interest (LifeAct::mRuby), and a 3′ UTR. Optionally, these constructs may also contain useful sequences such as introns, an internal ribosome entry site (IRES), a spliced leader 2 (SL2) sequence, or 2A peptide sequence.

Promoters must be empirically determined; however, the first 1–2 kb upstream of the gene start codon are usually sufficient for expression [38]. Promoters of specific genes with well-characterized expression profiles are commonly used for expression within distinct tissues [39]. A list of recommended promoters that have been validated by our lab is provided in (Table 3). In addition to the promoter's regulation, the 3′UTRs contain regulatory sequences, which affect mRNA stability and translation and must therefore be included in the construct [40]. We find that for most constructs, the well-established 3′ UTR of the *unc-54* gene is sufficient for stable gene expression. We utilize the last 29 bp of the gene (including the stop codon), in addition to the 700 bp after the stop codon. Alternatively, including the 3′ UTR of a gene-of-interest may more closely reflect the regulation of that particular gene.

Table 3
Tissue-specific promoters

Promoter	Sequence (bp upstream of start codon)	Tissue expression
vha-6p	1258 bp	Intestine
col-19p	2000 bp	Hypodermis (only late L4 and adulthood)
myo-3p	2001 bp	Body wall muscle
myo-2p	976 bp	Pharynx
rgef-1p	3338 bp	Pan-neuronal
hlh-17p	3000 bp	Cephalic sheath glia
sur-5p	3647 bp	All tissues

When creating transgenic *C. elegans* strains, considerations must be taken to promote gene expression level in the tissue of interest, as *C. elegans* nematodes possess the ability to silence the expression of foreign genetic material, especially for high-copy number transgenes within the germline [41]. The inclusion of introns within the sequence can reduce silencing and improve transgene expression, and we therefore recommend them for single-copy transgenes [42]. Inclusion of introns can be accomplished by inserting endogenous introns (i.e., genomic sequences) or introns foreign to the gene sequence. The optimal insertion site of foreign introns must be empirically determined, as they can potentially have negative effects on expression levels [43]. We find that constructs containing fluorescent protein genes with introns are stably expressed.

Lastly, an IRES, SL2, or 2A sequence may be utilized to allow for concurrent expression of two or more individual proteins from a single promoter construct [44, 45]. Combining these sequences with a fluorophore has been used for monitoring of fluorescent gene expression without fusion of the gene-of-interest to the fluorophore [46]. However, how well these multi-expression systems function across age is not well-characterized.

To assemble the MosSCI construct, we utilize an isothermal DNA assembly protocol [47]. For this method, we design primers to amplify all individual sequences within the MosSCI vector, the MosSCI vector, and a small portion of the upstream proximal sequence. This results in final products with overlapping regions, which are utilized to assemble all sequences in the correct order by isothermal assembly.

1. Utilize online resources such as Wormbase or Addgene to find the desired sequences.
2. Utilize software such as APE (A Plasmid Editor), SnapGene, or Benchling to assemble the sequences into a plasmid.

3. Design primers for each sequence to be amplified and then include 15–20 base pairs upstream of the primer within the in silico designed plasmid, to provide the necessary overlap for isothermal assembly.
4. Perform all individual PCR reactions following the polymerase manufacturer's guidelines. For Q5 Master Mix: Add 2.5 μL of both forward and reverse 10 μM primer to a PCR tube. Add 1–10 ng of the template DNA to be amplified. Add 25 μL of the Q5 hot start 2× master mix. Add water to a final volume of 50 μL and mix by pipetting. Run the following thermocycler protocol:
   (a) 30 s at 98 °C for initial denaturation.
   (b) 25–35 cycles of: 5–10 s at 98 °C, then 20 s at 50–72 °C (*see* **Note 3**) followed by 20–30 s/kb.
   (c) 2 min at 72 °C.
   (d) Hold at 4–10 °C.
5. Verify the product length by gel electrophoresis and purify the DNA according to the QIAGEN QIAquick PCR Purification Kit manufacturer's guidelines.
6. Assemble PCR fragments through an isothermal assembly reaction according to the manufacturer's guidelines:
   (a) Mix 0.03–0.5 pmols of vector and insert PCR fragments into a PCR tube with a 1:2 (vector:insert) molar ratio (*see* **Note 4**), totaling 10 μL.
   (b) Add 10 μL of NEBuilder reaction mix and pipette to mix thoroughly.
   (c) Incubate in a thermocycler for 60 min at 50 °C, then hold at 23 °C.
7. Transform the assembled plasmid with a transformation protocol. Here, we use a heat-shock transformation protocol:
   (a) Dilute isothermal assembly reaction 1:3 (reaction:water).
   (b) Add 5 μL of diluted reaction to 50 μL of competent cells (e.g., DH5a) thawed on ice.
   (c) Incubate on ice for 25 min.
   (d) Heat shock for 30 s at 42 °C, then immediately place on ice for 2 min.
   (e) Plate bacteria on an LB plate containing carbenicillin and incubate overnight at 37 °C.
8. Inoculate successful bacterial colonies into LB with 50 μg/mL carbenicillin and grow culture overnight to saturation.
9. Isolate the plasmid according to the QIAGEN QIAprep spin Miniprep Kit manufacturer's guidelines.
10. Sequence the plasmid with multiple primers to verify that all inserted sequences are present and correct.

3.4 C. elegans Microinjections and MosSCI Selection

Making *C. elegans* transgenic strains is a multistep, multi-day process. Injection pads and worms must be prepared at least 24 h prior to the day of injections. The injection mixture may be prepared in advance and should be adjusted to fit individual needs. We utilized mCherry fluorescent markers, but different tissue markers may be substituted for one of the markers in Table 2 to avoid similar fluorescence of both the MosSCI construct and a co-injection marker within the same tissue. While successful injections are dependent on the individual skill level and chance, we find that injecting 15–30 worms per construct usually results in creation of the MosSCI strain. Following injections, the progeny is iteratively screened for the presence of the MosSCI construct and absence of the co-injection markers. The screening process may require 2–6 generations of progeny to complete.

3.4.1 Preparation of Injection Pads

Injection pads are coverslips with dried agarose used to immobilize worms during injections. The pads must be made at least 24 h prior to injections for the agarose to dry. Once made, they can be stored indefinitely at room temperature. Care must be taken to not touch the agarose while handling the injection pad to avoid contamination.

1. Make 2% molten agarose.
2. Cut a pipette tip to allow for a bigger opening, and pipette a small amount of 2% molten agarose on the center of a 22 × 50 mm coverslip.
3. Place a second 22 × 50 mm coverslip on the molten agarose in perpendicular orientation to the first coverslip to flatten the agarose.
4. Allow ~1 min for the agarose to solidify, and then carefully slide off the top coverslip while leaving the agarose on the center of the bottom coverslip. If the agarose sticks to the top coverslip, flip over the coverslip and allow the agar to dry on it.
5. Repeat as necessary to make more injection pads.
6. Allow the flattened agarose pad to open-air dry overnight.
7. Once dry, stack injection pads in the same orientation and store in a coverslip case. Make a note of the orientation to ensure the correct side is used at a later time.

3.4.2 Worm Selection for Injections

Twenty-four hours prior to injecting, select L4 stage *unc* animals from the MosSCI strain, move them to a new NGM plate seeded with OP50, and store at 20 °C for 24 h to have a collection of day 1 adult animals to inject the following day. Alternatively, a standard bleaching protocol can be used to synchronize a population of worms (*see* Subheading 3.5 below) and timed to inject day 1 adult animals.

3.4.3 Preparing Needles for Injections

While needles can be pulled any time prior to injection and carefully stored in an empty petri plate on rows of mounting putty, the rest of the preparation must be performed immediately before injecting. The needles are loaded with a small volume of the DNA mixture. Loading the needles with DNA too far in advance can result in evaporation of some of the fluid, resulting in changes to the plasmid concentrations or evaporative loss of all fluid in the DNA mixture.

1. Prepare the injection mixture according to Table 2. (DNA can be mixed beforehand and stored at −20 °C.)
2. Use the micropipette puller to pull needles from the quartz capillary glass using setting 50, according to the P-2000 manufacturer's guidelines (a petri plate with mounting putty can serve as a holder for multiple needles so the fragile tips do not contact anything).
3. Pipette DNA mixture into the open end of needle. A small volume of liquid will fill the tip of the capillary tube. Allow capillary action to move the liquid to the needle end before adding more liquid to the open end. Repeat until ~1 μL of DNA mixture is loaded into each needle (Fig. 5a). Loading 2–4 needles at a time is recommended to have extra needles ready in the event that a needle clogs or breaks incorrectly.
4. Load a needle containing the plasmid mixture into the FemtoJet capillary holder, and secure it on the micromanipulator attached to the compound microscope. Set the FemtoJet settings to manual and the pressures to 1000 for pi and 333 for pc.
5. Set up a 22 × 22 mm coverslip on top of one end of a 22 × 50 mm coverslip, add a drop of oil on one edge of the top coverslip that is closest to the center of the larger bottom coverslip, and then spread the oil along the middle part of the edge so that the middle half of the edge is under oil.
6. Place the oil-covered coverslip on the stage of the compound microscope such that the edge of the coverslip is in the center of the field and focus on the coverslip edge using the 10× objective.
7. Switch to the 40× objective and get the center Z-position of the edge into focus.
8. Switch to the 10× objective and adjust the needle tip so that it is in the oil and in focus using the micromanipulator controls without adjusting the focus. Ensure that the needle tip is far away enough from the coverslip edge to avoid breaking the needle. If the needle holder must be adjusted to achieve the correct positioning, raise the needle away from the stage with micromanipulator controls, adjust the positioning manually, and then attempt to position the needle tip with the micromanipulator controls in order to avoid damaging the needle tip.

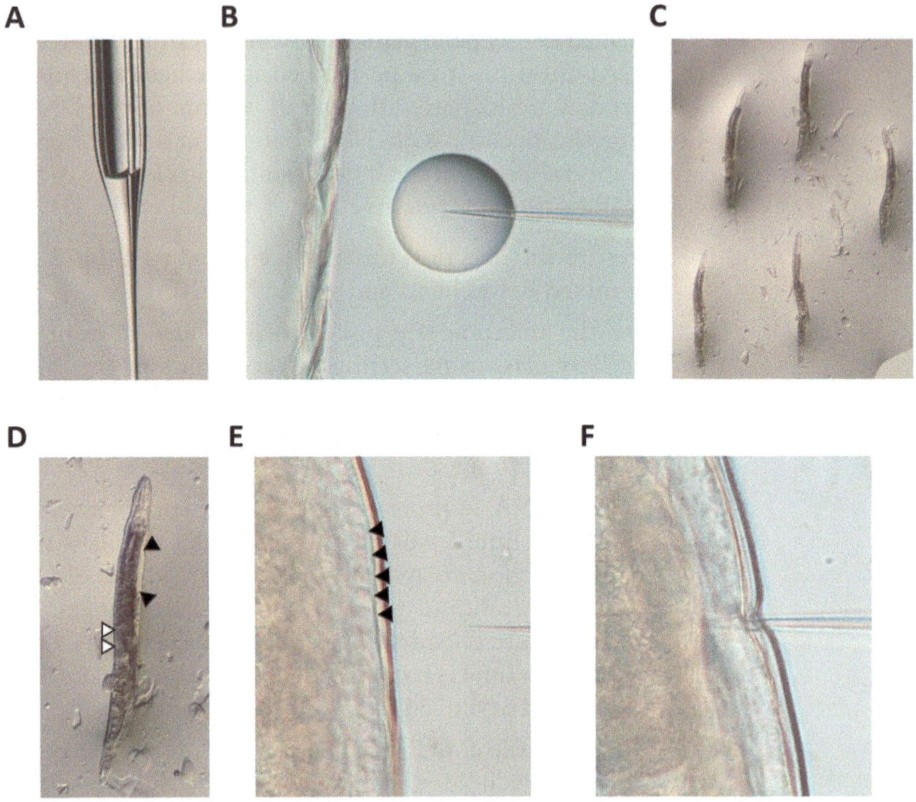

Fig. 5 Schematic for *C. elegans* microinjections. (**a**) Quartz needle loaded with adequate amount of DNA mixture. (**b**) Quartz needle after breaking the tip on the edge of the smaller coverslip and pressing "clean" to confirm the break by the formation of an aqueous droplet within the oil. (**c**) Animals properly lined up and immobilized on an oil patch on top of an agarose pad in preparation for microinjections. (**d**) Close-up view of an immobilized worm viewed through dissecting microscope. The black triangles mark the clear gonad positioned to the right side of the animal, and the white triangles mark the internal eggs positioned closer to the left side of the animal. (**e**) Immobilized worm viewed through compound microscope 40× DIC objective with needle in focus. Triangles mark nuclei "dimples" within outer edge of the gonad, indicating correct Z-position for injecting. (**f**) Immobilized worm viewed through compound microscope 40× DIC objective with injection needle tip positioned past the gonad

9. Change to the 40× objective and adjust the needle tip again so that it is in focus and close to the coverslip edge using the micromanipulator controls without adjusting the focus. Both the needle tip and coverslip edge center Z-position should be in focus.
10. Using the fine adjustments of the micromanipulator, break the needle tip by slowly moving it toward the coverslip edge. A small amount of liquid may come out and cling to the coverslip edge upon a successful break (*see* **Note 5**). Pressing inject or clean should release liquid and confirm that needle tip has broken (Fig. 5b).

11. Raise the needle using the micromanipulator Z-controls only, making sure not to change the X or Y settings so that the needle remains centered within the field of view when it is lowered back down in later steps.

3.4.4 Injecting Worms

Worms are immobilized for injections by adhering to the dried agarose pad that has been covered with a thin layer of injection oil. Excessive time on the agarose pad, even when covered by oil, can result in irreversible desiccation of the animal. Therefore, we recommend that animals be prepared for injections after the needle has been fully prepared (*see* previous Subheading 3.4.3).

1. Place an injection pad under the dissecting microscope with the dried agarose side up, and add a large drop of injection oil next to agarose pad. This drop will serve as the supply for further steps.
2. Adjust the dissecting microscope to get the agarose pad in focus by focusing on the edge of the pad (*see* **Note 6**).
3. Using a sterile worm pick, make a thin, square patch of oil on the agarose pad large enough to contain 4–6 worms.
4. Sterilize the pick and move 4–6 worms onto the oil patch from the synchronized population of day 1 *unc* MosSCI worms (*see* **Note 7**).
5. With a pick, manipulate the worms so that the gonad is visible on the side where the needle will be located, and pat down the worm to ensure that they are immobilized on the agarose under the oil. Correct positioning will also result in the vulva and eggs being positioned on the opposite side (Fig. 5c, d).
6. Move the injection pad to the compound microscope and focus on the worms using the 10× objective.
7. Switch to the 40× objective and get the center Z-position of either gonad in focus and at the center of the field of view. We recommend focusing on the widest part of the gonad to increase the likelihood of positioning the needle tip within the gonad in later steps. When focused on the center Z-position of the gonad, the large nuclei in the outer edge of the gonad will be visible as dimple-like structures through DIC microscopy (Fig. 5e).
8. Using the micromanipulator controls only, lower the needle until the tip is in focus and close to the gonad without contacting the worm (Fig. 5e).
9. Move the needle toward the gonad until it pierces the animal and the tip is within the gonad. Due to the force needed to puncture the cuticle, the needle tip may penetrate completely through the gonad (Fig. 5f). If this occurs, the needle must be retracted so that the tip is near the center before injecting the DNA.

10. Inject DNA into the gonad and move the needle out of the gonad. Injections can be accomplished by pressing the "inject" button on the FemtoJet, or alternatively, the "clean" button can be used if the needle break is small and requires more pressure. During the injection, fluid may travel up the gonad arm or a few eggs may be released from the vulva (*see* **Note 8**).
11. Raise the needle with the micromanipulator so that it is out of focus and out of the way, then switch to the 10× objective, and locate another worm to inject.
12. Repeat **steps 7–11** until all worms have been injected. Both gonads in a worm can be injected; however, doing so will increase the risk of excessive injections.
13. Raise the needle with the micromanipulator so it is completely out of the way and at least an inch above the coverslip.
14. Move the injection pad coverslip back to the dissecting microscope, and free the worms from the agarose pad by adding a small amount of M9 buffer to the patch of oil containing the worms. Excessive M9 may leak onto the unused agarose pad and reduce its useable space.
15. Sterilize a worm pick and move the worms from the M9/oil to a new NGM plate seeded with OP50 (*see* **Note 9**).
16. Repeat **steps 2–15** as necessary to inject more worms. If no more space is available on the agar pad, a new pad should be set up from **step 1**.
17. Allow injected worms to recover for 24 h at 20 °C, then maintain them at 25 °C.

3.4.5 Isolating MosSCI Transgenic Animals

After 2–3 days at 25 °C, a fluorescent dissecting microscope will be used to identify MosSCI integrants in the progeny of injected worms. While obtaining a MosSCI integrant without extra-chromosomal arrays in the F1 population is possible, obtaining animals with arrays is more likely. These animals may or may not have the MosSCI construct integrated into the genome and must be kept at 25 °C for screening of future progeny for loss of the array.

1. Under a fluorescence dissecting microscope, use white light to identify worms positive for the construct by looking for wild-type movement (non-*unc* worms). They will be larger and leave sinusoidal tracks in the bacteria.
2. Turn the white light off and turn on the fluorescence channel to determine if the worms identified in **step 1** possess an array by looking for fluorescence in the tissues corresponding to the co-injection markers utilized.

3. Animals positive for the construct (non-*unc*) and negative for the arrays (lack fluorescence from the co-injection markers) should be singled out and transferred to their own individual NGM plates seeded with OP50 and maintained at 25 °C.
   (a) If all non-*unc* animals are positive for arrays, the worms must be placed back at 25 °C until the next population of progeny have developed and are ready to be screened (*see* **step 1**).
   (b) If the plate has starved, worms can be chunked to a new NGM plate seeded with OP50 and screened once they have grown (*see* **step 1**).
4. Progeny of the individually selected animals from **step 3** (non-*unc*, array-negative) are then screened for homozygosity of the MosSCI construct and double-checked for array fluorescence and integration of the construct.
   (a) If all the offspring are non-*unc*, the original single animal selected is homozygous for the MosSCI construct integration. This new strain can now be confirmed using PCR genotyping and used for backcrosses or experiments. Higher magnification fluorescent imaging can be used to confirm lack of dimly expressed arrays.
   (b) If any *unc* worms are present on the plate, the original single animal selected was heterozygous, and 8–16 non-*unc* single worms must be transferred to individual plates for screening of the next progeny population.
   (c) If *unc* worms are present on every plate for more than two rounds of **step 4** screening, non-*unc* worms are likely arrays without co-injection markers and can be discarded. Injection of new animals is recommended if this is the case.

3.5 Synchronization of C. elegans Using Bleaching

A synchronized population of animals is required whenever age-related studies are performed. There exist many different methods for synchronization of animals, which can be as simple as hand-picking animals of a specific stage when a small number of animals are required or more involved like bleaching, when large numbers of animals are required. Bleaching is a method used to degrade the adult carcasses and isolate eggs from within the uterus. Because eggshells are more resilient to bleaching, a properly timed bleaching protocol will completely dissolve the animals while leaving the living embryos within the eggs intact. This protocol is a great method to collect a large population of synchronized animals. However, because eggs within a uterus can be 6–10 h apart in developmental stage, a tighter synchronization can be accomplished by L1-arresting animals. When eggs hatch in the absence of food, animals arrest in an early larval stage. Animals can remain in this larval stage for extended periods of time with limited effects on physiology [48].

1. Wash gravid worms (animals with a uterus full of eggs) off plates using 3–15 mL of M9 solution, swirling the plates to suspend the animals in the M9 (*see* **Note 10**).
2. Using a serological pipette, move the worm/M9 mix from plates into single 15 mL conical tubes. Multiple plates of the same strain may be collected to the same tube.
3. Centrifuge the worms at 1000 × *g* for 30 s and aspirate the supernatant, being mindful that adult worm pellets are loose. It is not necessary to wash off residual bacteria as the bleaching protocol will also eliminate bacteria. If there are large clumps of mold or bacteria, worms can be washed several times with M9.
4. Add 2–10 mL of freshly prepared bleaching solution to the tubes, using ~1 mL of bleach for every ~0.1 mL of animal pellet. Invert the worm and bleach mixture for ~5 min, or vigorously shake tubes to speed up the reaction. Regularly check the progress of the bleaching under a dissection microscope, as different strains may bleach at different rates. Continue to **step 5** once the adult carcasses have been fully dissolved and the eggs are suspended in the solution (*see* **Note 11**).
5. Spin the tubes at 1000 × *g* for 30 s (*see* **Note 12**), aspirate the supernatant, and add M9 solution up to 15 mL. Cap and shake the tubes to resuspend the egg pellets, and wash any residual bleach off the eggs, taking care to ensure entire egg pellet is resuspended.
6. Repeat **step 5** two times to ensure that all bleach has been washed from the eggs (*see* **Note 13**).
7. Spin the tubes at 1000 × *g* for 30 s, and then aspirate the supernatant and add M9 solution up to 10 mL, depending on the expected concentration of eggs.
8. Pipet egg/M9 mix onto plates to grow at 15–20 °C for experimentation. To approximate worm count, 5 μL of egg mixture can be placed on an agar plate or microscope slide to count the number of eggs under a microscope. Divide the count by 5 to get an approximation of number of eggs per 1 μL of volume. Averaging of 3 or more independent counts will improve accuracy of the egg count. To avoid starvation of worms for experiments, refer to Table 4 for recommended egg counts per plate.
9. Alternatively, animals can be L1-arrested for tighter synchronization of animals. To L1 arrest, eggs are resuspended in 10 mL of M9 and incubated in a rotator at 20 °C for up to 24 h. Wild-

Table 4
Recommendations for number of animals to plate

Plate size	Bacteria type	No. of animals to reach day 1 adulthood	No. of animals to reach L4 stage
60 mm	OP50	100–150	150–300
60 mm	HT115	70–100	120–200
100 mm	OP50	600–1000	1500–2000
100 mm	HT115	350–600	700–1300

type animals have shown no defects in physiology after L1 arresting for up to 48 h. The count of arrested L1 worms can be approximated as described in **steps 7–8**.

3.6 Aging of C. elegans for Age-Associated Studies

Since *C. elegans* are hermaphrodites and self-mate, maintaining a population of animals throughout age poses a challenge. Progeny must be eliminated to prevent contamination of the aging population by younger animals. When dealing with a small population of animals, adult animals can be moved away from their progeny by hand-picking adults onto new plates, similar to a life span assay and other survival protocols. However, this requires extensive manual labor and becomes much more difficult with protocols requiring a large population of worms. Another manual method is to collect animals from plates with M9 solution and allowing the adults to sediment in the liquid before aspiration of the upper fractions containing younger animals, which take longer to settle in a liquid. This method is applicable to larger populations of worms and does not require genetic or chemical manipulations. However, it requires extensive sample manipulation, is low throughput, may result in loss of animals, and has the greatest risk of being contaminated by younger animals.

To avoid manual or mechanical separation of adults from progeny, animals can be sterilized by genetic or chemical means. There exist several temperature sensitive mutants that do not produce progeny when grown at elevated temperatures, such as the *glp-1* and *glp-4* mutants, which lack a majority of their germlines [49, 50], or the sperm-deficient double-mutant CF512 strain [51]. However, these strains must be exposed to elevated temperatures and crossed with strains of interest. The simplest method for sterilization is through the use of chemicals, such as exposure to FUDR, which inhibits DNA replication [52]. For the imaging protocols described here, all methods of aging showed similar trends of data, although a minor level of variability was observed as expected: washing caused a slightly higher rate of deterioration of actin due to mechanical stress, while FUDR treatment caused the slowest deterioration of actin due to a complete lack of mechanical stress. However, these differences were not significant relative to the changes seen in the strains used. Representative data for each set of experiments are shown and described in figure legends. FUDR treatment can cause changes in life span and should not be used for sterilization for life span studies [53].

3.6.1 Manual Removal of Animals from Progeny with a Standard Pick

1. Prepare plates spotted with bacteria of interest (Subheading 3.1). We recommend using HT115 bacteria even in experiments not involving RNAi knockdown. Prepare enough plates for the experiments, based on the assumption that animals will be moved seven to ten times based on progeny development.
2. Synchronize animals of choice using methods described in Subheading 3.5, and plate animals onto spotted plates using criteria recommended in Table 4.
3. Incubate eggs at 20 °C for 3–4 days (65–96 h) until day 1 of adulthood, depending on the strain fecundity and speed of development.
4. At day 1, transfer animals onto separate plates. We recommend placing 20–25 animals on each plate.
5. For the first 5–7 days, animals should be moved away from progeny manually using a pick every day until progeny are no longer visible (*see* **Note 14**). After progeny are no longer visible, animals can be kept on plates and only moved if there is a risk of starvation.

3.6.2 Isolation of Adult Animals Using Washing and Sedimentation

1. Prepare plates spotted with bacteria of interest (Subheading 3.1). We recommend using HT115 bacteria even in experiments not involving RNAi knockdown. Prepare enough plates for experiments, based on the assumption that animals will be moved seven to ten times based on progeny development (*see* **Note 15**).
2. Synchronize animals of choice using methods described in Subheading 3.5, and plate animals onto spotted plates using criteria recommended in Table 4.
3. Incubate eggs at 20 °C for 3–4 days (65–96 h) until day 1 of adulthood, depending on the strain fecundity and speed of development.
4. Starting from day 2, animals are washed off plates using M9 and transferred into a 15 mL conical tube. Adult animals are then allowed to sediment for ~1–2 min. Immediately after an adult pellet is visible, aspirate off all M9/progeny mix, removing the top layer of the adult pellet to ensure all progeny are removed. If the number of animals is a major concern, do not remove the top later of adults, and instead perform two to three extra washes to ensure all progeny are removed.

5. Resuspend the pellet with M9 and repeat the sedimentation protocol in **step 4** two more times to ensure complete removal of progeny. Adult animals should be isolated through sedimentation every day until progeny are no longer visible.
6. After plating adult animals onto a new plate, inspect the plate under a microscope to ensure no progeny are present. Any progeny found on the plate should be manually removed.

3.6.3 Chemical Sterilization of Adult Animals

1. Prepare plates spotted with bacteria of interest (Subheading 3.1). We recommend using HT115 bacteria even in experiments not involving RNAi knockdown. Prepare two sets of plates.
2. Synchronize animals of choice using methods described in Subheading 3.5, and plate animals onto spotted plates using criteria recommended in Table 4.
3. On a second set of plates, add FUDR directly to the center of the bacterial lawn, and swirl the plates to ensure FUDR mix spreads evenly across the bacterial lawn. We recommend 100 μL for 60 mm plates and 500 μL for 100 mm plates. Let dry overnight.
4. Incubate eggs at 20 °C for 2 days (~56 h) to the L4 stage.
5. Wash L4 animals off plates using M9 into a 15 mL conical tube. Centrifuge at 1000 × *g* for 30 s to pellet and then aspirate the M9 solution.
6. Plate L4 animals onto the second set of plates containing FUDR. Animals can now be grown on FUDR plates until the desired age for experimentation. If there is a risk of starvation, animals can be moved onto a new FUDR-containing plate (*see* **Note 16**).

3.7 Visualization of the Actin Cytoskeleton in *C. elegans*

To visualize the actin cytoskeleton in *C. elegans*, we utilize transgenic animals in which LifeAct::mRuby is expressed using tissue-specific promoters. These constructs robustly and reliably label the actin cytoskeleton [17]. Here, we use fluorescent wide-field/compound microscopy to visualize the actin cytoskeleton in muscle, hypodermis, and intestine. In these cell types, the actin cytoskeleton is composed of bright and highly uniform structures, making confocal, super-resolution, or other more advanced microscopy unnecessary for general evaluation of these structures.

3.7.1 Preparation of Slides for Imaging

1. Prepare animals to the desired age following the protocol described in Subheading 3.6.
2. Prepare a fresh working stock of 100 mM sodium azide by dilution of 1 M stock 1:10 into M9 solution (*see* **Note 17**).
3. Place 10–18 μL of 100 mM sodium azide onto a glass microscope slide (*see* **Note 18**).

4. Manually pick 10–25 animals from NGM plates, and gently place them into the sodium azide solution on the glass slide.
5. Gently apply a cover slip onto the sample, taking care not to introduce any air bubbles or apply pressure onto the worms, as that may cause animals to be damaged.
6. The edges of the cover slip can be sealed with Valap, wax, or nail polish if desired, but for short-term imaging, sealing is not necessary (*see* **Note 19**).

3.7.2 Considerations for Wide-Field/Compound Microscopy

1. To reduce phototoxicity and photobleaching, any of the following strategies can be used:
   (a) Reduce excitation intensity. Most LED and laser sources have internal intensity control through the software. For those without internal intensity control, a manual attenuator can be used (for example, a manual attenuator can be installed into the Zeiss stand used in this protocol to reduce the intensity of the light source).
   (b) Reduce exposure time, but keep in mind that longer exposure time and lower intensity result in comparable image quality with less photo damage.
   (c) Reduce spatial or temporal resolution (e.g., increase the z-step size or reduce number of z-sections; increase the time interval duration).
2. To increase signal-to-noise ratio without altering light exposure, the following strategies can be used:
   (a) Apply binning to the camera. Increased binning will decrease spatial resolution but amplify signal.
   (b) Increase camera gain.
   (c) Reduce camera readout, if possible.
   (d) Use a filter with a broader spectral window, which will allow more light to pass through. However, keep in mind that this may increase autofluorescence or bleed-through between channels, so take care to properly understand the specimen used.

3.8 Quantitative Analyses of Actin Cytoskeletal Integrity in Worms

3.8.1 Quantification of Actin Thickness in Muscle Filaments Using ImageJ

Changes in actin cable thickness are an indirect readout of actin cytoskeletal health [17, 54]. Here, we describe a protocol adapted from yeast where ImageJ software can be used to objectively measure actin cable thickness in muscle. A line is drawn perpendicular to the actin filaments across the thickest part of the muscle cell, and the pixel intensity profile of this line is plotted. The width at the median of the peak (50% of the distance of the height of the peak) is measured to approximate actin cable thickness.

1. Acquire images of muscle actin with fluorescent LifeAct.
2. Open image in ImageJ.
3. If the image has more than one channel, ensure that the actin fluorescent channel is selected.
4. Set up ImageJ with the following parameters:
   (a) **Edit > Options > Input/Output**: Under Results Table Options, check Copy Row Numbers and leave other options unchecked.
   (b) **Edit > Options > Plot**: Set width and height equal to those of the image. Set Minimum Y to 0 and maximum Y to the Image Max, obtained from Analyze > Histogram.
   (c) **Analyze > Set Measurements**: Check Area, Display Label.
   (d) **Analyze > Set Scale**: Make sure that the spatial scale is set accurately.
5. Set the Line ROI tool to Straight Line mode (right click on the tool to change the mode). Double click on the Line ROI tool and set the Line Width to 3 and ensure that Spline Fit is unchecked.
6. Draw a line ROI perpendicular to the muscle actin filaments at the center of the muscle cell where the cell is thickest. Both ends of the line should extend into the outside of the cell while avoiding filaments within neighboring cells if possible (Fig. 6a).
7. **Analyze > Plot Profile**. A Plot window should appear showing Gray Value (pixel value) versus Distance.
8. If desired, the plot image can be saved (File > Save).
9. Cable thickness is defined as the width of the peak at half maximal height above background. To find the height, using the Line ROI tool, draw a vertical line from the maximum height of the peak to the background line. A background line can be determined by using the parts of the line plot that do not have fluorescence. Draw a horizontal line across the half maximal height (Fig. 6b).
10. Press T to add the horizontal line to the ROI Manager. **Analyze > Measure**. The Length value represents the thickness of the cable.

3.8.2 Quantification of Actin Quantity in the Hypodermis Using a Large Particle Biosorter

The actin cytoskeleton in the hypodermal syncytium shows a unique, star-like structure between day 2 and day 10 of adulthood [17]. Through use of a large particle flow cytometer, it is possible to quantify fluorescence intensities for thousands of worms [55]. We find that robust, quantitative measurements of the hypodermal actin structures can be made using a COPAS biosorter, as the measurements from this method reliably correlate with the decrease

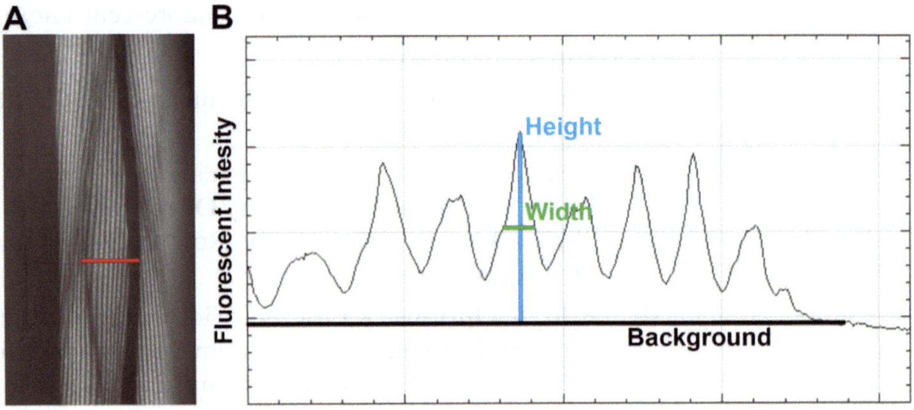

Fig. 6 Schematic for measurements of actin cable thickness. (**a**) Actin cytoskeleton of transgenic worms expressing LifeAct::mRuby under the *myo-3p* muscle-specific promoter. Transgenic worms were grown on EV and imaged at D1. (**b**) Image of intensity profile plot obtained from ImageJ for sample image in (**a**) using the line ROI drawn in red. The background level was determined from the line profile in the non-fluorescent region as drawn in black. The height of the peak is measured from the top of the peak to the background line as shown in blue. The width is measured across the peak at half of the maximum intensity

in actin cytoskeletal structures visualized upon RNAi knockdown of *act-1* (Fig. 7). For a thorough protocol on how to systematically use the COPAS biosorter, refer to [31]. Here, we briefly describe how to perform the biosorter analysis for hypodermal actin species.

1. Ensure that the sheath liquid and clean bottles are full and that all waste containers have sufficient space left.
2. Turn on the flow cytometer by turning on the air compressor, lasers, and instrument.
3. Open the FlowPilot software. The instrument should be starting up now.
4. Turn on the lasers within the software window by clicking "start."
5. Initiate argon laser control by hitting "run" on the pop-up window. Ensure that the laser reaches the proper power of 12 mW.
6. Hit "done" to close the pop-up window.
7. Check that the gauge pressures are correct (sheath 5.5–5.7; sample 6.0–6.6; sorter 3.1–3.3; clean 8.5–8.7), and adjust the pressure valves if necessary.
8. Check the fluidics and ensure that no air bubbles can be seen throughout the tubing.
9. Clean the tubing prior to use by flowing 3–5 mL of bleach solution through the sample collection cup by using the "acquire" feature. Remove excess bleach from the cup with a vacuum and rinse the sample cup with dH_2O. Repeat three times to remove all bleach from sample cup.

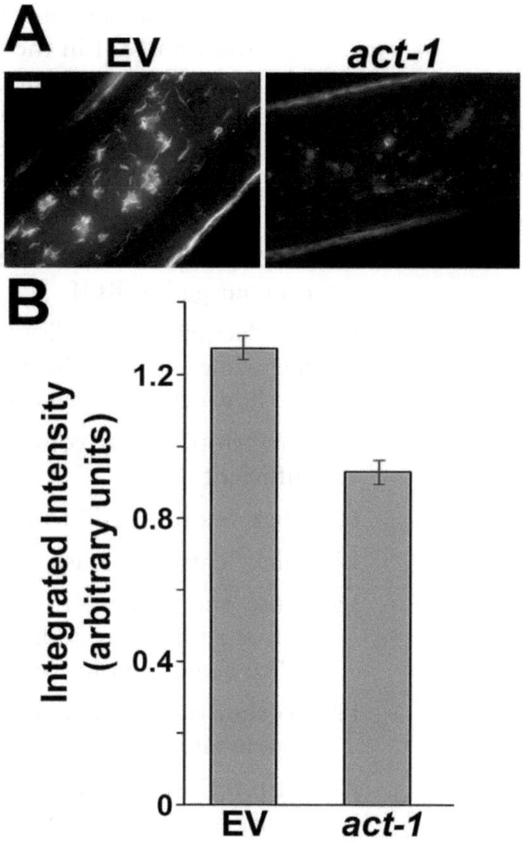

Fig. 7 Quantification of hypodermal actin. (**a**) Actin cytoskeleton of transgenic worms expressing LifeAct::mRuby under the *col-19p* hypodermis-specific promoter. Transgenic worms were grown on EV from hatch and moved onto either EV or *act-1* RNAi at the L4 stage of adulthood. Images were taken at D5 of adulthood. Scale bar is 5μm. (**b**) Quantification of (**a**) where >500 animals are imaged through a COPAS biosorter. Integrated fluorescence intensity is normalized to the integrated extinction of the animal. Data is plotted as mean ± standard deviation

10. After the final wash, run dH_2O through the sample line by using the "acquire" feature. Run 3–5 mL of dH_2O to ensure that all bleach has run out of the sample lines.
11. Remove the excess dH_2O from the sample cup with a vacuum and place the prepared worms/M9 sample into the cup.
12. Adjust the PMT level and TOF/EXT gating based on the conditions that allow for the most optimal imaging.
    (a) PMT level is adjusted by changing the numerical values for each channel PMT. For LifeAct::mRuby, adjust the numbers next to the "Red" under "PMT Control" to desired level to maximize fluorescent output without hitting saturation.

(b) To adjust the gate, **Region** > **Define Gate Region** and draw an ROI in the "TOF/EXT" gating panel to ensure that all worms of desired size are included in the gate and that all other materials are excluded. Some worms should be run through the sorter prior to experimental data collection to determine a guideline of where to draw the gate.

13. Click **Setup** > **Data Storage** > **Gated Only** to save data based on your gating ROI.
14. Hit "acquire," and allow all of the sample to run through the line while ensuring that not all the liquid is run as this will cause the flow cytometer to take in air.
15. Hit "stop" when your sample is low, or you have collected sufficient data.
16. Click "store gated" to save your gated data.
17. Click "erase" to erase the current data.
18. Rinse and aspirate the samples cup three times with dH_2O before adding any additional samples and repeating **steps 14–17** to acquire more data.
19. To shut down the system, repeat **steps 8** and **9** to wash the system after use.
20. Turn off the system by turning off the software, laser, instrument, and air compressor.
21. The saved data provides the integrated fluorescence intensity calculated along the animal and the integrated extinction (EXT) (estimated thickness) of the worm. We normalize the fluorescence intensity to size by dividing the red channel integrated fluorescence with the integrated extinction of the animal. The time of flight (TOF) (estimated length) is already incorporated into the integrated measurements, as both fluorescence and extinction measurements are integrated across the entire TOF.

3.9 Physiological Measurements of C. elegans Health

3.9.1 Life Span Measurements in C. elegans

Their short life span and ease of maintenance make *C. elegans* a great model organism for aging research. Moreover, the actin cytoskeleton shows a significant decline in form and function in multiple *C. elegans* tissues during aging, and premature breakdown of actin decreases life span, whereas protecting the actin cytoskeleton is sufficient to extend life span [17]. Therefore, we find life span measurements to be a reliable method of measuring the impact of cytoskeletal alterations on the physiological health of *C. elegans.*

1. Prepare plates spotted with bacteria of interest (*see* Subheading 3.1). We recommend using HT115 bacteria even in experiments not involving RNAi knockdown. Prepare approximately 80 plates per condition.

2. Synchronize animals of choice using methods described in Subheading 3.5 and place eggs/L1 onto plates using criteria suggested in Table 4. Incubate animals at 20 °C until day 1 of adulthood—approximately 3–4 days or ~65–90 h (growth at 15 °C is also acceptable).
3. Transfer 15 day 1 adults onto each of 8 seeded plates for every strain, for a total of 120 animals per condition. Take care not to transfer any progeny onto plates (*see* **Note 20**). Record the total number of live worms per strain per plate.
4. Continue to transfer adult animals onto a fresh plate every day until no progeny are detected on the plate. Typically, this requires 5–7 days (*see* **Note 21**).
5. Score the total number of live worms per strain per plate, daily. Animals that do not show any movement upon gentle touch at the head and tail are scored as dead. Animals showing extruded intestines, bagging, and desiccation at the sides of the plate or other unnatural causes of death are considered censored. All censored and dead animals should be removed from the plates to prevent counting them twice.
6. After worms have stopped producing progeny, score life spans every 1–2 days until all animals have been scored as dead or censored.

3.9.2 Thermotolerance in C. elegans

Thermotolerance assays are performed in *C. elegans* by placing animals in elevated temperatures post-development. Ambient temperatures for nematodes range from 15 to 20 °C, and thermal stress is induced at temperatures above 25 °C [56, 57]. Thermotolerance assays can be performed at temperatures ranging from 30 to 37 °C where animals exhibit major defects at this temperature and survival assays are completed within 24 h [56, 58]. Here, thermotolerance assays are performed by growth at 37 °C. We provide 37 °C as a simple experimental paradigm as most labs are already equipped with a 37 °C incubator for bacterial growth, and thus this protocol generally does not require purchase of additional equipment. However, it should be noted that thermotolerance assays are inherently variable due to the difficulties of maintaining a single, consistent temperature and experimental manipulations potentially altering strain variability in addition to survival. Additionally, minor shifts in temperature can have major differences in phenotype, and thus it is important to take all precautions possible to standardize a single assay within one's lab (a thorough guide to standardizing a protocol within a lab [59]).

1. Prepare plates spotted with bacteria of interest (*see* Subheading 3.1). We recommend using HT115 bacteria even in experiments not involving RNAi knockdown.

2. Synchronize animals of choice using methods described in Subheading 3.5 and grow animals to day 1 of adulthood by incubating animals at 20 °C for approximately 3–4 days (~65–90 h; *see* **Note 22**).
3. At day 1, animals are prepared by transferring them onto separate plates. Transfer ~10–15 animals onto each of 4–6 seeded plates for every strain, totaling ~60 animals per condition. This allows a manageable number of animals per plate for scoring.
4. Place plates into a 37 °C incubator (*see* **Note 23**).
5. Score animals every 2 h following hour 5 (*see* **Note 24**) until all worms have either been scored as dead or censored. Animals that completely lack movement after minor stimulation through gentle touch at the head and tail are scored as dead, whereas animals that exhibit extruded intestines, bagging, or other unnatural source of death are scored as censored.
6. Alternatively, if animals are placed at 34 °C instead of 37 °C, median thermotolerance is found at ~14 h in our incubators, so scoring can begin at hour 12 instead of hour 5, and is scored every 2 h from then.

4 Notes

1. We recommend growing OP50 bacteria at room temperature as we have noticed that growing bacteria at elevated temperatures can cause higher incidences of revertants of auxotrophy or other unwanted mutations and/or adaptations that cause more pathogenic growth. Moreover, once cultured, OP50 bacteria should be kept for a maximum of 1 week at 4 °C to avoid similar issues.
2. When dealing with animals with major defects in development or fecundity, we recommend chunking rather than picking animals. When chunking animals, a small piece of the agar is broken off the original plate containing worms and placed animal side down onto the bacterial lawn of a fresh NGM plate. Chunking can also be used as an alternative to picking animals for maintenance to decrease genetic drift, although the frequency of moving stocks will be increased due to the increased number of animals being placed on a plate.
3. For optimal annealing of the primers to the template, we recommend utilizing the "NEB Tm Calculator" with the correct polymerase selected. The annealing temperature must be calculated for the portion of the primer annealing to the template (i.e., not including the upstream overlap region).

4. For the isothermal assembly reaction, the manufacturer recommends a twofold molar excess (1:2, vector:insert ratio) for each insert, unless the insert is less than 200 bp. In such a case, a ratio of 1:5 (vector:insert) is recommended. Thus, the twofold molar excess can be adjusted to fivefold, depending on the insert fragment length. We generally use (vector length/insert length) as the fold excess of insert to add, without passing the fivefold excess recommendation. To simplify the calculation of total DNA required for each fragment, we recommend using the NEBioCalculator and setting the Vector DNA mass to 100 ng for the MosSCI vector.
5. A large break will result in excessive fluid to flow and will require a new needle to be broken. In contrast, a break too small will not release enough liquid and will increase the likelihood of clogging.
6. Several glass slides can be taped together at the edges and used as a platform for the injection pad. For our standard plates (~9 mL of volume in a 60 mm plate), we find that taping three slides together is the optimal height to reduce focus adjustment needed when switching between the injection pad and NGM plates.
7. Worms may desiccate if they are left on the agarose pads for too long. We recommend leaving worms on pads for no longer than 10–15 min (we usually avoid leaving worms on the pads for more than 5 min once they have adhered to the agarose beneath the oil patch). Therefore, the number of worms injected in a single round can be adjusted based on the user's speed of injecting.
8. Care must be taken not to inject excessive fluid into the gonad as the resulting increase in pressure forces the intestine out through the vulva. If the needle clogs, no fluid will be released from the needle. Clogs can be cleared by retracting the needle from the gonad and holding down the "clean" button for a few seconds while also monitoring the FemtoJet pressure level indicator that appears. If clogs do not clear and liquid does not flow from needle when pressing clean, use a new needle.
9. Using very dense OP50 can help worms stick better to a pick. If using bacteria to move worms from the agarose pads proves challenging, worms can also be pipetted from the M9/oil mix onto a new NGM plate. However, worms may adhere to the plastic of the tip, and this may result in loss of a few worms.
10. We recommend bleaching non-synchronized populations (e.g., chunked worms from Subheading 3.1), as successive rounds of bleaching can cause major genetic drift, especially in some mutant or transgenic lines. For example, successive

rounds of bleaching can result in a major decline in fecundity (sometimes even sterility) in the *sur-5p::hsf-1* animals used in this protocol.

11. Keeping eggs in the bleach solution for extended periods of time will damage the eggs and kill the embryos, so take care to regularly check the solution under a microscope until adult carcasses are gone and eggs are in solution.
12. Eggs can be centrifuged at faster speeds (e.g., 2500 × *g*) without disrupting egg integrity or organismal physiology.
13. It is essential that all bleach be washed from the tube if keeping eggs in the solution for extended periods of time (e.g., for L1 arresting). Eggs must be washed additional times if the residual smell of bleach remains in the tube. We found washing a total of 5× to be effective in removing all noticeable traces of bleach.
14. Animals can be moved every other day to save plates, but care must be taken to ensure that progeny do not exceed an age where they can no longer be differentiated from adults.
15. Prepare an excess number of initial plates with animals for this method, as each wash can lose anywhere from 10% to 30% of animals.
16. Day 1 adults can also be placed on FUDR plates, but if animals have developed eggs, these will hatch on the plate. FUDR will cause these animals to arrest, but care must be taken to ensure this does not cause the plate to starve, as larval animals consume significant amounts of bacteria. If animals need to be moved to a fresh plate after several days on FUDR to avoid starvation, it is not essential to put animals back onto an FUDR plate if reagents are limited. However, animals should be left on FUDR for at least 3–4 days before being taken off, to avoid progeny formation.
17. 100 mM sodium azide working solution can be kept at room temperature for several weeks. However, it will lose efficiency with time. For highly controlled experiments, use a fresh batch of sodium azide. We recommend keeping 1 M stocks at 4 °C in the dark for long-term storage.
18. The amount of sodium azide solution used on the slide is dependent on the size of the animals. A day 1 animal can generally survive with 10–13 μL of solution, while older animals will require 16–18 μL. The volume of sodium azide solution can be adjusted for smaller or larger variants to ensure that when the cover slip is placed, worms are not moving (too much liquid) or have not exploded (too little liquid). We recommend sealing the slides with Valap wax or nail polish to prevent liquid evaporation and eventual crushing of the worms.

19. We have noticed that the actin cytoskeleton starts to show defects if kept in sodium azide solution for more than 10 min. Therefore keep slides for no more than 5–10 min, and prepare fresh slides as needed. For slides being kept less than 10 min, we found that sealing the edges was unnecessary.
20. If life spans are being checked every day, some eggs and L1s can be transferred with the adults, as animals from eggs or L1 stage animals will not develop beyond the L3/L4 stage, which can still be distinguished from the adults on the next day. However, if life spans are only being moved every other day to conserve material, it is essential that no eggs or L1s be transferred to new life span plates, as 2 days may be sufficient for progeny to become indistinguishable from adults for some strains.
21. Generally, we recommend moving adults onto fresh plates every day until there are no progeny. However, if there are any limitations (e.g., time or cost of materials), the number of moves can be limited. For the first 3 days when egg-laying is at its maximum, we recommend moving animals every day. Starting from day 4 when egg-laying is already significantly reduced, animals can be moved every other day, although care should be taken to ensure that no progeny is moved onto new plates.
22. We have noticed a difference in thermotolerance measurements when animals are grown at 15 or 20 °C. Therefore, take care to be very consistent for growing animals in either temperature for all replicates of a single experiment.
23. We do not recommend stacking plates in an incubator as this will result in uneven temperature distribution across plates. The time spent opening the incubator and time spent examining plates outside of the incubator should also be minimized to decrease variability across conditions.
24. While scoring of thermotolerance can begin prior to hour 5, we notice minimal death prior to hour 7, and thus we believe scoring at earlier time points is not necessary. We have not seen significant death prior to hour 12 at 34 °C, and thus time points can start as late as hour 12.

Acknowledgments

This work was supported by 1K99AG065200-01A1 from the NIA to RHS. Several figures in this manuscript were originally published in [17]. We thank Dr. Liza Pon for careful review of the manuscript. We thank Dr. Andrew Dillin for technical assistance, valuable discussion, and providing the equipment required for this work, which was supported by 5R01AG055891-02 from the National Institute of Aging (NIA) and the Howard Hughes Medical Institute.

References

1. Berry AJ, Rinaldi MG, Graybill JR (1992) Use of high-dose fluconazole as salvage therapy for cryptococcal meningitis in patients with AIDS. Antimicrob Agents Chemother 36:690–692
2. Caviston JP, Holzbaur ELF (2006) Microtubule motors at the intersection of trafficking and transport. Trends Cell Biol 16:530–537
3. Higuchi R, Vevea JD, Swayne TC et al (2013) Actin dynamics affect mitochondrial quality control and aging in budding yeast. Curr Biol 23:2417–2422
4. Mooren OL, Galletta BJ, Cooper JA (2012) Roles for actin assembly in endocytosis. Annu Rev Biochem 81:661–686
5. Young JC, Barral JM, Ulrich Hartl F (2003) More than folding: localized functions of cytosolic chaperones. Trends Biochem Sci 28:541–547
6. Acsadi G, Dickson G, Love DR et al (1991) Human dystrophin expression in mdx mice after intramuscular injection of DNA constructs. Nature 352:815–818
7. Alim MA, Ma Q-L, Takeda K et al (2004) Demonstration of a role for alpha-synuclein as a functional microtubule-associated protein. J Alzheimers Dis 6:435–442; discussion 443-449
8. Alim MA, Hossain MS, Arima K et al (2002) Tubulin seeds alpha-synuclein fibril formation. J Biol Chem 277:2112–2117
9. Doyle T, Botstein D (1996) Movement of yeast cortical actin cytoskeleton visualized in vivo. Proc Natl Acad Sci U S A 93:3886–3891
10. Yamada S, Pokutta S, Drees F et al (2005) Deconstructing the cadherin-catenin-actin complex. Cell 123:889–901
11. Kamiyama D, Sekine S, Barsi-Rhyne B et al (2016) Versatile protein tagging in cells with split fluorescent protein. Nat Commun 7:1–9
12. Higuchi-Sanabria R, Swayne TC, Boldogh IR et al (2016) Live-cell imaging of mitochondria and the actin cytoskeleton in budding yeast. Methods Mol Biol 1365:25–62
13. Huckaba TM, Lipkin T, Pon LA (2006) Roles of type II myosin and a tropomyosin isoform in retrograde actin flow in budding yeast. J Cell Biol 175:957–969
14. Riedl J, Crevenna AH, Kessenbrock K et al (2008) Lifeact: a versatile marker to visualize F-actin. Nat Methods 5:605–607
15. Spracklen AJ, Fagan TN, Lovander KE et al (2014) The pros and cons of common actin labeling tools for visualizing actin dynamics during drosophila oogenesis. Dev Biol 393:209–226
16. Belin BJ, Goins LM, Mullins RD (2014) Comparative analysis of tools for live cell imaging of actin network architecture. BioArchitecture 4:189–202
17. Higuchi-Sanabria R, Paul Rd JW, Durieux J et al (2018) Spatial regulation of the actin cytoskeleton by HSF-1 during aging. Mol Biol Cell 29:2522–2527
18. Ono S, Pruyne D (2012) Biochemical and cell biological analysis of actin in the nematode *Caenorhabditis elegans.* Methods 56:11–17
19. Velarde N, Gunsalus KC, Piano F (2007) Diverse roles of actin in C. elegans early embryogenesis. BMC Dev Biol 7:142
20. Hochbaum D, Ferguson AA, Fisher AL (2010) Generation of Transgenic C elegans by Biolistic Transformation. J Vis Exp 42:e2090
21. Mello CC, Kramer JM, Stinchcomb D et al (1991) Efficient gene transfer in C.elegans: extrachromosomal maintenance and integration of transforming sequences. EMBO J 10:3959–3970
22. Praitis V, Casey E, Collar D et al (2001) Creation of low-copy integrated transgenic lines in Caenorhabditis elegans. Genetics 157:1217–1226
23. Evans TC (2006) Transformation and microinjection, WormBook http://www.wormbook.org/chapters/www_transformationmicroinjection/transformationmicroinjection.html
24. Frøkjaer-Jensen C, Davis MW, Hopkins CE et al (2008) Single-copy insertion of transgenes in Caenorhabditis elegans. Nat Genet 40:1375–1383
25. Kaymak E, Farley BM, Hay SA et al (2016) Efficient generation of transgenic reporter strains and analysis of expression patterns in Caenorhabditis elegans using library MosSCI. Dev Dyn 245:925–936
26. Mariol M-C, Walter L, Bellemin S et al (2013) A rapid protocol for integrating extrachromosomal arrays with high transmission rate into the C. elegans genome. J Vis Exp 82:e50773
27. Stinchcomb DT, Shaw JE, Carr SH et al (1985) Extrachromosomal DNA transformation of Caenorhabditis elegans. Mol Cell Biol 5:3484–3496
28. Yochem J, Herman RK (2003) Investigating C. elegans development through mosaic analysis. Development 130:4761–4768
29. Morley JF, Morimoto RI (2004) Regulation of longevity in Caenorhabditis elegans by heat shock factor and molecular chaperones. Mol Biol Cell 15:657–664

30. Baird NA, Douglas PM, Simic MS et al (2014) HSF-1-mediated cytoskeletal integrity determines thermotolerance and life span. Science 346:360–363
31. Bar-Ziv R, Frakes AE, Higuchi-Sanabria R et al (2020) Measurements of Physiological Stress Responses in C. Elegans. J Vis Exp 159: e61001
32. Rual J-F, Ceron J, Koreth J et al (2004) Toward improving *Caenorhabditis elegans* phenome mapping with an ORFeome-based RNAi library. Genome Res 14:2162–2168
33. Timmons L, Court DL, Fire A (2001) Ingestion of bacterially expressed dsRNAs can produce specific and potent genetic interference in *Caenorhabditis elegans*. Gene 263:103–112
34. Reinke SN, Hu X, Sykes BD et al (2010) *Caenorhabditis elegans* diet significantly affects metabolic profile, mitochondrial DNA levels, lifespan and brood size. Mol Genet Metab 100:274–282
35. Revtovich AV, Lee R, Kirienko NV (2019) Interplay between mitochondria and diet mediates pathogen and stress resistance in *Caenorhabditis elegans*. PLoS Genet 15:e1008011
36. Frøkjær-Jensen C, Davis MW, Ailion M et al (2012) Improved Mos1-mediated transgenesis in *C. elegans*. Nat Methods 9:117–118
37. Frokjaer-Jensen C, Davis MW, Sarov M et al (2014) Random and targeted transgene insertion in *C. elegans* using a modified Mosl transposon. Nat Methods 11:529–534
38. Okkema PG, Krause M (2005) Transcriptional regulation. WormBook:1–40
39. Riddle DL, Blumenthal T, Meyer BJ et al (1997) C. elegans II. Cold Spring Harbor Laboratory Press, Cold Spring Harbor (NY)
40. Mangone M, Manoharan AP, Thierry-Mieg D et al (2010) The landscape of C. elegans 3′UTRs. Science 329:432–435
41. Robert VJP, Sijen T, van Wolfswinkel J et al (2005) Chromatin and RNAi factors protect the C. elegans germline against repetitive sequences. Genes Dev 19:782–787
42. Crane MM, Sands B, Battaglia C et al (2019) In vivo measurements reveal a single 5′-intron is sufficient to increase protein expression level in *Caenorhabditis elegans*. Sci Rep 9:9192
43. Frøkjær-Jensen C, Jain N, Hansen L et al (2016) An abundant class of non-coding DNA can prevent stochastic gene silencing in the *C. elegans* germline. Cell 166:343–357
44. Ahier A, Jarriault S (2014) Simultaneous expression of multiple proteins under a single promoter in *Caenorhabditis elegans* via a versatile 2A-based toolkit. Genetics 196:605–613
45. Li D, Wang M (2012) Construction of a bicistronic vector for the co-expression of two genes in *Caenorhabditis elegans* using a newly identified IRES. BioTechniques 52:173–176
46. Wei X, Potter CJ, Luo L et al (2012) Controlling gene expression with the Q repressible binary expression system in *Caenorhabditis elegans*. Nat Methods 9:391–395
47. Gibson DG, Young L, Chuang R-Y et al (2009) Enzymatic assembly of DNA molecules up to several hundred kilobases. Nat Methods 6:343–345
48. Baugh LR (2013) To grow or not to grow: nutritional control of development during *Caenorhabditis elegans* L1 arrest. Genetics 194:539–555
49. Arantes-Oliveira N, Apfeld J, Dillin A et al (2002) Regulation of life-span by germ-line stem cells in *Caenorhabditis elegans*. Science 295:502–505
50. Kodoyianni V, Maine EM, Kimble J (1992) Molecular basis of loss-of-function mutations in the glp-1 gene of *Caenorhabditis elegans*. Mol Biol Cell 3:1199–1213
51. Garigan D, Hsu A-L, Fraser AG et al (2002) Genetic analysis of tissue aging in *Caenorhabditis elegans*: a role for heat-shock factor and bacterial proliferation. Genetics 161:1101–1112
52. Mitchell DH, Stiles JW, Santelli J et al (1979) Synchronous growth and aging of *Caenorhabditis elegans* in the presence of fluorodeoxyuridine. J Gerontol 34:28–36
53. Anderson EN, Corkins ME, Li J-C et al (2016) *C. elegans* lifespan extension by osmotic stress requires FUdR, base excision repair, FOXO, and sirtuins. Mech Ageing Dev 154:30–42
54. Higuchi-Sanabria R, Swayne TC, Boldogh IR et al (2016) Imaging of the actin cytoskeleton and mitochondria in fixed budding yeast cells. Methods Mol Biol 1365:63–81
55. Daniele JR, Esping DJ, Garcia G et al (2017) High-throughput characterization of region-specific mitochondrial function and morphology. Sci Rep 7:6749
56. Lithgow GJ, White TM, Hinerfeld DA et al (1994) Thermotolerance of a long-lived mutant of *Caenorhabditis elegans*. J Gerontol 49:B270–B276
57. Labbadia J, Morimoto RI (2015) The biology of proteostasis in aging and disease. Annu Rev Biochem 84:435–464
58. Park H-EH, Jung Y, Lee S-JV (2017) Survival assays using *Caenorhabditis elegans*. Mol Cells 40:90–99
59. Zevian SC, Yanowitz JL (2014) Methodological considerations for heat shock of the nematode *Caenorhabditis elegans*. Methods 68:450–457

Chapter 6

Imaging the Cytoskeleton in Living Plant Roots

Chenglin Chai, Sabrina Chin, and Elison B. Blancaflor

Abstract

For the past two decades, genetically encoded fluorescent proteins have emerged as the most popular method to image the plant cytoskeleton. Because fluorescent protein technology involves handling living plant cells, it is important to implement protocols that enable these delicate plant specimens to maintain optimal growth for the entire duration of the imaging experiment. To this end, we rely on a system that consists of a large coverslip coated with nutrient-supplemented agar. This agar-coverslip system is planted with surface-sterilized *Arabidopsis thaliana* seeds expressing cytoskeletal fluorescent protein reporters. The agar-coverslip system with planted seeds is then maintained in an environmentally controlled growth chamber. The entire setup is transferred onto the stage of a confocal microscope for imaging when roots of germinated seedlings reach a desired length. For plants with larger roots such as *Medicago truncatula*, the polymerized nutrient-supplemented agar is gently lifted or cut and used to secure pre-germinated seeds on the coverslip prior to root imaging. The agar-coverslip system we use for imaging the cytoskeleton in living roots along with general methods for expressing green fluorescent protein (GFP)-based cytoskeletal reporters in hairy roots of *Medicago truncatula* is described here.

Key words Actin, *Arabidopsis thaliana*, *Medicago truncatula*, Microtubules, Green fluorescent protein, Live cells, Root development

1 Introduction

The development of advanced light microscopes coupled with the introduction of genetically encoded fluorescent proteins has rapidly advanced our understanding of the plant cytoskeleton [1, 2]. The two major components of the cytoskeleton, microtubules and filamentous-actin (F-actin), can now be readily imaged in living plant cells using an array of green fluorescent protein (GFP) constructs under confocal microscopy (e.g., [3–9]). To ensure that data collected from plant specimens expressing GFP cytoskeletal reporters are reliable, it is important to develop simple methods for routine handling and mounting of plants for live cell imaging. Often times, live cell imaging is conducted on plant samples that are mounted between a coverslip and glass slide. Such practices can

Ray H. Gavin (ed.), *Cytoskeleton: Methods and Protocols*, Methods in Molecular Biology, vol. 2364, https://doi.org/10.1007/978-1-0716-1661-1_6,

lead to cell damage due to pressure from the coverslip or plant samples drying out during long-term imaging experiments. In this chapter, we describe procedures for handling plant seedlings for imaging the root cytoskeleton. We first outline a protocol for transforming F-actin and microtubule GFP markers in hairy roots of the model legume, *M. truncatula*. We then describe an agar-coverslip system for growing and mounting *A. thaliana* and *M. truncatula* seedlings/hairy roots to enable confocal imaging with minimal or no adverse consequences to root growth and cell integrity [10].

2 Materials

2.1 M. truncatula Seedling Preparation

1. *M. truncatula* seeds.
2. Concentrated sulfuric acid.
3. Autoclaved distilled water pre-chilled to 4 °C.
4. 30% bleach.
5. Falcon 50 mL conical centrifuge tubes.
6. Whatman filter paper circles (70 mm in diameter).
7. Sterile Petri dishes (90 × 15 mm).
8. Sterile extra deep Petri dishes (90 × 25 mm).
9. 95% ethanol.
10. 1% Triton X.
11. Agar.
12. 0.4% Gelzan (an agar substitute, if desired).
13. Personal protective equipment such as goggles, gloves, and lab coats.

2.2 Agrobacteria Preparation

1. *Agrobacterium rhizogenes* (strain ARqua1) transformed with GFP-based cytoskeletal constructs [3–5, 9].
2. Liquid lysogeny broth (LB) medium (commercially available) supplemented with 100 mg/L streptomycin and 50 mg/L kanamycin.
3. Polymerized LB agar supplemented with 100 mg/L streptomycin and 50 mg/L kanamycin.
4. Autoclaved distilled water.
5. Sterile cell spreader.
6. Sterile inoculating needles.

2.3 Hairy Root Transformation

1. Sterile forceps and scalpels.
2. Modified Fahraeus medium: 1 mM $CaCl_2$, 0.5 mM $MgSO_4$, 0.7 mM KH_2PO_4, 0.8 mM Na_2HPO_4, 50 μM FeEDTA,

0.5 mM NH_4NO_3, supplemented with 0.1 mg each of $MnSO_4$, $CuSO_4$, $ZnSO_4$, H_3BO_3, and Na_2MoO_4 per liter, adjust pH to 6.8 with NaOH.

3. Modified Fahraeus medium agar plates (round Petri dishes).
4. Modified Fahraeus medium agar plates (square Petri dishes, 120 × 120 mm).
5. 1.5 cm wide Millipore tape.
6. Black plastic wrap.
7. Growth chamber with controls for photoperiod and light intensity.
8. Fluorescence stereo microscope.

2.4 Live Cell Imaging of Roots and Agar Slabs for Root Mounting

1. Agar or Agargel (available commercially; *see* **Note 1**).
2. Murashige and Skoog (MS) basal salt mixture (available commercially).
3. Sterile 48 × 64 × 0.13–0.17 mm coverslips: sandwich each coverslip between sterile filter paper prior to autoclaving to prevent coverslips from sticking to each other.
4. Sterilized pointed end toothpicks.
5. 100 × 15 mm polystyrene sterile Petri dishes.
6. Sterilized filter paper.
7. 95% ethanol.
8. Bleach-Triton mix: 20% bleach-1% Triton-X.
9. 90 mm filter paper fitted in 100 × 15 mm polystyrene Petri dishes.
10. Confocal microscope.
11. Inverted microscope.

3 Methods

3.1 Generation of M. truncatula Hairy Roots Expressing Fluorescent Cytoskeletal Reporters

M. truncatula seeds have hard seed coats that can inhibit germination by blocking water and oxygen, which are required for germination. Seeds are usually scarified chemically or mechanically to promote fast and uniform germination.

3.1.1 Preparing Seedlings

1. Scarify *M. truncatula* seeds 4 days before transformation by placing them in a 50 mL Falcon tube, and mix with concentrated sulfuric acid of approximately thrice the volume of seeds. Put the mixture through a vortex (*see* **Note 2**). Repeat the vortex procedure at 2 min intervals for 6 min at room temperature.
2. Wash seeds three times with cold sterile distilled water in a laminar flow hood to remove residual sulfuric acid (*see* **Note 3**).

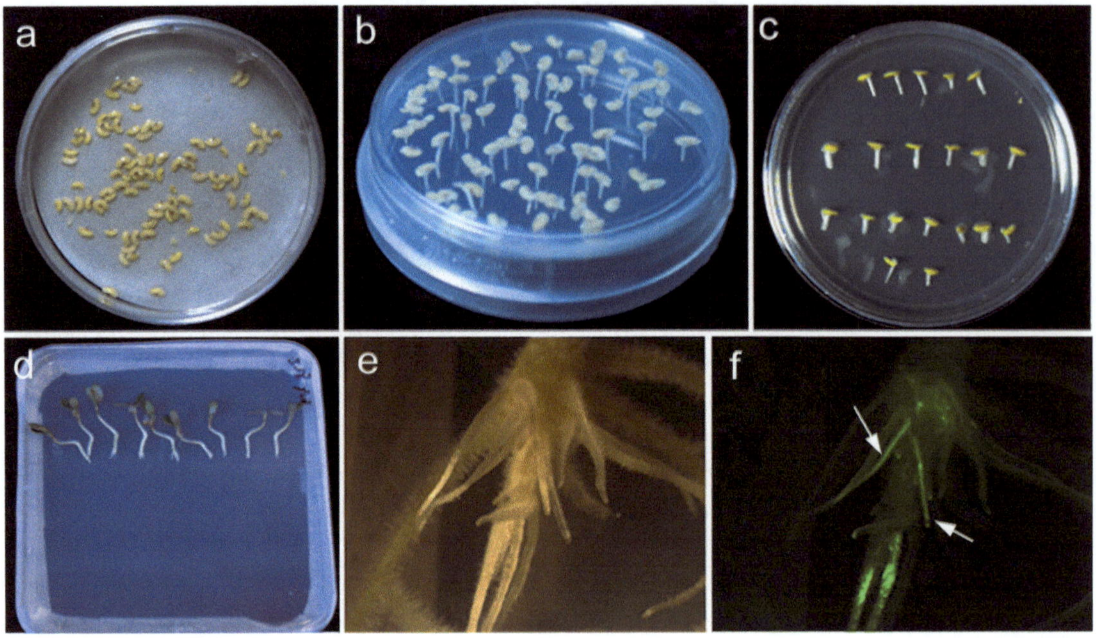

Fig. 1 Generating and screening cytoskeletal reporter transgenic hairy roots in *M. truncatula*. (**a**) Scarified seeds are incubated at 4 °C. (**b**) Germinating of seeds on an inverted Petri dish. (**c**) Co-cultivation of cut seedlings with *A. rhizogenes*. (**d**) Cultivation of inoculated seedlings on vertical plates for 2–3 weeks. (**e**) Hairy root formation viewed under bright field microscopy. (**f**) Fluorescent signal inform successfully transformed hairy roots (arrows, **e**)

3. Sterilize the scarified seed in 30% bleach for 10 min at room temperature.
4. Rinse the seeds five times with sterile distilled water. Incubate the seeds in sterile distilled water overnight at room temperature.
5. The next day, transfer the seeds onto a sterile Petri dish lined with sterile filter paper moistened with sterile distilled water (Fig. 1a). Seal the Petri dish with Parafilm, and wrap the dish with aluminum foil. Incubate at 4 °C in the dark for 2 days.
6. Approximately 16 h before transformation, plant the enlarged seeds on a deep Petri dish plated with 1% agar, and invert the plate (Fig. 1b; *see* **Note 4**). Place the plate in the dark at room temperature.

3.1.2 Agrobacteria Preparation

1. Three days before transformation, inoculate 5 mL liquid LB medium (*see* Subheading 2.2, **step 2**) with *Agrobacterium* harboring a construct of interest, and incubate in an incubator at 28 °C overnight at 250 rounds per minute (rpm).
2. The next day, spread 700 μL of overnight culture onto a LB plate containing antibiotic (*see* Subheading 2.2, **step 3**). Seal the plate with Parafilm, and incubate the inverted plate at 28 °C for 2 days.

3.1.3 Hairy Root Transformation and Transgenic Root Screening

1. Add 800 μL of sterile distilled water onto the plate and re-suspend the *Agrobacterium rhizogenes* using a triangular cell spreader (*see* **Note 5**).
2. The roots of the seedlings are about 2 cm long. Cut 1 cm of the root tip, and immediately dip the cut end into the bacterial culture for at least 5 s.
3. Place the cut seedlings horizontally on modified Fahraeus medium plates, and seal the plates with Parafilm (Fig. 1c). Incubate the plates in the dark at 20 °C overnight (*see* **Note 6**).
4. Place the seedlings vertically on square modified Fahraeus medium plates (Fig. 1d). Cover the bottom part of the plates with black plastic wrap.
5. Allow plants to grow in a growth chamber at 23 °C for 2 weeks at a 16 h photoperiod and a light intensity of 70 μE per s m^{-2} (*see* **Note 7**).
6. Transgenic hairy roots may be screened under a fluorescence stereo microscope 2 or 3 weeks after growing in a growth chamber (Fig. 1e, f). Transgenic roots can be marked on the plate using a Sharpie permanent marker pen (*see* **Note 8**).

3.2 Preparing* A. thaliana *Roots for Direct Live Cell Imaging

Most of the live cell imaging work on the plant cytoskeleton has been conducted using the model plant *A. thaliana*. This is because *A. thaliana* is readily transformed with various GFP cytoskeletal reporters and the smaller roots of *A. thaliana* make sample preparation for live cell imaging very convenient. However, a number of precautions have to be taken to ensure that growth of the root is not compromised by excessive physical handling of the seedling. If possible, imaging of live Arabidopsis roots should be done directly on the media where the seeds are germinated. Therefore, the method described here involves growing *A. thaliana* on a glass coverslip covered with MS media (Murashige and Skoog basal salt mixture). The coverslip can be removed from the growth chamber and placed directly on the microscope stage. This completely eliminates direct handling of the seedlings.

1. Surface sterilize the seeds by immersing them for 3 min each in 95% ethanol and the bleach-Triton mix. Rinse the seeds with sterile deionized water at least three times after bleach treatment. Transfer seeds to sterilized 90 mm filter paper fitted in 100 × 15 mm polystyrene Petri dishes, and spread the seeds evenly on the filter paper using a 1 mL pipette (Fig. 2a). Allow the seeds to dry in a laminar flow bench overnight. The seeds can be planted immediately after spreading onto the filter paper-lined polystyrene Petri dishes, but for long-term storage, they must be dried thoroughly prior to sealing with Parafilm.

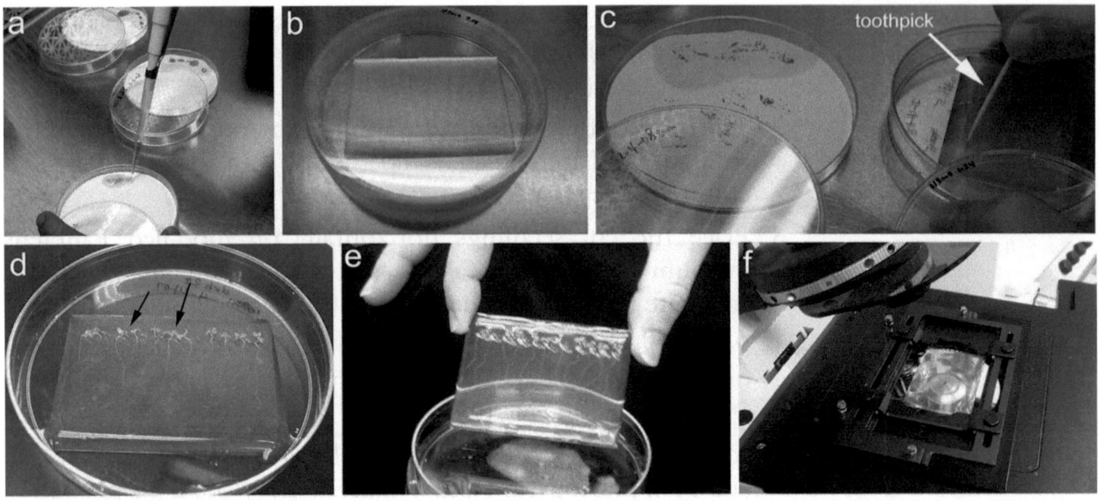

Fig. 2 *A. thaliana* seed sterilization and planting for live cell confocal imaging of the cytoskeleton. (**a**) Spreading sterilized seeds on filter paper. (**b**) Agar-coverslip setup for planting *A. thaliana* seeds. (**c**) Sterilized seeds can be picked individually using pointed toothpicks and planted directly on the agar-coverslip system. (**d**) Seven-day-old *A. thaliana* seedlings (arrows) growing on the agar-coverslip system. The agar-coverslip system with the growing seedlings can be easily picked up by gently pressing the bottom of the polystyrene Petri dish (**e**) and directly transferred to the stage of the confocal microscope for imaging (**f**)

2. Prior to seed sterilization, a coverslip-agar medium for germinating seeds must be prepared in advance. This system consists of sterilized coverslips with about a 1 mm layer of nutrient supplemented polymerized 0.5–1% agar. Alternatively, use 0.4% Gelzan for live cell imaging of root hairs. The coverslip agar system is placed inside 100 × 15 mm polystyrene Petri dishes, wrapped with Parafilm, and kept at 4 °C for future use (Fig. 2b).
3. To prepare the coverslip-agar system, autoclave 0.5× MS salts in 0.5–1% agar (v/v) supplemented with 0.5 mg/mL pyridoxine–HCl, 0.5 mg/mL nicotinic acid, and 1 mg/mL thiamine. The vitamins can be dissolved in water as 1000× stock, filter sterilized, and stored at 4 °C prior to use. Add 0.10 g/L myo-inositol, 0.5 g/L MES, and 1% sucrose to the MS vitamin solution. Adjust pH to 5.7 with 10 M KOH, add agar, and autoclave.
4. Sterilize 64 × 48 mm coverslips by autoclaving. Sandwich each coverslip between sterile filter papers prior to autoclaving to prevent coverslips from sticking to each other.
5. Use a pair of sterile forceps and place one coverslip in the center of a 9-cm-diameter round polystyrene Petri dish.
6. Carefully pipette 10 mL 1% (v/v) agar-MS solution onto the center of the coverslip until the entire coverslip is covered by

agar. Agar coverslip will be about 2 mm in height. The surface tension on the edge of the coverslip should hold the agar solution within the coverslip (*see* **Note 9**).

7. Allow the agar to polymerize under a laminar flow hood. This will take about ½ an hour (Fig. 2b; *see* **Note 10**).
8. Pick *A. thaliana* seeds individually with a sterile toothpick, and gently push the seed into the agar media so that the seed touches the bottom of the coverslip (Fig. 2c). This helps the roots grow straight inside the medium facilitating microscopic observation. Dried seeds can be easily picked up with a moist toothpick. This is done by immersing the tip of the toothpick into the agar and gently touching the seed.
9. After planting, seal the Petri dish with Parafilm, and transfer to a growth chamber. The Petri dish should be positioned at a 60° angle from the horizontal to allow the roots to direct their growth toward the surface of the coverslip. It takes about 3 to 4 days for the roots to be ready for imaging (Fig. 2d).
10. When the roots have reached the desired age for imaging, remove the coverslip containing the seedlings by gently pressing the bottom of the polystyrene Petri dish and lifting the coverslip making sure that the polymerized agar does not slide off (Fig. 2e).
11. Mount the coverslip directly on the stage of a confocal microscope for observation (Fig. 2f).

3.3 Mounting Plant Roots for Microscopy Using Agar Slabs

1. Place about two droplets of water on a glass slide (*see* **Note 11**).
2. Select four to six *A. thaliana* seedlings (ideal seedling length is approximately 15 mm. These are seedlings germinated on regular ½ MS 1% (v/v) agar) or two to three *M. truncatula* seedlings/hairy roots, and gently place them on the moistened 64 × 48 mm glass coverslip. For *A. thaliana*, space the seedlings about 5 mm apart. For *M. truncatula* seedlings or hairy roots, one seedling/root per coverslip should be sufficient.
3. Cut the agar slab into 50 × 30 mm to fit on top of the glass coverslip or directly lift the entire agar slab (Fig. 3a, b).
4. Gently transfer seedlings onto the surface of the exposed coverslip with fingers or forceps (Fig. 3c).
5. Secure the seedling by covering it with the agar (Fig. 3d). To do this, start by placing the agar support layer in the center of the coverslip, and gently spread it outward to prevent bubbling.
6. Transfer the coverslip system onto the stage of an inverted microscope for observations (*see* Fig. 2f). Representative images of actin and microtubule cytoskeleton of *A. thaliana* and *M. truncatula* roots using this process are shown in Fig. 3e, f.

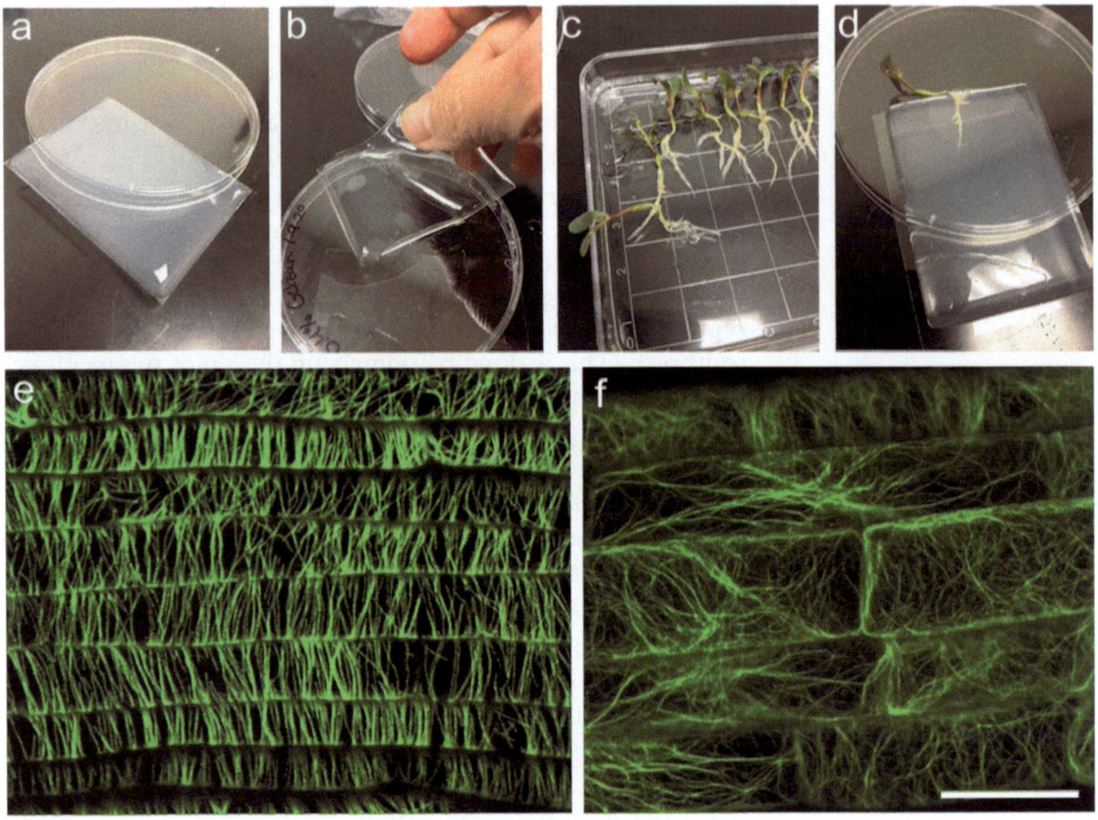

Fig. 3 Using the agar-coverslip system for mounting larger roots. (**a**) Coverslip coated with a 2 mm slab of nutrient-supplemented agar. (**b**) Lifting the agar slab from the coverslip. (**c**) Seedlings of *M. truncatula* expressing cytoskeletal GFP reporters. (**d**) Using the agar slab to secure one seedling of *M. truncatula* for imaging the root cytoskeleton. (**e**) Microtubules in living epidermal cells of *M. truncatula* hairy roots. (**f**) Filamentous-actin in epidermal cells of an *A. thaliana* root. Bar = 20μm

4 Notes

1. There are several companies that sell agar for plant culture, which can be used for preparing the agar-coverslip system. For imaging live *A. thaliana* roots, however, we found that Gelzan gives the most reliable images when layered on the coverslip.
2. Immediately vortexing the seeds right after adding sulfuric acid is important to prevent seeds from aggregating or sticking to the tube sides, which interferes with the efficacy of seed scarification.
3. After scarification, all subsequent steps should be conducted under sterile conditions.

   Scarified seeds are able to more efficiently absorb water allowing for better imbibition.

4. Inverting plates allows straight roots to emerge from germinating seeds for more efficient transplanting and transformation.
5. If the *Agrobacteria* solution dries up, re-suspend with 800 μL of sterile distilled water.
6. Incubating inoculated roots at a lower temperature (20 °C) can improve transformation efficiency.
7. The seedlings may fall off or be displaced during culture. It is important to check the plates 3 days after inoculation, and if necessary, reposition seedlings back to their original orientation.
8. If a slow-growing mutant is used for hairy root transformation, a longer time may be needed before it is ready for transgenic roots screening.
9. Surface tension of the molten agar will form a dome on the surface of the coverslip. Avoid pouring excess agar on the coverslip as this will break the dome and cause the agar to flow out of the coverslip. Furthermore, minimize vibrations and sudden movements as the agar solution can spill out of the coverslip. Alternatively, pour about the gel so that the entire Petri dish including the coverslip is covered with 2–3 mm for gel. This method ensures more uniform emergence of root hairs. Prior to imaging, cut the excess gel outlining the coverslip. Lift the entire coverslip and remove any excess gel prior to mounting on the microscope.
10. To store the agar coverslip, wrap the Petri dish with Parafilm and keep at 4 °C. Agar coverslip can be stored for months and be used as long as there is no contamination.
11. Add sterile distilled water to the Petri dishes so it does not completely submerge the seedlings. Using less water is better than placing too much water. Using too much water will overspread the seedlings. Seedlings will not desiccate once they are placed under the agar coverslip.

Acknowledgments

This work is supported by the National Aeronautics and Space Administration (NASA grants 80NSSC19K0129 and 80NSSC18K1462) and Noble Research Institute LLC.

References

1. Komis G, Novák D, Ovečka M, Šamajová O, Šamaj J (2018) Advances in imaging plant cell dynamics. Plant Physiol 176:80–93
2. Ovečka M, von Wangenheim D, Tomančák P, Šamajová O, Komis G, Šamaj J (2018) Multiscale imaging of plant development by light-sheet fluorescence microscopy. Nat Plants 4:639–650
3. Marc J, Granger CL, Brincat J, Fisher DD, Kao T, McCubbin AG, Cyr RJ (1998) A

GFP-MAP4 reporter gene for visualizing cortical microtubule rearrangements in living epidermal cells. Plant Cell 10:1927–1940

4. Wang Y-S, Yoo C-M, Blancaflor EB (2008) Improved imaging of actin filaments in transgenic Arabidopsis plants expressing a green fluorescent protein fusion to the C and N termini of the fimbrin actin binding domain 2. New Phytol 177:525–536
5. Dyachok J, Sparks JA, Liao F, Wang Y-S, Blancaflor EB (2014) Fluorescent protein-based reporters of the actin cytoskeleton in living plant cells: fluorophore variant, actin binding domain and promoter considerations. Cytoskeleton (Holboken) 71:311–327. https://doi.org/10.1002/cm.21174
6. Sparks JA, Kwon T, Renna L, Liao F, Brandizzi F, Blancaflor EB (2016) HLB1 is a tetratricopeptide repeat domain-containing protein that operates at the intersection of exocytic and endocytic pathways at the TGN/EE in Arabidopsis. Plant Cell 28:746–769. https://doi.org/10.1105/tpc.15.00794
7. Yoo C-M, Naramoto S, Sparks JA, Khan BR, Nakashima J, Fukuda H, Blancaflor EB (2018) Deletion analysis of AGD1 reveals domains crucial for plasma membrane localization and function in root hair polarity. J Cell Sci 131(2): jcs203828. https://doi.org/10.1242/jcs.203828
8. Kwon T, Sparks JA, Liao F, Blancaflor EB (2018) ERULUS is a plasma membrane-localized receptor-like kinase that specifies root hair growth by maintaining tip-focused cytoplasmic calcium oscillations. Plant Cell 30:1173–1177
9. Vidali L, Rounds CM, Hepler PK, Bezanilla M (2009) Lifeact-mGFP reveals a dynamic apical F-actin network in tip growing plant cells. PLoS One 4(5):e5744
10. Dyachok J, Paez-Garcia A, Yoo CM, Palanichelvam K, Blancaflor EB (2016) Fluorescence imaging of the cytoskeleton in plant roots. In: Gavin RH (ed) Cytoskeleton methods and protocols: methods in molecular biology series, 3rd edn. Humana Press, Totowa, New Jersey, pp 139–153

Chapter 7

Imaging of Actin Cytoskeleton in the Nematode *Caenorhabditis elegans*

Shoichiro Ono

Abstract

The nematode *Caenorhabditis elegans* is one of the major model organisms in cell and developmental biology. This organism is easy to culture in laboratories and suitable for microscopic investigation of the cytoskeleton. Because the worms are small and transparent, the actin cytoskeleton in many tissues and cells can be observed with appropriate visualization techniques without sectioning or dissection. This chapter describes the introduction to representative methods for imaging the actin cytoskeleton in *C. elegans* and a protocol for staining worms with fluorescent phalloidin.

Key words Actin filaments, *Caenorhabditis elegans*, Fluorescence microscopy, Nematodes, Phalloidin

1 Introduction

The nematode *Caenorhabditis elegans* was introduced as a model organism into genetics research by Brenner in 1970s [1] and has also been widely used in the fields of molecular, cell, and developmental biology. Tissue organization of an adult worm is relatively simple with ~1000 somatic cells [2, 3]. The small and transparent body of *C. elegans* is suitable for microscopic characterization of cytoskeletal structures in many tissues and cells. For example, *C. elegans* has been extensively used to study cell migration and invasion [4], cell division [5], embryonic morphogenesis [6], muscle sarcomere assembly [7, 8], and neuronal axon guidance [9]. *C. elegans* actin is biochemically similar to mammalian actins and plays central roles in these cell biological and developmental processes [10–12]. To visualize actin organization in *C. elegans*, three techniques are generally utilized: (1) immunofluorescent staining, (2) genetically encoded actin probes, and (3) phalloidin staining. Each technique is briefly described below, which is followed by a detailed description of the procedure for phalloidin staining.

Ray H. Gavin (ed.), *Cytoskeleton: Methods and Protocols*, Methods in Molecular Biology, vol. 2364,
https://doi.org/10.1007/978-1-0716-1661-1_7,

1.1 Immunofluorescent Staining of Actin

Because *C. elegans* actin is highly conserved with mammalian actin, several commercially available anti-actin antibodies can be used for immunofluorescent staining of actin. Our group has confirmed that a mouse monoclonal anti-actin antibody C4 (*see* **Note 1**) and a rabbit polyclonal anti-actin antibody (Cytoskeleton, Inc., catalog number: AAN01) can be used for immunofluorescent staining of actin in *C. elegans* [13–15]. The eggshell of embryos and the cuticle of the worms are highly impermeable and need to be disrupted during fixation. Several methods for immunofluorescent staining of *C. elegans* have been established [16–19], which can be applied to immunofluorescent staining for actin.

1.2 Live Imaging of Actin Filaments

The small and transparent body of *C. elegans* is suitable for noninvasive live imaging of the cytoskeleton in intact animals. Fluorescent probes for actin need to be expressed in tissues or cells of interest using specific promoters. Many strains expressing fluorescent actin probes have already been published (*see* Table 1) and can be obtained from *Caenorhabditis* Genetics Center (*see* **Note 2**) or researchers who developed them. Similarly to any other experimental systems, two major types of live actin probes have been utilized: (1) fluorescent proteins fused with small actin-binding peptides or protein fragments such as LifeAct [20], calponin-homology domain of utrophin (UtCH) [21], and actin-binding domain of *Drosophila* moesin (moesin ABD) [22] and (2) fluorescent proteins fused directly with actin. A fusion between actin and a fluorescent protein can cause dominant negative effects on the cytoskeleton when it is highly expressed [23]. Therefore, its expression levels should be kept low. For this reason, tagging of an endogenous actin gene with a fluorescent protein is not recommended. *C. elegans* has five actin genes that do not correspond to mammalian α, β, and γ isoforms [12]. Although functional differences among *C. elegans* actin isoforms remain unclear, differences in their expression patterns were reported: ACT-2 is expressed broadly in both muscle and non-muscle tissues [24], ACT-4 is predominantly expressed in the body wall muscle [25], and ACT-5 is specifically expressed in the intestine [26]. Therefore, if a direct fusion of actin with a fluorescent protein is planned, choosing an appropriate actin isoform may be important. Additionally, nanobody-based imaging could also be designed.

A variety of live imaging techniques can be applied to *C. elegans*, but technical details for microscopy are not described here. Except for early embryos, worms need to be immobilized during live imaging. Worms are often anesthetized using 10–25 mM sodium azide [27] or 0.01% tetramisole (with or without 0.1% tricaine) [28] and mounted on a pad of 2% agarose that is overlayed with a coverslip. Alternatively, a synthetic polymer, Pluronic F127, along with microbeads can be used to immobilize worms without anesthetic [29]. An immobilization medium based on this

Table 1
Examples of genetically encoded probes for actin

Probe	Fluorescent protein	Strain	Tissue(s)	CGC[a]	References
UtCH[b]	GFP	MG589	Embryos, gonad	Yes	[43]
LifeAct	GFP	FT1197	Epidermis	Yes	[44]
	mRuby	AGD1651	Body wall muscle	No	[45]
	mRuby	AGD1654	Epidermis	No	[45]
	mRuby	AGD1657	Intestine	No	[45]
	mRuby	AGD1679	Neuron	No	[45]
	mRuby	AGD1682	Pharynx	No	[45]
Moesin ABD[c]	GFP	CZ14748	Epidermis	Yes	[46]
	GFP	MAD36	Embryos	Yes	[47]
	GFP	NK887	Anchor cell	Yes	[48]
Actin (ACT-1)	GFP	UN1502	Spermatheca	No	[49]
Actin (ACT-2)	GFP	EU1382	Most tissues	Yes	[24]
Actin (ACT-4)	GFP	ON16	Body wall muscle	No	[37]
Actin (ACT-5)	GFP	ERT60	Intestine	Yes	[50]

[a]Availability at *Caenorhabditis* Genetics Center
[b]UtCH: calponin-homology domain of utrophin
[c]Moesin ABD: actin-binding domain of *Drosophila* moesin

method, NemaGel (InVivo Biosystems), is commercially available. Pluronic F127 is fluidic at low temperature (below 10 °C) but becomes a gel at room temperature. Microbeads maintain a constant distance between the two glass surfaces that apply gentle pressure to the worms. In advanced imaging experiments, microfluidic devices to immobilize worms could also be applied for imaging [30, 31].

1.3 Phalloidin Staining

Phalloidin, a phallotoxin from the death cap mushroom *Amanita phalloides*, specifically and tightly binds to filamentous actin [32]. Phalloidin conjugates with fluorescent dyes have been widely used to visualize actin filaments in tissues and cells from a wide variety of organisms [33] including *C. elegans* [34, 35]. Phalloidin staining is not very difficult, but insufficient permeabilization can cause poor staining. A protocol for staining worms with fluorescent phalloidin, which is based on the published procedure [36], is described below. The fluorescence of green fluorescent protein (GFP) is preserved with this method, such that co-localization of actin filaments and a GFP-fusion protein can be examined [15, 37, 38].

2 Materials

2.1 C. elegans Culture

1. *C. elegans* wild-type N2 strain and many other strains with mutations and/or genetic modifications can be obtained from *Caenorhabditis* Genetics Center (University of Minnesota; https://cgc.umn.edu/) (*see* **Note 2**).
2. *Escherichia coli* OP50, which is also available from *Caenorhabditis* Genetics Center.
3. Nematode Growth Medium (available commercially) agar plates [39].
4. 10 cm Petri dishes.

2.2 Buffers

1. M9 buffer: 22 mM KH_2PO_4, 42 mM Na_2HPO_4, 86 mM NaCl, 1 mM $MgSO_4$. Autoclave and store at room temperature.
2. 2× cytoskeleton buffer: 276 mM KCl, 6 mM $MgCl_2$, 4 mM EGTA, 20 mM MES-KOH, pH 6.1. Sterilize by filtration through a 0.22 mm filter, and store at 4 °C.
3. 1.28 M sucrose: sterilize by filtration through a 0.22 mm filter and store at 4 °C.
4. Phosphate-buffered saline (PBS)-TG: 137 mM NaCl, 2.7 mM KCl, 10 mM Na_2HPO_4, 1.8 mM KH_2PO_4, 0.5% Triton X-100, 30 mM glycine.

2.3 Fixative

1. Formaldehyde fixative: 4% formaldehyde, 1× cytoskeleton buffer, 0.32 M sucrose—immediately before use, mix 0.25 mL 16% formaldehyde solution, 0.5 mL 2× cytoskeleton buffer, and 0.25 mL 1.28 M sucrose.
2. Acetone: keep at −20 °C.

2.4 Fluorescently Labeled Phalloidin

1. 500× Tetramethylrhodamine (TMR)-labeled phalloidin: Dissolve TMR-phalloidin in methanol at 0.1 mg/mL, and store at −20 °C in small aliquots (*see* **Note 3**).

2.5 Mounting

1. Coverslips: Select an appropriate size of No. 1 coverslips depending on the volume of the sample.
2. 25 × 75 mm slide glasses.
3. Antifade Mountant (*see* **Note 4**).

3 Methods

3.1 C. elegans Culture

1. Worms can be cultured and maintained by following the standard methods as described in detail by Stiernagle [39]. Worms are generally grown on Nematode Growth Medium agar plates

using *Escherichia coli* OP50 as a food source. Although worms can be grown in liquid media, culturing worms on agar plates is much easier than liquid culture and provides sufficient quantities of worms for most microscopic observations. Typically, healthy worms grown on one or two 10 cm petri dishes before starvation are sufficient for one experiment (*see* **Note 5**).

2. Use worm cultures with no contamination of bacteria or fungi. If a culture is contaminated, follow decontamination protocols as described by Stiernagle [39].

3.2 Harvesting and Fixing Worms for Phalloidin Staining

1. Harvest worms from agar plates using 5–10 mL M9 buffer per plate, and collect in a 15 mL conical centrifuge tube. Worms may be sticky to certain types of plastic surfaces. In such a case, use glass pipettes and/or tubes.
2. Centrifuge at 400 × *g* for 2 min to pellet worms. Discard supernatant by aspiration.
3. Wash the worms by adding 15 mL M9 buffer, resuspending, pelleting, and discarding the supernatant. Repeat two times (total three washes) but leave a small amount (~0.5 mL) of the supernatant at the final wash.
4. Resuspend the worms in the remaining buffer and transfer to a 1.5 mL microcentrifuge tube (*see* **Note 6**).
5. Centrifuge at 400 × *g* for 2 min to pellet worms. Discard supernatant by pipetting.
6. Resuspend the worms in 1 mL 4% formaldehyde, 1× cytoskeleton buffer, 0.32 M sucrose, and incubate with gentle rocking for 30 min at room temperature.
7. Centrifuge at 400 × *g* for 2 min to pellet worms. Discard supernatant as much as possible by pipetting.
8. Loosen the worm pellet by tapping the bottom of the tube, add 1 mL cold acetone, mix well by inverting the tube, and incubate at −20 °C for 5 min (*see* **Note 7**).
9. Centrifuge at 400 × *g* for 2 min to pellet worms. Discard supernatant by pipetting.
10. Resuspend the worms in 1 mL PBS-TG, and incubate with gentle rocking for 10 min at room temperature.
11. Centrifuge at 400 × *g* for 2 min to pellet worms. Discard supernatant by pipetting.

3.3 Staining with TMR-Phalloidin

1. Mix 1 μL 500× TMR-phalloidin stock solution in 500 μL PBS-TG (*see* **Note 8**).
2. Resuspend the worms in 500 μL of the diluted TMR-phalloidin solution, and incubate with gentle rocking for 1 h at room temperature (*see* **Note 9**).

3. Centrifuge at 400 × *g* for 2 min to pellet worms. Discard supernatant by pipetting.
4. Resuspend the worms in 1 mL PBS-TG and incubate with gentle rocking for 10 min at room temperature.
5. Centrifuge at 400 × *g* for 2 min to pellet worms. Discard supernatant by pipetting.
6. Repeat the wash cycle two additional times. Leave a small amount of fluid after the last wash.
7. Add one drop of ProLong™ Diamond Antifade Mountant to a slide glass.
8. Resuspend the stained worms and transfer them onto ProLong™ Diamond Antifade Mountant. Mix with a slow swirling motion.
9. Overlay a coverslip and remove excess fluid with filter papers.
10. Cure overnight on a flat surface in the dark.

3.4 Observation and Interpretation

1. Observe the samples using a fluorescence microscope and record images (see sample images of wild-type worms in Fig. 1).
2. Compare experimental samples with control samples.
3. WormAtlas (www.wormatlas.org) and *C. elegans* Atlas [40] are useful references for worm anatomy.

4 Notes

1. Mouse monoclonal anti-actin antibody C4 was originally described in [41] and commercially available in the form of either ascites fluid or purified IgG from multiple manufacturers including MP Biomedicals, Thermo Fisher, Millipore Sigma, and Cell Signaling Technology.
2. *Caenorhabditis* Genetics Center (University of Minnesota; https://cgc.umn.edu/) collects and distributes *C. elegans* strains. Users are responsible to pay nominal fees for shipping and handling.
3. Phalloidin conjugates with various fluorophores are commercially available from multiple manufacturers.
4. We suggest ProLong™ Diamond Antifade Mountant. However, other mounting media can be used. Follow manufacturer's instructions for compatibility with fluorophores.
5. Typical cultures contain worms at mixed developmental stages. If worms at a particular developmental stage need to be examined, the culture should be synchronized as described by Porta-de-la-Riva et al. [42].

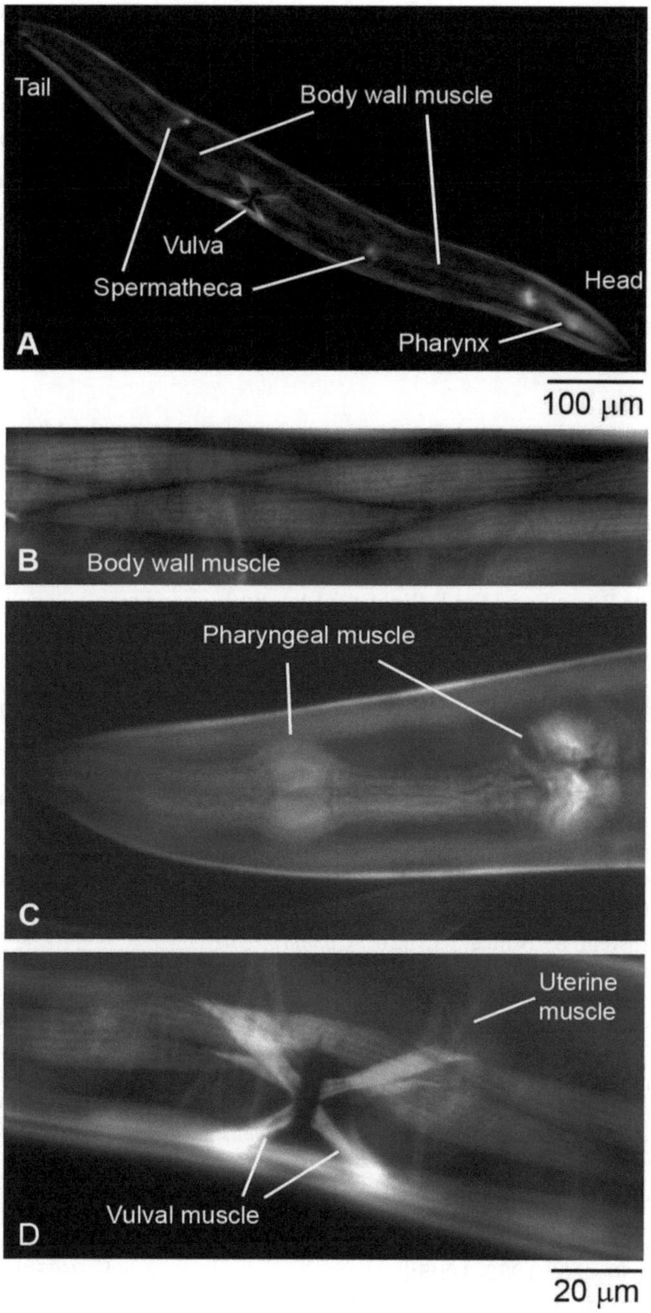

Fig. 1 Representative images of *C. elegans* stained with TMR-phalloidin. An entire view of an adult hermaphrodite (**a**) and magnified views of the body wall muscle (**b**), the pharyngeal muscle (**c**), and the vulval and uterine muscles (**d**) are shown. Bars: 100μm for **a** and 20μm for **b–d**

6. Worms may be sticky to certain microcentrifuge tubes. In our laboratory, Axygen® MaxyClear microcentrifuge tubes are preferred because of low retention of worms.

7. Exposure of the worms to acetone is important for successful staining. Clumping of the worms can occur when the worm pellets are not loosened enough or when too much salt, buffer, and formaldehyde are left before adding acetone, which can result in insufficient permeabilization.
8. Fixation of tissues and cells with methanol will significantly alter the pattern of phalloidin staining. However, carryover of a small amount of methanol from the TMR-phalloidin stock will not affect the staining.
9. Incubation can be extended to overnight at room temperature.

References

1. Brenner S (1974) The genetics of *Caenorhabditis elegans*. Genetics 77:71–94
2. White JG, Southgate E, Thomson JN, Brenner S (1986) The structure of the nervous system of the nematode *Caenorhabditis elegans*. Philos Trans R Soc Lond Ser B Biol Sci 314:1–340
3. Sulston JE, Schierenberg E, White JG, Thomson JN (1983) The embryonic cell lineage of the nematode *Caenorhabditis elegans*. Dev Biol 100:64–119
4. Sherwood DR, Plastino J (2018) Invading, leading and navigating cells in *Caenorhabditis elegans*: insights into cell movement *in vivo*. Genetics 208:53–78
5. Pintard L, Bowerman B (2019) Mitotic cell division in *Caenorhabditis elegans*. Genetics 211:35–73
6. Goldstein B, Nance J (2020) *Caenorhabditis elegans* gastrulation: a model for understanding how cells polarize change shape and journey toward the center of an embryo. Genetics 214:265–277
7. Ono S (2014) Regulation of structure and function of sarcomeric actin filaments in striated muscle of the nematode *Caenorhabditis elegans*. Anat Rec 297:1548–1559
8. Benian GM, Epstein HF (2011) *Caenorhabditis elegans* muscle: a genetic and molecular model for protein interactions in the heart. Circ Res 109:1082–1095
9. Chisholm AD, Hutter H, Jin Y, Wadsworth WG (2016) The genetics of axon guidance and axon regeneration in *Caenorhabditis elegans*. Genetics 204:849–882
10. Harris H, Tso M, Epstein HF (1977) Actin and myosin-linked calcium regulation in the nematode *Caenorhabditis elegans*: biochemical and structural properties of native filaments and purified proteins. Biochemistry 16:859–865
11. Ono S (1999) Purification and biochemical characterization of actin from *Caenorhabditis elegans*: its difference from rabbit muscle actin in the interaction with nematode ADF/cofilin. Cell Motil Cytoskeleton 43:128–136
12. Ono S, Pruyne D (2012) Biochemical and cell biological analysis of actin in the nematode *Caenorhabditis elegans*. Methods 56:11–17
13. Ono S, Baillie DL, Benian GM (1999) UNC-60B an ADF/cofilin family protein is required for proper assembly of actin into myofibrils in *Caenorhabditis elegans* body wall muscle. J Cell Biol 145:491–502
14. Ono K, Parast M, Alberico C, Benian GM, Ono S (2003) Specific requirement for two ADF/cofilin isoforms in distinct actin-dependent processes in *Caenorhabditis elegans*. J Cell Sci 116:2073–2085
15. Mohri K, Ono K, Yu R, Yamashiro S, Ono S (2006) Enhancement of actin-depolymerizing factor/cofilin-dependent actin disassembly by actin-interacting protein 1 is required for organized actin filament assembly in the *Caenorhabditis elegans* body wall muscle. Mol Biol Cell 17:2190–2199
16. Duerr JS (2006) Immunohistochemistry WormBook ed the *C elegans* research community WormBook. PubMed PMID: 18050446; PMCID: PMC4780882; https://doi.org/10.1895/wormbook.1.105.1
17. Finney M, Ruvkun G (1990) The *unc-86* gene product couples cell lineage and cell identity in *C elegans*. Cell 63:895–905
18. Shakes DC, Miller DM, Nonet ML (2012) Immunofluorescence microscopy. Methods Cell Biol 107:35–66
19. Wilson KJ, Qadota H, Benian GM (2012) Immunofluorescent localization of proteins in *Caenorhabditis elegans* muscle. Methods Mol Biol 798:171–181
20. Riedl J, Crevenna AH, Kessenbrock K, Yu JH, Neukirchen D, Bista M, Bradke F, Jenne D, Holak TA, Werb Z, Sixt M, Wedlich-Soldner R (2008) Lifeact: a versatile marker to visualize F-actin. Nat Methods 5:605–607

21. Burkel BM, von Dassow G, Bement WM (2007) Versatile fluorescent probes for actin filaments based on the actin-binding domain of utrophin. Cell Motil Cytoskeleton 64:822–832
22. Edwards KA, Demsky M, Montague RA, Weymouth N, Kiehart DP (1997) GFP-moesin illuminates actin cytoskeleton dynamics in living tissue and demonstrates cell shape changes during morphogenesis in drosophila. Dev Biol 191:103–117
23. Aizawa H, Sameshima M, Yahara I (1997) A green fluorescent protein-actin fusion protein dominantly inhibits cytokinesis cell spreading and locomotion in *Dictyostelium*. Cell Struct Funct 22:335–345
24. Willis JH, Munro E, Lyczak R, Bowerman B (2006) Conditional dominant mutations in the *Caenorhabditis elegans* gene *act-2* identify cytoplasmic and muscle roles for a redundant actin isoform. Mol Biol Cell 17:1051–1064
25. Stone S, Shaw JE (1993) A *Caenorhabditis elegans act-4::lacZ* fusion: use as a transformation marker and analysis of tissue-specific expression. Gene 131:167–173
26. MacQueen AJ, Baggett JJ, Perumov N, Bauer RA, Januszewski T, Schriefer L, Waddle JA (2005) ACT-5 is an essential *Caenorhabditis elegans* actin required for intestinal microvilli formation. Mol Biol Cell 16:3247–3259
27. Victor Ambros. (2006) Shaham S (ed) WormBook: Methods in Cell Biology (January 02, 2006), WormBook, ed. The C. elegans Research Community, WormBook, https://doi.org/10.1895/wormbook.1.49.1, http://www.wormbook.org.
28. McCarter J, Bartlett B, Dang T, Schedl T (1997) Soma-germ cell interactions in *Caenorhabditis elegans*: multiple events of hermaphrodite germline development require the somatic sheath and spermathecal lineages. Dev Biol 181:121–143
29. Dong L, Cornaglia M, Krishnamani G, Zhang J, Mouchiroud L, Lehnert T, Auwerx J, Gijs MAM (2018) Reversible and long-term immobilization in a hydrogel-microbead matrix for high-resolution imaging of *Caenorhabditis elegans* and other small organisms. PLoS One 13:e0193989
30. Hwang H, Barnes E, Matsunaga Y, Benian GM, Ono S, Lu H (2016) Muscle contraction phenotypic analysis enabled by optogenetics reveals functional relationships of sarcomere components in *Caenorhabditis elegans*. Sci Rep 6:19900
31. Krajniak J, Lu H (2010) Long-term high-resolution imaging and culture of C elegans in chip-gel hybrid microfluidic device for developmental studies. Lab Chip 10:1862–1868
32. Lengsfeld AM, Low I, Wieland T, Dancker P, Hasselbach W (1974) Interaction of phalloidin with actin. Proc Natl Acad Sci U S A 71:2803–2807
33. Wulf E, Deboben A, Bautz FA, Faulstich H, Wieland T (1979) Fluorescent phallotoxin a tool for the visualization of cellular actin. Proc Natl Acad Sci U S A 76:4498–4502
34. Strome S (1986) Fluorescence visualization of the distribution of microfilaments in gonads and early embryos of the nematode *Caenorhabditis elegans*. J Cell Biol 103:2241–2252
35. Waterston RH, Hirsh D, Lane TR (1984) Dominant mutations affecting muscle structure in *Caenorhabditis elegans* that map near the actin gene cluster. J Mol Biol 180:473–496
36. Ono S (2001) The *Caenorhabditis elegans unc-78* gene encodes a homologue of actin-interacting protein 1 required for organized assembly of muscle actin filaments. J Cell Biol 152:1313–1319
37. Hayashi Y, Ono K, Ono S (2019) Mutations in *Caenorhabditis elegans* actin which are equivalent to human cardiomyopathy mutations cause abnormal actin aggregation in nematode striated muscle. F1000Res 8:279
38. Barnes DE, Hwang H, Ono K, Lu H, Ono S (2016) Molecular evolution of troponin I and a role of its N-terminal extension in nematode locomotion. Cytoskeleton 73:117–130
39. Stiernagle T (2006) Maintenance of *C elegans*. WormBook ed The *C elegans* Research Community, PubMed PMID: 18050451; PMCID: PMC4781397 https://doi.org/10.1895/wormbook.1.101.1
40. Hall DH, Altun ZF (2008) *C elegans atlas*. Cold Spring Harbor Laboratory Press, Cold Spring Harbor, NY
41. Lessard JL (1988) Two monoclonal antibodies to actin: one muscle selective and one generally reactive. Cell Motil Cytoskeleton 10:349–362
42. Porta-de-la-Riva M, Fontrodona L, Villanueva A, Ceron J (2012) Basic *Caenorhabditis elegans* methods: synchronization and observation. J Vis Exp 10:e4019
43. Tse YC, Werner M, Longhini KM, Labbe JC, Goldstein B, Glotzer M (2012) RhoA activation during polarization and cytokinesis of the early *Caenorhabditis elegans* embryo is differentially dependent on NOP-1 and CYK-4. Mol Biol Cell 23:4020–4031
44. Zilberman Y, Abrams J, Anderson DC, Nance J (2017) Cdc42 regulates junctional actin but not cell polarization in the *Caenorhabditis elegans* epidermis. J Cell Biol 216:3729–3744
45. Higuchi-Sanabria R, Paul JW 3rd, Durieux J, Benitez C, Frankino PA, Tronnes SU, Garcia G, Daniele JR, Monshietehadi S, Dillin

A (2018) Spatial regulation of the actin cytoskeleton by HSF-1 during aging. Mol Biol Cell 29:2522–2527

46. Xu S, Chisholm AD (2011) A Gαq-Ca^{2+} signaling pathway promotes actin-mediated epidermal wound closure in *C elegans*. Curr Biol 21:1960–1967
47. Shivas JM, Skop AR (2012) Arp2/3 mediates early endosome dynamics necessary for the maintenance of PAR asymmetry in *Caenorhabditis elegans*. Mol Biol Cell 23:1917–1927
48. Schindler AJ, Sherwood DR (2011) The transcription factor HLH-2/E/daughterless regulates anchor cell invasion across basement membrane in *C elegans*. Dev Biol 357:380–391
49. Wirshing ACE, Cram EJ (2017) Myosin activity drives actomyosin bundle formation and organization in contractile cells of the *Caenorhabditis elegans* spermatheca. Mol Biol Cell 28:1937–1949
50. Szumowski SC, Estes KA, Popovich JJ, Botts MR, Sek G, Troemel ER (2016) Small GTPases promote actin coat formation on microsporidian pathogens traversing the apical membrane of *Caenorhabditis elegans* intestinal cells. Cell Microbiol 18:30–45

Chapter 8

Imaging of the Cytoskeleton Using Live and Fixed *Drosophila* Tissue Culture Cells

Derek A. Applewhite, Christine A. Lacy, Eric R. Griffis, and Omar A. Quintero-Carmona

Abstract

In recent years, the convergence of multiple technologies and experimental approaches has led to the expanded use of cultured *Drosophila* cells as a model system. Their ease of culture and maintenance, susceptibility to RNA interference, and imaging characteristics have led to extensive use in both traditional experimental approaches and high-throughput RNAi screens. Here we describe *Drosophila* S2 cell culture and preparation for live-cell and fixed-cell fluorescence microscopy and scanning electron microscopy.

Key words Fluorescence microscopy, S2 cells, Transfection, Microtubule, Actin, Electron microscopy

1 Introduction

The marriage of *Drosophila* tissue culture and cytoskeletal research has been a fruitful one. Numerous cytoskeletal proteins have been discovered using S2 cells [1], which is a testament to their practicality. Major advances in our understanding of cytoskeletal dynamics and function have been achieved using S2 cells, an indication of their versatility and strength as a model system [2–10]. *Drosophila* cells are cultured at room temperature in a medium that does not require buffering with CO_2. RNAi is both effective and easily administered [11]. *Drosophila* cells do not produce an interferon response to dsRNAs. Therefore, dsRNAs can be produced in vitro, eliminating the need to buy siRNAs. S2 cells readily incorporate dsRNAs when added to tissue culture medium. dsRNAs can also be generated against the 5′ and 3′ untranslated regions (UTRs) allowing for the depletion of endogenous target genes and testing of exogenous genes for their ability to rescue RNAi phenotypes. RNAi

Ray H. Gavin (ed.), *Cytoskeleton: Methods and Protocols*, Methods in Molecular Biology, vol. 2364,
https://doi.org/10.1007/978-1-0716-1661-1_8,

protocols can involve as few as 3 days, depending upon the turnover of a given protein, and frequently lead to substantial protein knockdown [12]. The *Drosophila* genome represents a pared down form of the mammalian genome, and while approximately 75% of human disease causing genes are retained in the *Drosophila* genome, there is far less functional redundancy [13–15]. Whereas the human or mouse genome may have three genes that are functionally interchangeable, the *Drosophila* genome will often have one. Collectively, these qualities have led to hundreds of RNAi screens of the full *Drosophila* genome at a fraction of the cost of screens preformed in mammalian tissue culture systems.

The usefulness of *Drosophila* S2 cells as tools for cytoskeletal research goes beyond their use in RNAi screens. S2 cells in particular adopt a flat, fried-egg-like morphology when plated on a concanavalin A (Con A)-coated surface [16]. This conformation makes them highly amenable to high-resolution microscopy [17], and their sessile nature eliminates the problems produced by motile cells. S2 cells form a circumferential actin-rich lamellipodium with characteristic fast actin dynamics where proteins such as the Arp2/3 complex, Capping Protein, and Ena/VASP proteins localize [5]. This lamellipodium is followed by a contractile network highly reminiscent of a lamella with its own characteristic slower actin dynamics and signature proteins such as tropomyosin, non-muscle myosin II, and alpha-actinin (Fig. 1, [18, 19]). This is even more evident as these actin-rich zones of S2 cells display similar phenotypes to those of migrating cells following RNAi depletion of these key actin-binding proteins [5, 20–22]. Thus, despite being non-motile, the behavior and dynamics of the actin-rich cell periphery of S2 cells mimics that of migrating cells. S2 cells have also been quite useful in studying microtubule dynamics. Our current understanding of spindle formation, in particular poleward flux, kinetochore attachments, and the role of molecular motors during mitosis owe much to experiments performed in S2 cells [16, 23–30]. In addition, the acentriolar interphase microtubule array offers a unique system to study with great detail the dynamics of the microtubule plus-end and minus-end as well as interactions between the actin and microtubule cytoskeletons [16, 31–36].

S2 cells are easily transfected with tagged protein (fluorescent or otherwise) but are naturally non-adherent and must be induced to attach to a substrate. If the cells have not been attached to a substrate, they will be washed away during fixation, permeabilization, and staining. There are two main coatings that will induce cell attachment, concanavalin A (Con A) and polylysine. These two coatings differ in how the cell responds to them. Con A is a lectin (sugar-binding protein) that is commonly purified from the jack bean (*Canavalia ensiformis*), and it presumably induces cell spreading in a Rac and Arp2/3-dependent manner as cells try to engulf

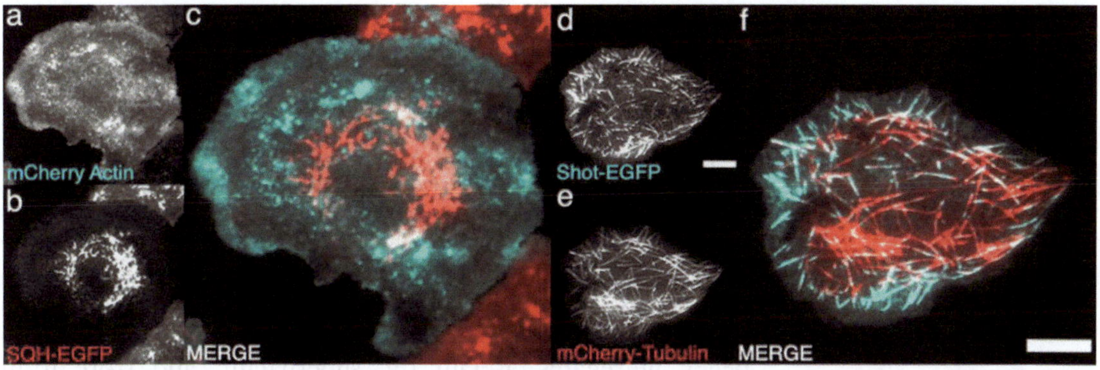

Fig. 1 Live imaging of cytoskeletal probes in *Drosophila* S2 cells. (**a**–**f**) Shown is a single time point from a live-cell image sequence of S2 cells imaged by TIRF-M. (**a**) An S2 cell co-expressing mCherry-Actin and (**b**) the regulatory chain of non-muscle myosin II (Spaghetti-Squash, Sqh), tagged with EGFP. (**c**) Shown at higher magnification, the merged image of (**a**) and (**b**) where mCherry actin pictured in (**a**) is in cyan and EGFP-Sqh pictured in (**b**) is in red. (**d**) An S2 cell co-expressing EGFP-Short stop (Shot) and (**e**) mCherry-α-Tubulin. (**f**) The merged image of (**d**) and (**e**) where Shot-EGFP pictured in (**d**) is shown in cyan and mCherry-α-Tubulin pictured in (**e**) is shown in red. Scale bars in both low and high magnification images are 10μm

the Con A-coated substrate [5]. Due to the strong inducement of spreading as well as the engagement of the phagocytic machinery to the basal cortex of the cell, many membrane trafficking events (including cytokinesis) are disrupted in Con A-plated cells, and these side effects should be considered when deciding whether to use Con A to induce adherence. Polylysine is a positively charged polymer of lysine that electrostatically interacts with negatively charged cell membranes and induces cell attachment and mild amounts of cell spreading. Polylysine is available as both L and D forms and available in a range of molecular weights. We have found that *Drosophila* cells need higher concentrations of polylysine than mammalian cells to attach to the substrate and they tend to be highly sensitive to the oxidizing effects of thimerosal. Therefore, it is not recommended to purchase premade polylysine that uses this chemical as a preservative.

2 Materials

2.1 Culturing and Transient Transfection of Drosophila *S2 Cells*

1. Schneider's *Drosophila* Medium (commercially available).
2. Fetal Bovine Serum, heat inactivated.
3. 100× Antibiotic/Antimycotic mix.
4. pMT/V5 His, pIZ/V5, and/or pAc5.1 vectors (available commercially).
5. $CuSO_4$: 100 mM aqueous solution.
6. FugeneHD (Promega).

7. Amaxa Kit V (Lonza) (optional).
8. Nucleofector Machine (Lonza) (optional).

2.2 Fixation and Staining of Drosophila S2 Cells for Immunofluorescence

1. 1 mg/mL polylysine: Dissolve Poly-D-Lysine or Poly-L-Lysine (70,000–150,000 average molecular weight) in 50 mM Tris–HCl, pH 8.0. Aliquot and freeze at −20°C.
2. 0.5 mg/mL concanavalin A (Con A): Dissolve Con A (Type IV-S, tissue culture grade) at 0.5 mg/mL in distilled water. Filter the Con A solution through a 0.22 μm filter into a fresh tube or centrifuge at 18,000 × *g* (14,000 RPM) for 5 min to pellet aggregates. Retain the supernatant, and store it in 0.5 mL aliquots in 1.5 mL tubes at −20°C.
3. 1 M Tris–HCl, pH 8.0.
4. #1.5 Glass coverslips, glass-bottom or optical plastic multi-well plates.
5. Humidified chambers.
6. Paraformaldehyde: EM Grade packed in ampoules or made fresh.
7. Glutaraldehyde: EM Grade.
8. PHEM Buffer: 10 mM EGTA, 2 mM $MgCl_2$, 60 mM PIPES, 25 mM HEPES, pH 6.9. This buffer can be made as a 2× stock solution, autoclaved, and stored until use.
9. Sodium borohydride: aqueous solution 1 mg/mL.
10. Saponin: 10 mg/mL aqueous stock solution.
11. Donkey or goat serum.
12. Bovine serum albumin.
13. Triton X-100.
14. Mounting medium: Dako, ProLong (Life Technologies), Fluoromount (Sigma), VectaShield (Vector Laboratories), or a solution of 90% glycerol, 20 mM sodium bicarbonate, 3% *n*-propyl gallate.
15. CoverGrip (Biotium) or nail polish.

2.3 Live-Cell Imaging

1. #1.5 glass-bottom dishes.
2. Titanium Step Drill Bit-1/4″ to 1–3/8′ Increments, 10 steps (Neiko) (optional).
3. 120 V 10-in. Drill Press (optional).
4. Norland Optical Adhesive 81 (Norland) (optional).
5. UV transilluminator (optional).
6. Cell-tak (Corning) (optional).
7. Polylysine (optional).

2.4 Fixation and Preparation of* Drosophila *S2 Cells for Scanning Electron Microscopy

1. Silver paint.
2. Double-sided conductive tape.
3. Critical point dryer.
4. Coverslip holder for critical point dryer.
5. 200 proof ethanol.
6. Sputter coater.

3 Methods

3.1 Culture of* Drosophila *S2 Cells (See Note 1)

1. Maintain *Drosophila* S2 cells at room temperature (preferably between 19 °C and 27 °C) and in semi-suspension in 25 cm or 75 cm flasks with plug-end caps.
2. Passage cells by pipetting them from the culture medium, making sure that the cells are adequately resuspended and that any cells lightly adhered to the tissue culture plastic are removed. Generally, a 1:4 split of the cells into fresh cell culture medium will yield a cell density that will require passage every 3–4 days. It is important to note that S2 cell viability decreases when the cell density falls below 5×10^5 cells/mL.

3.2 Transient Transfection of* Drosophila *S2 Cells (See Notes 2 and 3)

1. Plate S2 cells at 40–80% confluency in a 6- or 12-well tissue culture plate 15–30 min prior to transfection. Alternatively, plate cells a day in advance of transfection. This step is needed as the transfection protocol calls for the exchange of medium, and the cells will need this time to loosely adhere to the tissue culture plastic.
2. If using the FugeneHD transfection reagent (Promega) (*see* **Note 3**), prepare the DNA-FugeneHD complex by diluting the desired amount of DNA in water or serum-free Schneider's medium to a final volume of 100 μL. Next, add 6–8 μL of FugeneHD reagent directly to the diluted DNA with special care taken to avoid pipetting the reagent down sides of the microfuge tube. Incubate the DNA-FugeneHD mixture for 15–20 min at room temperature.
3. While the incubation is in progress, gently remove the cell culture medium from the tissue culture plate, taking precaution not to disturb the weakly adherent cells. Replace the cell culture medium with 900 μL of fresh cell culture medium. Add the 100 μL DNA-FugeneHD complex drop-wise to the 900 μL of medium containing the cells. The transfection is carried out overnight, and the medium is replaced with fresh medium the following day.
4. Alternatively, transiently transfect S2 cells by using electroporation systems such as Amaxa nucleofector from Lonza. The Amaxa nucleofector kits are tailored to specific cell lines. The manufacturer suggests Kit V for *Drosophila* S2.

5. The Amaxa kit provides a nucleofection solution and a supplement solution. Mix these two solutions at a ratio of 4.5:1 prior to use. The mixed nucleofector reagent can be stored at 4 °C for up to 3 months.
6. Prepare DNA for electroporation by diluting the desired amount of DNA, typically between 0.1 and 2 μg of DNA, in the nucleofector reagent to a final volume of 100 μL in a microfuge tube. This can be prepared and set aside until the cells are ready.
7. Determine the appropriate volume of cells and pellet them by gentle centrifugation at 300 × *g* at room temperature for 5–10 min.
8. Next, remove the supernatant (tissue culture medium) (*see* **Note 4**).
9. Resuspend the pellet of cells in the diluted DNA-nucleofector mixture, and transfer the mixture to the electroporation cuvette (*see* **Note 5**).
10. After electroporation, transfer cells to fresh medium, and allow them to recover prior to $CuSO_4$ induction if using pMT/His-5 vectors (*see* **Note 6**).

3.3 Fixation and Staining of Drosophila S2 Cells for Immunofluorescence (See Note 7)

3.3.1 Preparing the Surface of Coverslips (See Notes 8 and 9)

1. For flame cleaning, use fine-tipped forceps to grab an individual coverslip, and dip it into absolute ethanol. Allow most of the ethanol to run off the edge of the coverslip, and then wave the coverslip through an open flame. As the ethanol burns, continue to tilt and move the coverslip to prevent the ethanol from pooling in a single corner and superheating the glass.
2. Place flamed coverslips in a rack and rinse thoroughly in distilled water. Then place the coverslip rack over a heat block to allow them to dry thoroughly. Cover and store to block dust accumulation.

3.3.2 Con A Coating of Coverslips

1. Cover the bottom of a broad shallow chamber with Parafilm, and then place the cleaned coverslips onto the Parafilm.
2. For 22 × 22 mm coverslips, put a 50 μL drop of Con A on the center of the coverslip, and then evenly smear the drop over the surface of the glass.
3. For 18 × 18 mm coverslips, use a 35 μL drop.
4. Place the coverslips in an out-of-the-way location where they will not accumulate dust, and allow to dry overnight.

3.3.3 Con A Coating of 96-Well Plates

1. Dilute Con A to 0.05 mg/mL.
2. Add 80 μL of Con A to each well and allow to dry over at least 2 days.

3.3.4 Polylysine Coating of Coverslips

1. Place coverslips into the Parafilm-lined chamber.
2. Spread 50 μL of polylysine onto coverslips and let sit at room temperature for 3 h.
3. Rinse coverslips three times with distilled deionized water, and then aspirate excess water.
4. Allow the coverslips to dry, and then store at 4 °C until needed.

3.3.5 Coating of Coverslip-Bottom Imaging Chamber

1. Add a 100 μL drop of Con A or polylysine to the center of the coverslip, and spread it evenly over the surface.
2. Use the tip of a p100 pipette to remove any excessive fluid accumulation that might pool in the interface between the glass bottom and plastic side.
3. Allow Con A-coated chambers to dry overnight. Allow lysine-coated chambers to sit for 3 h before rinsing them three times with distilled-deionized water. Dry them by aspirating any leftover fluid and then store them at 4 °C.

*3.3.6 Plating of Cells (See **Note 10**)*

1. Resuspend cells that have been growing in a 96-well plate or 6-well dish.
2. Place prepared coverslips into a 6-well plate or 35 mm dish.
3. Place a drop of 100–200 μL of fresh medium onto the center of the coverslip.
4. Place a drop of the cell suspension into the fresh medium and pipette to mix. This step is very empirical. Allow the cells to settle to the surface and observe their density. Initially, add fewer cells to the coverslip, plate, or chamber than you think you will need. After 20 min the cells can be checked for density, and more cells can be added to achieve the desired density.
5. Add additional medium to the wells after the cells have attained the desired density and have begun to attach to the coverslip.
6. Initiate fixation, observation, or further manipulation after 1 h.

*3.3.7 Fixation and Staining (See **Note 11**)*

1. Remove the culture medium from each well of the 6-well plate or 35 mm dish, and fix cells for 10 min by adding in a solution of 0.3% glutaraldehyde, 4% formaldehyde in 1x PHEM supplemented with saponin (1 mg/mL).
2. Remove fixative and incubate cells with PHEM-T (1× PHEM + 0.5% Triton X-100) for 5 min.
3. Rinse cells in PHEM.
4. Reduce unreacted aldehydes in the sample by three treatments for 10 min with a solution of 1 mg/mL sodium borohydride dissolved in PHEM (make immediately before use).
5. Wash the coverslips three times in PHEM-Wash (1× PHEM + 0.1% Triton X-100)—5 min per wash.

6. Block nonspecific antibody binding sites and reduce background by incubating cells in a blocking solution (PHEM-Wash + 5% Boiled donkey serum) for 30 min to 1 h at room temperature (*see* **Note 12**).
7. Incubate cells in primary antibody for 1 h at room temperature or overnight at 4 °C if antibody staining is desired. Dilute antibodies to the appropriate concentration with blocking solution; determine concentrations empirically for each antibody. Dot 50 μL drops of antibody solution onto a Parafilm sheet in a humidified chamber, and place a coverslip on top of each drop. Ensure that the cell-adhering side of the coverslip makes contact with the drop of antibody solution.
8. Wash cells 4× 5 min in PHEM-Wash.
9. Perform secondary antibody incubation as described for primary antibody using fluorescent secondary antibodies and additions such as DAPI or fluorescent-labeled phalloidin diluted in blocking buffer.
10. Wash coverslips 4× 5 min in PHEM-Wash.
11. Rinse cells 1× in PHEM and 1× in distilled water to remove salts.
12. Wick away excess fluid from coverslips and then mount them (cell side down) on slides using 1 drop of a mounting media.
13. Use vacuum aspiration to remove any excess mounting media from around the edges of the coverslips, and then seal them with a hardening material such as nail polish, VALAP, or CoverGrip (Biotium).

3.4 Live-Cell Imaging (See Note 13)

1. Plate S2 cells on glass-bottom dishes (*see* **Note 14**) that have been coated with a lectin solution, concanavalin A (con A). Con A should be applied to the glass surface and allowed to incubate for 1–2 min.
2. Remove con A, carefully ensuring that the reagent is completely removed, and allow the dishes to dry before the addition of cell culture medium containing cells. Allow 30–60 min for S2 cells to fully attach and spread on the con A-coated surface.
3. Alternatively, plate S2 plate on polylysine or Cell-tak (Corning Cat. No. 354240)-coated glass. See manufacturer's instructions for coating glass-bottom dishes with Cell-tak.
4. Perform imaging in Schneider's Insect Medium supplemented with 10% heat-inactivated FBS and antibiotic. Imaging is conveniently carried out at room temperature without CO_2.

3.5 Fixation and Preparation of *Drosophila* S2 Cells for Scanning Electron Microscopy

3.5.1 Fixation (See Note 15)

1. Equilibrate fixative to the same temperature as the cell culture. Remove the cell medium from the coverslips, and replace it with a solution of 4% glutaraldehyde with 4% paraformaldehyde in 0.1 M phosphate-buffered saline.
2. Incubate the sample in the fixative for at least 30 min at room temperature.
3. Store samples for a maximum of 2 weeks at 4 °C if desired.

3.5.2 Dehydration and Critical Point Drying (See Note 16)

1. Wash the sample three times for 5 min in distilled water to remove fixative.
2. Exchange distilled water for 30% ethanol for 5 min, and continue through dehydration with 5 min steps in 50%, 70%, 90%, 100% ethanol, using three washes at each ethanol concentration.
3. Under 100% ethanol, stack coverslips in a CPD holder with a washer between each.
4. Fill the CPD chamber with enough dehydrated ethanol to cover the holder. Place the holder in the chamber. Operate the CPD as directed by the manufacturer's instructions.

3.5.3 Mounting and Sputter Coating (See Note 17)

1. Remove dry coverslips from the CPD chamber, and adhere onto SEM stubs with double-sided conductive tape.
2. To further facilitate conductivity, apply silver paint to create a bridge from the surface of the coverslip to the conductive tape or SEM stub.
3. After the silver paint has dried, place the samples in a sputter coater chamber.
4. Operate the sputter coater according to manufacturer's instructions.
5. Image the coverslips in a SEM.

4 Notes

1. S2 cells (Fig. 2) can be maintained in either Schneider's *Drosophila* medium (available from many suppliers) supplemented with 10% FBS and antibiotics or serum free media that are designed for SF9 cell culture. However, live-cell imaging should be performed in Schneider's medium as serum-free media both produce background fluorescence and contain trace amounts of divalent ions, which tend to induce expression from pMT/V5-His vectors. S2 do not require CO_2, and they are tolerant of and perform well at high cell densities. Cells should be passaged when they become confluent. S2 cells are

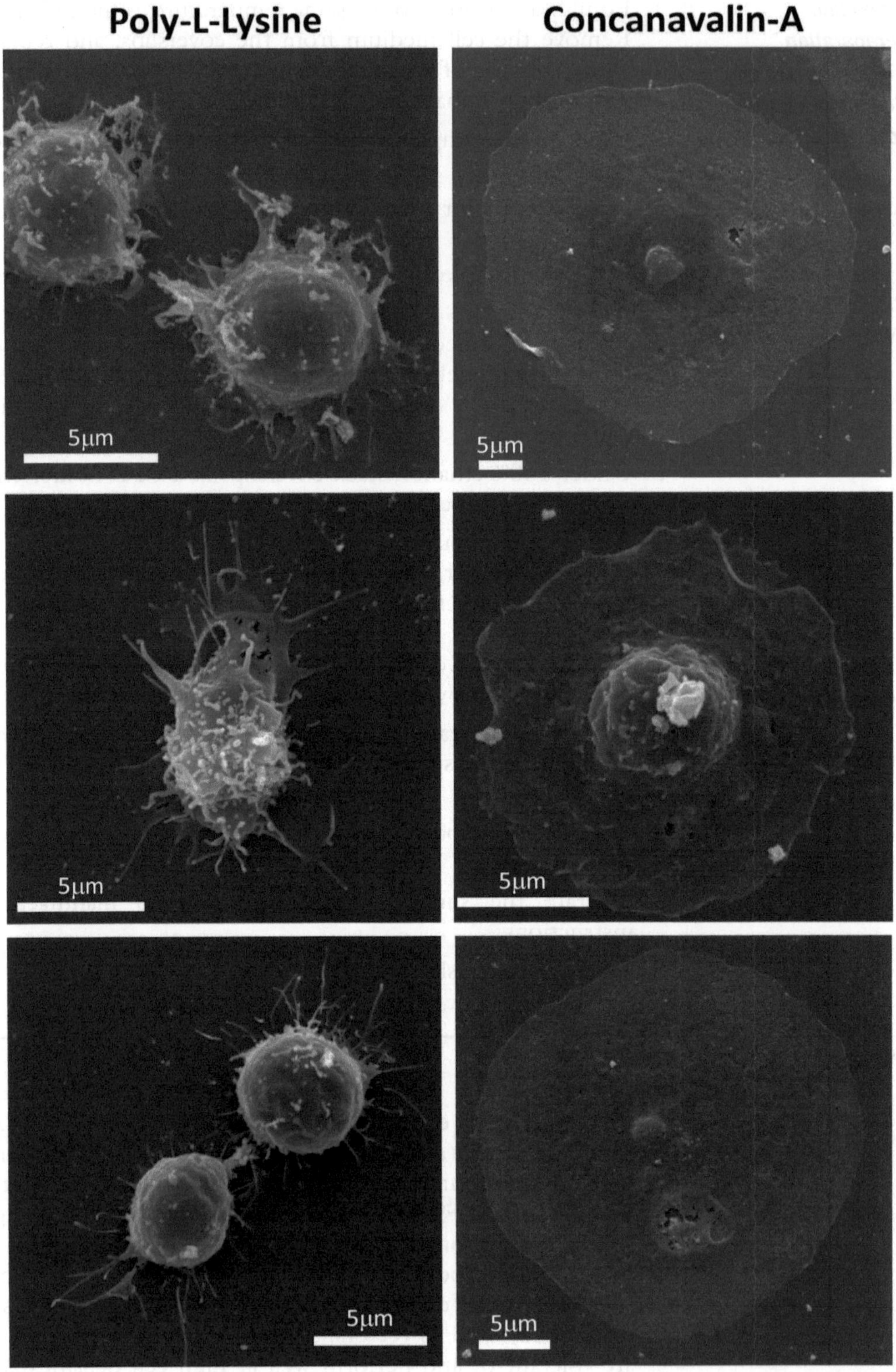

Fig. 2 Scanning electron micrography of *Drosophila* S2 cells. Shown are representative images of S2 cells allowed to adhere to glass coverslips coated with either Poly-L-Lysine, 1mg/mL for 3 hours. Con-A, 0.5mg/mL

loosely adherent on tissue culture plastic that has been treated for cell culture and thus do not require trypsin/EDTA treatments.

2. The amount of plasmid DNA transfected depends on the cytoskeletal protein being imaged and the *Drosophila* expression vector used. pMT/V5 His vectors that are copper inducible are typically transfected at a range of 0.1–2 μg DNA. An advantage of using pMT/V5-His vectors is that the amount of protein expressed can be titrated with the amount of $CuSO_4$ supplemented in the tissue culture medium prior to imaging. Successful imaging experiments have been carried out using vectors that have constitutively active promoters such as the pIZ/V5 vector, which uses the OpIE2 promoter, and the pAc5.1/V5 which uses the *Drosophila* actin 5C promoter. However, it is often best to titrate the amount of DNA transfected in order to achieve optimal protein expression levels with these vectors. It is also possible to transfect in vectors containing the UAS promoter, which are usually generated for producing transgenic flies. To work with UAS vectors, one can co-transfect a pMT-Gal4 to allow for inducible expression of the UAS promoter. For live-cell imaging of fluorescently tagged cytoskeletal proteins, good results have been achieved by an overnight induction of 25–600 μM $CuSO_4$. It is particularly important to use the low end of this range of $CuSO_4$ when imaging microtubule plus-end tracking proteins (+TIPS) that are sensitive to mis-localization as a consequence of overexpression. Expression of other cytoskeletal proteins such as tubulin, actin, or actin-binding proteins that are less sensitive to the artifacts generated by overexpression can be achieved by overnight induction at the higher end of this range.

   Transient transfection of cytoskeletal proteins is generally carried out 1–2 days prior to live-cell imaging. The exact timing is highly dependent on the vector and the expression of the protein of interest, and an effort should be made to optimize this timing. For visualizing mitosis, it is usually best to wait 3 or more days before imaging. Mitosis is a rare event in a cell's lifetime, and only 5% of cells will be in mitosis at any point in time. If the transfection efficiency is 50%, a culture will only have 2.5% of cells transfected and in mitosis at any point during imaging. For this reason, stably transfected cells are usually preferred for studies of mitosis.

Fig. 2 (continued) overnight until dry.. Cells were allowed to adhere for 1 h prior to fixation and processing for SEM. While the size and surface morphology for cells was somewhat variable, nearly all cells grown on concanavalin A-coated coverslips displayed a smooth dorsal surface morphology with few projections, had a smooth edge morphology, and were well-spread. Scale bars are 5μm

3. We have achieved good results using FugeneHD transfection reagent (Promega). It is always best to follow the manufacture's instructions while preparing both the plasmid DNA and cells for electroporation. Lonza recommends 2×10^6 cells per transfection and from 0.1 to 2 μg of DNA per transfection.
4. It is important to remove as much tissue culture medium as possible without disturbing the pellet as residual media can interfere with the electroporation.
5. The electroporation machine provided by the Lanza system contains a preset S2 cell protocol, G-030 (or G-30 for Nucleofector I device).
6. Electroporation does lead to cell death, but in general if the protocol is adhered to, the amount of cell death does not interfere with the progress of the experiment. Very high transfection efficiencies can be achieved with electroporation; however, transfection efficiency is highly dependent on plasmid DNA quality and the health of the cells prior to electroporation.
7. Drosophila S2 cells cannot be fixed and stained like mammalian cells. They are nonadherent and must be induced to adhere to coverslips or multiwall plates; they do not tolerate low-density cultures and therefore must be seeded carefully in order to function optimally. And they do not tolerate hypertonic buffers and therefore require specific buffers for washing and fixation.
8. In order to plate cells on glass, the hydrophobic coating found on most coverslips must be removed. There are several ways to do this: plasma cleaning, hydrogen peroxide + sulfuric acid soaking, and flame cleaning. Flame cleaning does not produce any chemical waste that requires storage and disposal or requires a special piece of equipment.
9. S2 cells are only weakly adherent to glass surfaces, which must be functionalized in order to keep cells immobilized for fixation and staining.
10. S2 cells are extremely sensitive to plating density. If plated at too low of a concentration, they will not form extensive microtubule networks. Additionally, if one wishes to visualize mitotic cells, a too low density will make it very difficult to observe enough cells. However, plating at a too high density will inhibit their spreading and lead to cells stacking upon each other, resulting in out of focus noise in images. It is advisable to empirically determine how many cells to plate for any given application.
11. Many existing protocols for cell fixation work well for S2 cells. Several methods may need to be tested to find one that is compatible with both fixing the structure you wish to observe

and that preserves the antigenicity of the molecules you wish to localize. Fixatives that work for S2 cells include −80 °C methanol, a mixture of methanol and formaldehyde (90% methanol, 3.2% formaldehyde, 5 mM sodium bicarbonate, pH 9 (from a 0.5 M stock) chilled to −80 °C prior to fixation), 4–10% formaldehyde, and ice-cold acetone. The protocol in Subheading 3.3.5 uses a common fixation method that contains a mild detergent to give simultaneous fixation and permeabilization. Simultaneous fixation/permeabilization works well for many antigens and structures as the detergent speeds up the access of the fixative while allowing some loosely bound protein to leave the cells, thereby reducing background. Alternatively, fixation followed by permeabilization or rapid fixation (10–60 s) followed by permeabilization and then 10 min of a second fixation can be utilized as well.

12. We use donkey serum and the highly cross-subtracted donkey secondary antibodies from Jackson Immunoresearch) or 1× TBS + 0.1% Triton X-100 + 2% BSA).

13. Imaging is performed on an inverted microscope. Both the actin and microtubule cytoskeletons are sensitive to phototoxicity. Exposure to excess fluorescent light can quickly lead to a slowing or altering of the natural filament dynamics. Imaging should be performed to minimize the fluorescent light exposure. Both spinning-disk confocal and total internal reflection (TIRF) microscopy lend themselves to limiting the amount of light exposure while providing a good signal-to-noise ratio ideal for imaging proteins that are expressed at low levels. The speed of image acquisition by both these types of microscopy is also ideal for capturing the dynamics of the actin and microtubule cytoskeletons of live cells.

14. Glass bottom dishes can be made using a step drill bit (Neiko Titanium Step Drill Bit- 1/4″ to 1–3/8″ Increments, 10 steps) installed in an upright drill press (SKIL 3320-01 120 V 10-in. Drill Press); however, commercial dishes with their superior flatness are highly recommended for TIRF. The bit is used to bore a circular opening in the bottom of a 35 mm plastic dish 1.5 to 2 cm in diameter. Coverslips can then be glued to the dish, using any number of quick-drying plastic compatible adhesives. We prefer an ultraviolet curing adhesive available from Norland (Norland Optical Adhesive 81) which quickly cures following a brief exposure (5–10 min) to ultraviolet light from a transilluminator.

15. Chemical fixation halts deterioration of the sample by crosslinking proteins. This crosslinking also provides some support for dehydration and critical point drying. Chemical fixatives are hazardous. Gloves and a fume hood should be used during fixation.

16. Samples must be dry to withstand the high vacuum of a scanning electron microscope. Air-drying creates enormous surface tension that causes samples to crumple losing their original surface shapes. To preserve a sample's structure, dehydration in ethanol and critical point drying (CPD) in carbon dioxide are used to dry the sample. During CDP 100% ethanol, which is miscible with water and carbon dioxide, is exchanged for liquid carbon dioxide. Heat and pressure are applied to the sample to reach the point where liquid carbon dioxide changes into gaseous carbon dioxide (the critical point). The sample is now dry. Having never been exposed to surface tension, the original structures are preserved.
17. Scanning electron microscopy (SEM) requires conductivity throughout the system. Biological samples lose their conductivity once water is removed. Mounting samples on aluminum stubs facilitate conductivity from the SEM stage to the sample and provide a stable surface. Conductivity must continue across the surface of the sample. This requires the deposition of a thin layer of metal. Sputter coating uses a heavy inert gas (typically Argon) to create a plasma field around a metal target. The plasma field erodes the target so that metal atoms are ejected onto the surface of the sample.

References

1. Schneider I (1972) Cell lines derived from late embryonic stages of *Drosophila melanogaster*. J Embryol Exp Morphol 27:353–365
2. Somma MP, Fasulo B, Cenc G, Cundari E, Gatti M (2002) Molecular dissection of cytokinesis by RNAi interference in drosophila tissue culture cells. Mol Biol Cell 13:2448–2460
3. Pearson AM, Baksa K, Rämet M, Protas M, McKee M, Brown D, Ezekowitz RA (2003) Identification of cytoskeletal regulatory proteins required for efficient phagocytosis in drosophila. Microbes Infect 10:815–824
4. Kiger AA, Baum B, Jones S, Jones MR, Coulson A, Echeverri C, Perrimon N (2003) A functional genomic analysis of cell morphology using RNA interference. J Biol 2:27
5. Rogers SL, Wiedemann U, Stuurman N, Vale RD (2003) Molecular requirements for actin-based lamella formation in drosophila S2 cells. J Cell Biol 162:1079–1088
6. Eggert US, Kiger AA, Richter C, Perlman ZE, Perrimon N, Mitchison TJ, Field CM (2004) Parallel chemical genetic and genome-wide RNAi screens identify cytokinesis inhibitors and targets. PLoS Biol 12:e379
7. Goshima G, Wollman R, Goodwin SS, Zhang N, Scholey JM, Vale RD, Stuurman N (2007) Genes required for mitotic spindle assembly in drosophila S2 cells. Science 316:417–421
8. D'Ambrosio MV, Vale RD (2010) A whole genome RNAi screen of drosophila S2 cell spreading performed using automated computational image analysis. J Cell Biol 191:471–479
9. Moutinho-Pereira S, Stuurman N, Afonso O, Hornsveld M, Aguiar P, Goshima G, Vale RD, Maiato H (2013) Genes involved in centrosome-independent mitotic spindle assembly in drosophila S2 cells. Proc Natl Acad Sci U S A 110:19808–19813
10. Toret CP, D'Ambrosio MV, Vale RD, Simon MA, Nelson WJ (2014) A genome-wide screen identifies conserved protein hubs required for cadherin-mediated cell-cell adhesion. J Cell Biol 201:265–279
11. Caplen NJ, Fleenor J, Fire A, Morgan RA (2000) dsRNA-mediated gene silencing in cultured drosophila cells: a tissue culture model for the analysis of RNA interference. Gene 252:95–105
12. Rogers SL, Rogers GC (2008) Culture of drosophila S2 cells and their use for RNAi-mediated loss-of-function studies and immunofluorescence microscopy. Nat Protoc 3:606–611

13. Adams MD et al (2000) The genome sequence of *Drosophila melanogaster.* Science 287:2185–2195
14. Reiter LT, Potocki L, Chien S, Gribskov M, Bier E (2001) A systematic analysis of human disease-associated gene sequences in *Drosophila melanogaster.* Genome Res 11:1114–1125
15. Beir E (2005) Drosophila, the golden bug, emerges as a tool for human genetics. Nat Rev Genet 39:715–720
16. Rogers SL, Rogers GC, Sharp DJ, Vale RD (2002) Drosophila EB1 is important for proper assembly, dynamics, and positioning of the mitotic spindle. J Cell Biol 158:873–884
17. Kner P, Chhun BB, Griffis ER, Winoto L, Gustafsson MG (2009) Super-resolution video microscopy of live cells by structured illumination. Nat Methods 6:339–342
18. Iwasa JH, Mullins RD (2007) Spatial and temporal relationships between actin-filament nucleation, capping, and disassembly. Curr Biol 17:395–406
19. Uehara R, Goshima G, Mabuchi I, Vale RD, Spudich JA, Griffis ER (2010) Determinants of myosin II cortical localization during cytokinesis. Curr Biol 20:1080–1085
20. Biyasheva A, Svitkina T, Kunda P, Baum B, Borisy G (2004) Cascade pathway of filopodia formation downstream of SCAR. J Cell Sci 117:837–884
21. Kim JH, Cho A, Yin H, Schafer DA, Mouneimne G, Simpson KJ, Nguyen KV, Brugge JS, Montell DJ (2011) Psidin, a conserved protein that regulates protrusion dynamics and cell migration. Genens Dev 25:730–741
22. Bai SW, Herrera-Abreu MT, Rohn JL, Racine V, Tajadura V, Suryavanshi N, Bechtel S, Wiemann S, Baum B, Ridley AJ (2011) Identification and characterization of a set of conserved and new regulators of cytoskeletal organization, cell morphology and migration. BMC Biol 9:54
23. Maiat H, Sampaio P, Lemos CL, Findlay J, Carmena M, Earnshaw WC, Sunkel CE (2002) MAST/orbit has a role in microtubule-kinetochore attachment and is essential for chromosome alignment and maintenance of spindle bipolarity. J Cell Biol 157:749–760
24. Logarinho E, Bousbaa H, Dias JM, Lopes C, Amorim I, Antunes-Martins A, Sunkel CE (2004) Different spindle checkpoint proteins monitor microtubule attachment and tension at kinetochores in drosophila cells. J Cell Sci 117:1757–1771
25. Maiato H, Rieder CL, Khodjakov A (2004) Kinetochore-driven formation of kinetochore fibers contribute to spindle assembly during animal mitosis. J Cell Biol 167:831–840
26. Goshima G, Nédélec F, Vale RD (2005) Mechanisms for focusing mitotic spindle poles by minus end-directed motor proteins. J Cell Biol 171:229–240
27. Maiato H, Hergert PJ, Moutinho-Pereira S, Dong Y, Vandenbeldt KJ, Rieder CL, McEwe BF (2006) The ultrastructure of the kinetochore and kinetochore fiber in drosophila somatic cells. Chromosoma 115:469–480
28. Griffis ER, Stuurman N, Vale RD (2007) Spindly, a novel protein essential for silencing the spindle assembly checkpoint, recruits dynein to the kinetochore. J Cell Biol 117:1005–1015
29. Zhang D, Rogers GC, Buster DW, Sharp DJ (2007) Three microtubule severing enzymes contribute to the "Pacman-flux" machinery that moves chromosomes. J Cell Biol 177:231–242
30. Maresca TJ, Salmon ED (2009) Intrakinetochore stretch is associated with changes in kinetochore phosphorylation and spindle assembly checkpoint activity. J Cell Biol 184:373–381
31. Rogers GC, Rusan NM, Peifer M, Rogers SL (2008) A multicomponent assembly pathway contributes to the formation of acentrosomal microtubule arrays in interphase drosophila cells. Mol Biol Cell 19:3163–3178
32. Rogers SL, Wiedemann U, Häcker U, Turck C, Vale RD (2004) Drosophila RhoGEF2 associates with microtubule plus ends in an EB1-dependent manner. Curr Biol 14:1827–1833
33. Mennella V, Rogers GC, Rogers SL, Buster DW, Vale RD, Sharp DJ (2005) Functionally distinct kinesin-13 family members cooperate to regulate microtubule dynamics during interphase. Nat Cell Biol 7:235–245
34. Goodwin SS, Vale RD (2010) Patronin regulates the microtubule network by protecting microtubule minus ends. Cell 143:263–274
35. Rothenberg ME, Rogers SL, Vale RD, Jan LY, Jan YN (2003) Drosophila pod-1 crosslinks both actin and microtubules and controls the targeting of axons. Neuron 39:779–791
36. Applewhite DA, Grode KD, Keller D, Zadeh AD, Slep KC, Rogers SL (2010) The spectraplakin short stop is an actin-microtubule crosslinker that contributes to organization of the microtubule network. Mol Biol Cell 21:1714–1724

Part II

Cytoskeleton Dynamics

Chapter 9

Influencing the Actin Dynamics in Plant Cells by Jasplakinolide, Chondramides, Phalloidin, Cytochalasins, and Latrunculins

Andreas Holzinger

Abstract

This chapter presents an overview of the most common F-actin influencing substances, used to study actin dynamics in living plant cells for studies on morphogenesis, motility, organelle movement, apoptosis, or abiotic stress. These substances can be divided into two major subclasses—F-actin-stabilizing and F-actin-polymerizing substances like jasplakinolide and chondramides and F-actin-severing compounds like cytochalasins and latrunculins. Jasplakinolide, which may have anti-cancer activities, was originally isolated from a marine sponge and can now be synthesized and has become commercially available, which is responsible for its wide distribution as membrane-permeable F-actin-stabilizing and F-actin-polymerizing agent. Recently an acyclic derivate of jasplakinolide was isolated. Cytochalasins, derived from fungi, show an F-actin-severing function, and many derivatives are commercially available (A, B, C, D, E, H, J), also making it a widely used compound for F-actin disruption. The same can be stated for latrunculins (A, B), derived from Red Sea sponges; however the mode of action is different by binding to G-actin and inhibiting incorporation into the filament. In the case of swinholide, isolated from red algae or the cyanobacterium *Nostoc*, a stable complex with actin dimers is formed, resulting in severing F-actin.

For influencing F-actin dynamics in plant cells, only membrane permeable drugs are useful in a broad range. We, however, introduce also the phallotoxins and synthetic derivatives thereof, as they are widely used to visualize F-actin in fixed cells. A particular uptake mechanism has been shown for hepatocytes but has also been described in siphonal giant algae. The focus is set on F-actin dynamics in plant cells where alterations in cytoplasmic streaming can be particularly well studied; moreover fluorescence methods for phalloidin- and antibody staining as well as techniques for immunoelectron microscopy are explained.

Key words Actin filaments, Chondramides, Cytochalasins, Depsipeptide, Jasplakinolide, Latrunculin, Phalloidin, Phallotoxin, Swinholide

1 Introduction

Filamentous actin (F-actin), as a major structural cytoskeletal component, has been targeted by inhibitor drugs for decades, and the importance of an undisturbed turnover has been established as prerequisite for living functions in plant cells [1]. Recently the

Ray H. Gavin (ed.), *Cytoskeleton: Methods and Protocols*, Methods in Molecular Biology, vol. 2364,
https://doi.org/10.1007/978-1-0716-1661-1_9,

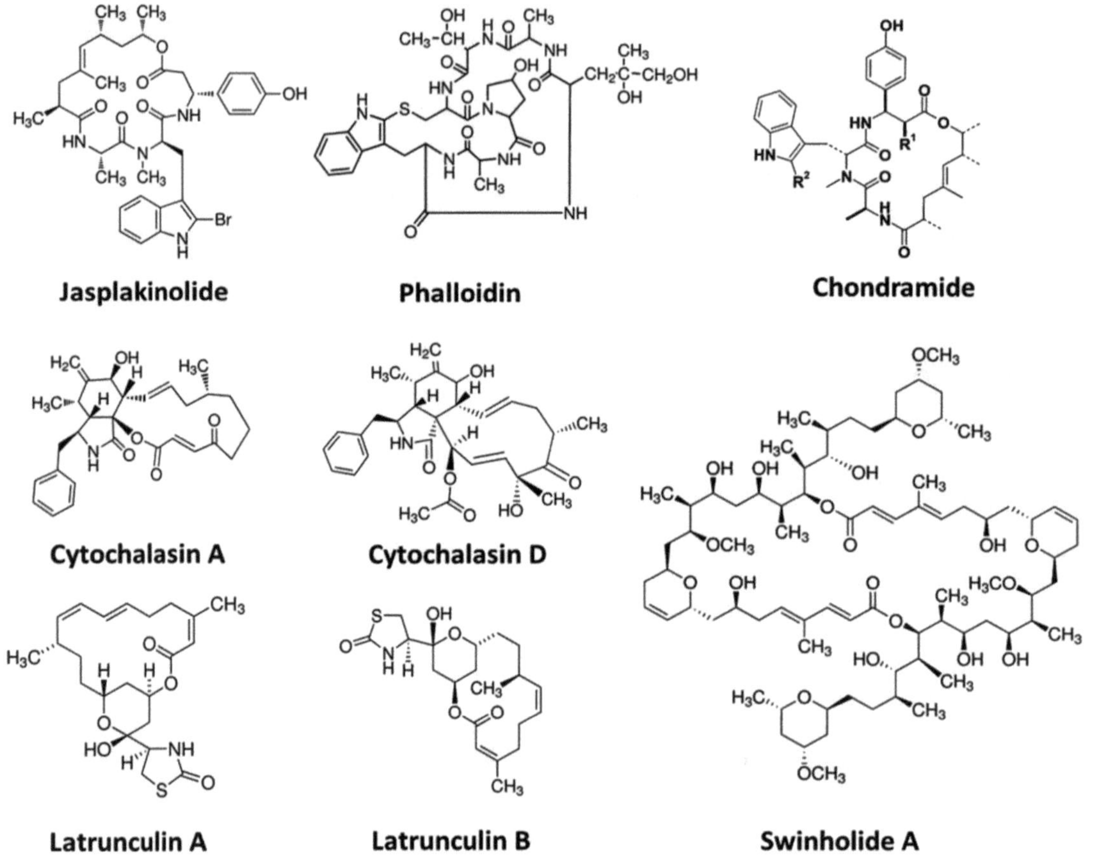

Fig. 1 Molecular structures of a selection of F-actin-stabilizing/F-actin-polymerizing drugs (jasplakinolide, phalloidin, chondramide) and F-actin-disrupting drugs (cytochalasin A, cytochalasin D, latrunculin A, latrunculin B, swinholide A)

first near-atomic structure of plant actin has been published using cryogenic electron microscopy and a real-space helical reconstruction of *Zea mays* pollen actin (ZMPA) at a resolution of 3.9 Å [2]. Yet, a comprehensive overview of the different categories of F-actin-perturbing substances is still missing. We will therefore give a short overview on (A) actin-stabilizing and actin-polymerizing substances including jasplakinolide, chondramide, and phalloidin and actin-depolymerizing substances like cytochalasin, latrunculin, and swinholide (Fig. 1).

The actin system is involved in different cellular processes of plants like exo- and endocytosis, organelle motions and maintenance of organelle distribution, motility, cell division, cell shape formation, and cytoplasmic streaming [1]. The near-atomic structure of ZMPA has been compared with rabbit skeletal muscle actin and jasplakinolide-stabilized *Plasmodium falciparum* actin, where the D-loop of subdomain 2 differs from the animal actin but resembles the jasplakinolide-stabilized actin, suggesting that plant actin filaments are more stable than their mammalian counterparts [2]. Several substances are known to influence actin dynamics by altering intracellular actin organization and have been used extensively for cell biological research [3].

While the mechanism of action is different for the described substances, the effect on the cells might be quite similar—a disruption of the actin-related functions. The present article does not cover drugs or inhibitors acting only indirectly on F-actin polymerization (i.e., ATP-synthase blockers like ATPase blockers *N*-ethylmaleimide (NEM) and 2,3-butanedione monoxime (BDM) [4]) or inhibitors interacting with actin-binding proteins like the Arp2/3 complex (i.e., wiskostatin, [5]).

1.1 F-Actin-Stabilizing and F-Actin-Polymerizing Substances

Actin stabilization is generated mostly by three different compounds: phalloidin [6, 7], jasplakinolide [8, 9], and chondramide [10, 11]. While all three substances have the capacities to stabilize F-actin, phalloidin is not membrane permeable, whereas jasplakinolide and chondramides readily enter cells [12, 13]. Therefore, phalloidin is used in cell biological research mainly for visualization of F-actin after fluorescence labeling of the compound in fixed tissues [14]. In contrast, the commercially available jasplakinolide can be used for F-actin stabilization in living cells due to its membrane permeability [15].

1.1.1 Jasplakinolide

The common feature of these actin-stabilizing substances is a cyclic depsipeptide (Fig. 1), which is a polymeric compound containing both amino acids and hydroxy acids, joined by peptide and ester bounds. Chemically, jasplakinolide is a cyclo-depsipeptide containing a tripeptide moiety linked to a polyketide chain [16, 17] (Fig. 1). Jasplakinolide is also called jaspamide [17–20]. For jasplakinolide also F-actin-polymerizing capacities were described [21–25]. Originally, jasplakinolide was isolated from the marine sponge *Jaspis* sp. collected at Fiji or the Palau islands [16, 18]. The sponge was authenticated as *Jaspis johnstoni* [17]; however taxonomic difficulties were pointed out [26], and the compound can also be isolated from other sponge genera [26, 27]. Only recently, a new acyclic cytotoxic jasplakinolide derivative has been isolated from the marine sponge *Jaspis splendens* [28]. Detailed procedures for jasplakinolide extraction can be found in [29].

Different approaches were made to synthesize this substance [30, 31] or generate nonpeptide mimetics [32]; finally Jasplakinolide became commercially available from Molecular Probes in the late 1990s. An enantioselective total synthesis of (+)-jasplakinolide has been described [33].

The biological activities of jasplakinolide are described as anthelminthic, antifungal, insecticidal, and selective antimicrobial [16, 18, 34]. In vitro investigations have elucidated the actin-polymerizing and actin-stabilizing capacities of jasplakinolide

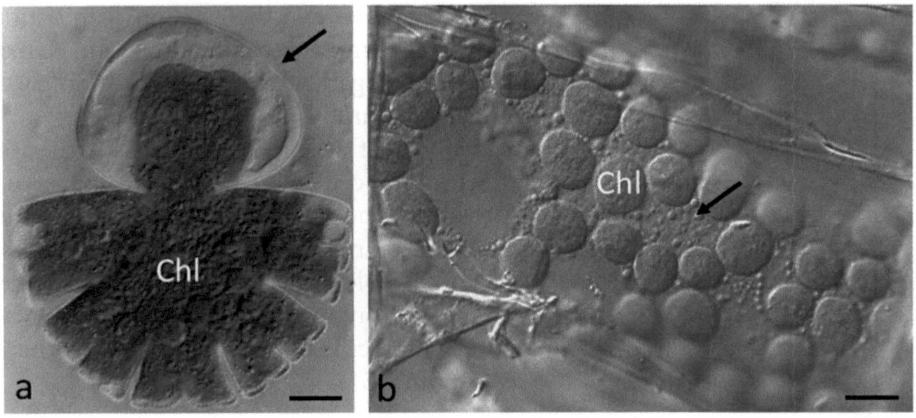

Fig. 2 Light microscopy of F-actin perturbation. (**a**) Effect of jasplakinolide treatment on developing *Micrasterias denticulata* cell (Subheading 3.1) and young developmental stage treated with 3μM jasplakinolide for 0.5 h and allowed to recover for 3 h in nutrient solution; young semi-cell exhibits complete loss of normal cell pattern (arrow), chloroplast (chl). (**b**) Effect of 5μm latrunculin B on *Oxyria digyna* mesophyll cell; cytoplasmic streaming is inhibited, between the chloroplasts (chl) granular cytoplasmic portions are visible. Bars = (**a**) 20μm, (**b**) 5μm. ((**a**) Reprinted from Ref. 22 with permission. "Copyright Wiley-Liss, Inc., a subsidiary of John Wiley & Sons, Inc." (**b**) Reprinted from Ref. 58 with permission of "G. Thieme Verlag KG, Stuttgart")

[9, 21, 25]. In plant systems the unicellular green alga *Micrasterias* [22], green algae *Acetabularia*, *Pseudobryopsis*, and *Nitella* [13], fucoid (brown algal) Zygotes of *Silvetia compressa* and *Pelvetia compressa* [35], *Allium* bulb scale cells and *Sinapis* root hairs [13], tobacco BY-2 cells [36], *Lilium* pollen tubes [37], and *Papaver rhoeas* pollen tubes [38] have been tested for jasplakinolide reaction.

Jasplakinolide causes a tremendous effect already visible at the light microscopic level by malformation of cells after recovery in the development of *Micrasterias* [22] (Fig. 2a). Inhibition or retardation of cell development occurs; during recovery from drug treatment, the cells develop a malformed pattern [22]. In pollen tubes the normal bidirectional cytoplasmic streaming is altered; instead a rotary streaming is observed in the swollen apex [37]. Moreover, altering F-actin levels or dynamics by jasplakinolide plays a functional role in initiating programmed cell death in *Papaver* pollen, triggering a caspase-3-like activity [38]. Polarity establishment is severely changed in fucoid zygotes [35]. When applied during mitosis, binucleated cells are generated as a consequence of jasplakinolide treatment [22].

Jasplakinolide specifically targets F-actin, other cytoskeletal components like desmin or β-tubulin were not influenced by jasplakinolide [39], microtubule-dependent processes are not affected [40], and the distribution of microtubules and microtubule-dependent processes were not altered in green algae [22].

1.1.2 Chondramides

Another group of substances with actin-stabilizing and actin-polymerizing capacities are chondramides [9, 41]. These depsipeptides were originally obtained from myxobacteria of *Chondromyces crocatus* [10, 42], Chemically they are very similar to jasplakinolide but contain instead of an 19-membered ring an 18-membered macrocyclic ring [43]. Until to date they are underrepresented due to their limited availability. However, synthetic chondramides were produced recently [43] and may become commercially available. In the future, chondramides may also be used for cancer therapy [44, 45].

1.2 Actin-Depolymerizing Substances

A prerequisite for an F-actin-perturbing drug is it's membrane permeability. Cytochalasins as well as latrunculins represent the most widely used F-actin-sequestering drugs [46].

1.2.1 Cytochalasins

Cytochalasins have been shown to act on cytoplasmic streaming in plant cells [47–50]. A vast number of different cytochalasins (A, B, C, D, E, H, J) have been described, with different inhibitory concentrations and chemical properties, occasionally termed cytochalasans [51]. It is generally accepted that the cytochalasins slow the rate of filament polymerization by inhibiting the rate of elongation [52]. This action is caused by the high affinity binding of cytochalasins to the barbed (plus end) of F-actin; thus the monomer addition is prevented by a "capping" mechanism [46, 52]. Due to the prevention of G-actin incorporation into the filament, a net depolymerization is provoked [46]. While the main target is the F-actin cytoskeleton, cytochalasins A and B also inhibit monosaccharide transport across the plasma membrane [7].

Cytochalasins are fungal metabolites, independently discovered and isolated from distinct fungal species (cytochalasin A and B derived from *Helminthosporium dematioideum*; cytochalasin C and D from *Metarhizium anisopliae*) by Aldridge et al. [53]. Previously, Rothweiler and Tamm [54] termed the substance "phomin" as it was detected in *Phomas* sp. (*Fungi imperfecti*, strain S 298). Chemically cytochalasins are a diverse group of polyketide-amino acid hybrid metabolites [51]. Various cytochalasins have a wide range of biological activities, some of which not directly related to actin binding. Some cytochalasins (e.g., cytochalasin B) interfere with the monosaccharide transport systems by binding to high-affinity sites on glucose-transporter proteins and may interfere with hormones.

1.2.2 Latrunculins

Latrunculins (Fig. 1) provide another useful drug for the study of actin polymerization with a complementary mode of action when compared to cytochalasin [55]. The main mechanism of latrunculin action is in an interaction with actin monomers (G-actin) in order to avoid actin polymerization [55]. It has been shown by an in vitro assay that latrunculins bind to purified G-actin resulting in a non-polymerizable 1:1 complex [8]. Latrunculin A is a toxin purified from the Red Sea sponge *Latrunculia magnifica* [55, 56]. It has been used in a variety of plant cells inhibiting pollen germination and pollen tube growth [50] and induced plant dwarfism [57]. Both Latrunculins arrested cytoplasmic streaming after disrupting the subcortical actin bundles [49]. In a mesophyll cell of the high alpine plant *Oxyria digyna*, latrunculin B arrested cytoplasmic streaming (Fig. 2b) [58]. Latrunculin B has been used to depolymerize actin in the green alga *Chlamydomonas reinhardtii* [59, 60]; in higher concentrations it can lead to the formation of ring-like structures [59]. Latrunculin B has a diminished binding capacity when compared to latrunculin A [6]. The function of actin depolymerization by latrunculin B may have a more wide-ranging effect than previously thought as recently demonstrated by a proteomic study, where not only a downregulation of several proteins involved in actin organization and dynamics was observed but also a change in phytohormone dynamics by reducing the abundance of abscisic acid [61]. Complex interaction patterns upon drought stress were seen in *Arabidopsis*, where a stress-induced protein (AFL1) made FM4-64 uptake as a marker for endocytosis less sensitive to the actin-disrupting agents latrunculin B as well as cytochalasin D [62].

1.2.3 Swinholide

Swinholide A (Fig. 1) was isolated from the marine sponge *Theonella swinhoei*, collected from the Red Sea [8, 63]. This polyketide was long time suspected to be produced by symbiotic bacteria, and recently it was reported to be produced by an axenic strain of a terrestrial cyanobacterium *Nostoc* sp. [64]. It is membrane permeable and acting as an F-actin-disrupting toxin that stabilizes actin dimers and thus severs F-actin. Chemically it consists of a 44-carbon ring dimeric dilactone macrolide with a twofold axis of symmetry [65–67]. The effects of swinholide A on the actin cytoskeleton and cell morphology are similar to latrunculins rather than of cytochalasins [65].

1.3 Visualization of Inhibitory Effects

The function of these drugs can be either studied by arresting or inhibiting cytoplasmic streaming (Fig. 2b) [58]. The velocity of cytoplasmic streaming can provide useful insights in the function of the different drugs [49]. Visualization of the changes in F-actin is usually performed by phalloidin staining [49, 59, 66]. It has been shown in the charophyte green alga *Nitella pseudoflabellata* that thick F-actin bundles may not appear influenced by, e.g., cytochalasin A and H (Fig. 3), despite that the cytoplasmic streaming is arrested. In *Chlamydomonas reinhardtii* a time-dependent loss of F-actin has been visualized after latrunculin A treatment; prolonged treatments (45 min) led to the formation of F-actin rings [59] (Fig. 4). After jasplakinolide or chondramide treatment, an altered

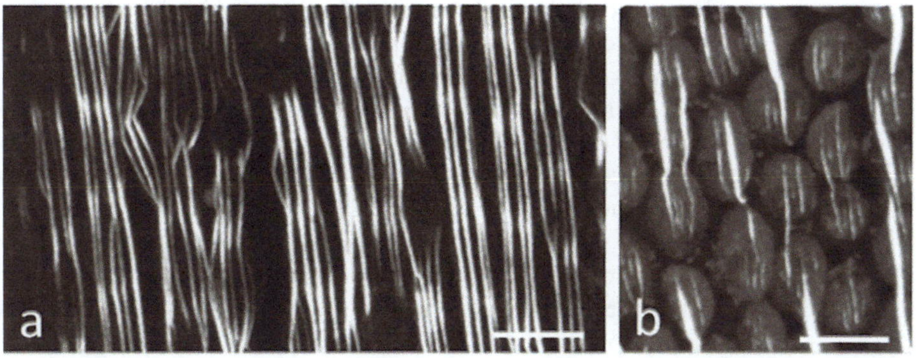

Fig. 3 Subcortical F-actin bundles in internodal cells of *Nitella pseudoflabellata* treated with (**a**) 1μM cytochalasin A, incubated for 1 day and (**b**) 50μM cytochalasin H, incubated for 2 days. Despite these concentrations lead to an arresting of cytoplasmic streaming, the F-actin bundles are clearly visible after phalloidin staining with the perfusion method (*see* Subheading 3.2.5) clear cytoplasm. Bar = 5 μm (Reprinted from Ref. 49 with permission of Oxford University Press)

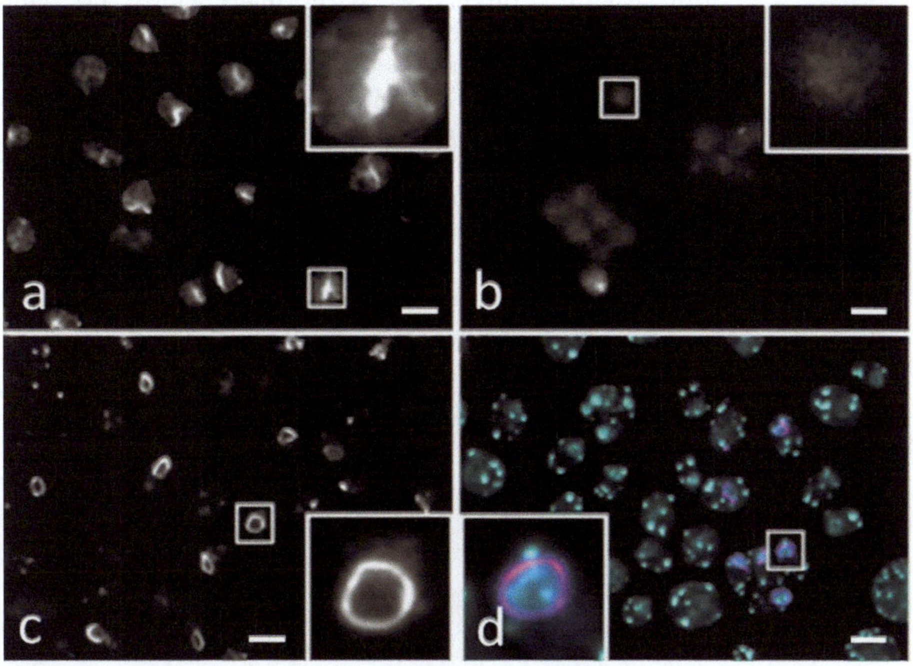

Fig. 4 Effects of different concentrations of latrunculin B on *Chlamydomonas* CC-125 wild-type cells. (**a**) Vegetative wild-type cells stained with Atto 488 phalloidin, (**b**) filamentous actin signal disappears after 10 min treatment with 10μM latrunculin B, (**c**) filamentous actin ring structures form in many cells treated for 45 min with 10μM latrunculin B, (**d**) actin ring structures (magenta) co-localize around the nucleus (DAPI in cyan), cells treated for 45 min in 10μM latrunculin B. The insets are magnified details of the respective panels (indicated by the small squares). The details of the staining procedure are described in Subheading 3.2.4. [59] Bars = 10 μm. (Reprinted from Ref. 59 with permission of the authors © 2019 Craig et al.)

F-actin distribution was found, characterized by a patchy appearance of cortical actin [68], the formation of actin dots [69], or a disruption of the actin cytoskeleton [13, 39]. Also abiotic stresses like desiccation have been successfully visualized by phalloidin staining [70], making this a universal tool.

As competitive binding inhibition of phalloidin to F-actin by jasplakinolide is described [21], and decreased FITC-phalloidin labeling is noted as a consequence of jasplakinolide treatment in MDCK cells [71], the problems visualizing jasplakinolide effects by fluorescently labeled phalloidin are obvious. Moreover, phalloidin and jasplakinolide have the same effect when applied to living *Dictyostelium* cells, namely, the formation of actin aggregates [24].

In order to avoid the difficulties mentioned, only electron microscopy appears to be a sufficient technique for detection of jasplakinolide effects on F-actin [22, 72].

At electron micrographs tangential sections of F-actin can easily be recognized by the filament diameter of about 4–5 nm. In organisms like *Micrasterias*, where normally almost no bundling of F-actin occurs, and thus F-actin is hardly found at electron microscopical images, alterations causing extensive bundling of actin are easily detected [22] (Fig. 5). In addition, immunodetection of the altered F-actin system after jasplakinolide treatment is successful [68] and allows a comprehensive picture to be drawn in combination with other techniques.

Herein different procedures shall be described for the detection of actin filament aggregates generated by application of jasplakinolide. Due to the advantages of electron microscopy and/or

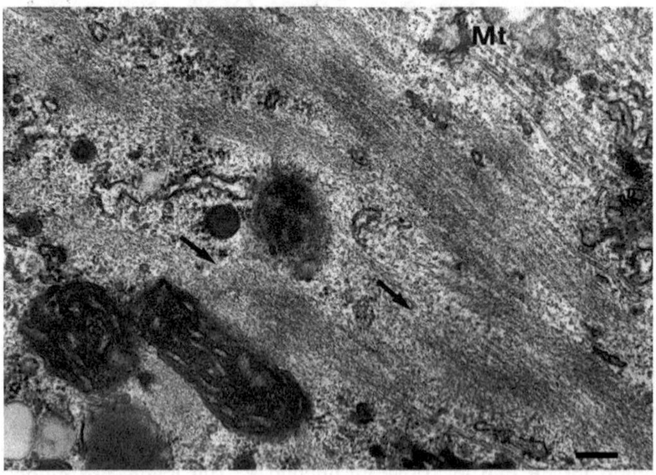

Fig. 5 Detail of the ultrastructure of *Micrasterias denticulata* cell treated with 3μM jasplakinolide for 2 h and fixed for electron microscopy according to Subheading 3.2.1; dense accumulations of F-actin are visible in the cytoplasm. Bar = 0.2μm. (Reprinted from Ref. 22 with permission "Copyright Wiley-Liss, Inc., a subsidiary of John Wiley & Sons, Inc.")

immunomethods, these will be mentioned more in detail; however different protocols for phalloidin stainings will be given as they have been used for some organisms in combination with jasplakinolide treatment [24, 68, 71].

2 Materials

2.1 Cell Cultures

1. *Micrasterias denticulata* Bréb. is a unicellular green alga used for light microscopy, electron microscopy [22], and immunoelectron microscopy because alterations in cell shape caused by drug treatments are easily detected. The cells are grown in culture medium under semisterile conditions in a light/dark regime of 14/10 h at 20 °C.
2. *Chlamydomonas reinhardtii* CC-125 (wild type).
3. Characean green algae (*Chara* sp., *Nitella* sp.), fresh collected or culture grown material.
4. *Acetabularia acetabulum*.
5. Stably transformed *Arabidopsis thaliana* expressing a fusion construct of GFP with the actin-binding domain 2 (ABD2) of the plant actin-binding protein fimbrin [73, 74].
6. Tris acetate phosphate (TAP) liquid medium (commercially available).
7. Tris acetate phosphate (TAP) agar plates.

2.2 Chemicals, Buffers, Solutions

2.2.1 Actin-Perturbing Drugs and Solvents

1. Jasplakinolide, Mol. Wt. 709.68: 100 μg diluted upon arrival in methanol to achieve 10 mM stock solution and kept refrigerated at −20 °C (*see* **Note 1**).
2. Chondramides: although the original isolate from myxobacteria [10, 42] is not commercially available, a synthetic analog, condramide, has been produced [43].
3. Cytochalasins: cytochalasin A, MW 477.59; cytochalasin B, MW 479.61; cytochalasin C, MW 507.62; cytochalasin D, MW 507.62; cytochalasin E, MW 495.56; cytochalasin H, MW 493.63.
4. Latrunculin (LAT): LAT A, MW 421.55; LAT B, MW 395.51.
5. Swinholide A, MW 1389.87.
6. Phalloidin MW 788.87.

2.2.2 Electron Microscopy

1. Cacodylic buffer: 50 mM cacodylic acid sodium salt, pH 7.2.
2. Cacodylate-glutaraldehyde: 50 mM cacodylic buffer, 1% glutaraldehyde, pH 7.2.
3. Cacodylate-osmium: 50 mM cacodylic acid buffer, 1% osmium tetraoxide, 0.8%, potassium hexacyanoferrate III.

4. Phosphate buffer-fixative: 25 mM sodium hydrogen phosphate, 25 mM potassium dihydrogen phosphate, 1% glutaraldehyde, 1% osmium tetroxide, pH 6.2.
5. 2% Aqueous uranyl acetate dihydrate.
6. Dehydrating ethanol series: 15%, 30%, 40%, 50%, 60%, 70%, 80%, 90%, 95%, 100%.
7. Propylene oxide-ethanol mixture: 1,2-propylene oxide and ethanol (1:1 v/v).
8. Propylene oxide-embedding resin mixture: mix (1:1 v/v) 1,2-propylene oxide and embedding resin (Embed 812, Araldite 502, Spurr's, or others).
9. Copper grids.
10. 3% Formvar 1595 E in chloroform.
11. 10% uranyl acetate.

2.2.3 Immunoelectron Microscopy

1. 10% Lecithin dissolved in chloroform.
2. Liquid nitrogen.
3. Acetone-tannic acid: acetone containing 0.1% tannic acid (v/v).
4. Acetone-osmium: acetone containing 2% osmium tetroxide and 0.05% uranyl acetate.
5. Embedding medium: LR white resin, medium grade.
6. TBS (Tris-buffered saline): 50 mM Tris–HCl, pH 7.5, 150 mM NaCl.
7. 2% Blocking solution: TBS containing 1% BSA (fraction V), 1% acetylated BSA, 01% Tween-20.
8. 10% Blocking solution: 10% bovine fetal calf serum.
   Primary antisera/antibodies (e.g., N 350 monoclonal anti-actin, Amersham).
9. 10 nm colloidal gold-conjugated secondary antibody appropriate for the primary antibody.

2.2.4 Fluorescence Microscopy

1. MT-stabilizing buffer (MTSB): 50 mM PIPES buffer, 50 mM $MgSO_4$, 50 mM EGTA, pH 6.9.
2. MTSB/DMSO solution: 4.6 mL MTSB buffer mixed with 0.5 mL of DMSO.
3. MTSB/DMSO fixative: 4% freshly prepared paraformaldehyde in MTSB.
4. Paraformaldehyde-HEPES fixative: 4% fresh paraformaldehyde in 7.5 mM HEPES pH 7.4.
5. PBS (phosphate-buffered saline): 137 mM sodium chloride, 2.7 mM potassium chloride, 4.3 mM sodium hydrogen phosphate, 1.4 mM potassium dihydrogen phosphate.

6. TBS (Tris-buffered saline): 50 mM Tris–HCl, pH 7.5, 150 mM NaCl.
7. 7.5 mM HEPES pH 7.4.
8. Triton solution: 0.1% Triton X-100 in PBS.
9. Blocking solution: 2% bovine serum albumin (BSA).
10. Anti-actin monoclonal antibody N 350 (Amersham) (*see* **Note 2**).
11. Anti-actin monoclonal antibody, clone C4. (ICN, Costa Mesa, CA).
12. Phalloidin: 0.4 mg/mL FITC, Alexa or Atto conjugated phalloidin Alexa phalloidin (Molecular Probes, Leiden, The Netherlands; prepare from a 6.6 mM stock solution in methanol) at a concentration of 0.16 mM.
13. Mounting medium: phenylenediamine.
14. Mounting medium: Fluoromount-GTM.
15. Texas red conjugated anti-mouse IgM antibody.
16. Steedman's wax: PEG 400 distearate and 1-hexadecanol mixed in proportions 9:1 (w/w) [78].
17. Dehydrating ethanol series in PBS: PBS, 15%, 30%, 40%, 50%, 60%, 70%, 80%, 90%, 95%, 100%.
18. Rehydrating ethanol series in PBS: 100%, 95%, 90%, 80%, 70%, 60%, 50%, 40%, 30%, 15%, PBS.
19. Isotonic perfusion solution: 200 mM sucrose, 70 mM KCl, 4.5 m*M* $MgCl_2$, 5 mM ethyleneglycoltetraacetic acid (EGTA), 1.48 mM $CaCl_2$, 10 mM piperazine-*N*,*N*′-bis (2-ethanesulfonic acid) (PIPES, pH 7.0).
20. Cell wall digest solution: 1% hemicellulose dissolved in 0.5 M EGTA, 0.4 M mannitol, 1% Triton X-100, 0.3 mM phenylmethylsulfonyl fluoride (dissolved in MTSB).

3 Methods

3.1 Treatment and Recovery Experiments

1. Incubate cells (from the abovementioned plant cell types or tissue, *see* Subheading 2.1) in 1 mL of appropriate culture medium containing appropriate inhibitory drug concentrations for time periods of 0.5, 1, 2, 3, 4, 24, 48 h, and observe the effect on cytoplasmic streaming (Fig. 2b; *see* **Note 3**).
2. Wash several times with culture medium, and allow developing in culture medium for up to 24 h. This is regarded as "recovery experiment."
3. Investigate the cells from the recovery experiment under a regular light microscope or fix for electron microscopy (*see* Subheading 3.2.1).

3.2 Visualization of F-Actin After Drug Treatment

3.2.1 Chemical Fixation/Electron Microscopy of Algal Cells after Holzinger and Meindl [22]. The Procedure Described Herein Has First Been Used for the Unicellular Green Alga Micrasterias [75]

1. Fix recovered cells (Subheading 3.1, **step 2**) and untreated control cells in cacodylate-glutaraldehyde for 15 min (*see* **Note 4**).
2. Wash cells three times for 5 min each in cacodylic buffer.
3. Postfix cells in cacodylate-osmium for 2 h.
4. Wash cells, as in **step 2**, in distilled water.
5. Incubate cells in 2% uranyl acetate for 2 h (*see* **Note 5**).
6. Wash cells, as in **step 2**, in distilled water.
7. Use a cold dehydrating ethanol series to dehydrate the cells for 15 min per concentration except for 100% ethanol, which requires 30 min (*see* **Note 6**).
8. Transfer the cells to the propylene oxide-ethanol mixture, allow to equilibrate for 10 min, and transfer to propylene oxide.
9. Transfer the cells to the propylene oxide-embedding resin mixture (*see* **Note** 7).
10. Rotate the cells in this mixture for 48 h in order to allow the propylene oxide to evaporate and a sufficient penetration of the cells with the resin to be achieved.
11. Transfer the cells to freshly prepared resin in aluminum dishes, and orientate them by the use of eyelashes connected to holders.
12. Incubate in a desiccator for 4 days.
13. Polymerize the resin for 24 h at 60 °C.
14. Select cells and section at an ultramicrotome.
15. Collect sections on Formvar-coated copper grids (*see* **Note 8**).
16. Counterstain the sections with 2% uranyl acetate and Reynold's lead citrate (*see* **Note 9**).
17. Investigate and photograph at a transmission electron microscope at 60–80 kV.

3.2.2 Immunoelectron Microscopy

The following fixation technique has been carried out for *Micrasterias* for the first time by Meindl et al. [76]. The recipe for actin detection by means of immuno-gold localization is based on Holzinger et al. [77].

1. Prepare gold or aluminum specimen holders with bed depths of 100–300 μm by dipping in lecithin.
2. Collect treated cells to be transferred to specimen holders. Large cells like *Micrasterias* (200 μm in diameter) can be easily collected under a stereo microscope by wrapping with cotton fibers.

3. Use high-pressure freezing to fix treated cells (High Pressure Freezing Machine HPM 010, Boeckeler Instruments Inc., Tucson/AZ, USA). *See* **Note 10** for other methods.
4. Collect and store samples under liquid nitrogen.
5. Transfer samples to a freeze substitution device (Reichert-Jung, cs auto or LEICA EM AFS, Leica Microsystems GmbH, Vienna, Austria).
6. Substitute samples at −80 °C for 24 h in acetone-tannic acid solution.
7. Wash several times with acetone.
8. Substitute samples at −80 °C for 24 h in acetone-osmium solution (*see* **Note 11**).
9. Allow samples to reach −30 °C within 5 h.
10. Continue substitution at −30 °C for 10 h.
11. Allow samples to reach room temperature (20 °C) within 5 h.
12. Remove osmium tetroxide/uranyl acetate solution and rinse several times with acetone.
13. Change acetone for ethanol by several rinses.
14. Transfer samples to LR white in aluminum dishes and cover with cellophane foil.
15. Allow samples to infiltrate in a desiccator for 24 h.
16. Polymerize under UV light at room temperature for 24 h.
17. Prepare sections for electron microscopy (compare Subheading 3.2.1) and transfer to Formvar-coated gold or gilded-copper grids.
18. Incubate in 2% blocking solution for 30 min up to 1 h. A blocking step with 10% blocking solution may be applied if blocking is not sufficient.
19. Transfer into primary antibody against actin. Purified antibody may be diluted in blocking solution. Incubation should last for 1.5 h at room temperature or for up to 24 h at 4 °C.
20. Wash four times for 15 min each in TBS by transferring grids to 50 μL droplets of TBS.
21. Incubate in 10 nm gold (*see* **Note 12**) labeled secondary antibody diluted in blocking solution (*see* **Note 13**).
22. Wash by rinsing with TBS, followed by incubation in droplets of TBS for 2 min, followed by a brief rinse with distilled water.
23. Counterstain with uranyl acetate and lead citrate, if necessary (*see* Subheading 3.2.1).
24. Investigate at a transmission electron microscope at 80 kV.

3.2.3 Immunofluorescence Microscopy of Latrunculin B-Treated Maize Roots After Baluska et al. [57, 78]

1. Grow maize root tips for 48 h.
2. Cut apical segments (6–8 mm) and transfer into MTSB/DMSO solution for 15 min.
3. Fix for 1 h in MTSB/DMSO fixative.
4. Dehydrate in increasing ethanol series in PBS.
5. Embed in Steedman's wax by stepwise infiltration in proportions 2:1, 1:1, and 1:2 (v/v) for 2 h each step at 37 °C (*see* Subheading 2.2.4, **step 14**).
6. Infiltrate in pure wax under vacuum, using three changes in pure wax.
7. Polymerize wax at room temperature.
8. Cut longitudinal sections with a thickness between 4 μm and 8 μm.
9. Mount sections on poly-L-lysine-coated slides.
10. Dewax the sections in ethanol.
11. Rehydrate in ethanol/PBS series and incubate in MTSB for 45 min.
12. Digest cell walls in digest solution (*see* Subheading 2.2.4, **step 18**).
13. Incubate in blocking solution for 30 min.
14. Incubate in anti-actin monoclonal antibody (clone C4, ICN, Costa Mesa, CA) for 2 h at room temperature.
15. Rinse three times for 15 min each in MTSB.
16. Incubate in FITC-conjugated anti-mouse IgGs (diluted 1:100–1:200) for 1 h at room temperature.
17. Remove secondary antibody and mount in phenylenediamine.
18. Examine at a confocal laser scanning microscope at excitation with an argon laser at 488 nm.

3.2.4 Immunofluorescence Microscopy of Latrunculin B-Treated Chlamydomonas Cells After Craig et al. [59] and Craig and Avasthi [66]

1. Grow two inoculating loops of *Chlamydomonas* CC-125 mating type + cells, taken from a Tris Acetate Phosphate (TAP) agar plate, in 2 mL of TAP liquid media on a roller drum (40 rpm) overnight (~16 h) in growth lighting at room temperature.
2. Use a hydrophobic marker to draw a circle on the coverslip where you pipette the cells and media used in the experiment. Add 200 μL of poly-L-lysine at room temperature to the hydrophobic circle, and after 10 min, wash in distilled water; poly-L-lysine will fall off the coverslip when submerged. Quickly place coverslip on Parafilm and remove excess moisture.
3. Select healthy cells by centrifuging 1 mL of cell culture at 1800 rpm (203 × *g*) for 1.5 min, discard the supernatant, and resuspend cells in 600 μL fresh TAP media. Slowly pipette

media up and down a few times; healthy cells migrate to the top of the culture after about 10 min. Take 200 μL of resuspended cells from the top of the culture to poly-L-lysine-coated coverslips for 5 min, and shield samples from light to prevent phototaxis swimming to the light.

4. After 5 min, tilt off liquid from the coverslips and replace with 200 μL paraformaldehyde-HEPES fixative (Subheading 2.2.4, **item 4**), and incubate for 15 min at room temperature.
5. Remove fixative solution with Kimwipe and place coverslips in a Columbia jar containing 1× PBS, and wash for 3 min.
6. For cell permeabilization, submerge coverslips in a Columbia jar containing 80% pre-cooled acetone (diluted in water and stored at −20 °C) and then incubate for 5 min at −20 °C.
7. Quickly transfer coverslips into a second Columbia jar containing 100% pre-cooled acetone and incubate for another 5 min at −20 °C.
8. Place coverslips back on Parafilm and allow them to air dry for a minimum of 2 min.
9. Rehydrate cells by transferring coverslips to a Columbia jar containing 1× PBS, and incubate for 5 min.
10. Stain coverslips with Atto 488 Phalloidin for 16 min in the dark (shorter than recommended staining time significantly reduces background and increases the signal to noise ratio).
11. Tilt off staining solution, and wash cells by transferring coverslips to a Columbia jar containing 1× PBS for 5 min.
12. Remove excess liquid from the coverslip with a Kimwipe and mount coverslips with self-sealing Fluoromount-GTM.
13. Capture 0.3 μm step Z-stacks in bright-field and wide-field fluorescence channels (GFP filter set) using a Nikon Eclipse Ti-S equipped with a QIMAGING QICAM. Deconvolve fluorescence images using Huygens Essential deconvolution software and format in ImageJ.

3.2.5 Perfusion of Nitella Internodal Cells for FITC-Phalloidin Staining of F-Actin After Foissner and Wasteneys [49] (Fig. 3; See ***Note 14****)*

1. Place an internodal cell on the cover slip bottom of a perfusion chamber and press into vacuum grease lines; place small reservoirs with grooves over the grease and firmly press down without damaging the cells.
2. Cover the central portion of the cell between the reservoirs with silicon fluid to avoid evaporation.
3. Bath the ends of the cells with isotonic perfusion solution.
4. Reduce the turgor pressure of the cells in a hypertonic solution.
5. Cut the ends with small scissors, and allow the perfusion solution to enter the cells.

6. After 1 min replace the perfusion solution by an F-actin staining solution containing Alexa-phalloidin.
7. After 20 min examine at the CLSM (*see* Subheading 3.2.3).

4 Notes

1. Jasplakinolide is also dissolvable in DMSO [34].
2. The choice of the anti-actin antibody is very important. The epitope of the N 350 (a mouse monoclonal IgM antibody directed against chicken gizzard actin) binding site is obviously not changed by jasplakinolide [13], whereas binding sites for other actin antibodies (e.g., clone C4—a mouse monoclonal IgG antibody directed against chicken gizzard actin, ICN) seem to be altered substantially which prevents visualization of F-actin after jasplakinolide treatment [13]. This antibody may be used without problems in combination with latrunculin B treatment [61].
3. The concentrations of the inhibitory drugs may vary drastically. Jasplakinolide was found to be effective at 1.5 μM to 3 μM in producing F-actin aggregates for *Micrasterias denticulata* [22]. Concentrations above 250 μM of jasplakinolide were lethal. The concentrations of jasplakinolide needed for other cells have to be found empirically. Variations in the literature range from as low as 100 nM to 1 μM, 2.5 μM to 10 μM [13, 35, 68, 79]. However they should be in the same range as described herein, since the effective concentrations correlate with data obtained from the literature [24, 34, 71]. Latrunculins had different lowest effective concentrations in *Micrasterias denticulata*: chondramide A 20 μ*M*, chondramide B 15 μM, chondramide C 5 μM, chondramide D 10 μM [41]. In contrast with human tumor cell lines, 3–85 nM chondramide effectively inhibited proliferation [11]. Cytochalasins [49] found streaming-arresting concentration in *Nitella pseudoflabellata* between 1 and 200 μM. The lowest concentrations to arrest streaming were at 1 μM in cytochalasins A and E, 30 μM in cytochalasin H, 60 μM cytochalasin D, and 200–220 μM in cytochalasins C, J, B. These authors distinguish between streaming arresting (i.e., fully arresting streaming within 1 h, Fig. 3a, b) and streaming inhibiting (i.e., reducing streaming velocity by 15–80% within 1 h) concentrations. In another green alga *Xanthidium armatum*, cytochalasin D arrested cytoplasmic streaming at concentrations below 10 μM.

   In contrast, latrunculins have lower effective concentrations; latrunculin B may cause severe effects even at extremely low concentrations in the pM to nM range [48]. Several authors use it in higher concentration in the μM range (10 μM [57]: 5 μM, 10 μM [58, 59]. In the case of latrunculin B, the duration of the treatment might also be critical, whereas 10 min leads to a disruption of actin filaments; prolonged treatment of 45 and 120 min causes actin rings in *Chlamydomonas* [59] (*see* Fig. 4). Moreover, a combined treatment with latrunculins and cytochalasins rapidly arrested cytoplasmic streaming even at concentrations that had only mild effects on the streaming rate when used separately [49]. Swinholide has not been extensively applied to plant cells; only IC 50 values for human nasopharynx cancer cells have been determined at 6 nM [80].

4. For fixation, we use a Balzers Fixomat (not commercially available) consisting of sintered glass suction filters connected via a valve to a pump station, allowing the removal of solutions by underpressure while cells remain in the filter. Moreover, the temperature can be adjusted via a cooling bath. Alternatively, the cells can be fixed in glass dishes, ideally in the shape of a hemisphere. In this case it is best to remove only the solutions with a pipette and not to transfer the cells. For other organisms more specialized methods might be necessary, for example, embedding of the specimens in agarose prior to the fixation procedure.
5. During incubation in uranyl acetate solution, it is necessary to keep the cells in darkness, by covering the filter or dish with aluminum foil.
6. During the whole procedure of dehydration, the temperature should be kept at 6 °C. At the steps 50% and 70% ethanol, temperature might be lowered to 4 °C. In case of fixation in round glass dishes, they should be kept over ice. At the step of 100% ethanol, the temperature should be increased to room temperature.
7. Very good results are obtained with a 1:1 mixture of glycid ether 100 (a substance equivalent to Epon 812) and MNA. Prior to use DMP must be added (5 droplets to 10 mL of resin). However, a mixture of Embed-812/Araldite 502/DDSA will also give adequate results (2,2-dimethoxypropane or benzyldimethylamine must be added prior to their use).

   When different hardness has to be achieved, Spurr's resin may be the best choice: As several components of the original mixture are not deliverable any more, we now use the kit "Agar Low Viscosity Resin (LV)" by Agar Scientific Ltd. Essex CM24 8DA, England.
8. Coating of grids with Formvar may not be necessary for all objects. We use the following procedure: A light microscopic slide is cleaned with lens tissue, tipped into 0.3% Formvar dissolved in chloroform. The resulting Formvar film is cut at

the edges of the slide and allowed to float on distilled water. Grids are placed on the film which is then removed with Parafilm.

9. The time period needed for counterstaining depends on the staining already achieved during the fixation/staining procedure and is typically 5–30 min for uranyl acetate and 1–5 min for Reynold's lead citrate.
10. For large cells like *Micrasterias* (about 200 μm in diameter), high-pressure freezing has been found the only appropriate technique for freeze fixation. The Leica EMPACT high-pressure freezer (Leica Microsystems GmbH, Vienna, Austria) is an alternative to the HPM 01. However, for small and less vacuolated cells, plunge freezing which might be achieved with relatively simple equipment will also give reasonable results.
11. Osmium tetroxide dissolves well in acetone. Uranyl acetate is best sonicated to dissolve it. For other organisms it might be appropriate to use substitutes for osmium tetroxide and uranyl acetate.
12. Gold particles with 10 nm diameter are used commonly. The usage of smaller gold particles might enhance the accuracy of the detected locus.
13. The actual dilution factor has to be found empirically for each system, but it should be in the range of 1:50–1:200.
14. As mentioned in the introduction, the use of fluorescently labeled phalloidin is problematic due to competitive binding with jasplakinolide [22, 25]. It has been reported that simultaneous addition of jasplakinolide and FITC-phalloidin (1:1) did not result in F-actin labeling, whereas addition of the inactive analogue jasplakinolide B and FITC-phalloidin resulted in staining as strong as in controls [13].

Acknowledgments

I would like to thank Dr. Ilse Foissner, University of Salzburg, for providing the images in Fig. 3. This chapter was supported by Austrian Science Fund (FWF) grants I 1951-B16 and P 34181 to A.H.

References

1. Szymanski D, Staiger CJ (2019) The actin cytoskeleton: functional arrays for cytoplasmic organization and cell shape control. Plant Physiol 176:106–118. https://doi.org/10.1104/pp.17.01519
2. Ren Z, Zhang Y, Zhang Y et al (2018) Cyro-EM structure of actin filaments from *Zea mays* pollen. Plant Cell 31(12):2855–2867. https://doi.org/10.1105/tpc.18.00973
3. Allingham JS, Klenchin VA, Rayment I (2006) Actin-targeting natural products: structures, properties and mechanisms of action. Cellular Mol Life Sci 63(18):2119–2134. https://doi.org/10.1007/s00018-006-6157-9

4. Oertel A, Holzinger A, Lütz-Meindl U (2003) Involvement of myosin in intracellular motility and cytomorphogenesis in *Micrasterias*. Cell Biol Int 27(12):977–986. https://doi.org/10.1016/j.cellbi.2003.07.004
5. Guerriero CJ, Weisz OA (2007) N-WASP inhibitor wiskostatin nonselectively perturbs membrane transport by decreasing cellular ATP levels. Am J Physiol Cell Physiol 292(4): C1562–C1566. https://doi.org/10.1152/ajpcell.00426.2006
6. Lynen F, Wieland U (1937) Über die Giftstoffe des Knollenblätterpilzes IV. Justus Liebigs Ann Chem 533(1):93–117. https://doi.org/10.1002/jlac.19385330105
7. Cooper JA (1987) Effects of cytochalasin and phalloidin on actin. J Cell Biol 105 (4):1473–1478. https://doi.org/10.1083/jcb.105.4.1473
8. Spector I, Braet F, Shochet NR et al (1999) New anti-actin drugs in the study of the organization and function of the actin cytoskeleton. Microsc Res Tech 47(1):18–37. https://doi.org/10.1002/(SICI)1097-0029(19991001)47:1<18:AID-JEMT3>3.0.CO;2-E
9. Bubb MR, Spector I, Beyer BB et al (2000) Effects of jasplakinolide on the kinetics of actin polymerization. An explanation for certain in vivo observations. J Biol Chem 275 (7):5163–5170
10. Kunze B, Jansen R, Sasse F et al (1995) Chondramides A-D, new antifungal and cytostatic Depsipeptides from *Chondromyces crocatus* (Myxobacteria). Production, Physico-chemical and biological properties. J Antibiot 48 (11):1262–1266. https://doi.org/10.7164/antibiotics.48.1262
11. Sasse F, Kunze B, Gronewold TMA et al (1998) The Chondramides: cytostatic agents from Myxobacteria acting on the actin cytoskeleton. J Natl Cancer Inst 90 (20):1559–1563. https://doi.org/10.1093/jnci/90.20.1559
12. Waldmann H, Hu T, Renner S et al (2008) Totalsynthese von Chondramid C und Bindung an F-Aktin. Angew Chem 120 (34):6573–6577. https://doi.org/10.1002/ange.200801010
13. Sawitzky H, Liebe S, Willingale-Theune J et al (1999) The anti-proliferative agent jasplakinolide rearranges the actin cytoskeleton of plant cells. Eur J Cell Biol 78(6):424–433. https://doi.org/10.1016/S0171-9335(99)80085-5
14. Peterson JR, Mitchison TJ (2002) Small molecules, big impact: a history of chemical inhibitors and the cytoskeleton. Chem Biol 9 (12):1275–1285. https://doi.org/10.1016/S1074-5521(02)00284-3
15. Matthews JB, Smith JA, Hrnjez BJ (1997) Effects of F-actin stabilization or disassembly on epithelial cl- secretion and Na-K-2Cl cotransport. Am J Physiol Cell Physiol 272 (1):C254–C262
16. Crews P, Manes LV, Boehler M (1986) Jasplakinolide, a cyclodepsipeptide from the marine sponge, *Jaspis* sp. Tetrahedron Lett 27 (25):2797–2800. https://doi.org/10.1016/S0040-4039(00)84645-6
17. Braekman JC, Daloze D, Moussiaux B et al (1987) Jaspamide from the marine sponge *Jaspis johnstoni*. J Nat Prod 50(5):994–995. https://doi.org/10.1021/np50053a048
18. Zabriskie TM, Klocke JA, Ireland CM et al (1986) Jaspamide, a modified peptide from a *Jaspis* sponge, with insecticidal and antifungal activity. J Am Chem Soc 108(11):3123–3124. https://doi.org/10.1021/ja00271a062
19. Fabian I, Halperin D, Lefter S et al (1999) Alteration of actin organization by Jaspamide inhibits ruffling, but not phagocytosis or oxidative burst, in HL-60 cells and human monocytes. Blood 93(11):3994–4005
20. Terracciano S, Bruno I, D'Amico E et al (2008) Synthetic and pharmacological studies on new simplified analogues of the potent actin-targeting Jaspamide. Bioorg Med Chem 16(13):6580–6588. https://doi.org/10.1016/j.bmc.2008.05.019
21. Bubb MR, Senderowicz AM, Sausville EA et al (1994) Jasplakinolide, a cytotoxic natural product, induces actin polymerization and competitively inhibits the binding of phalloidin to F-actin. J Biol Chem 269 (21):14869–14871
22. Holzinger A, Meindl U (1997) Jasplakinolide, a novel actin targeting peptide, inhibits cell growth and induces actin filament polymerization in the green alga *Micrasterias*. Cell Motil Cytoskeleton 38(4):365–372. https://doi.org/10.1002/(SICI)1097-0169(1997)38:4<365:AID-CM6>3.0.CO;2-2
23. Sheikh S, Gratzer WB, Pinder JC et al (1997) Actin polymerisation regulates integrin-mediated adhesion as well as rigidity of neutrophils. Biochem Biophys Res Commun 238 (3):910–915. https://doi.org/10.1006/bbrc.1997.7407
24. Lee E, Shelden EA, Knecht DA (1998) Formation of F-actin aggregates in cells treated with actin stabilizing drugs. Cell Motil Cytoskeleton 39(2):122–133. https://doi.org/10.1002/(SICI)1097-0169(1998)39:2<122:AID-CM3>3.0.CO;2-8
25. Visegrády B, Lorinczy D, Hild G et al (2005) A simple model for the cooperative stabilisation of actin filaments by phalloidin and

jasplakinolide. FEBS Lett 579(1):6–10. https://doi.org/10.1016/j.febslet.2004.11.023

26. Murray LM, Johnson A, Diaz MC et al (1997) Geographic variation in the tropical marine sponge *Jaspis cf. johnstoni*: an unexpected source of new terpene-benzenoids. J Org Chem 62:5638–5641
27. Fabian I, Shur I, Bleiberg I et al (1995) Growth modulation and differentiation of acute myeloid leukemia cells by jaspamide. Exp Hematol 23(7):583–587
28. Ebada SS, Müller WEG, Lin W et al (2019) New acyclic cytotoxic jasplakinolide derivative from the marine sponge *Jaspis splendens*. Mar Drugs 17:100. https://doi.org/10.3390/md17020100
29. Ayscough KR, Stryker J, Pokala N et al (1997) High rates of actin filament turnover in budding yeast and roles for actin in establishment and maintenance of cell polarity revealed using the actin inhibitor latrunculin-a. J Cell Biol 137(2):399–416
30. Chu KS, Negrete GR, Konopelski JP (1991) Asymmetric total synthesis of (+)-jasplakinolide. J Org Chem 56(17):5196–5202. https://doi.org/10.1021/jo00017a037
31. Inman W, Crews P (1989) Novel marine sponge-derived amino acids. 8. Conformational analysis of jasplakinolide. J Am Chem Soc 111(8):2822–2829. https://doi.org/10.1021/ja00190a013
32. Kahn M, Nakanishi H, Su T et al (1991) Design and synthesis of nonpeptide mimetics of jaspamide. Int J Pept Protein Res 38(4):324–334
33. Ghosh AK, Moon DK (2007) Enantioselective total synthesis of +−jasplakinolide. Org Lett 9(12):2425–2427. https://doi.org/10.1021/ol070855h
34. Scott VR, Boehme R, Matthews TR (1988) New class of antifungal agents: jasplakinolide, a cyclodepsipeptide from the marine sponge, *Jaspis* species. Antimicrob Agents Chemother 32(8):1154–1157
35. Hable WE, Miller NR, Kropf DL (2003) Polarity establishment requires dynamic actin in fucoid zygotes. Protoplasma 221(3-4):193–204. https://doi.org/10.1007/s00709-002-0081-0
36. Ou GS, Chen ZL, Yuan M (2002) Jasplakinolide reversibly disrupts actin filaments in suspension-cultured tobacco BY-2 cells. Protoplasma 219(3-4):168–175. https://doi.org/10.1007/s007090200018
37. Cárdenas L, Lovy-Wheeler A, Wilsen KL et al (2005) Actin polymerization promotes the reversal of streaming in the apex of pollen tubes. Cell Motil Cytoskeleton 61(2):112–127. https://doi.org/10.1002/cm.20068
38. Thomas SG, Huang S, Li S et al (2006) Actin depolymerization is sufficient to induce programmed cell death in self-incompatible pollen. J Cell Biol 174(2):221–229. https://doi.org/10.1083/jcb.200604011
39. Senderowicz AM, Kaur G, Sainz E et al (1995) Jasplakinolide's inhibition of the growth of prostate carcinoma cells in vitro with disruption of the actin cytoskeleton. J Natl Cancer Inst 87(1):46–51
40. da Costa SR, Yarber FA, Zhang L et al (1998) Microtubules facilitate the stimulated secretion of beta-hexosaminidase in lacrimal acinar cells. J Cell Sci 111(Pt 9):1267–1276
41. Holzinger A, Lütz-Meindl U (2001) Chondramides, novel cyclodepsipeptides from myxobacteria, influence cell development and induce actin filament polymerization in the green alga *Micrasterias*. Cell Motil Cytoskeleton 48(2):87–95. https://doi.org/10.1002/1097-0169(200102)48:2<87:AID-CM1000>3.0.CO;2-C
42. Jansen R, Kunze B, Reichenbach H et al (1996) Chondramides A-D, new cytostatic and antifungal cyclodepsipeptides from *Chondromyces crocatus* (myxobacteria): isolation and structure elucidation. Liebigs Ann 1996:285–290
43. Ma CI, Diraviyam K, Maier ME et al (2013) Synthetic chondramide a analogues stabilize filamentous actin and block invasion by *Toxoplasma gondii*. J Nat Prod 76(9):1565–1572. https://doi.org/10.1021/np400196w
44. Foerster F, Braig S, Moser C et al (2014) Targeting the actin cytoskeleton: selective antitumor action via trapping PKCε. Cell Death Disease 5:e1398. https://doi.org/10.1038/cddis.2014.363
45. Menhofer MH, Kubisch R, Schreiner L et al (2014) The actin targeting compound Chondramide inhibits breast cancer metastasis via reduction of cellular contractility. PLoS One 9(11):e112542. https://doi.org/10.1371/journal.pone.0112542
46. Hussey PJ, Ketelaar T, Deeks MJ (2006) Control of the actin cytoskeleton in plant cell growth. Ann Review Plant Biol 57:109–125. https://doi.org/10.1146/annurev.arplant.57.032905.105206

47. Url T, Höftberger M, Meindl U (1993) Cytochalasin B influences dictyosomal vesicle production and morphogenesis in the desmid *Euastrum*. J Phycol 29(5):667–674. https://doi.org/10.1111/j.0022-3646.1993.00667.x
48. Höftberger M, Lütz-Meindl U (1999) Septum formation in the desmid *Xanthidium* (Chlorophyta): effects of cytochalasin D and latrunculin B suggest the involvement of actin microfilaments. J Phycol 35(4):768–777. https://doi.org/10.1046/j.1529-8817.1999.3540768.x
49. Foissner I, Wasteneys GO (2007) Wide-ranging effects of eight cytochalasins and latrunculin a and B on intracellular motility and actin filament reorganization in characean internodal cells. Plant Cell Physiol 48 (4):585–597. https://doi.org/10.1093/pcp/pcm030
50. Gibbon BC, Kovar DR, Staiger CJ (1999) Latrunculin B has different effects on pollen germination and tube growth. Plant Cell 11 (12):2349–2363
51. Scherlach K, Boettger D, Remme N et al (2010) The chemistry and biology of cytochalasans. Nat Prod Rep 27(6):869–886. https://doi.org/10.1039/b903913a
52. Bonder EM (1986) Cytochalasin B slows but does not prevent monomer addition at the barbed end of the actin filament. J Cell Biol 102(1):282–288. https://doi.org/10.1083/jcb.102.1.282
53. Aldridge DC, Armstrong JJ, Speake RN et al (1967) The cytochalasins, a new class of biologically active mould metabolites. Chem Commun (London) 1:26. https://doi.org/10.1039/c19670000026
54. Rothweiler W, Tamm C (1966) Isolation and structure of Phomin. Experientia 22 (11):750–752. https://doi.org/10.1007/BF01901360
55. Coué M, Brenner SL, Spector I et al (1987) Inhibition of actin polymerization by latrunculin a. FEBS Lett 213(2):316–318. https://doi.org/10.1016/0014-5793(87)81513-2
56. Spector I, Shochet N, Kashman Y et al (1983) Latrunculins: novel marine toxins that disrupt microfilament organization in cultured cells. Science 219(4584):493–495. https://doi.org/10.1126/science.6681676
57. Baluska F, Jasik J, Edelmann HG et al (2001) Latrunculin B-induced plant dwarfism: plant cell elongation is F-actin-dependent. Dev Biol 231(1):113–124. https://doi.org/10.1006/dbio.2000.0115
58. Holzinger A, Wasteneys GO, Lütz C (2007) Investigating cytoskeletal function in chloroplast protrusion formation in the arctic-alpine plant *Oxyria digyna*. Plant Biol (Stuttg) 9 (3):400–410. https://doi.org/10.1055/s-2006-924727
59. Craig EW, Mueller DM, Bigge BM et al (2019) The elusive actin cytoskeleton of a green alga expressing both conventional and divergent actins. Mol Biol Cell 30(22):2827–2837. https://doi.org/10.1091/mbc.E19-03-0141
60. Onishi M, Umen JG, Cross FR et al (2020) Cleavage-furrow formation without F-actin in *Chlamydomonas*. Proc Natl Acad Sci U S A 117 (31):18511–18520. https://doi.org/10.1073/pnas.1920337117
61. Takáč T, Bekešová S, Šamaj J (2017) Actin depolymerization-induced changes in proteome of *Arabidopsis* roots. J Proteome 153:89–99. https://doi.org/10.1016/j.jprot.2016.06.010
62. Kumar MN, Bau Y-C, Longkumer T et al (2019) Low water potential and At14a-Like1 (AFL1) effects on endocytosis and actin filament organization. Plant Physiol 179 (4):1594–1607. https://doi.org/10.1104/pp.18.01314
63. Carmely S, Kashman Y (1985) Structure of swinholide-a, a new macrolide from the marine sponge *Theonella swinhoei*. Tetrahedron Lett 26(4):511–514. https://doi.org/10.1016/S0040-4039(00)61925-1
64. Humisto A, Jokela J, Liu L et al (2018) The swinholide biosynthesis gene cluster from a terrestrial cyanobacterium, *Nostoc* sp. strain UHCC 0450. Appl Environ Microbiol 84(3): e02321–e02317. https://doi.org/10.1128/AEM.02321-17
65. Bubb MR, Spector I, Bershadsky AD et al (1995) Swinholide a is a microfilament disrupting marine toxin that stabilizes actin dimers and severs actin filaments. J Biol Chem 270 (8):3463–3466
66. Craig E, Avasthi P (2019) Visualizing filamentous actin in *Chlamydomonas reinhardtii*. Bio-Protoc 9:e3274
67. Klenchin VA, King R, Tanaka J et al (2005) Structural basis of swinholide a binding to actin. Chem Biol 12(3):287–291. https://doi.org/10.1016/j.chembiol.2005.02.011
68. Posey SC, Bierer BE (1999) Actin stabilization by jasplakinolide enhances apoptosis induced by cytokine deprivation. J Biol Chem 274 (7):4259–4265

69. Braet F, Spector I, de Zanger R et al (1998) A novel structure involved in the formation of liver endothelial cell fenestrae revealed by using the actin inhibitor misakinolide. Proc Natl Acad Sci U S A 95(23):13635–13640
70. Blaas K, Holzinger A (2017) F-actin reorganization upon de- and rehydration in the aeroterrestrial green alga *Klebsormidium crenulatum*. Micron 98:34–38. https://doi.org/10.1016/j.micron.2017.03.012
71. Shurety W, Stewart NL, Stow JL (1998) Fluid-phase markers in the basolateral endocytic pathway accumulate in response to the actin assembly-promoting drug Jasplakinolide. Mol Biol Cell 9(4):957–975
72. Shaw MK, Tilney LG (1999) Induction of an acrosomal process in *Toxoplasma gondii*: visualization of actin filaments in a protozoan parasite. Proc Natl Acad Sci U S A 96 (16):9095–9099
73. Sheahan MB, Staiger CJ, Rose RJ et al (2004) A green fluorescent protein fusion to actin-binding domain 2 of *Arabidopsis* fimbrin highlights new features of a dynamic actin cytoskeleton in live plant cells. Plant Physiol 136 (4):3968–3978. https://doi.org/10.1104/pp.104.049411
74. Voigt B, Timmers ACJ, Samaj J, Müller J, Baluska F, Menzel D (2005) GFP-FABD2 fusion construct allows in vivo visualization of the dynamic actin cytoskeleton in all cells of *Arabidopsis* seedlings. Eur J Cell Biol 84 (6):595–608. https://doi.org/10.1016/j.ejcb.2004.11.011
75. Meindl U (1990) Effects of temperature on cytomorphogenesis and ultrastructure of *Micrasterias denticulata* Bréb. Protoplasma 157:3–18
76. Meindl U, Lancelle S, Hepler PK (1992) Vesicle production and fusion during lobe formation in *Micrasterias* visualized by high-pressure freeze fixation. Protoplasma 170 (3-4):104–114. https://doi.org/10.1007/BF01378786
77. Holzinger A, de Ruijter N, Emons AM et al (1999) Spectrin-like proteins in green algae (Desmidiaceae). Cell Biol Int 23(5):335–344. https://doi.org/10.1006/cbir.1999.0365
78. Baluska F, Parker JS, Barlow PW (1992) Specific patterns of cortical and endoplasmic microtubules associated with cell growth and tissue differentiation in roots of maize (*Zea mays* L.). J Cell Sci 103(1):191–200
79. Chichili GR, Rodgers W (2007) Clustering of membrane raft proteins by the actin cytoskeleton. J Biol Chem 282(50):36682–36691. https://doi.org/10.1074/jbc.M702959200
80. de Marino S, Festa C, D'Auria MV et al (2011) Swinholide J, a potent cytotoxin from the marine sponge *Theonella swinhoei*. Mar Drugs 9(6):1133–1141. https://doi.org/10.3390/md9061133

Chapter 10

Analysis of Morphogenesis and Flagellar Assembly During Spermatogenesis in Planarian Flatworms

Labib Rouhana, Tracy Chong, and Phillip A. Newmark

Abstract

Spermatogenesis is one of the most dramatic cellular differentiation events observed in animals. In particular, spermiogenesis (the final stage of spermatogenesis) involves extensive shedding of cytoplasmic organelles, dramatic nuclear rearrangements, and assembly of long flagellar structures. In planarian flatworms, the spherical nucleus present in round spermatids elongates to produce the filamentous nucleus of mature sperm. Newly formed cortical microtubules participate in cytoskeletal rearrangements observed during spermiogenesis and remain present in sperm. In addition, a pair of flagella assemble at one end of each spermatid in a process that likely involves de novo formation of centrioles. This chapter includes a brief introduction to planarian spermatogenesis and current tools for the analysis of molecular players in this process. Step-by-step protocols for isolating and imaging spermatogenic cells are provided with enough detail to be carried out by newcomers to the field who would like to study this unique organism in the laboratory.

Key words Spermatogenesis, Spermiogenesis, Flagella, Manchette, Microtubules, Planarian, Platyhelminthes

1 Introduction

The post-meiotic events that take place during spermatogenesis showcase some of the most dramatic cellular differentiation processes present in animal development. During this process, which is defined as spermiogenesis, spermatids of most animal species develop one or more flagella and undergo extensive cytoskeletal rearrangements to pack the cell nucleus into a compact sperm head. The mechanisms of flagellar assembly during spermiogenesis are largely conserved across animal species and apply many of the same molecular processes present during cilia formation in the soma [1, 2]. On the other hand, cytoskeletal rearrangements that take place during spermiogenesis give rise to a remarkable range of diverse morphologies, even in closely related phyla [3]. Because of

Ray H. Gavin (ed.), *Cytoskeleton: Methods and Protocols*, Methods in Molecular Biology, vol. 2364, https://doi.org/10.1007/978-1-0716-1661-1_10,

the analogy with somatic processes, more is known about flagellar assembly than about the mechanisms that drive structural changes during sperm head formation. However, ongoing studies suggest that molecular activities that drive sperm head development are also highly conserved in distant species, but differences in the extent of their activity and presence result in the vastly diverse morphologies observed throughout the animal kingdom. For example, a perinuclear array of microtubules required for cytoskeletal rearrangements of the sperm head (i.e., the manchette) is only present briefly during spermatid elongation in mice [4, 5]. In contrast, similar arrays of cortical microtubules appear during spermatid elongation and remain present in sperm of species that produce sperm with filamentous nuclear structures, such as the planarian *Schmidtea mediterranea* and other Platyhelminthes [6–11]. Current experimental techniques allow the analysis of molecular mechanisms underlying the cytoskeletal rearrangements observed in planarian spermatogenesis, facilitating our understanding of processes involved in development of this rapidly evolving cell type.

Protocols for inducing systemic RNA interference (RNAi) have facilitated identification of genes that regulate stem cell dynamics, regeneration, and establishment of developmental polarity in planarian flatworms [12–15]. The use of this organism as a model has also increased due to availability of genomic and transcriptomic resources [16–21], including single-cell RNAseq data for thousands of cells [22–26], as well as from optimized protocols for in situ hybridization that allow assessment of gene expression at the cellular level in whole-mount samples [27–29]. More recently, an increasing number of researchers have turned to planarians to uncover mechanisms involved in motile cilia assembly and function [30–32]. Planarians offer important advantages for the analysis of cilia. First, specified cells in these organisms, including those that possess motile cilia in the epidermis and protonephridia, undergo continuous turnover and replacement by differentiating somatic stem cells. This allows for structural analysis of cilia in vivo after RNAi is induced to disrupt expression of specific genes of interest. Second, mitotic cells in planarians lack centrosomes, and centrioles are formed during terminal differentiation of ciliated cells [33], which allows for analysis of the entire process of ciliary assembly de novo. Finally, disruption of essential ciliary activity has no immediate direct effect upon planarian viability, but it does result in physiological readouts that are visible to the naked eye, such as abnormal motility and bloating [34–37]. Detailed protocols for analysis of ciliary structures in the planarian soma are available [31, 32] and, although not covered in this chapter, are inherent to the analysis of sperm flagella.

Although most of the current work using planarians as research organisms is performed using asexual strains, the same phenotypic readouts of ciliary dysfunction in the soma (abnormal motility and

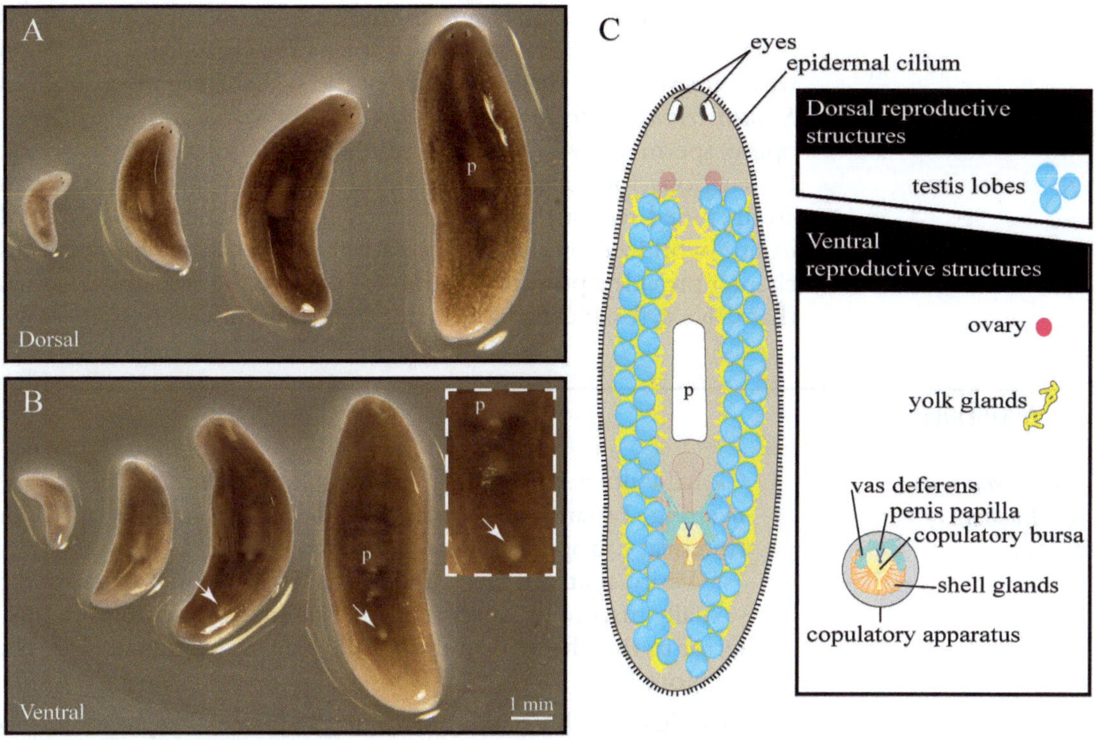

Fig. 1 Dorsal (**a**) and ventral (**b**) views of sexual planarians of different sizes (sp. *Schmidtea mediterranea*). The gonopore (arrows) is an external indicator of sexual maturity that is visible ventrally, posterior to the pharynx (p). Inset in (**b**) shows twofold magnified view. (**c**) Illustration depicting photoreceptors (eyes), epidermal cilia, and the anatomy of the hermaphroditic reproductive system of mature sexual planarians. Legend boxes on the right indicate dorsal and ventral structures that develop post-embryonically in sexual planarians

bloating) are observed in sexual strains [38, 39]. A constitutively sexual laboratory strain of *S. mediterranea* (Fig. 1a, b) [40] is routinely used for analysis of gene function during germline development [41]. Sexually mature planarians commit much of their body to the development of sperm, which is readily observed using nuclear dyes to stain testis lobes present dorsolaterally throughout these animals (Fig. 1c) [42, 43]. Planarian spermiogenesis involves dramatic elongation of the nucleus and head, as well as assembly of a pair of flagella, giving rise to biflagellate sperm with a filamentous nucleus [7, 44–46]. Evidence suggests that the dramatic elongation of developing sperm in planarians may result from the extended presence of manchette-like structures throughout the final stages of spermiogenesis; these structures remain as a ring of cortical microtubules surrounding the nucleus of sperm [11]. Since RNAi in planarians is systemic, and development of the germline occurs post-embryonically [42], functional contributions of cytoskeletal components can be analyzed simultaneously in somatic cells and developing gametes of sexual planarians. This approach is exemplified in recent publications [38, 39, 47].

This chapter includes simple step-by-step instructions to assess progression of spermatogenesis in planarian flatworms. We begin with directions for the husbandry of the sexual laboratory strain of *S. mediterranea*, followed by protocols for visualization of developing sperm in whole-mount samples and isolated testis lobes. Instructions for surgical release and structural analysis of sperm are included to conclude the chapter. Combination of RNAi and in situ hybridization techniques with the methods described in this chapter provide a powerful option for uncovering conserved factors that drive cytoskeletal changes during sperm differentiation.

2 Materials

2.1 Planarians and Husbandry Medium

1. Clonal cultures of sexual planarians (sp. *Schmidtea mediterranea*, Fig. 1a) are available from academic laboratories in North America, Europe, and Asia (*see* **Note 1**).

   Verify that samples used to analyze spermatogenesis are sexually mature by observing the presence of a gonopore posterior to the pharynx on the ventral side of the animal (Fig. 1b, *see* **Note 2**).
2. Montjuïc salts husbandry medium [48]: Use ultrapure water to make separate filtered stocks of 5 M NaCl, 1 M $CaCl_2$, 1 M $MgSO_4$, 1 M $MgCl_2$, and 1 M KCl. A **5× Montjuïc salts** concentrated stock solution is prepared by adding 1.6 mL of 5 M NaCl, 5 mL of 1 M $CaCl_2$, 5 mL of 1 M $MgSO_4$, 0.5 mL of 1 M $MgCl_2$, 0.5 mL of 1 M KCl, and 0.504 g of $NaHCO_3$ per liter of ultrapure water. Routinely, the components for 9 L of a 5× stock solution are mixed in a 10-L carboy for 20 min, before adding ~6.3 mL of 2 N HCl to adjust the pH and stirring overnight. After overnight stirring and verification of a pH in the range of 6.8–7.2, the 5× stock is diluted to a working concentration of **0.75×** (e.g., 2.25 L of 5× Montjuïc salts added to 12.75 L of ultrapure water in a 15 L carboy) and stored at room temperature (*see* **Notes 3** and **4**).
3. Large plastic containers (e.g., Ziploc 40 oz. or 72 oz. plastic container, BPA-free).
4. Organic calf liver (e.g., Golden Forest calf liver, Fremont Beef Company, Fremont, NE) stored at −80 °C as chunks or minced aliquots (*see* **Note 5**).
5. Disposable plastic transfer pipettes.
6. 500 mL wash bottles.
7. Disposable paper towels.

2.2 Whole-Mount Sample Fixation and DAPI Staining

1. Phosphate-buffered saline (PBS): A 10× PBS stock solution is prepared by adding 80 g of NaCl, 2 g of KCl, 14.4 g of Na_2HPO_4, 2.4 g of KH_2HPO_4, and ultrapure water to 1 L.

Adjust to pH 7.4 with HCl and autoclave. Dilute to 1× PBS working solution using ultrapure water. Store at room temperature.

2. PBSTx: phosphate-buffered saline (PBS) supplemented with 0.3% Triton X-100 – Use 3 mL of Triton X-100 per liter of 1× PBS and stir until completely dissolved. Store at room temperature.
3. 5 mL microcentrifuge tubes.
4. *N*-Acetyl-L-cysteine (NAC) solution: 8% NAC in PBS.
5. PBSTx fixative solution: PBSTx containing 4% final concentration of formaldehyde. Prepared fresh by diluting 36.5–38% formaldehyde stock in PBSTx.
6. Formamide bleaching solution as per King and Newmark [28], made by combining 9 mL of ultrapure water, 0.5 mL of formamide, 0.25 mL of 20× saline sodium citrate (SSC, commercially available), and 0.4 mL of 30% hydrogen peroxide (*see* **Note 6**).
7. 1 mg/mL DAPI stock solution: Dissolve 10 mg of 4′,6-′-diamidino-2-phenylindole dihydrochloride (DAPI) in 10 mL of ultrapure water. Stock solution can be stored at 4 °C for 6–12 months.
8. Mounting solution: glycerol diluted to 80% with PBS.
9. Microscope slides and 24 × 40 mm coverslips.
10. Non-hardening modeling clay.
11. Fluorescence stereomicroscope with ultraviolet light wavelength excitation and filter capabilities (*see* **Note 7**).
12. Confocal microscope.

2.3 Isolation of Testis Lobes and Labeling of Developing Sperm Nuclei and Flagella

1. Maceration solution [49]: 7.4 mL ultrapure water, 1 mL of glycerol, 1.5 mL methanol, and 0.1 mL of glacial acetic acid. Make fresh before use.
2. 5 mL microcentrifuge tubes.
3. Falcon 60 mm × 15 mm or 35 mm × 10 mm bacteriological Petri dish or equivalent.
4. Graduated 200 μL pipette tips and P200 micropipette.
5. Phosphate-buffered saline (PBS).
6. PBSTx: PBS containing 0.3% Triton X-100.
7. PBS fixative solution: PBS containing 4% final concentration of formaldehyde, freshly prepared by diluting a commercially available 36.5–38% formaldehyde stock with PBS.
8. Blocking solution: PBSTx containing 1% bovine serum albumin and 0.45% fish gelatin (40–50% stock solution, available commercially). Store at 4 °C for up to 2 weeks.

9. Primary antibody solution: blocking solution supplemented with 1:400 dilution of monoclonal anti-α-tubulin antibody (clone B-5-1-2; Cat. No. T5168, Sigma Aldrich, St. Louis, MO) or alternative.
10. Secondary antibody solution: blocking solution supplemented with 1:400 dilution of Alexa Fluor 488-conjugated goat-anti-mouse IgG.
11. DAPI.
12. Mounting solution: glycerol diluted to 80% in PBS.
13. Superfrost Plus microscope slides and appropriate coverslips.

2.4 Visualization of Individual Sperm by Surgical Release from Vas Deferens

1. Phosphate-buffered saline (PBS).
2. PBSTx.
3. PBS fixative solution.
4. Blocking solution.
5. Primary antibody solution.
6. Secondary antibody solution.
7. DAPI.
8. Mounting solution.
9. Superfrost Plus microscope slides and slide covers.

3 Methods

3.1 Handling and Maintenance of Planarian Colonies

1. Clean plastic husbandry containers by rinsing and scrubbing twice with tap water, deionized water, and ultrapure water (*see* **Note 8**). Dry containers using disposable paper towels. Planarians can be kept in the same container for long periods of time. However, the containers must be cleaned once every 1–2 weeks or after every feeding session.
2. Upon receipt or collection of samples, transfer gradually into 0.75× Montjuïc salts husbandry medium. Use a 1:1 dilution of the solution in which planarians are found and 0.75× Montjuïc salts for storage during the first week. If planarian morphology looks normal (Fig. 1a), transfer planarians to 0.75× Montjuïc salts husbandry medium. For studies of spermatogenesis, verify the presence of a gonopore in ~1 cm or larger planarians (Fig. 1b). The gonopore indicates that planarians are sexual and have progressed through the post-embryonic process of reproductive system development (Fig. 1c, *see* **Note 2**).
3. Maintain planarians under dark conditions as much as possible and leave container lids slightly open to allow for aeration. Sexual *S. mediterranea* cultures are maintained between 17 °C and 19 °C. Temperature adjustments may be needed to mimic the natural habitat for other strains or species.

4. Planarians should be fed at least once every 2 weeks to maintain their size; starvation periods for longer than 2 weeks will lead to de-growth and loss of reproductive structures. A gonopore will likely not be present in planarians that are smaller than 0.8 mm (Fig. 1b). If this is the case, increase feeding frequency to once or twice per week (with at least 3 days between feedings) to drive growth and development of reproductive structures.
5. To feed, place a piece of organic calf liver ~1 cm^3 in size into the husbandry container housing planarians and allow planarians to eat for 1–2 h at room temperature and away from direct light. If food is completely eaten, then increase liver quantity in subsequent feedings. Planarians do not often eat floating liver, so remove air bubbles from liver that fails to sink to the bottom of the container.
6. Clean containers (as in **step 1**) after feeding and replace husbandry medium. A squirt bottle can be used to rinse out planarians, which can also be placed in a clean Petri dish while cleaning the container and replacing old medium.
7. To increase population size under laboratory husbandry, cut planarians transversely anterior and posterior to the pharynx with a clean scalpel. Clean after an hour and then allow to regenerate for 2 weeks before reinstating feedings (*see* **Note 9**).
8. Planarians are usually not fed the week prior to experimentation in order to avoid artifacts related to contamination or mitotic increase after feeding. Similarly, planarians that have been recently amputated or injured will lose their reproductive structures; therefore, it is best to allow them to heal, regenerate, and grow before using them in experiments.

3.2 Whole-Mount Analysis of Testis Lobes by DAPI Staining

Spermiogenesis in planarians involves dramatic elongation of the spermatid head and nucleus. Sexual planarians can be fixed and analyzed by whole-mount staining with nuclear fluorescent dyes such as DAPI or Hoechst, which reveal the presence and structure of sperm and sperm precursors in planarian testis lobes under fluorescence microscopy. This process can be combined with established protocols for RNAi [15] to assess gene function during spermatogenesis.

1. Transfer up to six sexual planarians to a 5 mL centrifuge tube using a transfer pipette (*see* **Note 10**). The use planarians of at least 0.9 cm length and with visible gonopores (Fig. 1b) is needed to avoid artifacts caused by use of immature sexual animals.
2. Remove excess husbandry medium by decanting or using a 1 mL micropipette, and rinse twice with clean husbandry medium using a squirt bottle or by gently shaking the capped tube by hand for 5 s.

3. Remove husbandry medium, add 5 mL of a freshly prepared (within 15 min prior to use) solution of NAC, and gently rock planarians in the tubes horizontally for 10 min (*see* **Note 11**). Planarians are euthanized and external mucous removed by NAC. It is crucial to use PBS (not PBSTx) and to agitate gently by hand if samples begin to clump during this step.
4. Replace NAC solution with 5 mL of PBSTx fixative solution, and incubate tubes horizontally on a rocking platform at 4 °C for 1 h. Monitor every 15 min, and agitate briefly by hand if samples begin to clump.
5. Bring samples back to room temperature, remove fixative solution, and wash twice with 5 mL of PBSTx for 5 min on a rocking platform.
6. Bleach samples by incubating in formamide bleaching solution for 45 min at room temperature under a fluorescent white light.
7. Wash samples twice for 5 min each in PBSTx.
8. Replace PBSTx solution with PBSTx containing 1 μg/mL of DAPI, and incubate horizontally on a rocking platform for 2 h at room temperature or overnight at 4 °C.
9. Wash samples at room temperature 3× 10 min each with 5 mL of PBSTx.
10. Replace PBSTx with mounting solution.
11. Use a transfer pipette to place samples on microscope slides, and position them so that their ventral side faces the surface of slide (*see* **Note 12**).
12. Add non-hardening modeling clay to the corners of a slide cover and place on top of the samples, flatten, and seal the slide cover with clear nail polish (*see* **Notes 13** and **14**). Allow nail polish seal to dry.
13. Proceed to image under low magnification (Fig. 2a-a′, b) or confocal fluorescence microscopy (Fig. 2c-c′). This procedure can be combined with RNAi to analyze genetic requirements for sperm development (Fig. 3a, b). In addition, markers of specific stages during sperm development have been identified [42, 43, 50] and can be used for detailed phenotypic characterization through in situ hybridization (*see*, e.g., [38, 51, 52]). Plasmids for these markers are available through Addgene (Watertown, MA) and include *nanos* (germline stem cells; Addgene ID: 26601), *germinal histone H4* (spermatogonia; ID: 26600), *Smed-tkn-1* (spermatocytes; ID: 44686), and *Smed-pka* (spermatids; ID: 44667).

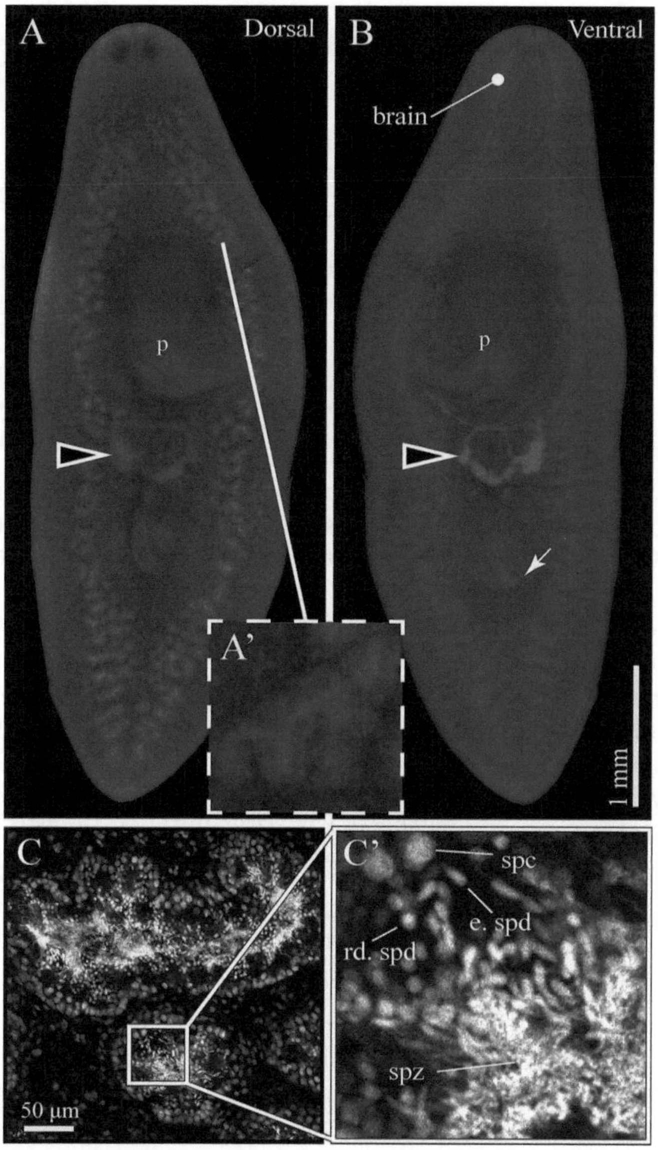

Fig. 2 (**a** and **b**) Dorsal (**a**) and ventral (**b**) views of a whole-mount sample stained with DAPI reveal structures with high cellular density in sexual planarians, such as the pharynx (p), brain, penis papilla (arrow), vas deferens (arrowhead), and testis lobes (inset; threefold magnification). (**c**-**c**′) Confocal section of the inner region of testis lobes in (**a**) reveals cells at progressive steps of spermatogenesis, including spermatocytes (spc), round spermatids (rd. spd), elongating spermatids (e. spd), and spermatozoa (spz), which are labeled in (**c**′; fivefold magnification)

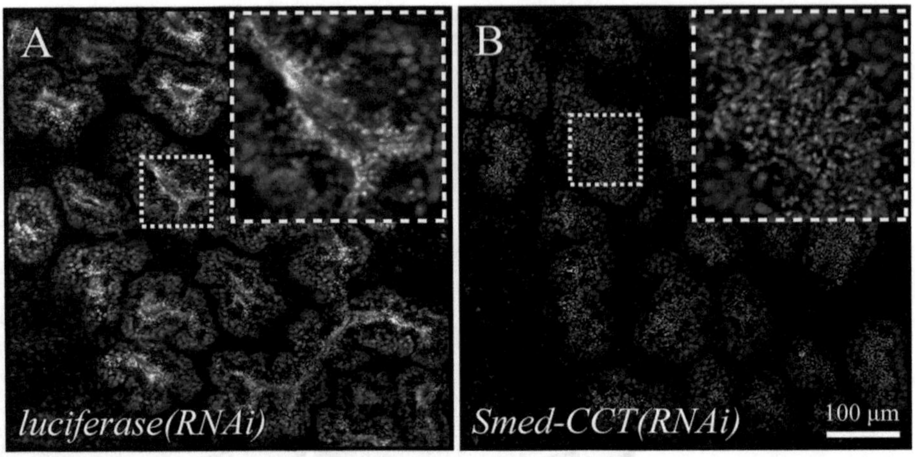

Fig. 3 Single confocal sections of whole-mount DAPI-stained (**a**) control (*luciferase(RNAi)*) and (**b**) testis-specific chaperonin-containing TCP-1 subunits knockdown planarians (*Smed-CCT(RNAi)*) [47] reveal testis lobes with normal and disrupted spermatid elongation, respectively. Insets show threefold magnified views

3.3 Analysis of Spermatid Elongation and Flagellar Development in Isolated Planarian Testis Lobes

It is possible to separate testis lobes from other planarian tissues by maceration, which allows assessment of structural changes in developing sperm in more detail than in whole-mount samples. This protocol allows visualization of spermatid head nuclear elongation and development of flagella under confocal microscopy. Sexual planarians longer than 1 cm are recommended, as these are most likely to have large numbers of testis lobes actively producing sperm under normal conditions.

1. Transfer individual sexual planarians (>1 cm in length) to 5 mL microcentrifuge tubes, one planarian per tube.
2. Remove carry over husbandry medium, and add 5 mL of maceration solution.
3. Incubate sample in maceration solution for 2 h in a nutating mixer, tube rotator, or rocking platform with moderate speed (~1 cycle per 3 s).
4. Decant contents of the tube into a small Petri dish while leaving large tissue in the tube.
5. Cut a graduated P200 tip at the 10 μL mark (Fig. 4a-a′), and use it to transfer testis lobes onto positively charged microscope slides. Testis lobes can be found by placing the Petri dish with the contents of the maceration under a dissecting microscope (Fig. 4b) and locating white clumps of ~0.1–0.2 mm in diameter (Fig. 4c).
6. Allow testis lobes to dry by placing slide on benchtop overnight (*see* **Note 15**).
7. Wash slides by placing them in a microscope slide staining jar filled with PBS for 10 min.

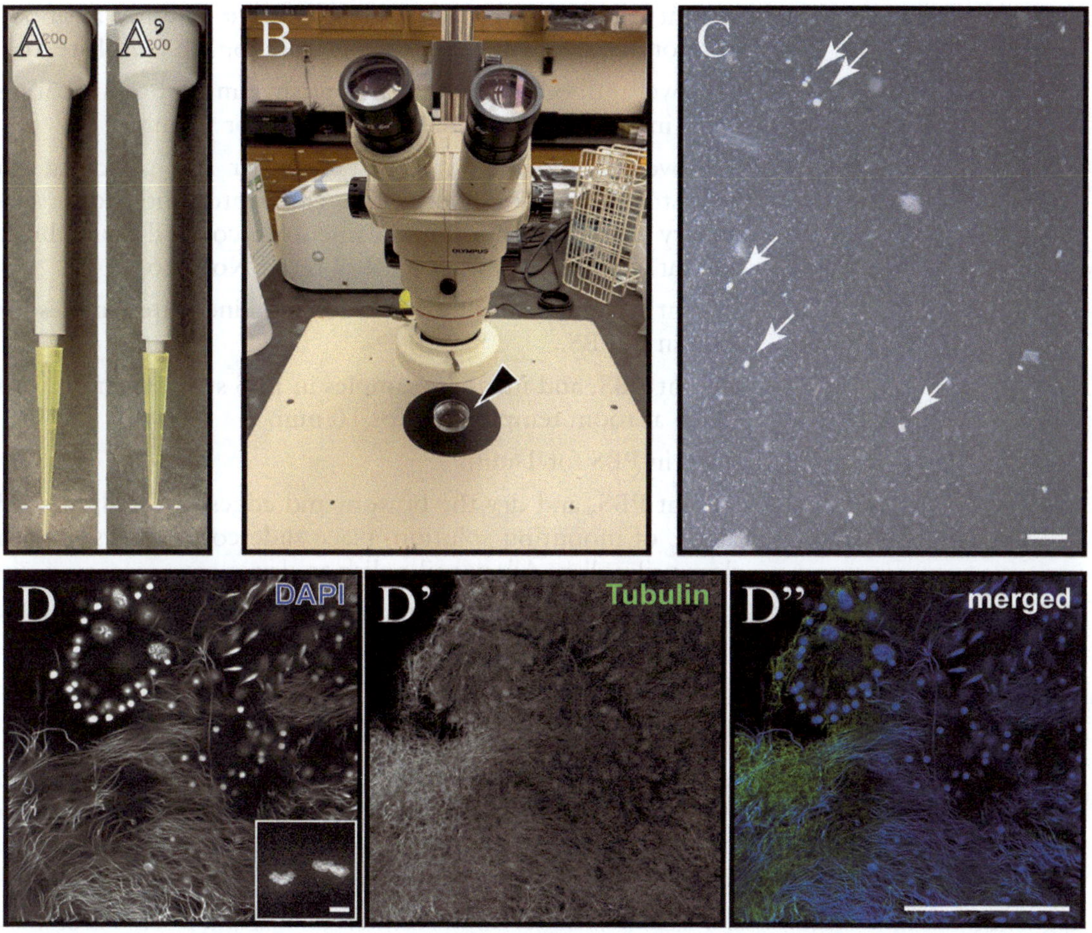

Fig. 4 Analysis of nuclear structures and flagella in testis lobes isolated by maceration. (**a-a**′) A 200μL pipette tip (**a**) cut at the 10μL mark (dashed line; **a**′) is used to collect testis lobes from maceration. (**b**) A Petri dish (arrowhead) holds the contents of the maceration mixture to be observed under a dissecting microscope. (**c**) Testis lobes appear as bright clumps of 0.1–0.2 mm diameter (arrows) in a field of single cells and larger macerated tissue. (**d-d**″) Single confocal sections of isolated testis lobes are observed under a 10× objective (inset) and imaged under a 60× oil immersion objective. Detection of DAPI staining reveals nuclear structures (**d**, blue in **d**″), whereas anti-α-Tubulin staining reveals flagella (**d**′, green in **d**″). Scale bars: 1 mm in (**c**), 0.2 mm in inset, and 50μm in (**d**″)

8. Dry the bottom and edges of the slide with a paper towel, and place on bench top.
9. Cover the top of the slide with PBS fixative solution, and incubate for 15 min.
10. Decant fixative solution, and incubate slides in a staining jar filled with PBS for 1 min.
11. Return slides to the benchtop, and cover with PBSTx. Incubate flat for 15 min.
12. Decant PBST, and cover the sample with blocking solution. Incubate on benchtop for 30 min.

13. Decant blocking solution, and add anti-α-Tubulin antibody solution. Incubate at room temperature for 2 h (*see* **Note 16**).
14. Remove antibody solution, and wash sample by incubating slides in a staining jar filled with PBSTx for 15 min.
15. Remove the slides from the staining jar and place flat on benchtop. Then, add fluorophore-conjugated anti-mouse secondary antibody solution, and incubate covered from direct light at room temperature for 1–2 h (*see* **Note 16**).
16. Decant secondary antibody solution, and incubate samples for 15 min in PBS.
17. Decant PBS, and incubate samples in PBS supplemented with DAPI at room temperature for 10 min.
18. Wash in PBS for 1 min.
19. Decant PBS, and dry the bottom and edges of the slide. Add 10 μL of mounting solution, place slide cover, and seal with clear nail polish. Allow nail polish to dry.
20. Proceed to fluorescence confocal microscopy. Find samples using 10× or 20× objective (Fig. 4d, inset), and image using higher magnification objectives (60× or above; Fig. 4d-d″). Slides can be stored briefly at 4 °C for future analysis.

3.4 Visualization of Individual Sperm by Surgical Release from Vas Deferens

Because of their length, abundance, and density, it is difficult to visualize and perform detailed measurements of individual sperm in whole-mount samples and isolated testis lobes. An approach to examine individual sperm in their entirety is to amputate planarians at the location of the vasa deferentia and release sperm onto slides.

1. Transfer a single sexual planarian (>1.2 cm in length) onto a positively charged microscope slide and remove carry-over husbandry medium, leaving only 5–10 μL behind.
2. While on the slide, amputate planarians immediately posterior to the pharynx using a glass microscope slide cover (Fig. 5a), and drag posterior fragment away from the anterior (Fig. 5a′). The goal is to amputate the planarian at the location of vasa deferentia (Fig. 5b) in order to release sperm onto the slide (*see* **Note 17**).
3. Remove planarian fragments and add a 10 μL drop of PBS onto the slide. Add a slide cover to proceed with direct observation by phase-contrast microscopy (Fig. 5c) or proceed with the following steps for staining and observation under fluorescence microscopy.
4. Allow slide to dry on a benchtop overnight.
5. Follow procedures for fixation, staining, and imaging (as in Subheading 3.3, **steps 7–20**). Planarian sperm are distinguished by their long and filamentous nucleus, which is readily observed after DAPI staining (Fig. 5d).

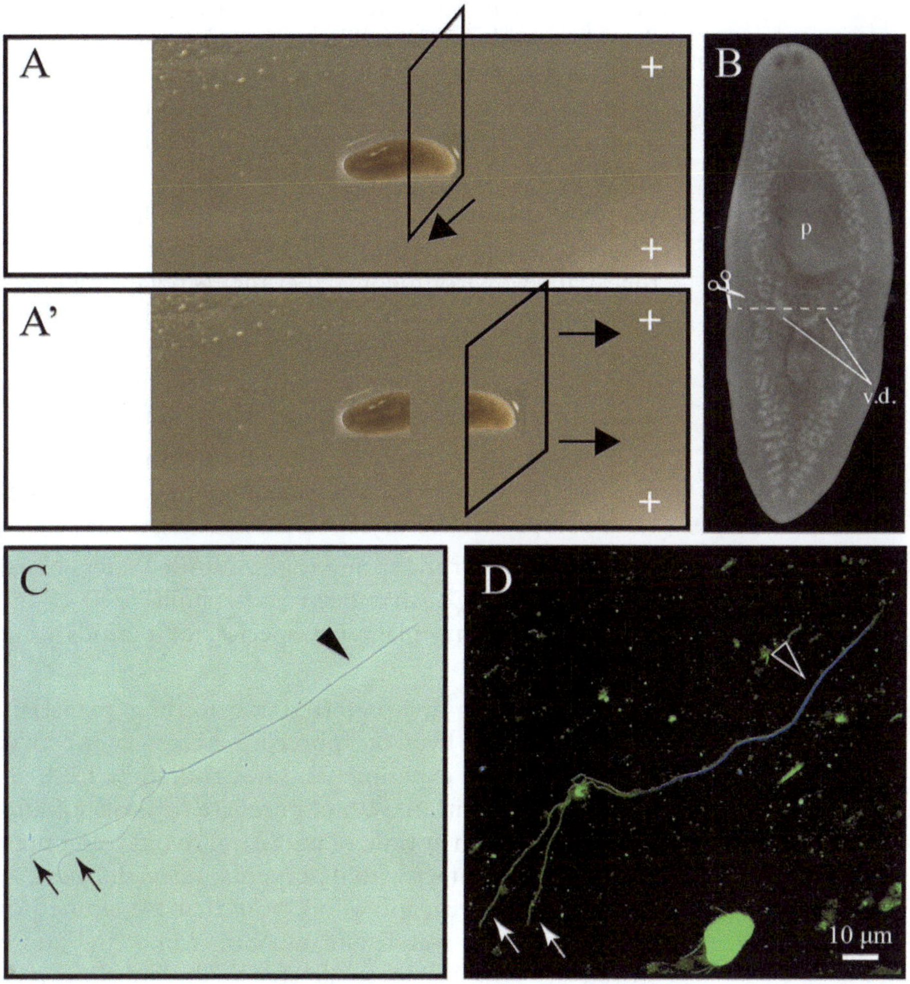

Fig. 5 Illustration depicting the release of sperm from sexually mature planarians by transverse amputation posterior to the pharynx (**a**) and removal of body fragments (**a'**). (**b**) The amputation (dashed line) is posterior to the pharynx (p) and targets the vasa deferentia (v.d.), in which sperm accumulate, as shown in a sample stained with DAPI. (**c**) Planarian spermatozoon imaged under phase-contrast microscopy using a Nikon Eclipse E2000 (Ph2 setting, 60× objective). (**d**) Planarian spermatozoon stained with DAPI (blue) and anti-α-Tubulin (green) imaged using a Nikon C2+ confocal microscope (60× oil immersion objective, maximum intensity of z-stack sections). Position of sperm nucleus (arrowheads) and flagella (arrows) is indicated in (**c**) and (**d**)

4 Notes

1. Sexual laboratory lines of *S. mediterranea* were derived from amputation and regeneration of samples originally collected from Sardinia in 1999 by Dr. Maria Pala [53]. This species of planarian flatworms is endemic to islands and coastal areas of the Mediterranean Sea [54], but other species of sexual planarians can be collected from the wild in different regions of the world by placing perforated plastic containers with a small piece

of liver as bait in shaded regions of freshwater ecosystems. Samples caught in the wild can be maintained in the laboratory in Montjuïc salts (*see* **Note 3**) or commercially available pasteurized spring water (*see* **Note 4**). It is important to mimic source temperature and maintain planarians in the dark as much as possible.

2. We have observed sexual planarians collected in the wild become asexual, shrink, and/or fission once removed from their native environment. Experiments regarding development of the germline will not be reliable if animals show signs of transitioning into an asexual state (e.g., by inducing fission or losing the gonopore) or if health of the animal seems compromised (blisters and head regression are signs of this).
3. The use of ultrapure deionized water (resistivity of 18.2 MΩ cm at 25 °C) is recommended. Carboys are autoclaved empty once every 6–12 months to avoid contamination.
4. Ice Mountain natural spring water (Nestlé Waters North America, Stamford, CT) has been successfully used as husbandry medium for some planarian species for periods longer than 6 months.
5. Chunks of liver are preferred for feeding planarian stocks, whereas minced liver or "puree" is necessary for mixing with dsRNA in RNAi experiments. Liver should be fresh and either minced or cut within 24 h of purchase (if fresh) or after partial thawing if frozen at time of purchase. Avoid using utensils that have been in contact with detergents. Instead, have a dedicated set of trays, knives, and a food mill that are cleaned only with water, ultrapure water, and ethanol. Minced aliquots can be stored at −80 °C in small sterile Petri dishes and chunks wrapped in aluminum foil.
6. To avoid a violent reaction, prepare the formamide bleaching solution by mixing the ultrapure water and SSC first, then adding formamide, and the hydrogen peroxide last. Scale up proportionally as needed.
7. We have successfully adapted dissecting microscopes with NIGHTSEA ultraviolet light and filter sets (Catalog No. SFA-LFS-UV, Electron Microscopy Sciences, Hatfield, PA) and a Moticam 1080 camera (Motic, Hong Kong) for low-magnification imaging of DAPI-stained whole-mount samples.
8. Planarians are very sensitive to detergents and bleach. Use gloves while cleaning husbandry containers to avoid depositing soap or other harsh chemicals that may be on your skin. Never clean husbandry containers with soap.

9. Cutting can be done by placing the planarian in a Petri dish with or without husbandry medium. However, the preferred method to generate clean cuts is to immobilize planarians by placing them on a double layer of Whatman paper slightly dampened with husbandry medium placed on top of an aluminum plate that sits on an ice bucket. After cuts are made, return planarian fragments to a container with clean husbandry medium, rinsing them off the Whatman paper with a transfer pipette or wash bottle.
10. Special care must be taken not to damage planarians during the transfer, for any lesion will likely result in sample degradation during NAC treatment. Decanting while using a squirt bottle with husbandry medium to generate flow is a safe method. Another precautionary step is to slightly enlarge the entry of a transfer pipette by cutting the tip. To move planarians using a transfer pipette, first detach planarians from the surface by squirting husbandry medium on them, suction into pipette as they float, and eject as quickly as possible into tube.
11. The speed setting of the rocking platform is particularly important at this step. Planarians will break down with too much agitation in NAC solution. Conversely, surface mucus will persist on samples without enough agitation, which restricts the quality of imaging in future steps. As a starting point, a GeneMate Variable Rocker (Item No. R-3200-1; VWR, Radnor, PA) is set to go through eight rocking cycles per minute in which the platform travels 8 cm vertically per half cycle. It is also beneficial to leave ~10% air space in the tube, which creates a bubble to move planarians in waves while rocking.
12. Testis lobes are difficult to image from the ventral side, so it is crucial to mount samples with the ventral side facing the slide and the dorsal side facing the coverslip. In order to determine the orientation of the samples, look for the eyes at the anterior tip of the sample. If DAPI staining of the brain is visible over the eyes (Fig. 2b), then the sample should be flipped using a small bristle brush.
13. The clay prevents the samples from being damaged but will not attach to a surface that is directly covered with mounting solution. Therefore, it is important to include enough mounting solution to prevent samples from drying out, but not so much that it gets in the way of attaching the coverslip. It may be necessary to clean around the samples with a moist Kimwipe, followed by dry one, or to transfer samples to new slides.
14. Flattening the samples leads to improvements during imaging. To flatten, place the slide assembly upside-down (slide cover at the bottom, then sample in mounting medium protected by

clay spacers, and slide on top) on a Kimwipe, and allow to sit on the benchtop for a couple of hours. Moderate weight may be added to augment flattening.

15. The drops of maceration solution may appear incompletely evaporated after overnight incubation, or even 2 or 3 days after transfer onto the slides. Nevertheless, testes lobes will be attached to the surface by this time and it is safe to proceed. Make sure to pipette solutions from the side, and not directly onto the sample, in subsequent steps to avoid detachment.
16. A piece of Parafilm the size of a rectangular coverslip can be placed on top of the slide to diminish loss of solution during long incubations.
17. The vasa deferentia are not visible to the naked eye, so performing a second amputation (about 1 mm posterior to the initial one) and applying pressure from the top while dragging fragments to the end of the slide may enhance sperm release. Attempt this procedure on multiple individuals, as even large sexually mature planarians are depleted of sperm in the vas deferens at times.

Acknowledgments

Special thanks to Sydney Lesko and Donovan Christman for assistance during development of these protocols. The authors are supported by NIH awards R15 HD082754 (LR) and R01 HD043403 (PAN). PAN is an investigator of the Howard Hughes Medical Institute.

References

1. Ishikawa M, Marshall WF (2011) Ciliogenesis: building the cell's antenna. Nat Rev Mol Cell Biol 12:222–234
2. King SM (2016) Axonemal Dynein Arms. Cold Spring Harb Perspect Biol 8(11): a028100
3. Pitnick S, Hosken DJ, Birkhead TR (2009) Sperm morphological diversity. In: Birkhead TR, Hosken DJ, Pitnick S (eds) Sperm Biology: An evolutionary perspective. Academic Press, Cambridge, MA, pp 69–149
4. Kierszenbaum AL (2002) Intramanchette transport (IMT): managing the making of the spermatid head, centrosome, and tail. Mol Reprod Dev 63:1–4
5. Lehti MS, Sironen A (2016) Formation and function of the manchette and flagellum during spermatogenesis. Reproduction 151:R43–R54
6. Thomas MB, Henley C (1971) Substructure of the cortical singlet microtubules in spermatozoa of Macrostomum (Platyhelminthes, Turbellaria) as revealed by negative staining. Biol Bull 141(3):592–601
7. Silveira M, Porter KR (1964) The spermatozoids of flatworms and their microtubular systems. Protoplasma 59:240–265
8. Justine JL (1998) Spermatozoa as phylogenetic characters for the Eucestoda. J Parasitol 84:385–408
9. Justine JL, Lambert A, Mattei X (1985) Spermatozoon ultrastructure and phylogenetic relationships in the monogeneans (Platyhelminthes). Int J Parasitol 15:601–608
10. Justine JL, Poddubnaya LG (2018) Spermiogenesis and spermatozoon ultrastructure in basal polyopisthocotylean monogeneans,

Hexabothriidae and Chimaericolidae, and their significance for the phylogeny of the Monogenea. Parasite 25:7

11. Harrath AH, Alwasel S, Zghal F et al (2012) Ultrastructure of spermatogenesis and mature spermatozoon of the freshwater planarian Schmidtea mediterranea (Platyhelminthes, Paludicola). C R Biol 335:87–95
12. Shibata N, Agata K (2018) RNA interference in planarians: feeding and injection of synthetic dsRNA. Methods Mol Biol 1774:455–466
13. Newmark PA, Reddien PW, Cebria F et al (2003) Ingestion of bacterially expressed double-stranded RNA inhibits gene expression in planarians. Proc Natl Acad Sci U S A 100 (Suppl 1):11861–11865
14. Sanchez Alvarado A, Newmark PA (1999) Double-stranded RNA specifically disrupts gene expression during planarian regeneration. Proc Natl Acad Sci U S A 96:5049–5054
15. Rouhana L, Weiss JA, Forsthoefel DJ et al (2013) RNA interference by feeding in vitro-synthesized double-stranded RNA to planarians: methodology and dynamics. Dev Dyn 242:718–730
16. Brandl H, Moon H, Vila-Farré M et al (2016) PlanMine--a mineable resource of planarian biology and biodiversity. Nucleic Acids Res 44:D764–D773
17. Grohme MA, Schloissnig S, Rozanski A et al (2018) The genome of Schmidtea mediterranea and the evolution of core cellular mechanisms. Nature 554:56–61
18. Robb SM, Gotting K, Ross E et al (2015) SmedGD 2.0: the Schmidtea mediterranea genome database. Genesis 53:535–546
19. Robb SMC, Ross E, Alvarado AS (2008) SmedGD: the Schmidtea mediterranea genome database. Nucleic Acids Res 36: D599–D606
20. Rozanski A, Moon H, Brandl H et al (2019) PlanMine 3.0-improvements to a mineable resource of flatworm biology and biodiversity. Nucleic Acids Res 47:D812–D820
21. An Y, Kawaguchi A, Zhao C et al (2018) Draft genome of Dugesia japonica provides insights into conserved regulatory elements of the brain restriction gene nou-darake in planarians. Zool Lett 4:24
22. Wurtzel O, Cote LE, Poirier A et al (2015) A generic and cell-type-specific wound response precedes regeneration in planarians. Dev Cell 35:632–645
23. Fincher CT, Wurtzel O, de Hoog T et al (2018) Cell type transcriptome atlas for the planarian Schmidtea mediterranea. Science 360:eaaq1736
24. Plass M, Solana J, Wolf FA et al (2018) Cell type atlas and lineage tree of a whole complex animal by single-cell transcriptomics. Science 360(6391):eaaq1723
25. Molinaro MA, Pearson BJ (2016) In silico lineage tracing through single cell transcriptomics identifies a neural stem cell population in planarians. Genome Biol 17:87
26. Zeng A, Li H, Gao X et al (2018) Prospectively isolated Tetraspanin(+) Neoblasts are adult pluripotent stem cells underlying Planaria regeneration. Cell 173(7):1593–1608
27. Rybak-Wolf A, Solana J (2014) Whole-mount in situ hybridization using DIG-labeled probes in planarian. Methods Mol Biol 1211:41–51
28. King RS, Newmark PA (2013) In situ hybridization protocol for enhanced detection of gene expression in the planarian Schmidtea mediterranea. BMC Dev Biol 13:8
29. Pearson BJ, Eisenhoffer GT, Gurley KA et al (2009) Formaldehyde-based whole-mount in situ hybridization method for planarians. Dev Dyn 238:443–450
30. Rompolas P, Patel-King RS, King SM (2009) Schmidtea mediterranea: a model system for analysis of motile cilia. Methods Cell Biol 93:81–98
31. Rompolas P, Azimzadeh J, Marshall WF et al (2013) Analysis of ciliary assembly and function in planaria. Methods Enzymol 525:245–264
32. King SM, Patel-King RS (2016) Planaria as a model system for the analysis of ciliary assembly and motility. Methods Mol Biol 1454:245–254
33. Azimzadeh J, Wong ML, Downhour DM et al (2012) Centrosome loss in the evolution of planarians. Science 335:461–463
34. Thi-Kim Vu H, Rink JC, McKinney SA et al (2015) Stem cells and fluid flow drive cyst formation in an invertebrate excretory organ. Elife 4:e07405
35. Rink JC, Thi-Kim Vu H, Sanchez Alvarado A (2011) The maintenance and regeneration of the planarian excretory system are regulated by EGFR signaling. Development 138:3769–3780
36. Rink JC, Gurley KA, Elliott SA et al (2009) Planarian Hh signaling regulates regeneration polarity and links Hh pathway evolution to cilia. Science 326:1406–1410
37. Scimone ML, Srivastava M, Bell GW et al (2011) A regulatory program for excretory system regeneration in planarians. Development 138:4387–4398
38. Magley RA, Rouhana L (2019) Tau tubulin kinase is required for spermatogenesis and

development of motile cilia in planarian flatworms. Mol Biol Cell 30(17):2155–2170

39. Lesko SL, Rouhana L (2020) Dynein assembly factor with WD repeat domains 1 (DAW1) is required for the function of motile cilia in the planarian Schmidtea mediterranea. Develop Growth Differ 62(6):423–437
40. Zayas RM, Hernández A, Habermann B et al (2005) The planarian Schmidtea mediterranea as a model for epigenetic germ cell specification: analysis of ESTs from the hermaphroditic strain. Proc Natl Acad Sci U S A 102:18491–18496
41. Issigonis M, Newmark PA (2019) From worm to germ: germ cell development and regeneration in planarians. Curr Top Dev Biol 135:127–153
42. Wang Y, Zayas RM, Guo T et al (2007) nanos function is essential for development and regeneration of planarian germ cells. Proc Natl Acad Sci U S A 104:5901–5906
43. Wang Y, Stary JM, Wilhelm JE et al (2010) A functional genomic screen in planarians identifies novel regulators of germ cell development. Genes Dev 24:2081–2092
44. Ishida S, Yamashita Y, Teshirogi W (1991) Analytical studies of the ultrastructure and movement of spermatozoa of freshwater triclads. In: Turbellarian biology. Developments in hydrobiology, vol 69. Springer, Dordretch, pp 95–104
45. Hyman LH (1925) The reproductive system and other characters of Planaria dorotocephala Woodworth trans. Amer Microsc Soc 44:51–89
46. Stevens NM (1904) On the germ cells and the embryology of Planaria simplissima. Proc Acad Natl Sci U S A 56:208–220
47. Counts JT, Hester TM, Rouhana L (2017) Genetic expansion of chaperonin-containing TCP-1 (CCT/TRiC) complex subunits yields testis-specific isoforms required for spermatogenesis in planarian flatworms. Mol Reprod Dev 84:1271–1284
48. Cebria F, Newmark PA (2005) Planarian homologs of netrin and netrin receptor are required for proper regeneration of the central nervous system and the maintenance of nervous system architecture. Development 132:3691–3703
49. Baguñá J, Romero R (1981) Quantitative analysis of cell types during growth, degrowth and regeneration in the planarians Dugesia mediterranea and Dugesia tigrina. Hydrobiologia 84:181–194
50. Chong T, Stary JM, Wang Y et al (2011) Molecular markers to characterize the hermaphroditic reproductive system of the planarian Schmidtea mediterranea. BMC Dev Biol 11:69
51. Iyer H, Issigonis M, Sharma PP et al (2016) A premeiotic function for boule in the planarian Schmidtea mediterranea. Proc Natl Acad Sci U S A 113:E3509–E3518
52. Steiner JK, Tasaki J, Rouhana L (2016) Germline defects caused by Smed-boule RNA-interference reveal that egg capsule deposition occurs independently of fertilization, ovulation, mating, or the presence of gametes in planarian flatworms. PLoS Genet 12: e1006030
53. Chong T. (2012) A planarian's journey from Sardinia to the Midwest. The Node. https://thenode.biologists.com/a-planarian%E2%80%99s-journey-from-sardinia-to-the-midwest/research/
54. Lazaro EM, Harrath AH, Stocchino GA et al (2011) Schmidtea mediterranea phylogeography: an old species surviving on a few Mediterranean islands? BMC Evol Biol 11:274

Chapter 11

An In Vitro Model System to Test Mechano-Microbiological Interactions Between Bacteria and Host Cells

Luis Carlos Santos, Emilia Laura Munteanu, and Nicolas Biais

Abstract

The aim of this chapter is to present an innovative technique to visualize changes of the F-actin cytoskeleton in response to locally applied force. We developed an in vitro system that combines micromanipulation of force by magnetic tweezers with simultaneous live cell fluorescence microscopy. We applied pulling forces to magnetic beads coated with the *Neisseria gonorrhoeae* Type IV pili in the same order of magnitude than the forces generated by live bacteria. We saw quick and robust F-actin accumulation in individual cells at the sites where pulling forces were applied. Using the magnetic tweezers, we were able to mimic the local response of the F-actin cytoskeleton to bacteria-generated forces. In this chapter, we describe our magnetic tweezers system and show how to control it in order to study cellular responses to force.

Key words Magnetic tweezers, *Neisseria*, F-Actin, Type 4 pili, Magnetic beads, Cytoskeleton

1 Introduction

The ability of eukaryotic cells to adapt their shape and sense their environment through their cytoskeleton has been a subject of scientific inquiry for a very long time. Comprised of actin microfilaments, microtubules, and intermediate filaments, the cytoskeleton plays a crucial role both in the interior and exterior of individual cells. On the one hand, the cytoskeleton in conjunction with cell membranes organizes the interior of cells and maintains cell shape by controlling intracellular trafficking and cell division [1–3]. On the other hand, it connects cells to their chemical and physical surroundings, namely, to the extracellular matrix and to other cells [4, 5]. Different extracellular matrix proteins transmitting forces to the interior of the cell will lead to different responses from the cytoskeleton [6–8].

The impact of physical force on biological outcomes, usually referred to as mechanobiology, is a key component of many landmark cellular mechanisms such as motility, development,

Ray H. Gavin (ed.), *Cytoskeleton: Methods and Protocols*, Methods in Molecular Biology, vol. 2364,
https://doi.org/10.1007/978-1-0716-1661-1_11,

differentiation, metabolism, and cancer [9–13]. Some of these force mechanisms are conserved across multiple domains of life, representing perhaps one of nature's most remarkable examples of convergent evolution at a molecular level [14]. Consider, for example, the elongation-retraction cycles that drive the directed locomotion of single eukaryotic cells in substrate rigidity-dependent migration [15, 16]. Using traction force microscopy and computational simulations, recent studies revealed that some individual bacteria powered by long thin appendages called Type IV pili use a similar mechanism in twitching motility and microcolony formation, deploying these dynamic filaments that polymerize and depolymerize, thus pulling the cell forward at its front [17–20]. Recent evidence shows that certain bacteria also use mechanical forces as a way to induce cellular changes in host cells with implications on bacterial pathogenesis [21–25]. Bacteria have not ceased to coevolve with their host, and the study of the interactions between bacteria and host cells has often illuminated certain chemical aspects of the cytoskeleton (e.g., ActA from *Listeria monocytogenes* [26], toxin from enteropathogenic *Escherichia coli* [27, 28]).

The bacterium *Neisseria gonorrhoeae*, the causative agent of gonorrhea, has emerged in recent years as a paradigm for the study of mechano-microbiological interaction [29–31]. Forces generated by cycles of elongation and retraction of Type IV pili—long thin polymers emanating from the surface of *Neisseria gonorrhoeae*—have been shown to have a dramatic impact on the outcome of the interactions between bacteria and human cells [21, 32]. In contrast, in vitro model studies show that bacteria that do not have the ability to exert pulling forces (including mutants that are missing ATPase PilT, the protein responsible for force generation) exhibit reduced level of invasion compared to their force bearing counterparts [21]. Most importantly, bacteria-generated force is required to remodel the host cell cytoskeleton [21, 33]—a hallmark of infection in vitro as well as in vivo.

How the eukaryotic cytoskeleton responds to force has been the subject of many experimental studies. Central to this question are technologies that allow exerting a controlled force on a single cell. Techniques enabling the application of local force at a subcellular scale typically involve magnetic or optical tweezers [34, 35]. Newly developed biophysical techniques involving AFM, lithography, and micropillars have also been employed to exert and measure global forces on an entire cell [36–38]. Previous studies have used optical tweezers to investigate the impact of forces in the range of hundreds of picoNewton (pN) on remodeling of the cytoskeleton [39–41]. These techniques have provided unprecedented insight on the response of the eukaryotic cytoskeleton to the magnitude and speed of the local force applied, and on the nature of the physical link to the cytoskeleton. Thus, unraveling the physical cross talk between eukaryotic cells and their environment is a complex subject of tremendous importance with potential applications ranging from cancer to bioengineering.

To better understand the role of local force in eukaryotic cells, we explored the F-actin response to Type IV pili-mediated force interactions that occur between bacteria from the *Neisseria* species and human skin epithelial cells. Forces generated by *Neisseria gonorrhoeae* bacteria are in the nanoNewton (nN) range and have been shown to induce rearrangements of the eukaryotic cytoskeleton [30, 42]. We developed an innovative and robust technique by combining magnetic tweezers with live cell fluorescence microscopy. We applied local forces to human skin epithelial cells through magnetic beads coated with Type IV pili from *Neisseria gonorrhoeae* and simultaneously visualized F-actin dynamics. The tweezers create an electromagnetic field gradient that pulls on the pili-coated magnetic beads, which in turn pull on the cell that the beads are attached to. Here, we describe the magnetic tweezers built in our laboratory to study the response of the F-actin cytoskeleton to local pulling forces exerted on the cell body. Our results mimic the early steps of infection of skin epithelial cells with bacteria from the *Neisseria* species. The techniques presented here should be easily implemented for other pili-bearing bacterial species interacting with eukaryotic host cells.

2 Materials

2.1 Live Cell Imaging of F-Actin in Human Epithelial Cells

2.1.1 Cell Culture and DNA Transfection

1. Human skin epithelial cell line A-431 (ATCC CRL-1555).
2. Mammalian expression plasmid DNA vector (pEGFP-N1) encoding GFP-tagged tractin—an F-actin-associated protein [43].
3. Transfection reagent: Fugene HD (commercially available).
4. Dulbecco's Modified Eagle Medium without phenol red (commercially available) supplemented with 10% heat-inactivated fetal bovine serum, FBS (commercially available).
5. OPTI-MEM reduced serum medium with HEPES and sodium bicarbonate (commercially available).
6. CO_2-independent medium (serum-free, commercially available).
7. 0.05% trypsin solution (commercially available).
8. Phosphate-buffered saline (PBS), commercially available.
9. Glass-bottom cell imaging dishes, round bottom, 50 mm diameter, 7 mm low-height, glass coverslip #0.

10. 60 mm tissue culture plates.
11. Incubator with controlled atmosphere at 37 °C with 5% CO_2.

2.1.2 Time-Lapse Fluorescence Microscopy

1. Inverted microscope, equipped with a 63×/1.40 NA Oil, Plan Apo Chroma objective (Zeiss).
2. Electron-multiplying charge-coupled device (EMCCD) digital camera PRO-EM 512 (Princeton Instruments, NJ).
3. 100 W high-pressure mercury bulb with power supply (BH2-RFL-T3, Olympus).
4. Environmental control chamber to maintain temperature of the sample stable at 37 °C (assembled in-house).
5. Digital image acquisition software: Micromanager (NIH, MD).
6. Digital image processing software: Fiji/ImageJ (NIH, MD).
7. Model-design software: MATLAB (MathWorks, MA).

2.2 Preparing Magnetic Beads Coated with Bacteria Pili

2.2.1 Purifying Pili from Neisseria gonorrhoeae

1. *Neisseria gonorrhoeae* clone MS11 (wild type).
2. 100 mm Petri dishes.
3. Gonococcal broth (GCB) agar (available commercially).
4. CHES buffer, pH 9.5 [2-(cyclohexylamino)ethanesulfonic acid] (available commercially).
5. MES buffer, pH 5.0 [2-(*N*-morpholino)ethanesulfonic acid] (available commercially).
6. Micro Ultracentrifuge (RC-M120EX, Sorvall, CT), and fixed angle rotor (RP100AT-289).

2.2.2 Coating Beads with Purified Pili

1. Superparamagnetic beads (Life Technologies, CA): uniform, monosized, 1 μm (Dynabeads MyOne) beads and 3 μm (Dynabeads M-270) in diameter beads, composed of highly cross-linked polystyrene with evenly distributed magnetic material, with a carboxylic acid surface chemistry.
2. EDC [1-(3-dimethylaminopropyl)-3-ethylcarbodiimide hydrochloride] (Alfa Aesar, MA).
3. MES buffer, pH 5.0 [2-(*N*-morpholino)ethanesulfonic acid].
4. 0.01 M NaOH.
5. Phosphate-buffered saline (PBS), available commercially.
6. 50 mM Tris–HCl buffer, pH 7.4.
7. BSA: 0.1% bovine serum albumin in PBS.
8. Sonicator Branson #1800 (Branson Ultrasonics, CT).

2.3 Magnetic Tweezers Setup for Applying Local Force to Eukaryotic Cells

2.3.1 Components of the Magnetic Tweezers

1. XYZ 3D micromanipulator MP-285 (Sutter Instrument, CA).
2. Programmable DC power supply and operating software (BK Precision, CA).

2.3.2 Force Calibration

1. Polystyrene beads (2 μm diameter) (Polysciences, Inc., PA).
2. Dimethylpolysiloxane (silicone oil) (viscosity $\mu = 12{,}500$ centipoise) (Sigma, CO).
3. Poly-L-lysine hydrobromide MW 30,000–70,000 (Sigma, CO).

3 Methods

3.1 Preparations for Live Cell Imaging of F-Actin in Human Epithelial Cells

Here we describe a detailed protocol to use live fluorescence microscopy for high-resolution visualization of fluorescently tagged tractin—a protein that specifically binds to F-actin—in human epithelial cells (*see* **Note 1**).

3.1.1 Cell Culture and DNA Transfection

We visualized actin dynamics in live cells by time-lapse image acquisition of cells expressing GFP-tagged tractin—an F-actin-associated protein [43]. With this purpose, we introduced a DNA vector carrying the GFP-tagged tractin sequence into A-431 cells—a human epithelial cell line. The CMV promoter and enhancer sequence allow strong and clear visual detection of the tractin-GFP reporter as early as 12 h after transfection, and during the whole course of the experiment. To introduce the tractin-GFP vector into A-431 cells, we used the transfection reagent Fugene HD. This method of transfection involves a non-liposomal reagent that achieves a good balance between transfection efficiency—high percentage of cells expressing the fluorescent protein—and low cytotoxicity, low percentage of dead cells due to toxicity of the transfection reagent (*see* **Note 2**).

The steps below describe the optimal transfection protocol that we followed to perform the set of experiments presented in this chapter. All cell culture handling should be performed in sterile conditions inside a laminar flow biosafety cabinet. For optimal cell performance, A-431 cells should have been subcultured for at least 1 week, or equivalent to two passages, in 60 mm tissue culture plates, in full medium (DMEM, supplemented with 10% FBS), and in a 37 °C/5% CO_2 atmosphere. Follow standard subculture protocol using 0.05% trypsin solution (TrypLE Express recommended) to detach cells from the culture plate.

1. 48 h before imaging: Prepare one glass-bottom low wall imaging dish for each experimental condition you want to test (e.g., different types of beads, or pili, or force applied, etc.). Seed 5×10^5 cells per imaging dish with 3 mL DMEM/10% FBS culture medium, and incubate the cells for 24 h (*see* **Note 3**).
2. 24 h before imaging: In a sterile tube, add 80 μL of room temperature OPTI-MEM medium, and then add 20 μL of room temperature Fugene HD for a total volume of 100 μL and incubate at room temperature for no longer than 5 min. The Fugene HD reagent should be added directly into the OPTI-MEM medium, and do not allow undiluted Fugene HD reagent to contact the sides of the tube.
3. In a new sterile tube, add 5 μg of DNA and then add OPTI-MEM medium to a total volume of 100 uL. The concentration of the DNA stock solution used should be 1–2 μg/μL to avoid dilution of the DNA/Fugene complexes formed in the next step (*see* **Note 4**).
4. Add the Fugene/OPTI-MEM to the DNA/OPTI-MEM solution tube, mix immediately by gently pipetting up and down, and incubate at room temperature for no longer than 15 min to allow formation of the DNA/Fugene complexes.
5. Slowly, and dropwise, add the 200 μL DNA/Fugene mix to the cell dish (containing 3 mL of full medium), and mix by gently swirling the dish.
6. Return the dish to the incubator.
7. After at least 12 h of incubation, assess cell viability and transfection efficiency by briefly examining the level of fluorescence intensity and appropriate subcellular localization of the fluorescent protein (tractin-GFP localizes to F-actin). The tractin-GFP protein is expected to be visible as early as 12 h post-transfection, and the expression peak occurs at 24–48 h post-transfection. If the transfection was successful, proceed to cell imaging.

3.1.2 Time-Lapse Fluorescence Microscopy

Epifluorescence microscopy is a powerful tool to visualize the dynamics of tractin-GFP over time in live cells. We used an inverted microscope, with a 63x oil objective, brightfield illumination, and halogen light source for epifluorescence illumination. Here we provide guidelines to setting up live cell imaging by time-lapse fluorescence microscopy (*see* **Note 5**).

1. Turn on all microscope components. Stabilization of the halogen light source may take a few minutes.
2. Microscope temperature stabilization: Turn on the local heater attached to the microscope's environmental chamber

approximately 30 min before image acquisition to make sure all components of the microscope are at a stable temperature at the start of the experiment (*see* **Note 6**).

3. Carefully place the imaging dish on the microscope stage, turn on brightfield illumination, and focus on the cell monolayer.
4. Adjust the brightfield image quality in order to have the sharpest contrast of the cell edges.
5. Adjust the image brightness by controlling the amount of incident light allowed by the diaphragm, and by adding/removing neutral density filters in the light path (*see* **Note 7**).
6. Switch to epifluorescent light, and quickly scan the dish to find a region of interest (ROI) that includes a transfected cell, i.e., GFP-expressing cell, with one or more beads on top. Be swift with this ROI search in order to minimize photobleaching (exposure time-dependent fading) (*see* **Note 8**).
7. Once you have selected the ROI, switch to "camera view," and adjust the exposure time on both channels (brightfield and epifluorescence). The aim here is to maximize signal-to-noise ratio while simultaneously minimizing photobleaching and phototoxicity. Choose the shortest exposure time possible that provides an optimal balance between these factors.

3.2 Coating Magnetic Beads with Bacteria Pili

The following protocols involving bacteria handling, streaking, and lawning must be performed under sterile conditions and appropriate safety precautions.

3.2.1 Purifying Pili from *Neisseria gonorrhoeae*

1. Prepare two lawns of *N. gonorrhoeae* MS11 wild type (WT) using two 10 cm plates of Gonococcal broth (GCB) agar.
2. Incubate bacteria lawns at 37 °C in 5% CO_2 atmosphere for 16–20 h. This incubating time window is critical to obtaining good quality pili with high yields. In general, each 10 cm Petri dish of lawn bacteria yields 0.1 mg of pili.
3. Harvest the bacteria lawns using a polyester swab, and resuspend in 1 mL of 50 mM CHES buffer (pH 9.5).
4. Vortex the bacteria suspension for 2 min, and then centrifuge bacteria bodies at 18,000 × *g* for 5 min (~14,000 rpm using a fixed rotor tabletop centrifuge).
5. Collect the supernatant and ultracentrifuge it at 100,000 × *g* for 90 min. The ultracentrifugation step must be performed in suitable ultracentrifuge microtubes. The tubes should be well balanced (<10 μg weighting error) and carefully loaded onto the ultracentrifuge rotor.
6. After centrifugation, carefully remove the tubes from the rotor without disturbing the pili pellet and discard the supernatant. The pili pellet is almost invisible to the eye.

7. Gently resuspend the pili pellet in 1 mL of 50 mM MES buffer (pH 5.0).

3.2.2 Coating Beads with Purified Pili

In the steps below, we describe one of the protocols we optimized for coating beads with purified pili from *N. gonorrhoeae* (*see* Subheading 3.3.1). To replace the solution in which the beads are suspended, simply place the tubes containing the beads in contact with a magnet for 2 min, and then carefully remove the liquid supernatant.

Surface Activation of Beads with EDC

1. Wash 100 μL of beads twice with 1 mL of 0.01 M NaOH for 10 min with gentle rotation.
2. Repeat wash three times this time with 100 μL deionized water.
3. Dissolve EDC in cold, deionized water to 50 mg/mL.
4. Add 100 μL of EDC solution to the beads. Vortex to ensure good mixing.
5. Incubate for 30 min at room temperature with gentle rotation.
6. Wash once with cold, deionized water and once with 50 mM MES (pH 5.0) as quickly as possible to avoid hydrolysis of the activated carboxylic acid groups.
7. The beads are now activated and ready for coating with a ligand containing primary amine groups, in this case the purified pili (*see* **Note 9**).

Immobilization of Ligand After Activation

1. Remove the wash solution used in **step 5** above. Add 60 μL of pili in 50 mM MES (pH 5.0) to the activated beads.
2. Add 40 μL of 50 mM MES (pH 5.0) to the final volume of 100 μL. Vortex to ensure good mixing.
3. Incubate for 30 min at room temperature with gentle rotation.
4. Wash the pili-coated beads as described below.

Quenching and Washing of Pili-Coated Beads

All immobilization procedures require washing of the pili-coated beads to remove excess pili and to block unreacted surface.

1. Quench the non-reacted activated carboxylic acid groups, and incubate the pili-coated beads with 1 mL 50 mM Tris–HCl pH 7.4 for 15 min at room temperature with gentle rotation.
2. Wash the pili-coated beads four times with 1 mL PBS.
3. Block by adding 0.1% BSA when it does not interfere with downstream applications of the beads.
4. Resuspend the pili-coated beads in deionized water to the desired concentration. The pili-coated beads are now ready for use. Store coated beads at 4 °C with gentle rotation.

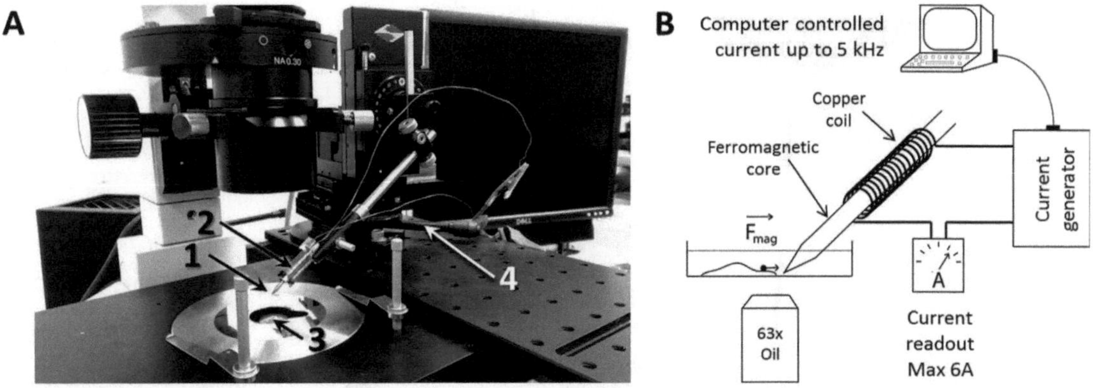

Fig. 1 (**a**) Magnetic tweezers and fluorescence microscopy system used in our lab. (1) Ferromagnetic core tip. (2) Two coiled copper wires. (3) Oil objective (63×). (4) Micromanipulator. (**b**) Schematic representation of the magnetic tweezers and imaging setup. The computer-controlled current generator is connected to a power amplifier (not depicted)

3.3 Magnetic Tweezers Setup for Applying Local Force to a Eukaryotic Cell

Here we present the magnetic tweezers setup and protocol developed in our lab for studying local F-actin cytoskeleton response to force (Fig. 1). The magnetic tweezers setup we implemented features a solenoid electromagnet as a magnetic field source that is mounted on a microscope stage in order to be brought in close proximity to the magnetic beads attached to cells (Fig. 1).

3.3.1 Components of the Magnetic Tweezers

Using a cylindrical holder made of a nonmagnetic material, a solenoid electromagnet was handcrafted by coiling a copper wire on the outside of that holder. Typically, two to three layers of wire are coiled. Several inches of wire were left at the beginning and end in order to connect the solenoid to the electric current source. A rod-shaped ferromagnetic core was fixed with plastic screws inside of the holder. The core should not touch the inside walls of the holder. This space creates an insulating layer of air that minimizes the heat transfer from the copper wire when an electric current is applied to the solenoid. One end of the rod was machined to be a sharp tip. The role of the ferromagnetic core is to amplify the magnetic field produced by the solenoid electromagnet. The rod's sharp tip creates a high gradient in the magnetic field close to its tip, necessary to apply high forces to the magnetic beads in the sample field. With the use of two electronic connectors (we use alligator-banana end), the beginning and the end of the copper coil are connected to an electric current source. Typically, 1 to 3 amperes (A) of continuous current was applied to the solenoid to mimic the range in magnitude of the pulling forces exerted by bacteria in the *Neisseria* species.

3.3.2 Force Calibration

Force magnitude depends on the magnitude of the magnetic field, the size of the bead, and the material the bead is made of. In turn, the magnitude of the magnetic field depends on the geometry of

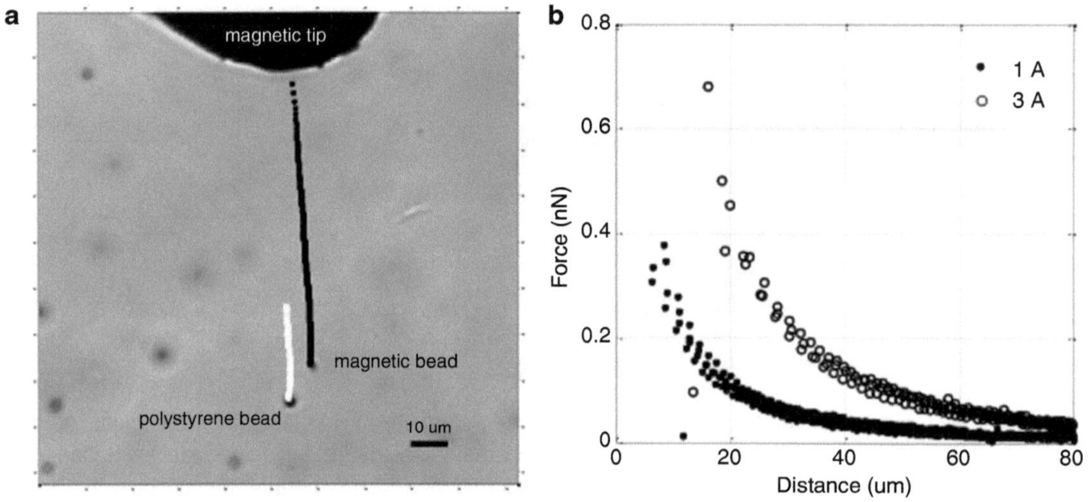

Fig. 2 (**a**) Brightfield micrograph depicting the magnetic tip at the top of the display image and several beads in silicone oil. Trajectories of a magnetic bead and a polystyrene bead are shown (acquisition rate = 10 Hz). The magnetic bead (1μm diameter, trajectory in black dots) is subject to magnetic force. The polystyrene bead (2μm diameter, trajectory in white dots) is only subject to local flow in the oil. Scale bar = 10μm. (**b**) Calibration curves of "force vs. distance." Distance represents the position of the bead relative to the tip. Values represent three beads for each of two amplitudes of applied current: 1 A (solid dots) and 3 A (open circles). The force was calculated using Stokes' law $F = 6 \times \pi \times \mu \times r \times v$ ($\mu = 12{,}500$ centipoise, $r = 0.5\mu m$, v was calculated from bead displacement between subsequent frames corrected for nonmagnetic-induced flow)

the magnetic tip, the amplitude of the electric current applied to the solenoid, and the distance between the tip of the magnetic core and the bead. In our setup, given the high variability in the shape of the magnetic tip and of the complex spatial geometry between the bead and the magnetic tip positions, it is not possible to estimate the magnetic forces mathematically. Therefore, we performed a calibration procedure to obtain force vs distance curves for each type of magnetic beads used in our experiment (Fig. 2).

Preparing a Mixture of Magnetic and Nonmagnetic Beads for Force Calibration

1. Add a small volume of 100–200 μL of silicone oil to the center of a new glass-bottom imaging dish (*see* **Note 10**).
2. In two separate 1.5 mL tubes, prepare silicone oil suspensions of magnetic and nonmagnetic polystyrene beads (*see* **Note 11**).
3. Wash 5 μL of magnetic beads (Myo-270) in 500 μL deionized water, sonicate for 1 min, gather beads on the bottom of the tube with a magnet, and carefully discard the supernatant. Resuspend the washed beads in 500 μL silicone oil by adding the oil to the tube and mix vigorously with a pipette tip. To further homogenize the solution, sonicate the bead-silicone oil mixture for 1 h in a heated water bath. The beads do not disperse easily in the viscous oil and repeating the sonication two to three times is often necessary.

4. Polystyrene (nonmagnetic) beads: Wash 10 μL of polystyrene beads in 1 mL deionized water, vortex briefly, centrifuge for 2 min at 2,400 × *g* in a tabletop centrifuge, and discard the supernatant. Resuspend the beads in 500 μL silicone oil as described above for the magnetic beads.
5. Mix 50 μL of the magnetic beads with the polystyrene beads to a 3:1 ratio (magnetic/polystyrene).
6. Add the mixture of magnetic and polystyrene beads to the clear silicone oil previously set on the imaging dish (*see* **Note 12**).

Manipulating Tweezers for Force Calibration

1. Mount the imaging dish on the pre-heated microscope stage (set environmental chamber temperature in advance to 37 °C).
2. Manually position the magnetic tip over the bead-silicone mixture in the dish.
3. Using the XYZ micromanipulator, slowly lower the tweezers tip until it touches the silicone in the dish.
4. Switch to camera view and, using the micromanipulator, slowly lower the tweezers until it is 1–5 μm away from the surface of the dish. Steer it carefully so that the end of its tip shows on the top edge of the displayed image field of view (Fig. 2a) (*see* **Note 13**).
5. Allow the silicone oil to settle for 10 min as the tip manipulation typically causes local flow.
6. Position the magnetic tweezers tip at the edge of a field of view with five to ten magnetic and polystyrene beads (Fig. 2a).
7. Using BK Precision software, set up the current you wish to apply for your calibration experiment.
8. Using Micromanager, set up the Multi-Dimensional acquisition parameters according to your experimental aims.
9. Start image acquisition by clicking "Acquire" on the Micromanager acquisition panel, and subsequently turn on the tweezers by clicking "Run" in the BK software control panel.
10. Tracking of beads is performed using ImageJ plugin "Particle Tracker," and force analysis is performed using MATLAB.

Qualitative In Situ Monitoring of the Magnetic Force

1. During image acquisition, keep visual track of the beads of interest.
2. Correct for focus drift of the sample by choosing fiduciary structures on the glass surface as position references.
3. Maintain the location of the tip of the tweezers the same in all experiments involving cells (*see* **Note 14**).

Plotting "Force vs. Distance" Curves

The motion of the magnetic bead in the silicone oil obeys Stokes' law of particle movement in viscous fluids:

$$F = 6 \times \pi \times \mu \times r \times v,$$

where μ is the dynamic viscosity of the silicone oil, r is the bead radius, and v is the speed of the bead.

Local flow in the oil can be induced by the magnetic beads that are moving toward the magnetic tip, or by changes in temperature due to the heating of the magnetic tip, and/or by mechanical vibrations. Any motion detected on a polystyrene nonmagnetic bead is attributed to local flow in the oil. Therefore, to correct for any motion induced by local flow in the oil, track the motion of nearby polystyrene beads and subtract it from the motion of the magnetic bead.

Local speed of the magnetic bead is calculated for each time point as $v = \Delta d/\Delta t$, where Δd is the corrected displacement of the magnetic bead between the current and the subsequent frame and Δt is the time interval between frames. In the calibration shown in Fig. 2, $\Delta t = 0.1$ s (corresponding to a 10 Hz acquisition rate).

The force on the magnetic bead at each time point is then calculated using Stokes' law above. Figure 2b shows two calibration curves of "force vs. distance" from the tip for magnetic beads of 1 μm in diameter (MyOne) and for two amperages applied on the electromagnet 1A and 3A. We can apply forces up to several nN at a distance of 5–10 μm away from the magnetic tip.

Controlling the Amplitude and Speed of Force Pulling on Beads

Magnetic beads subject to a magnetic field are pulled with a force, F, in the direction of the magnetic field. The electric current applied to the electromagnetic coil can be programmed to generate forces up to a few nN, applied in pulses with variable duration and frequency controlled by the BK Precision software. By controlling the magnitude and duration of the pulling force, we control the speed at which the bead moves toward the magnetic tip and, hence, the speed of force pulling on a bead attached to a cell or other substrate.

Here, we demonstrate the robustness and versatility of force control of our magnetic tweezers system. To visualize local forces generated by the magnetic tweezers pulling on the magnetic beads, we attached magnetic beads coated with bacteria pili to an array of flexible micropillars. The bending angle of a micropillar in response to a pulling force is proportional to the force applied. The micropillar array was made according to a protocol described elsewhere [44]. Briefly, the micropillar array is made using a lithography-fabricated silicone master, which was then applied to a mixture of polyacrylamide and cross-linking agent followed by heat curing. The micropillar array was mounted on a glass-bottom imaging dish, the micropillars were coated with poly-L-lysine, and the dish was filled with PBS buffer. Then, pili-coated magnetic beads were

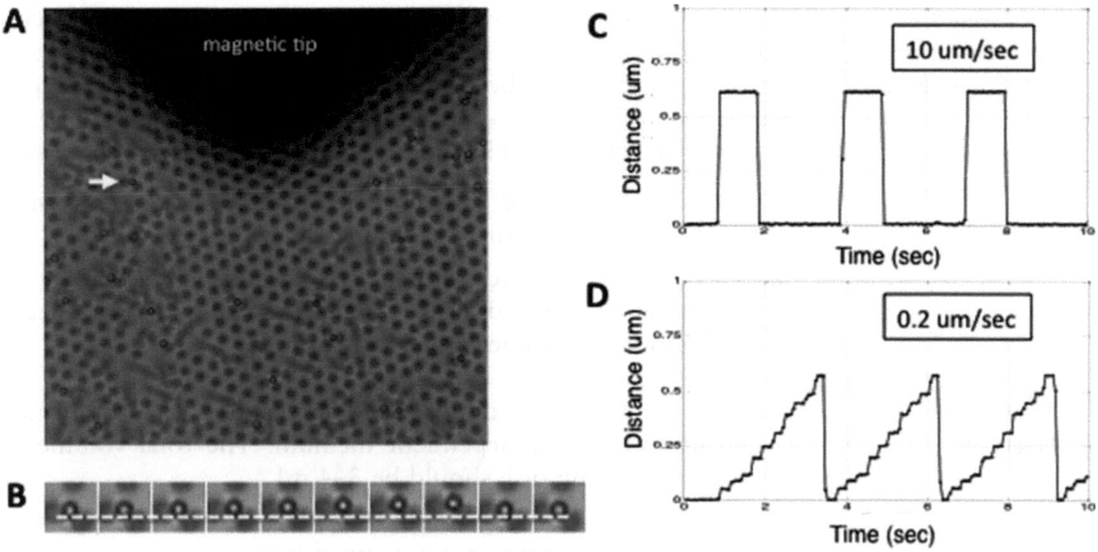

Fig. 3 Magnetic beads attached to an array of flexible polyacrylamide micropillars are subject to cycles of increasing magnetic force. (**a**) The magnetic tip is positioned 5–10μm above the micropillar tops (pillars are spaced 3μm center to center). One bead (white arrow) was tracked during two different cycles shown in (**c** and **d**). (**b**) Time series montage of images of bead tracked in (**a**) (white arrow). Dashed line represents initial position before force was applied. (**c**) Bead displacement during cycles of "fast pulling." One cycle comprised the following steps: 1 s at 0 A, 1 s at 3 A, and 1 s at 0 A. The electric current applied to the electromagnet at the second phase produced a sudden magnetic force on the bead that pulled the bead at a high speed of 10μm/s. (**d**) Bead displacement during cycles of "slow pulling" by gradually increasing the magnetic force. The cycle applied to the electromagnet to produce this force was comprised of ten short current pulses of 0.2 s in duration and increasing amplitude from 0 A to 3 A. The average speed at which bead moved during one cycle was 0.2μm/s

added to the micropillar array. As shown in Fig. 3a, the beads attached strongly to the poly-L-lysine-coated surface.

In Fig. 3b, d we show two examples of a bead pulling at different speeds.

1. A series of 1 s current pulses with a current amplitude of 3 A was applied to the electromagnet. This cycle induced a sudden increase in the magnetic force on the attached bead as indicated by an increasing bending angle of the micropillar, and consequently a high speed of pulling on the substrate (Fig. 3b).
2. A second series of repeating cycles is shown in Fig. 3d. Each cycle was comprised of ten short current pulses of 0.2 s and with increasing amplitude from 0 A to 3 A. As a result, the force exerted on the bead gradually increased, as indicated by an increasing bending angle of the micropillar, and the magnetic bead pulls on the substrate with a slower speed compared with example 1. The maximum value of the current during one cycle is chosen based on the desired maximum force exerted by the bead on the substrate (or cell body).

3.3.3 Tweezers Manipulation: Applying Force to Pili-Coated Beads

We used the magnetic tweezers to study the response of the actin cytoskeleton of A-431 cells to force applied on *N. gonorrhoeae* pili-coated 3 μm diameter beads (Fig. 3). Force exerted by magnetic tweezers on pili-coated beads mimics the force that bacteria exert on epithelial cells during infection. Prior to starting the protocol:

1. Prepare an imaging dish containing A-431 cells expressing tractin-GFP as described in Subheading 3.1.1.
2. Turn on in advance all the microscope components as described in Subheading 3.1.2, and prepare pili-coated beads as described in Subheading 3.2.1.

Set Beads onto Epithelial Cells

1. Change the cell medium from full medium to pre-warmed serum-free CO_2-independent medium. The total volume of the cell imaging dish should be 3–4 mL.
2. Remove the bead suspension from the 4 °C rotation apparatus and warm them up to room temperature.
3. Sonicate the beads for 1 min to break up eventual aggregates and ensure a uniform suspension of single beads.
4. Remove the cell imaging dish from the 37 °C incubator and pipette 1 μL bead suspension onto the cells. Perform this in sterile conditions (*see* **Note 15**).
5. Return the cells to the 37 °C incubator and let the beads settle for 30 min.

Position Tweezers over the Cells

1. Remove the cell imaging dish with beads from the incubator and mount it on the microscope stage and onto the 63× oil objective. Use the metal clamps attached to the stage to hold the dish in place.
2. Remove the lid of the cell imaging dish and manually bring the magnetic tweezers into place by setting its needle tip above the center of the dish.
3. Using the XYZ micromanipulator, lower the tweezers tip until it touches the medium in the dish. Follow the tweezer manipulation guidelines from the force calibration in Subheading 3.3.2.
4. Following the guidelines from Subheading 3.1.2, focus on a region of interest (ROI) that includes a tractin-GFP-expressing cell with several beads on top (*see* **Note 16**).
5. Under brightfield illumination, switch to camera view, and, using the micromanipulator, slowly lower the tweezers until it is 1–5 μm away from the cell monolayer. Steer it carefully so that the end of its tip shows on the top edge of the displayed image field of view without touching the cells (Fig. 4a).

A
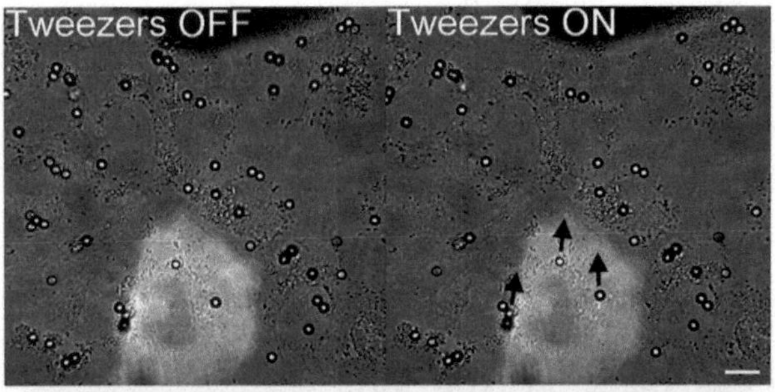

B

Current, *I* (A)	Pulse duration per cycle* (s)
0	1
1	0.5
1.5	0.5
2	0.5

* Total duration of each cycle is 2.5s. Cycles repeat continuously throughout the experiment.

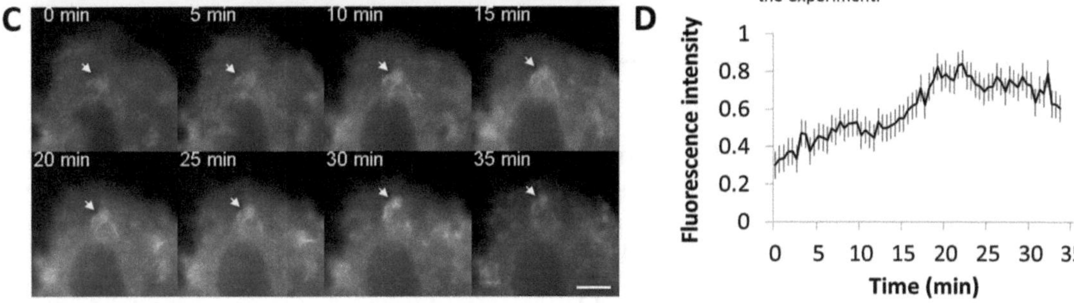

Fig. 4 Pili-coated beads pulled by magnetic tweezers trigger local F-actin accumulation. (**a**) Simultaneous brightfield and fluorescence image acquisition of live epithelial cells. Magnetic beads (high contrast spheres) and tip of the magnetic tweezers (dark field on center top of the image) are visible through the brightfield channel. Floating beads abandon the field of view after current is turned on. Beads attached to the epithelium remain in the field of view. Black arrows (right image, Tweezers ON) point to three beads attached to one cell expressing actin-GFP (center bottom of the image, visible through the fluorescent channel). (**b**) Magnetic force gradient is generated by applying repeated cycles of current pulses of increasing intensity. The table depicts one cycle ranging from no current (0 A) up to 2 A. (**c**) Visualization of actin-GFP in response to local magnetic forces generated by bead pulling. White arrows point to bead location at different time points. Scale bar = 10μm. (**d**) Mean fluorescence intensity of actin-GFP accumulated at the sites of pulling beads such as shown in (**c**). Error bars represent standard error of the mean of three beads

Time-Lapse Acquisition and "Current vs. Time" Parameters

1. Using BK Precision software, set up the Current vs. Time program you wish to run for your experiment (*see* **Note 17**).
2. Using Micromanager, set up the Multi-Dimensional time-lapse acquisition parameters according to your experimental aims (*see* **Note 18**).
3. Start image acquisition by clicking "Acquire" on the Micromanager acquisition panel, and immediately after the first frame is captured, turn on the tweezers by clicking "Run" in the BK software control panel (*see* **Note 19**).

Qualitative In Situ Monitoring of Pulling Forces

1. After turning on the tweezers, some beads abandon the field of view, either because they were not adhering to the cells or because the force generated by the tweezers overwhelmed the bead-cell adhesion strength and pulled the beads away.
2. Correct for focus drift of the sample by choosing fiduciary structures on the glass surface as position references.

Quantitative Analysis of Force-Dependent F-Actin Accumulation

1. Using ImageJ, measure the fluorescence intensity of the area occupied by one bead (signal) over time by drawing a circle around that area in each frame, and then using the function "measure" from the "analyze" menu.
2. Similarly, measure the fluorescence background (noise) of another area of the cell that is not occupied by a bead.
3. For each frame, subtract the noise from the signal to obtain the corrected fluorescence signal, and normalize it to the background for each frame (*see* **Note 20**).

4 Notes

1. Why tractin? Traditional vectors expressing GFP-tagged actin monomers (g-actin) have a drawback: the tendency to suppress assembly of actin filaments (F-actin). An alternative approach is to use GFP-tagged proteins that bind to F-actin, such as tractin [43]. This approach does not interfere with actin polymerization allowing F-actin assembly rates and actin-driven forces to remain unaffected. Furthermore, the signal from actin-GFP represents both g-actin and F-actin, thus yielding a higher fluorescence background. In contrast, tractin-GFP has a specific affinity for F-actin, so it has a higher signal-to-noise ratio than actin-GFP.
2. Other transfection reagents such as Lipofectamine are also available, but often exhibit significantly higher cytotoxicity, and increased fluorescence background compared to Fugene HD. The transfection protocol described here leads to minimal cell death, and it does not require removal/replacement of the culture medium following transfection.
3. 50 mm imaging dishes with low walls (7 mm height) are suitable for micromanipulation because they provide easy access to the tweezers setup on the microscope stage, without interfering with image acquisition.
4. The amount of DNA used per dish corresponds to a final DNA/Fugene (μg/μL) ratio of 1:4. This ratio should be optimized for each plasmid and cell line. Suboptimal deviations from this ratio will result in decreased transfection efficiency and increased cell death.
5. High-quality image acquisition is critical in quantitative fluorescence microscopy. Protein subcellular localization can only be accurately tracked and measured if all steps in the cell preparation protocol, DNA plasmid purification, and transfection are optimized and reproducible. Slight variations in the image acquisition settings of the microscope and of the imaging software dramatically impact the quality of results. Thus,

optimal image acquisition settings should be saved and reused each time for different experimental runs.

6. This step is critical to avoid focus drift during image acquisition. Due to temperature sensitivity of the materials comprising the microscope stage, objective, and oil, small temperature variations result in relatively large distortion of those materials.
7. Density filters are placed in the light path between the light source and the sample to decrease the amount of light illuminating the sample by partially blocking/reflecting incident light and, thus, help to prevent phototoxicity. This is particularly important for long-term acquisition experiments.
8. It is important to always minimize photobleaching for purposes of fluorescence intensity quantification analyses.
9. Activated beads cannot be stored and you should proceed directly to the following step.
10. To pipette the very viscous silicone oil, cut off the end of a 1000 μL pipette tip and set the pipette to aspirate 500 μL. Allow the silicone oil to settle for 10 min at 37 °C.
11. The two types of beads should have different diameter to be easily identified in the microscope.
12. The final bead-silicone oil mixture should yield about five to ten beads in the microscope field of view of the 63× oil objective, as shown in Fig. 2a.
13. Practice makes perfect! Get comfortable with the XYZ micromanipulator before performing this experiment. Adjust the sensitivity of the micromanipulator to "fine" as the tip moves closer to the surface of the dish. Move the magnetic tip slowly to minimize local flow in the oil.
14. The magnetic force exerted on the beads depends exponentially on the distance of the beads of interest to the tip (Fig. 2b).
15. You may choose to pipette more beads according to the concentration of your bead suspension. Do this on different regions of the glass-bottom center of the dish in order to have beads uniformly spread out and thus increase chance of having single beads interacting with a tractin-GFP cell.
16. Beads that do not adhere to the cell surface usually display Brownian-like motion and tend to promptly abandon the image field of view as soon as an electromagnetic current is applied.
17. Take into account your tweezers force calibration curve so that the applied force falls into the biologically meaningful range of your experiment. As shown in Fig. 4b, we ran continuous cycles of a four-step force gradient, as defined by the increasingly stronger current values applied.

18. In Fig. 4, we acquired a 35-min image sequence, and tracked tractin-GFP in the fluorescent channel with a 30-s interval between frames. We only captured images in the brightfield channel every 5 min to minimize phototoxicity.
19. As shown in Fig. 4a (left image "Tweezers OFF"), we captured the first frame with the tweezers off, then turned on the tweezers, and 30 s later captured the second frame with the "Tweezers ON."
20. The graph in Fig. 3d represents the variation over time of the corrected and normalized mean fluorescence signal of three beads.

References

1. McMahon HT, Boucrot E (2015) Membrane curvature at a glance. J Cell Sci 128:1065–1070
2. Roux A, Cappello G, Cartaud J, Prost J, Goud B, Bassereau P (2002) A minimal system allowing tubulation with molecular motors pulling on giant liposomes. Proc Natl Acad Sci U S A 99:5394–5399
3. Pollard TD (2010) Mechanics of cytokinesis in eukaryotes. Curr Opin Cell Biol 22:50–56
4. Chicurel ME, Chen CS, Ingber DE (1998) Cellular control lies in the balance of forces. Curr Opin Cell Biol 10:232–239
5. Mammoto T, Mammoto A, Ingber DE (2013) Mechanobiology and developmental control. Annu Rev Cell Dev Biol 29:27–61
6. Roca-Cusachs P, del Rio A, Puklin-Faucher E, Gauthier NC, Biais N, Sheetz MP (2013) Integrin-dependent force transmission to the extracellular matrix by α-actinin triggers adhesion maturation. Proc Natl Acad Sci U S A 110: E1361–E1370
7. Kundu AK, Putnam AJ (2006) Vitronectin and collagen I differentially regulate osteogenesis in mesenchymal stem cells. Biochem Biophys Res Commun 347:347–357
8. Vogel V, Sheetz M (2006) Local force and geometry sensing regulate cell functions. Nat Rev Mol Cell Biol 7:265–275
9. Vogel V, Sheetz MP (2009) Cell fate regulation by coupling mechanical cycles to biochemical signaling pathways. Curr Opin Cell Biol 21:38–46
10. Farge E (2003) Mechanical induction of twist in the drosophila foregut/Stomodeal primordium. Curr Biol 13:1365–1377
11. Engler AJ, Sen S, Sweeney HL, Discher DE (2006) Matrix elasticity directs stem cell lineage specification. Cell 126:677–689
12. Nisticò P, Di Modugno F, Spada S, Bissell MJ (2014) β 1 and β 4 integrins : from breast development to clinical practice. Breast Cancer Res 16(5):459
13. Park JS, Burckhardt CJ, Lazcano R, Solis LM, Isogai T, Li L et al (2020) Mechanical regulation of glycolysis via cytoskeleton architecture. Nature 578:621–626. https://doi.org/10.1038/s41586-020-1998-1
14. Paluch EK, Nelson CM, Biais N, Fabry B, Moeller J, Pruitt BL et al (2015) Mechanotransduction: use the force(s). BMC Biol 13:1–14. https://doi.org/10.1186/s12915-015-0150-4
15. Giannone G, Dubin-Thaler BJ, Döbereiner HG, Kieffer N, Bresnick AR, Sheetz MP (2004) Periodic lamellipodial contractions correlate with rearward actin waves. Cell 116:431–443
16. Meshel AS, Wei Q, Adelstein RS, Sheetz MP (2005) Basic mechanism of three-dimensional collagen fibre transport by fibroblasts. Nat Cell Biol 7:157–164
17. Sabass B, Koch MD, Liu G, Stone HA, Shaevitz JW, Schwarz US (2017) Force generation by groups of migrating bacteria. Proc Natl Acad Sci U S A 114:7266–7271
18. Simsek AN, Braeutigam A, Koch MD, Shaevitz JW, Huang Y, Gompper G et al (2019) Substrate-rigidity dependent migration of an idealized twitching bacterium. Soft Matter 15:6224–6236. https://doi.org/10.1039/C9SM00541B
19. Craig L, Forest KT, Maier B (2019) Type IV pili: dynamics, biophysics and functional consequences. Nat Rev Microbiol 17:429–440. https://doi.org/10.1038/s41579-019-0195-4
20. Pönisch W, Eckenrode KB, Alzurqa K, Nasrollahi H, Weber C, Zaburdaev V et al

(2018) Pili mediated intercellular forces shape heterogeneous bacterial microcolonies prior to multicellular differentiation. Sci Rep 8:1–10

21. Lee SW, Higashi DL, Snyder A, Merz AJ, Potter L, So M (2005) PilT is required for PI (3,4,5)P3-mediated crosstalk between Neisseria gonorrhoeae and epithelial cells. Cell Microbiol 7:1271–1284
22. Bieber D, Ramer SW, Wu YC, Murray JW, Tobe T, Fernandez R et al (1998) Type IV pili, transient bacterial aggregates, and virulence of enteropathogenic Escherichia coli. Science 280:2114–2118
23. Dufrêne YF, Persat A (2020) Mechanomicrobiology: how bacteria sense and respond to forces. Nat Rev Microbiol 18:227–240. https://doi.org/10.1038/s41579-019-0314-2
24. Harper CE, Hernandez CJ (2020) Cell biomechanics and mechanobiology in bacteria: challenges and opportunities. APL Bioeng 4:021501
25. Viela F, Prystopiuk V, Leprince A, Mahillon J, Speziale P, Pietrocola G et al (2019) Binding of staphylococcus aureus protein a to von willebrand factor is regulated by mechanical force. MBio 10:321–323
26. Lambrechts A, Gevaert K, Cossart P, Vandekerckhove J, Van Troys M (2008) Listeria comet tails: the actin-based motility machinery at work. Trends Cell Biol 18:220–227
27. Cleary J, Lai L-C, Shaw RK, Straatman-Iwanowska A, Donnenberg MS, Frankel G et al (2004) Enteropathogenic Escherichia coli (EPEC) adhesion to intestinal epithelial cells: role of bundle-forming pili (BFP), EspA filaments and intimin. Microbiology 150:527–538
28. Caron E, Crepin VF, Simpson N, Knutton S, Garmendia J, Frankel G (2006) Subversion of actin dynamics by EPEC and EHEC. Curr Opin Microbiol 9:40–45
29. Merz AJ, So M (1997) Attachment of piliated, Opa- and Opc- gonococci and meningococci to epithelial cells elicits cortical actin rearrangements and clustering of tyrosine-phosphorylated proteins. Infect Immun 65:4341–4349
30. Merz AJ, Enns CA, So M (1999) Type IV pili of pathogenic Neisseriae elicit cortical plaque formation in epithelial cells. Mol Microbiol 32:1316–1332
31. Merz AJ, So M (2000) Interactions of pathogenic Neisseriae with epithelial cell membranes. Annu Rev Cell Dev Biol 16:423–457
32. Howie HL, Glogauer M, So M (2005) The N. gonorrhoeae type IV pilus stimulates mechanosensitive pathways and Cytoprotection through a pilT-dependent mechanism. PLoS Biol 3:e100
33. Higashi DL, Zhang GH, Biais N, Myers LR, Weyand NJ, Elliott DA et al (2009) Influence of type IV pilus retraction on the architecture of the Neisseria gonorrhoeae-infected cell cortex. Microbiology 155:4084–4092
34. Tanase M, Biais N, Sheetz M (2007) Magnetic tweezers in cell biology. Trends Cell Biol 2:116–118
35. Kuo SC, Sheetz MP (1992) Optical tweezers in cell biology. Trends Cell Biol 2:116–118
36. Webster KD, Crow A, Fletcher DA (2011) An AFM-based stiffness clamp for dynamic control of rigidity. PLoS One 6:1–7
37. Desprat N, Richert A, Simeon J, Asnacios A (2005) Creep function of a single living cell. Biophys J 88:2224–2233
38. Fouchard J, Bimbard C, Bufi N, Durand-Smet P, Proag A, Richert A et al (2014) Three-dimensional cell body shape dictates the onset of traction force generation and growth of focal adhesions. Proc Natl Acad Sci U S A 111:13075–13080
39. Jiang G, Sheetz MP (2003) Two-piconewton slip bond between fibronectin and the cytoskeleton depends on talin. Nature 424:334–337
40. Dai J, Sheetz MP (1999) Membrane Tether Formation from Blebbing Cells. Biophys J 77:3363–3370
41. Jiang G, Huang AH, Cai Y, Tanase M, Sheetz MP (2006) Rigidity sensing at the leading edge through alphavbeta3 integrins and RPTPalpha. Biophys J 90:1804–1809
42. Higashi DL, Biais N, Weyand NJ, Agellon A, Sisko JL, Brown LM et al (2011) N. elongata produces type IV pili that mediate interspecies gene transfer with N. gonorrhoeae. PLoS one 6:7
43. Johnson HW, Schell MJ (2010) Neuronal IP3 3-kinase is an F-actin–bundling protein: role in dendritic targeting and regulation of spine morphology. Mol Biol Cell 20:5166–5180
44. Biais N, Higashi D, So M, Ladoux B (2012) Techniques to measure pilus retraction forces. Methods Mol Biol 799:197–216

Chapter 12

Purification of Cytoskeletal Proteins by Fast Protein Liquid Chromatography (FPLC) Using an *ÄKTA Start System*

Michael Gotesman, Yueyang Wang, Sharath C. Madasu, and Carter A. Mitchell

Abstract

Actin, myosin, and tubulin are ubiquitous components of the fibrous network known as the cytoskeleton. Cytoskeletal proteins are involved in a plethora of intracellular processes such as maintenance of cellular organization, organelle translocation, and various nuclear roles including chromosome separation during mitosis. Early methods for protein extraction primarily relied on the salting-out method which was performed in conjunction with biochemical assays. Since the advent of recombinant molecular biology, protein tagging has been coupled with chromatography to obtain highly purified proteins required for sensitive assays. This chapter provides a general standard operating procedure (SOP) for using the *ÄKTA™ Start System* controlled by UNICORN software for fast protein liquid chromatography (FPLC) of 6× his-tagged cytoskeletal proteins. The protocol can readily be modified for affinity and non-affinity purification techniques using the various ÄKTA™ Chromatography Systems.

Key words Protein purification, Salting-out, Recombinant DNA technology, Affinity purification, UNICORN software, Cytiva Life Sciences

1 Introduction

Actin, myosin, and tubulin are ubiquitous components of the fibrous network known as the cytoskeleton. The importance of actin and myosin in cellular processes within non-muscle cells began to emerge with the identification of actin [1] and myosin [2] in non-muscle cells and the subsequent development of the concept of the cytoskeleton. Myosins are generally referred to as conventional if they are filament-forming and unconventional if they are non-filament-forming. Both conventional and unconventional myosins are ubiquitous in eukaryotic cells, and conventional myosin is an integral fibrous component of the cytoskeleton. Eukaryotic cells contain multiple unconventional myosins that are

Ray H. Gavin (ed.), *Cytoskeleton: Methods and Protocols*, Methods in Molecular Biology, vol. 2364,
https://doi.org/10.1007/978-1-0716-1661-1_12,

involved with diverse organelle function. Some protists are known to contain a family of unconventional myosins [3]. The actomyosin complex is important in a plethora of functions such as muscle contraction, cell organization and division, and organelle trafficking [4–10]. Tubulin is the major constituent of microtubules, which are better appreciated for their involvement in chromosome segregation [11, 12] but may also be involved in cellular processes such as osmoregulation [13]. Diverse proteins form 10-nm-diameter cytoskeleton fibers known as intermediate filaments, which are well noted for their role in nuclear envelope stabilization and other nuclear processes [14].

Traditionally, cytoskeletal proteins were primarily purified by the salting-out method [15, 16] or in conjunction with ATPase activity assays [1] (*see* **Note 1**). Such purification methods can be useful for identification of cytoskeletal proteins but may be too crude for more detailed characterization [17] as required in X-ray crystallography [18], biochemical assays [19], or microscopy studies that can visualize single molecules [20]. Antibody-based purification [21, 22], which relies on an antibody to bind to a specific epitope, can be used to purify cytoskeletal proteins [9, 21, 23]. However, the aforementioned technique may not be practical due to the inhibitory cost of antibody production for large-scale purification. With the advent of modern recombinant technology, the field of protein purification has been greatly advanced by the ability to clone a purification tag(s) specifically for a protein of interest [19, 24] such as polyhistidine [25] or maltose-binding protein tag, the latter of which may even aid in proper folding of target proteins [26]. Traditional salting-out protein purification can be further clarified using chromatography, such as ion-exchange chromatography, as performed to purify the first unconventional myosin in *Acanthamoeba castellanii* [2]. Chromatography in general offers many advantages for protein purification such as compatibility with both affinity and non-affinity purification techniques, reusability of columns, enhancement of purity, and scalability [27, 28].

This chapter provides standard operating procedures (SOP) for 6× his-tagged protein purification using the *ÄKTA Start* Fast Protein Liquid Chromatography System controlled by UNICORN software (*see* **Notes 2** and **3**). The protocols are useful for purification of proteins by chromatography through HisTrap HP column installed on an *ÄKTA Start System* and can readily be readily be adapted to the purification method of choice. The ÄKTA™ Chromatography System has been used to purify cytoskeletal proteins [20, 29, 30] and analyze cytoskeleton dynamics [31, 32], as well as purification of proteins that interact with the cytoskeletal system [33–35].

2 Materials and Equipment

1. *ÄKTA Start System* (Fig. 1).
2. Column (e.g., HisTrap HP).
3. Glass bottles and flasks (1 L).
4. Filtration glassware.
5. Membrane filters (0.2 μm).
6. Syringes (5, 50 mL).
7. Tubes (1.5, 15, 50 mL).
8. Ultrapure water.
9. 20% ethanol.
10. 10% (w/v) sodium azide.
11. 0.5 M disodium phosphate.
12. 1 M monosodium phosphate.
13. Sodium chloride.
14. Imidazole.
15. 1 M hydrogen chloride.
16. 1 M sodium hydroxide.
17. 0.5 M EDTA.

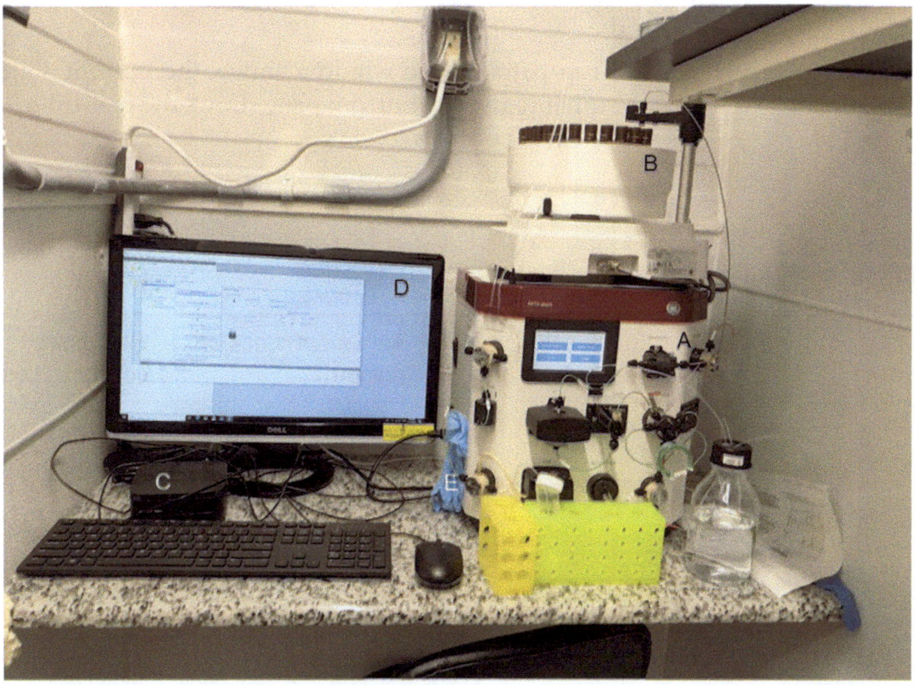

Fig. 1 FPLC setup (**a**–**e**): (**a**) ÄKTA Start, (**b**) fraction collector, (**c**) computer, (**d**) control screen, (**e**) gloves; *see* **Note 4**

3 Methods

3.1 Prepare Reagents (See Notes 4 and 5)

1. Make 1 L of 0.02% sodium azide solution in a glass bottle by diluting 2 mL of 10% sodium azide stock with 998 mL of ultrapure water.
2. Make Buffer A (lysis buffer) and Buffer B (elution buffer) in 1 L glass bottles according to compositions shown in Tables 1 and 2.
3. Prepare a suitable HisTrap HP Affinity Column (*see* **Note** 7).

3.2 Engage the ÄKTA Start Instrument (See Note 4)

1. Place Buffer A and Buffer B on top of the ÄKTA Start instrument (Fig. 2).
2. Locate inlet tubing A and B (Fig. 2). Transfer tubing A to Buffer A bottle and tubing B to Buffer B bottle.
3. Turn on the PC, and then the ÄKTA Start machine, and wait a few minutes for the instrument to initialize.
4. Double click the UNICORN start icon on the desktop to open the software. Select the following modules: *System Control*, *Method Editor*, and *Evaluation*. *System Control* is used to start, view, and control a run. *Method Editor* provides flexibility to create or edit the chromatography methods. *Evaluation* module is used to manage and evaluate results from chromatography runs.
5. In *System Control*, click "Connect" to connect the instrument to the PC.
6. The top pane of the window shows the current state of the instrument, and the bottom pane shows the real-time flow path and manual controls.

Table 1
Buffer A (lysis buffer)

Component	Formula	Stock solution or reagent	Amount of stock or reagent	Final buffer concentration
Disodium phosphate	Na_2HPO_4	0.5 M	94 mL	47 mM
Monosodium phosphate	NaH_2PO_4	1 M	3 mL	3 mM
Sodium chloride	NaCl	58.44 g/mol	17.53 g	300 mM
Sodium azide	NaN_3	10%	2 mL	0.02%
Ultrapure water	H_2O	–	Top up to 1 L	–
Total	–	–	*1 L*	

Table 2
Buffer B (elution buffer; *see* Note 6)

Component	Formula	Stock solution or reagent	Amount of stock solution or reagent	Final buffer concentration
Disodium phosphate	Na_2HPO_4	0.5 M	94 mL	47 mM
Monosodium phosphate	NaH_2PO_4	1 M	3 mL	3 mM
Sodium chloride	NaCl	58.44 g/mol	17.53 g	300 mM
Imidazole	$C_3H_4N_2$	68.08 g/mol	34.04 g	500 mM
Sodium azide	NaN_3	10%	2 mL	0.02%
Ultrapure water	H_2O	–	Top up to 1 L	–
Hydrogen chloride	HCl	1 M	A few drops	–
Total	–	–	*1 L*	

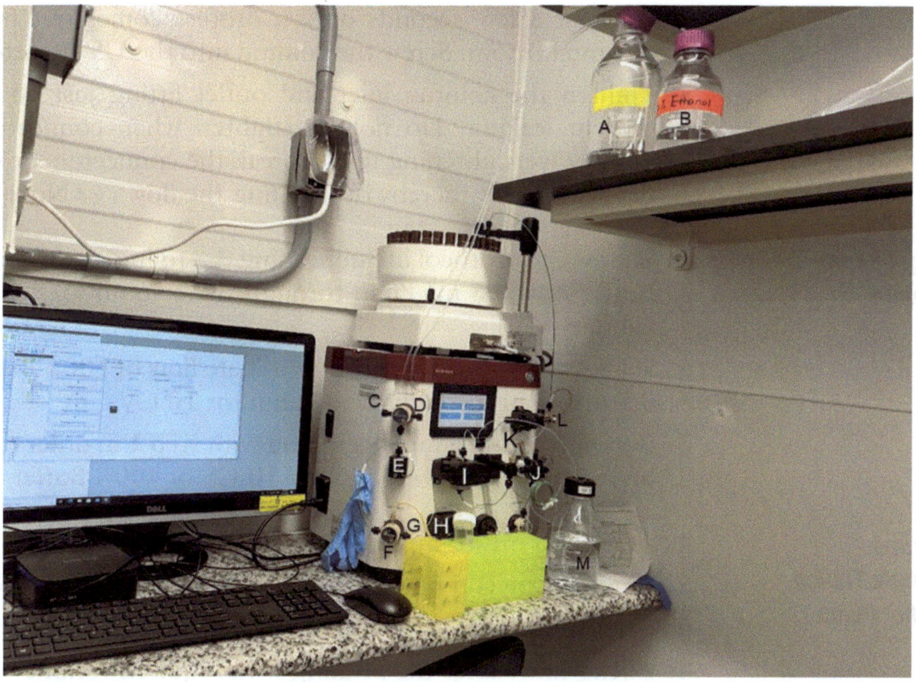

Fig. 2 Instrument modules with annotated parts (**a**–**m**): (**a**) Bottle A; (**b**) Bottle B; (**c**) Bottle A attachment to buffer valve; (**d**) Bottle B attachment to buffer valve; (**e**) Mixer; (**f**) Sample valve; (**g**) Sample valve attachment port; (**h**) Pump; (**i**) UV monitor; (**j**) Injection valve; (**k**) Column attachment; *see* **Note 16**; (**l**) Outlet valve; (**m**) Waste bottle

3.3 Priming the Pump

1. Place all waste tubing, labeled as W1, W2, and W3, in a 500 mL glass bottle.
2. In *System Control*, select "Manual Run." Set the flow rate at 0.5 mL/min.
3. Set %B as 50% at the mixer and click the wash valve to initiate flow of 50% Buffer A and 50% Buffer B at 0.5 mL/min through the system.
4. Wait until the pump is primed with 10 mL of the buffers.
5. Set %B as 0% at the mixer to initiate 10 mL flow of 100% Buffer A at 0.5 mL/min through the system.

3.4 Install/Change a Chromatography Column on the ÄKTA Start System (See Note 8)

1. Remove the column stoppers; mount the column on the union connector if the column type requires a union.
2. Remove the tube connector from the injection valve part 1 by unscrewing the knurled fastener.
3. As a droplet emerges from the tubing, touch the tubing to the column head and begin to thread the fitting in. Leave a slight looseness of threads, which allows liquid to escape around the fitting and prevents pressure buildup in the column.
4. Wait for a few seconds until a droplet comes out from the column bottom. Screw the column into the UV detector inlet.
5. Tighten the column inlet and outlet fitting just enough to prevent leaking. Do not overtighten when connecting columns. Overtightening might break the connectors or squeeze the tubing and thereby obstructing the flow (*see* **Note 9**).

3.5 Equilibrate the Column

1. Follow the protocol in Subheading 3.3 to initiate flow of 100% Buffer A at 0.5 mL/min through the column for 10 column volume (CV).
2. Set %B as 100% at the mixer to initiate flow of 100% Buffer B at 0.5 mL/min through the column for 10 CV.
3. Repeat this procedure beginning with **step 1** in order to equilibrate the column with another 10 CV of 100% Buffer A.

3.6 Prime the Sample Tubing Using the Pump

1. Transfer 5 mL of Buffer A into a 15 mL conical tube.
2. Immerse the end of the sample valve inlet tubing in the tube.
3. Click the sample valve and waste valve to divert the flow.
4. Click "End" after the priming with required volume of buffer has been completed.
5. Carefully transfer the sample valve inlet tubing from the buffer tube to the sample tube. Make sure no air bubbles enter the tubing (*see* **Note 10**).

3.7 Prime the Sample Loop Before Injecting Sample

1. Fill a syringe with 5 mL of Buffer A.
2. Tap the syringe with a finger to move air bubbles to the top. Then push gently on the plunger to push out air bubbles.
3. Make sure that the injection valve is set to position: Load Sample.
4. Connect the syringe to the injection valve port 3.
5. Load 1.5 mL of the buffer into the sample loop.
6. Repeat all steps in Subheading 3.7 using at least 5× the loop volume before loading the sample (*see* **Note 11**).

3.8 Write Methods in UNICORN Start (See Note 12)

1. In the UNICORN software, navigate to the *Method Editor* window.
2. Under the *File* menu, select "New Method."
3. In the pop-up window, select a pre-defined method (affinity, anion exchange, cation exchange, desalting, gel filtration) for the desired type of chromatography, and then click OK.
4. Select "Method Settings" in the left pane.
5. If a GE Life Sciences prepacked column is used, find and select the column in the Column Type dropdown menu. Properties of the column will be automatically displayed.
6. If the column is not listed, select "user desired column" in the Column Type dropdown menu. Manually enter the column volume, pressure limit, and flow rate, as specified by the manufacturer.
7. Select "Prime and Equilibration" in the left pane.
8. Check the prime-the-pump box.
9. Set the equilibration volume as 5 CV.
10. In *Advanced Options*, select "Use the same flow rate as in Method Settings" to equilibrate the column with the default flow rate.
11. Set the start concentration of Buffer "B" as 0%.
12. Select "reset UV monitor (Auto Zero)" to zero the detector before loading the sample.
13. Select "Sample Application" in the left pane.
14. Select either "loop/super loop or pump" for applying the sample.
15. Enter the sample volume. If the pump will be used to apply the sample, enter a volume at least 5 mL less than the actual volume to prevent the introduction of air into the system.
16. In *Advanced Options*, enter the desired flow rate. Capture methods such as ion exchange or affinity separations usually achieve a higher binding efficiency at lower flow rates.

17. Select either "fractions" or "no fractions" for collecting the flow through.
18. Enter the desired volume per fraction if fractionation is desired.
19. The flow through will be collected in the "W3" waste tubing if fractionation is not selected.
20. Select "Wash out unbound" in the left pane.
21. Enter the desired wash volume.
22. Enter the desired flow rate in *Advanced Options.*
23. Set the start concentration of Buffer "B" as 0%.
24. Select either "fractions" or "no fractions" for collecting the wash.
25. Enter the desired volume per fraction if enabling fractionations.
26. The wash will be collected in the W3 waste tubing if fractionation is not selected.
27. Select "Elution and Fractionation" from the left pane.
28. Select either "isocratic elution" (constant elution strength) or "gradient elution" (increasing elution strength).
29. If using isocratic elution, enter the target concentration of Buffer B and desired elution volume.
30. If using gradient elution, enter the start and target concentration of Buffer B and the desired gradient volume.
31. Select "Add more steps" to add step segment(s) during the gradient elution.
32. Select "Enable Fractionation" to collect eluates in 1.5 or 15 mL tubes in the fraction collector carousel. Set the fractionation type as "fixed volume fractionation." Enter the desired fractionation volume.
33. In *Advanced Options*, enter the desired flow rate.
34. Select "Prime and Equilibration" from the left pane.
35. Set the equilibration volume as 5 CV.
36. In *Advanced Options*, select "Use the same flow rate as in Method Settings" to equilibrate the column with the default flow rate.
37. Set the start concentration of Buffer B as 0%.
38. Select "reset UV monitor" (Auto Zero) to zero the detector for subsequent runs.
39. Under the *File* menu, select "Save" and name the method. Save the method in a designated folder.

40. In *System Control*, click "Method Run" and select the saved method to perform the purification.
41. Make sure enough tubes are placed on the fraction collector carousel and the drip outlet is aligned with the tube before starting the run.

3.9 Equipment Shutdown and Short-Term Storage (See Note 13)

1. After completion of the final run of the day, transfer inlet tubing A and B to a flask of degassed 0.02% sodium azide solution.
2. In *System Control*, select "Manual Run." Set the flow rate at 0.5 mL/min.
3. Set %B as 50% at the mixer and click the wash valve to initiate flow of the sodium azide solution at 0.5 mL/min through the system.
4. Pump 20 mL of the sodium azide solution through the system.
5. End the run and leave the system filled with the sodium azide solution during the storage period.
6. Close UNICORN start software. Turn off the ÄKTA Start instrument and then the PC.

3.10 Equipment Shutdown and Long-Term Storage (See Note 14)

1. After completion of the System Short-Term Storage method (*see* Subheading 3.9), transfer inlet tubing A and B to a flask of degassed 20% ethanol.
2. Follow the protocol in Subheading 3.9 (**steps 2–5**) and start a manual run. Pump 20 mL of 20% ethanol through the system. Use the same run parameters recommended for the short-term storage procedure.
3. Pause the run when 10 mL of the ethanol has been pumped through the system.
4. Disconnect the tubing from the column.
5. Use the union mounted on the tubing connected to the UV flow cell to reconnect the flow path between the injection valve and the UV monitor (Fig. 2, i).
6. Screw the column stoppers back onto the column.
7. Continue the run until 20 mL of 20% ethanol is pumped through the system.
8. End the run and leave the system filled with 20% ethanol during the storage period.
9. Close UNICORN start software. Turn off the *ÄKTA Start* instrument, and then turn off the PC (*see* **Note 15**).
10. Store the column filled with 20% ethanol at 4 °C.

4 Notes

1. Salting-out refers to the use of ammonium sulfate precipitation which can gently precipitate proteins of interest from a cytosolic mixture using gentle fractionation techniques. Although it is considered crude, ammonium sulfate precipitation does significantly reduce downstream burdens in purification processes.
2. Cytiva Life Sciences, formerly part of General Electric, has recently been acquired by Danaher and rebranded as a standalone operating company within the Life Sciences. Cytiva offers several options for UNICORN software-controlled protein purification under the ÄKTA™ Chromatography Systems label. This chapter is based on the manual for the *ÄKTA™ Start System*, which is our recommendation for beginners. Additional information about the various options can be found on Cytiva's website (last accessed 10.23.2020): https://www.cytivalifesciences.com/en/us/shop/chromatography/chromatography-systems/akta-start-p-05773.
3. The type of column (such as size exclusion, hydrophobicity, and affinity-based) and/or tag used for purification is outside the realm of this chapter and should be worked out empirically. Although this chapter describes the procedure for a 6× His-tag purification, more helpful information regarding tag (s) selection can be found in refs. 24–28.
4. Always wear gloves during preparation of reagents as well as operation and maintenance of *ÄKTA Start*.
5. Prepare fresh reagents shortly before use. Solutions made from the highest-grade reagents and ultrapure water should be passed through a 0.2 μm filter and degassed thoroughly in an ultrasonic bath shortly before their use.
6. Buffer B serves to elute the trapped His-tag proteins from the Ni-NTA (HisTrap HP) column. Prior to using Buffer B, it must be pH balanced by adding a few drops of HCl.
7. HisTrap HP columns can be purchased directly from Cytiva Life Sciences; we recommend starting with Cat. # 17524701 which provides five (×) 1 mL HisTrap HP columns.
8. In order to avoid introducing air bubbles into the column and flow path, make a drop-to-drop connection prior to inserting the threaded fitting into its position.
9. Avoid applying excessive forces when connecting plastic fasteners in order to prevent damaging the threads of the fitting.

10. Avoid air bubbles in the flow path in order to prevent spurious readouts from the UV detector and damage to the column and pumps.
11. Samples should be clarified by centrifugation and filter sterilization immediately before introducing them into the flow path.
12. To better control the *ÄKTA Start instrument*, assemble all programmed steps into a method. Steps vary for a given column and protein of interest but typically include *Prime and Equilibration*, *Sample Application*, *Wash Out Unbound*, and *Elution and Fractionation*.
13. If the system will be unused for 1–4 days, prepare for short-term storage by filling the system with 20% ethanol or 0.02% sodium azide to prevent bacterial growth.
14. If the system will be unused for more than 4 days, prepare the system for long-term storage.
15. Keep the pump cover open when not using the system to enhance the lifetime of the pump tubing.
16. Figure 2 shows a 1 mL HisTrap HP column attached to the column holder located immediately below the letter K.

Acknowledgments

Protocols in this chapter are based on research performed at Ibex Biosciences, LLC, and was partially funded by a 2020 PSC-CUNY Adjunct/CET Professional Development Fund Award granted to M.G. The authors would like to thank the amazing CEOs at Ibex Biosciences, LLC, Michael J. Karlin, Esq, and Murat Croci, for their support in writing this manuscript.

References

1. Ishikawa H, Bischoff R, Holtzer H (1969) Formation of arrowhead complexes with heavy meromyosin in a variety of cell types. J Cell Biol 43(2):312–328
2. Pollard TD, Korn ED (1973) *Acanthamoeba* myosin I isolation from *Acanthamoeba castellanii* of an enzyme similar to muscle myosin. J Biol Chem 10:4682–4690. PMID: 4268863
3. Williams SA, Gavin RH (2005) Myosin genes in *Tetrahymena*. Cell Motil Cytoskeleton 61:237–243
4. Szent-Györgyi AG (2004) The early history of the biochemistry of muscle contraction. J Gen Physiol 123:631–641
5. Betapudi V (2014) Life without double-headed non-muscle myosin II motor proteins. Front Chem 7:45. https://doi.org/10.3389/fchem.2014.00045. PMID: 25072053
6. Huxley AF, Niedergerke R (1954) Structural changes in muscle during contraction: interference microscopy of living muscle fibres. Nature 173:971–973
7. Sandquist JC, Kita AM, Bement WM (2011) And the dead shall rise: actin and myosin return to the spindle. Dev Cell 21:410–419
8. Basciano PA, King-Smith C (2002) Actin-dependent, retrograde motility of surface-attached beads and aggregating pigment

granules in dissociated teleost retinal pigment epithelial cells. Pigment Cell Res 15:184–191
9. Hosein RE, Gavin RH (2007) Myo1 localizes to phagosomes, some of which traffic to the nucleus in a Myo1-dependent manner in *Tetrahymena thermophila*. Cell Motil Cytoskeleton 64:926–935
10. Gotesman M, Hosein RE, Gavin RH (2010) A FERM domain in a class XIV myosin interacts with actin and tubulin and localizes to the cytoskeleton, phagosomes, and nucleus in *Tetrahymena thermophila*. Cytoskeleton 67 (2):90–101. https://doi.org/10.1002/cm.20426. PMID: 20169533
11. Lakshmi RB, Nair VM, Manna T (2018) Regulators of spindle microtubules and their mechanisms: living together matters. IUBMB Life 70:101–111
12. Monda JK, Cheeseman IM (2018) The kinetochore-microtubule interface at a glance. J Cell Sci 131(16). https://doi.org/10.1242/jcs.214577. PMID: 30115751; PMCID: PMC612773
13. Korolyev E, Steinberg-Neifach O, Eshel D (2005) Mutations in the yeast kinesin-like Cin8p are alleviated by osmotic support. FEMS Microbiol Lett 1:379–383
14. Dechat T, Adam SA, Taimen P, Shimi T, Goldman RD (2010) Nuclear lamins. Cold Spring Harb Perspect Biol 2:a000547. https://doi.org/10.1101/cshperspect.a000547
15. Kühne W (1864) Untersuchungen uber das Protoplasma und die Contractilitat. W. Engelmann, Leipzig
16. Halliburton WD (1887) On muscle-plasma. J Physiol 8:133
17. Oosawa F (2018) Discovery of myosin I and Pollard-san. Biophys Rev 10:1481–1482. https://doi.org/10.1007/s12551-018-0485-5
18. Koubassova NA, Tsaturyan AK (2011) Molecular mechanism of actin-myosin motor in muscle. Biochemistry (Mosc) 76:1484–1506. https://doi.org/10.1134/S0006297911130086
19. Yang Y, Baboolal TG, Siththanandan V, Chen M, Walker ML, Knight PJ, Peckham M, Sellers JR (2009) A FERM domain autoregulates Drosophila myosin 7a activity. Proc Natl Acad Sci U S A 106:4189–4194. https://doi.org/10.1073/pnas.080868210
20. Fujita K, Ohmachi M, Ikezaki K, Yanagida T, Iwaki M (2019) Direct visualization of human myosin II force generation using DNA origami-based thick filaments. Commun Biol 2:1–1
21. Gotesman M, Menanteau-Ledouble S, El-Matbouli M (2016) Proteomic analysis of cytoskeleton proteins in fish. In: Gavin RH (ed) Methods in molecular biology: cytoskeleton methods and protocols, vol 1365. Humana Press, New York, NY, pp 357–372
22. Fitzgerald J, Leonard P, Darcy E, Sharma S, O'Kennedy R (2017) Immunoaffinity chromatography: conceptds and applications. In: Walls D, Loughran (eds) Methods in molecular biology: protein chromatography, vol 1485. Humana Press, New York, NY, pp 27–51
23. Gotesman M, Hosein RE, Gavin RH (2011) MyTH4, independent of its companion FERM domain, affects the organization of an intramacronuclear microtubule array and is involved in elongation of the macronucleus in *Tetrahymena thermophila*. Cytoskeleton 68:220–236
24. Loughran ST, Walls D (2017) Tagging recombinant proteins to enhance solubility and aid purification. In: Walls D, Loughran (eds) Methods in molecular biology: protein chromatography, vol 1485. Humana Press, New York, NY, pp 131–156
25. Loughran ST, Bree RT, Walls D (2017) Purification of polyhistidine-tagged proteins. In: Walls D, Loughran (eds) Methods in molecular biology: protein chromatography, vol 1485. Humana Press, New York, NY, pp 275–303. https://doi.org/10.1007/978-1-4939-6412-3_14
26. Lebendiker M, Danieli T (2017) Purification of proteins fused to maltose-binding protein. In: Walls D, Loughran (eds) Methods in molecular biology: protein chromatography, vol 1485. Humana Press, New York, NY, pp 257–273
27. Bonner PLR (2018) Chromatography and the strategy of protein purification. In: Protein purification, 2nd edn. Taylor & Francis, London, pp 31–57
28. Bonner PLR (2018) Affinity-based procedures used to purify proteins. In: Protein purification, 2nd edn. Taylor & Francis, London, pp 145–175
29. Tenney A, Price M, Macatangay J, Hatt D, Kline T, Berezuk MA (2016) Partial characterization of an interaction between kinesin associated protein 3 (Kap3) of kinesin-2 and the actin cytoskeleton. Austin Biochem 1:1003
30. Galkin VE, Orlova A, Lukoyanova N, Wriggers W, Egelman EH (2001) Actin depolymerizing factor stabilizes an existing state of F-actin and can change the tilt of F-actin subunits. J Cell Biol 153:75–86
31. Herm-Götz A, Weiss S, Stratmann R, Fujita-Becker S, Ruff C, Meyhöfer E, Soldati T,

Manstein DJ, Geeves MA, Soldati D (2002) Toxoplasma gondii myosin A and its light chain: a fast, single-headed, plus-end-directed motor. EMBO J 21:2149–2158

32. Altman D, Sweeney HL, Spudich JA (2004) The mechanism of myosin VI translocation and its load-induced anchoring. Cell 116:737–749
33. Senju Y, Kalimeri M, Koskela EV, Somerharju P, Zhao H, Vattulainen I, Lappalainen P (2017) Mechanistic principles underlying regulation of the actin cytoskeleton by phosphoinositides. Proc Natl Acad Sci U S A 114:E8977–E8986
34. Khatau SB, Bloom RJ, Bajpai S, Razafsky D, Zang S, Giri A, Wu PH, Marchand J, Celedon A, Hale CM, Sun S (2012) The distinct roles of the nucleus and nucleus-cytoskeleton connections in three-dimensional cell migration. Sci Rep 2:1
35. Mentel M, Ionescu AE, Puscalau-Girtu I, Helm MS, Badea RA, Rizzoli SO, Szedlacsek SE (2018) WDR1 is a novel EYA3 substrate and its dephosphorylation induces modifications of the cellular actin cytoskeleton. Sci Rep 8:1–6

Chapter 13

The Cilioprotist Cytoskeleton, a Model for Understanding How Cell Architecture and Pattern Are Specified: Recent Discoveries from Ciliates and Comparable Model Systems

Linda A. Hufnagel

Abstract

The cytoskeletons of eukaryotic, cilioprotist microorganisms are complex, highly patterned, and diverse, reflecting the varied and elaborate swimming, feeding, reproductive, and sensory behaviors of the multitude of cilioprotist species that inhabit the aquatic environment. In the past 10–20 years, many new discoveries and technologies have helped to advance our understanding of how cytoskeletal organelles are assembled in many different eukaryotic model systems, in relation to the construction and modification of overall cellular architecture and function. Microtubule organizing centers, particularly basal bodies and centrioles, have continued to reveal their central roles in architectural engineering of the eukaryotic cell, including in the cilioprotists. This review calls attention to (1) published resources that illuminate what is known of the cilioprotist cytoskeleton; (2) recent studies on cilioprotists and other model organisms that raise specific questions regarding whether basal body- and centriole-associated nucleic acids, both DNA and RNA, should continue to be considered when seeking to employ cilioprotists as model systems for cytoskeletal research; and (3) new, mainly imaging, technologies that have already proven useful for, but also promise to enhance, future cytoskeletal research on cilioprotists.

Key words Basal body, Centriole, Cilioprotists, Kinetosome, Infraciliature, DNA, RNA, *Tetrahymena*, *Paramecium*, *Euplotes*

This review is dedicated to Janine Beisson, upon her untimely departure from this world at the age of 89. Prof. Beisson was committed throughout her life to research on the cilioprotist cytoskeleton, pattern formation, and non-Mendelian inheritance. She also has had profound influence on her many collaborators and colleagues, resulting in numerous discoveries concerning ciliate architecture and its assembly, only a small sample of which could be mentioned herein. This review is also dedicated to my mentors as follows: Reuben Torch, who introduced me to the amazing diversity and complexity of the cilioprotists, the value of experimental cytology, and the importance of attention to detail, in both observation and writing; John Preer, who from the outset of our collaboration treated me as a colleague and not as a student and who unconditionally supported my quest for new knowledge about ciliates; and Joseph Gall, who also lent me his unconditional support, to my pursuit of a career as a cell biologist.

Ray H. Gavin (ed.), *Cytoskeleton: Methods and Protocols*, Methods in Molecular Biology, vol. 2364,
https://doi.org/10.1007/978-1-0716-1661-1_13,

1 Introduction

For those of us who conduct or have conducted research on the ultrastructure and reproduction of the ciliated, eukaryotic microorganisms known as "cilioprotists" or "ciliates," such as *Paramecium*, *Tetrahymena*, *Stentor*, or *Euplotes*, or who have simply become students or collectors of cilioprotist lore, the term "cytoskeleton" has a very special meaning. For, in the microscopic cilioprotists, we encounter the arguably most sophisticated, patterned, and diversified cytoskeleton to be found in eukaryotic cells [1, 2]. Not only does the ciliate cytoskeleton carry out the many functions that cytoskeletons perform in all other eukaryotic cells (such as cell motility and re-establishment of mature cell architecture and form following cytokinesis and fertilization), but it also supports the highly organized and repetitive patterning of the ciliature that underlies the diverse motilities, sensory reception [3], and feeding behaviors that are emblematic of the diverse hunter-gathering cilioprotist tribes inhabiting the microbial world [4, 5]. Furthermore, the cilioprotist cytoskeleton is most certainly the armature for a unique type of epigenetic inheritance, referred to as "perpetuation of pattern" by Nanny [6], "cytotaxis" by Sonneborn and many others [7, 8], or "structural inheritance" [9, 10], first discovered in the ciliates but possibly occurring in other eukaryotic cells as well, as suggested by Beisson and Sonneborn [11], Bornens [12], and others. Thus, cilioprotists have long been viewed as attractive model organisms from which to gain insight into the many roles of the cytoskeleton in eukaryotic cellular processes [10, 13, 14]. This is especially true because the genomes of *Tetrahymena* [15, 16], *Paramecium* [17, 18], *Stentor* [19], and *Oxytricha* [20] have been completely sequenced and have already revealed many generalizable secrets, such as the properties of telomeres, and gene amplification (*see* Gall, 1986, for an early summary of genomic discoveries in cilioprotists [21]). Furthermore, especially for *Tetrahymena*, a tool kit of genetic and molecular biological methods is readily available [16, 22].

Within the past 10–20 years, many new discoveries have advanced our understanding of the assembly of cytoskeletal organelles, particularly in relation to their role in determining cell architecture and function. Some of these discoveries have emanated from research on cilioprotists, but many have resulted from non-cilioprotist model organisms. Time, space, and the number of published works do not allow all relevant topics or publications to be included in this review. Therefore, I will attempt to explore a few of the most significant advances, and raise questions and suggest approaches that could be applied to ciliate model organisms. This review will not attempt to cover all possible topics related to ciliates as models for cytoskeletal research, but will seek to pull together a variety of published resources that can help to define some of the questions that remain unanswered or unresolved, for example, regarding the question of whether nucleic acids are found within or associated with ciliate kinetosomes (aka basal bodies), and if so what role(s) they might have in cytoskeletal assembly.

2 Basal Bodies

At the many hearts of the cilioprotist cytoskeleton are its cylindrical kinetosomes, which serve as seeds for and support of the numerous motile cilia that aid the organism in swimming and feeding. The arrangements of these kinetosomes reflect the arrangements of cilia into somatic rows for motility and oral rows that sweep prey microbes into their buccal cavities. As a type of microtubule organizing center (MTOC), ciliate kinetosomes also appear to be central to the formation of several components of the ciliate cytoskeleton. Since kinetosomes share numerous structural, functional, and molecular features with the basal bodies found in many other types of ciliated and flagellated cells, including cylindrical shape and ninefold symmetry, hereafter ciliate kinetosomes will be referred to as "basal bodies" or "BB." For recent summaries of basal body properties in a typical and much analyzed cilioprotist, *Tetrahymena*, *see* Wloga and Frankel [23] and Bayless et al. [24]. For *Paramecium*, ultrastructural details of ciliary basal body structure, arrangements, and cytoskeletal associations were beautifully presented by Ehret and McArdle [25]; important aspects of *Paramecium tetraurelia* basal body ultrastructure have more recently been shown and discussed by Tassin et al. [26]. For comparison with basal body structure in another protist model cell, the green alga *Chlamydomonas reinhardtii*, *see* Dutcher and O'Toole [27].

Ciliate basal bodies share an ancestry, basic structural organization, and protein composition with centrioles, organelles that are usually found at the poles of the mitotic spindle and in close association with the nucleus of most eukaryotic cells, but can become basal bodies to support the growth of motile and/or sensory cilia or flagella in many types of cells [28, 29]. Centrioles/basal bodies (C/BBs) apparently arose early in the evolution of the eukaryotic cell and are almost universally found in various genomic branches of eukaryotes [29]. However, in some eukaryotic groups and cell types, including higher plants, higher fungi, human mature oocytes, mouse spermatozoa, and certain cancer cells, C/BBs are lacking [30, 31]. A number of recent authors, including Chapman et al. [32], Zhou and Roy [33], Jana et al. [34], and Sathanathan et al. [35], have provided useful diagrams that illustrate the detailed structure of centrioles and ciliary/

flagellar basal bodies, as well as their central role in many of the functions of eukaryotic cells, including those of protists, mammals, and humans.

C/BBs have many functions in human biology, including the selective contribution of sperm centrioles to human embryonic development, discussed by Avidor-Reiss et al. [31], and the activities of basal bodies during development and functioning of cilia and cilia-derived cell structures. Therefore, they have been implicated in a variety of human dysfunctions (such as male factor infertility [31]) that affect human reproduction, cellular aging [36], sensory functions including vision, respiration (e.g., Kartagener syndrome [37]), and sperm cell motility. C/BB malfunction also appears to underlie many developmental disorders and cancers [38]. Davis and Katsanis [39], Oh and Katsanis [40], and Wheatley [41] have published recent reviews that further address this topic.

Nearly 50 years ago, Fulton [42] thoroughly and elegantly reviewed what was understood at that time regarding the structural, functional, and molecular properties of C/BBs, but much more has been learned since then. In 2000, Chapman, Dolan, and Margulis [32] also addressed in great detail the properties of C/BBs, particularly in relation to opposing theories of their evolutionary origin, either from the integration of bacterial endosymbionts or through gradual evolution from simpler cellular components, called "direct filiation" by Bornens and Azimzadeh [43]. However, since 2000, many new discoveries have emerged, as a result of the development of new microscopical, genetic, and molecular biological technologies, and the emergence of new model systems. For example, in 2019, the structural organization and protein arrangements of C/BBs (and procentrioles; *see* definition below) in the unicellular biflagellate *Chlamydomonas* were characterized at 23 angstrom (and better) resolution through the application of cryoelectron tomography, by Li et al. [44]. When this technology was combined with reverse (subtractive) genetics, the precise structural locations of several specific centriolar and centriole-associated proteins were elegantly determined. The ability of cryoelectron tomography [45, 46] to capture the precise locations of specific macromolecules in cells, without the need for pretreatment with fixatives, embedding media, and other reagents, suggests that this approach to characterizing cell architecture and macromolecular arrangements, and for following changes in cell architecture at the molecular level, will be very important in the future (for another relevant example, *see* Subheading 4).

Through genome sequencing and other types of analysis, many conserved C/BB proteins have been identified. These include the alpha-, beta-, gamma-, delta-, and epsilon-tubulins and SAS-6 [47]. Alpha- and beta-tubulins are major components of the microtubule triplets in all C/BBs. SAS-6 is a protein required for centriole assembly [48]. *Paramecium*, with its approximately 4000 basal bodies, and well-developed genetics, has proven to be a useful system for identifying important basal body proteins and determining their roles in C/BB function [49, 50]. For example, in studies utilizing *Paramecium*, Ruiz et al. [51] showed that gamma-tubulin is required for basal body duplication, Garreau de Loubresse et al. [52] confirmed earlier findings from *Chlamydomonas* that delta-tubulin has a role in the formation of the C-tubule of the basal body triplet microtubules, and Dupuis-Williams et al. [50], using immunolocalization at the light and electron microscopic levels, confirmed localization of the epsilon-tubulin within the basal body microtubule triplets and its requirement for basal body duplication. By placing the *Tetrahymena* gene coding for gamma-tubulin (GTU1) under control of an inducible-repressible promoter, Shang et al. [53] found that gamma-tubulin knockdown led to the gradual disappearance of basal bodies and disruption of normal cortical morphology; however, when gamma-tubulin was re-expressed, the normal patterned structure of the cell cortex was restored, suggesting to the authors that the pattern can form de novo. Their observations also raise questions regarding whether the basal bodies, or the gamma-tubulin basal body "starter" protein, possess the ultimate blueprints for cortical patterning in cilioprotists, or whether the blueprints are stored elsewhere in the cell.

Another conserved C/BB protein is Poc1, also required for C/BB assembly. In studies on *Tetrahymena*, Pearson et al. [54] used fluorescence immunocytochemistry, electron microscopy, and gene knockout methods to confirm findings from other model systems that Poc1 is located in the basal body cylinder walls and at the site of assembly of nascent basal bodies. A new finding from their study was that Poc1 is also a stable component of the basal body cartwheel structure. In addition, the findings in this study appeared to shed light on the activities of Poc1 important for assembly and stability of the basal body. Studies similar to this, using cilioprotists to further analyze the roles of conserved as well as unique proteins required for basal body assembly and the correct assembly of associated cytoskeletal structures (*see* Subheading 3), should continue to reveal new information applicable to other organisms and to human health.

3 Infraciliature

Surrounding and attached to the basal bodies of cilioprotists, the basal bodies of other types of ciliated and flagellated cells, and the centrioles of the mitotic apparatus is a more or less complex cytoskeletal architecture (in 2011, Carvalho-Santos et al. [29] published useful summary diagrams). In the case of the mitotic apparatus, the centriole plus its most closely associated cytoskeleton is known as the "centrosome" [28]. In recent years, because of its important role in mitotic spindle formation and related human diseases, much attention has been given to the composition and behavior of centrosomes in vertebrate, invertebrate, and protist model systems. In 2018, Graf [55] published a useful comparative review of centrosomes in eukaryotes. Model organisms that have been useful for centrosomal research include *Drosophila* [56, 57], *Naegleria* [58], and the so-called "green yeast" *Chlamydomonas* [59]. The role of centrosomes in the formation of the mitotic spindle and microtubular asters has been analyzed exhaustively; besides studies on whole cells, this topic has been investigated in cellular extracts and homogenates. Of particular interest here are studies using cell-free preparations from oocytes of *Xenopus* [60] (c.f. Ito et al., 2020) and *Spisula* (c.f. Zackroff et al., 1976 [61]; Alliegro et al., 2008 [62]; and Alliegro and Alliegro, 2013 [63]); one outcome of such in vitro studies has been the discovery of centrosomal RNA, discussed below.

In cilioprotists, such as *Paramecium*, *Tetrahymena*, and the hypotrich ciliate *Paraurostyla*, the basal bodies and associated cytoskeletal elements comprise an architecture known as the "infraciliature" [64], located mainly near the cell surface (within the cell "cortex," i.e., "cortical"). The cortical architecture of the model ciliate *Tetrahymena* has been amply described by Frankel [65] and Wloga and Frankel [23]. Besides the basal bodies themselves, there are a number of other structural elements, including a submembranous layer termed the "epiplasm" [66], striated fibers (e.g., "kinetodesmal fibers," a type of ciliary rootlet) [1, 67], postciliary and transverse ribbons [1], and the infraciliary lattice (ICL), a polygonal, contractile network of microfibrillar bundles located at the level of the proximal ends of the ciliary basal bodies (described in *Paramecium* by Beisson et al. [68]). Microtubules constitute an important structural component of the basal bodies, as well as the postciliary and transverse ribbons [1]; microtubules, composed of proteins known as "tubulins," are highly conserved structures common to most eukaryotic cells, and have long been considered to be among the most important structural components determining the architecture and remodeling of eukaryotic cells.

Unlike the cytoskeleton of other eukaryotic cells, the cilioprotist infraciliature is uniquely composed of repetitive units that are centered on one or two basal bodies and are asymmetrically organized; these are known as "cortical units" [23] or "kinetids" [1]. As a result of cortical unit organization, cilioprotists exhibit a distinct cellular patterning that includes an anteroposterior axis and left-right asymmetry, described in detail by Frankel [69, 70] and more recently by Iftode and Fleury-Aubusson [9] and Soares et al. [2]. Ciliates asexually reproduce through a process of transverse fission, in which, in the simplest cases, the anteroposterior rows of somatic cilia and their basal bodies are transected, and in the anterior daughter (called "proter"), new basal bodies are added within and to the posterior end of each row, while in the posterior daughter (called "opisthe"), new basal bodies are added within and to the anterior end of each row. In this way, each ciliary row is propagated to the daughter cells. As addressed by Sonneborn [7], in 1963, and somewhat later in several important papers by Nanny [4, 6, 71], extensive observations on dividing cells of many different cilioprotist species have shown that the patterns of ciliation on daughter cells are a direct result of the patterns of ciliation on parent cells, even in cases where some cilia are grouped into complex arrangements for feeding, such as within the buccal cavity (aka gullet) of *Paramecium* [72, 73], or for swimming, such as the cirri of *Euplotes* [74]. Experimental surgical alterations of the cell surface architecture in a variety of cilioprotists, for example, *Stentor* [75], *Euplotes* [76, 77] and *Paramecium* [78], have confirmed the presence of an epigenetic mechanism of inheritance of cell patterning called "cytotaxis" [7] (*see* summaries by Nanney [71] and Beisson [79]). However, the experimental results have also suggested an underlying influence of nuclear genes in returning experimentally altered cell architecture to normal over several generations of cell division [76, 77, 80].

In recent years, the advent of new technologies in combination with gene sequencing has permitted the use of *Tetrahymena* in studies to better understand the role of the infraciliature in positioning and stabilizing basal bodies in relation to effects of ciliary beating forces [81, 82]. This topic is particularly relevant medically because of the role of cilia in transporting mucus in the human respiratory airways. In studies on *Tetrahymena*, Junker et al. [82] recently combined light microscopic immunofluorescence analysis with EM tomography to show that glycylation of basal body-associated microtubules has an important architectural role in positioning of basal bodies and therefore the organization of the ciliary rows.

4 Basal Body/Centriole Assembly

To appreciate the mystique held by basal bodies and centrioles, one must be aware that they appear to be "self-replicating" organelles. But unlike membrane-limited self-replicating organelles like mitochondria and chloroplasts, they do not "divide" in two and are not surrounded by their own membranes. Instead, basal bodies and centrioles usually reproduce by a type of "budding," in which a new basal body or centriole (called "procentriole") assembles in close association with, and at right angles to, a preexisting basal body or centriole. This process was first described at the ultrastructural level in *Paramecium* by Dippell [83] (1968), who observed that a component called the "cartwheel" assembled earlier than other basal body structures, during the formation of a new basal body.

In ciliates, self-replication of basal bodies was recognized very early, through the extensive observations that Chatton, Lwoff, and their co-workers carried out on the asexual reproduction of many different ciliate species, following fixation and staining with silver [84, 85]. Their observations led Lwoff [86], in 1950, to propose that basal bodies of ciliates might contain genetic information, later referred to by Nanny as corticogenes [71] and interpreted to mean the recognized genetic material, DNA (discussed in more detail below), although one consequence of the ability of basal bodies in ciliates to self-replicate is that certain aspects of cortical patterning in ciliates appeared to some to be under epigenetic control. Examples include (1) the inheritance of kinety (ciliary row) number in *Paramecium* (Sonneborn, 1963 [7]) and *Tetrahymena* (Nanney, 1968, 1975, 1977 [4, 87, 6]; Frankel, 1979 [80]); (2) the inheritance of experimentally rotated kineties in *Tetrahymena* (Ng and Frankel, 1977 [88]); and (3) the occurrence of intraclonal dimorphism in *Euplotes vannus*, a special phenomenon in which, during cell division, the anterior daughter cell develops a different number of caudal cirri (clusters of cilia used for cell motility) than the posterior daughter (Hufnagel and Torch, 1967 [74]). In the last case, the number of caudal cirri on a daughter cell was found to be dependent on where on the parent cell surface the daughter cell cirri are assembled, reflecting local influences by the parental cell on cortical patterning in daughter cells.

The steps in assembly of C/BBs have received detailed scrutiny since Dippell's early ultrastructural analysis in *Paramecium* [83]. Much of the research has been carried out on non-cilioprotist cells, for example, the ultrastructural studies of Johnson and Rosenbaum (1992) on *Chlamydomonas* centrioles [89]. In 2010, Azimzadeh and Marshall provided a useful review of genes and conserved proteins required for centriole assembly, as well as non-conserved proteins that may be involved in specific structural and functional relationships between centrioles and the cytoskeletons of diverse organisms [90]. They also focused on the mechanisms determining ninefold symmetry, control of centriole length, and steps in centriole maturation. Much more has been learned since 2010. For example, in 2018, Guichard et al. detailed much of what is now known about the steps in formation of the cartwheel, the "seed" for C/BB assembly, based on studies on a variety of model systems including *Chlamydomonas*, *Drosophila*, and human cells [91]. Analysis of cartwheel proteins and the use of cryoelectron tomography have allowed Guichard and his collaborators (and others) to dissect many details of cartwheel assembly. Recently, their group developed its own version of a relatively new super-resolution imaging approach that enables molecular resolution imaging at the light microscopic level. By combining expansion microscopy (ExM) [92–95] with single-molecule localization microscopy (SMLM) [96, 97], they have been able to localize fluorescent labels at the molecular level and have recently begun to apply their methods to the analysis of centrioles [98].

Other recent reviews that address the molecular events in C/BB assembly are available [99–102]. Loncarek and Bettencourt-Dias, in 2018, addressed research that illuminates how diverse C/BB number, size, and organization are established [99]. Gonczy and Hatsopoulos, in 2019, provided a summary of steps in formation and elimination of centrioles, and very significantly, called attention to recent use of advanced super-resolution microscopy methods to transform our understanding of C/BB assembly by providing "nanometer-scale precision" [100]. They also called attention to the need to even more precisely localize centriolar proteins, and suggested that the use of "super-resolution time-lapse microscopy of centriole assembly in a cellular context," presumably in living cells, will accomplish this mission.

In another recent review, Arslanhan et al., in 2020, addressed the structure and regulation of the mammalian centrosome/cilium complex in relation to the cell cycle, and pointed out various problems inherent in analyzing these activities, one of which is the small number of organelles in most cells [101]; this is a problem that is not encountered in cilioprotists! Of particular interest here, Arslanhan et al. summarized various studies that have used proximity-based labeling methods (BioID) to analyze protein interactions in relation to assembly-related functions of the centrosome/cilium. They explained the methods of proximity labeling, which utilize enzymes that label neighboring proteins with biotin or peroxidase, and suggested that such studies, when correlated with super-resolution imaging, will help to explain the roles of the >1700 unique mammalian proteins known to participate in the assembly and functioning of centrioles, their satellites, and associated cilia.

In another up-to-date summary, Sullenberger et al., in 2020, addressed the structure and composition of mature human centrioles and centrosomes, including so-called pericentriolar material (PCM) (*see* Subheading 7) [102]. They called attention to the fact that proliferating cells simultaneously contain three distinct types of centrioles: pro-centrioles, daughter centrioles, and mother centrioles. They differ in age, structure, and biochemical function, and represent maturation stages that take place over three consecutive cell cycles. Sullenberger et al. covered in great detail the extensive published research regarding the biochemical events during centriole formation and maturation, and suggested that answers to the many questions that remain will depend on a combination of improved genetic and imaging tools.

Another experimental approach to C/BB assembly in mammals, including humans, was reported in 2015, by Wong et al. [103], who conducted experiments on NIH/3T3 mouse embryonic fibroblasts and HCT-116 human colon carcinoma cells, in which they developed and applied an inhibitor ("centrinone") of Polo-like kinase 4 (Plk4), a protein that controls centriole assembly. When normal 3T3 cells were treated with centrinone, they were irreversibly arrested "in a senescence-like G_1 state." In contrast, colon carcinoma cells with normal or experimentally enhanced centrosome numbers "could proliferate indefinitely after centrosome loss" induced by centrinone treatment. Furthermore, the centrinone-induced loss of centrosomes in the cancer cells was reversible; after centrinone washout, the cancer cell lines returned to their predrug centrosome numbers. Clearly, experimental reports like this raise many questions. One question is whether or not centrinone induced the loss of a centriole precursor (invisible by the labeling methods used) in the cancer cells. Another question is whether centrinone would affect the proliferation of basal bodies in cilioprotist cells, and whether the results of centrinone treatment would provide insight about basal body assembly in ciliates, particularly regarding the role of a Plk4-like homologue in this model system.

5 De Novo Assembly of Basal Bodies

In 2007, Satir et al. [104] pointed out that C/BBs do not always arise in close proximity to existing C/BBs by a semiconservative process of budding. In many cases, they appear to condense *en masse* from a fibrogranular matrix, as seen in the formation of an "anarchic field" of basal bodies during formation of buccal cilia in the cilioprotists (e.g., Kaneda and Hansen, 1974 [105]). Thus, while self-replication is often the source of new C/BBs, de novo formation, in the absence of preexisting centrioles or basal bodies, also can occur [30]. Other well-known examples of de novo C/BB formation occur during the transition of the amoeboid stage to the flagellated stage in the protist *Naegleria* [49, 58, 106, 107] and during meiosis in the fern *Marsilea* [108]. In the case of *Naegleria*, this intriguing phenomenon was subjected to repeated microscopic investigation at higher and higher resolutions, in hopes of finding a visible centriolar "seed," but it has only been recently, with a more molecular approach, that evidence for such a seed has been uncovered [58, 109]. Using light microscopic immunocytochemistry, these investigators have revealed, in interphase cells of *Naegleria*, a single transacylase-containing precursor complex (GMPp) that appears to precede the formation of a transient "cell cycle-dependent duplicating complex" (GMP complex; containing γ-tubulin, pericentrin-like protein, and myosin II), from which a pair of basal bodies is formed during the differentiation of the amoeboid stage into the flagellated stage. Whether the GMPp itself assembles de novo is not clear. But these findings in *Naegleria* suggest the need to examine whether GMP- or GMPp-like complexes exist in cilioprotists.

Other challenges to the dependence on preexisting C/BBs for the formation of new C/BBs came from experiments testing the need for centrioles during cell cycle progression in cultured mammalian cells, and observations on the rapid formation of multiple basal bodies in multiciliated mammalian epithelial cells (summarized by Meunier and Spasskey, in 2016 [110]). Earlier observations had appeared to show in these systems that centrioles could form without the participation of preexisting centrioles. However, Rieder et al., in 2001, found that when centrioles in cultured mammalian cells were removed by surgery during the S phase, or laser ablation during the G2 phase, the cells progressed through mitosis but were arrested in G1, suggesting that all of the functions of the centrioles were not restored [111]. In 2016, Meunier and Spasskey [110] suggested that more recent evidence, derived from super-resolution microscopy and genetic manipulation [112, 113], as well as live cell imaging [114], lends support to the idea that preexisting centrioles transmit structural information to newly forming C/BBs, even when a direct structural association is not readily apparent. Here we see how newer advances in microscopy are helping to revise our understanding of cytoskeletal mechanisms related to C/BB function. Furthermore, these new discoveries are showing how the concept of cytotaxis, developed through studies on cilioprotists, may extend to other systems as well, including to mammalian cells.

6 Cilioprotist Cell Patterning and Architecture

Understanding the architecture and assembly of basal bodies/centrioles and their associated cytoskeletal elements requires detailed knowledge of the ultrastructure and macromolecular composition of the entire cell cytoskeleton. Cilioprotists have contributed significantly to this knowledge. Proteins of the ciliate infraciliature and their locations have been identified through studies on detergent-extracted cells, and isolated "pellicles" and oral apparatuses, in combination with a variety of techniques for labeling various macromolecular components. Pellicles are fragments of the cell surface enriched for cytoskeletal structures, including basal bodies (for an example from *Paramecium*, *see* Hufnagel, 1966, 1969 [115, 116]). The extensive contributions of Norman Williams, his wife Ruth Jaeckel-Williams, and his many students and collaborators, on pellicular proteins from the somatic and oral infraciliature of *Tetrahymena*, were summarized in 2017, by Frankel and Buhse [117]. Abundant information has been gathered concerning the types of proteins present and their properties and their structural locations. In 1980, Gavin reported that as many as 138 different polypeptides have been identified in the *Tetrahymena* oral infraciliature ("oral apparatus") alone [118].

In *Tetrahymena*, *Paramecium*, and other ciliates, the main cytoskeletal *structural* proteins (which we can think of as "timbers") are evolutionarily conserved; they include *tubulins* [119], *actin* [120], *myosins* [121], and *tetrins* [122]. Also included in this category are proteins that form a thin submembranous layer, the epiplasm, including *epiplasmins* [123–125] and *articulins* [126–129]. In *Paramecium*, another cortical cytoskeletal structure, the contractile infraciliary lattice (ICL), is composed mainly of *centrin* or centrin-like polypeptides [68, 130]. Cilioprotists also share with other members of the infrakingdom Alveolata the presence of a family of proteins called the *alveolins* [131]. The alveolins contain a repeat motif (EKIVEVP) and may be related to other cortical proteins of ciliates that contain repeat motifs, including the epiplasmins, articulins, and *plateins*, a protein family identified in *Euplotes aediculatus* by Kloetzel et al. [132]. Various studies, including an early immunofluorescence analysis of morphogenesis in the cilioprotist *Paraurostyla* by Fleury et al. [64], in which a monoclonal antibody (CTR 210) raised against centrosomes isolated from human lymphoblastic cells was used, have shown that many of the timber proteins of the ciliate infraciliature are members of multigene families also found in mammals, including humans [124, 130]. It will be interesting to see to what extent the "nails and screws" that fasten these timbers together are also evolutionarily conserved, and how they may have diversified to perform localized cytoskeletal assembly functions in the cilioprotists.

There have been many inquiries into the role of centrioles and basal bodies in the nucleation and assembly of other cytoskeletal organelles, such as the mitotic spindle (see below); this is true also for cytoskeletal assembly in cilioprotists. For example, in the experiments using *Paramecium* mentioned earlier, Beisson et al. [68] investigated the role of basal bodies in nucleating the assembly of the ICL, using immunocytochemistry following experimental disassembly of the ICL. The results suggested that newly developing ICL organization is controlled by preexisting basal body patterning rather than genetic information. However, as mentioned earlier for surgical experiments on cilioprotists, there was also evidence for an underlying genetic influence on patterning. It is expected that further studies will resolve this apparent conflict.

In another example of how cilioprotists can contribute to understanding the role of C/BBs in cytoskeleton assembly and thus cell architecture, in 1993, Fleury et al. [64] used the cross-reacting antibody mentioned earlier (CTR210) to investigate the potential role of cytoskeletal proteins in re-establishing patterning in dedifferentiated cells of the cilioprotist, *Paraurostyla weissei*. In the future, such experimental studies, using cross-reacting antibodies to conserved C/BB proteins and other proteins implicated in centrosomal functions in other organisms, will likely reveal the usefulness of cilioprotists in elucidating the role of C/BBs in cell patterning and architecture.

Recent studies that identify proteins important in docking and positioning of basal bodies in multiciliated animal cells provide another opportunity to use cross-reacting antibodies to dissect cell patterning in cilioprotists. Experiments on *Xenopus* and zebrafish, by Epting et al. [133], reveal molecular tools that may be usefully applied to cilioprotists, in order to identify proteins and biochemical pathways responsible for basal body migration, interaction with actin fibrils, docking at the cell surface, and spacing. Studies on rat tracheal epithelial cells, by Hegan et al. [134], in which rotational positioning of cilia was altered in Myo1d knockout (KO) cells, provide another such example.

In studies on *Paramecium*, Iftode et al. [135] demonstrated how fluorescent antibodies specific for certain cytoskeletal structural proteins, in conjunction with scanning confocal microscopy, can be used to follow the participation of specific proteins in establishing rotational positioning of basal bodies. More recently, the roles of two conserved proteins (OFD1 and VFL3) in positioning and anchoring of basal bodies in *Paramecium* were examined by Bengueddach et al. [136]. A combination of gene silencing, immunolabeling, and light and electron microscopy was used. The results revealed a conserved role for OFD1 in anchoring of motile cilia and a role for VFL3 in establishing rotational asymmetry of the basal body. Rotational positioning is an early event in the development of the oral ciliature of cilioprotists, when a rotationally disarrayed field of basal bodies is transformed into orderly rows [69]. KO mutants that selectively affect ciliate oral morphogenesis could be identified through their effects on both swimming and feeding. These may be useful for investigating the role of less abundant proteins during cytoskeletal assembly in cilioprotists.

In 2016, Galati et al. [137] summarized the literature concerning the importance of multiciliary arrays and their correct assembly, for the generation of proper hydrodynamic flow in the respiratory tract, brain ventricles, and elsewhere. They also demonstrated that automated 3D image analysis can be used to quantify basal body patterning in whole, fixed, and fluorescently labeled *Tetrahymena* cells, using widefield and confocal imaging. Parameters that were quantified included BB frequency during the cell cycle, spacing between BBs, and angular orientation of BBs. In addition, they successfully demonstrated that label intensity averaging can be used to align several labeled proteins relative to each other; this can be utilized to reconstruct macromolecular assemblies and follow their changes during the cytoskeletal events responsible for basal body positioning and growth of associated cortical structures. Many interesting observations resulted from the image analysis study of Galati et al. that cannot be adequately covered here. Most importantly for the purpose of this review, their methods were applied to staged, fixed cells; as Galati et al. reminded us, immobilizing live cells for "extended periods of time has deleterious effects on cellular physiology," and the *Tetrahymena* cell cycle is about 180 min. Looking into the future, it would be desirable to find the means to overcome this problem and adapt their methods to living, immobilized cilioprotist cells, followed throughout the cell cycle.

7 Pericentriolar Material (PCM)

Detailed ultrastructural analysis of the architecture of centrioles and centrosomes has led to recognition of the presence in some cells of a special structure surrounding the centriole, known as the "pericentriolar material" (PCM) [138]. Proteins found in the PCM (aka "centromatrix") include gamma-tubulin and pericentrin [139]. A detailed molecular comparison of the PCM with material directly surrounding the basal bodies of ciliates seems warranted.

8 Nucleic Acids and Basal Bodies/Centrioles: DNA

Because both centrioles and basal bodies appeared in early studies to be self-reproducing, the idea that they might, like nuclei, contain genetic material in the form of DNA was entertained by some biologists, including the Nobel Prize winner, Andre Lwoff, in his influential 1950 book, *Problems of Morphogenesis in Ciliates* [86]. However, later Lwoff appears to have set aside the concepts of inheritance based on genes in basal bodies (or on ordered cell structure) in his 1962 book *Biological Order* [140], in which he focused his attention on the ability of chromosomal DNA and messenger RNA to explain the diverse aspects of inheritance and evolution in living cells and viruses.

Nevertheless, partly through Lwoff's influence and partly because it was becoming more and more evident that other self-reproducing organelles, namely, mitochondria and chloroplasts, contain DNA genes, the hunt was on for DNA in basal bodies and centrioles. In 1962, Rampton reported his autoradiographic analysis following incorporation of tritiated thymidine into living *Tetrahymena pyriformis* cells, which included appropriate enzyme controls [141]. In Rampton's light microscopical studies, the results were negative for the presence of DNA in basal bodies, but Rampton's results were later open to criticism, because his experiments were not sensitive enough to reveal the DNA now known to be present in mitochondria or the amount of DNA earlier reported by Seaman (1959) to be present in basal body fractions isolated from *Tetrahymena pyriformis* [142]. It should be noted that Rampton also did not obtain positive evidence for RNA in *Tetrahymena*'s cytoplasm! This is a good example of the old adage: "Absence of evidence is not necessarily evidence of absence!"

Autoradiographic studies on *Tetrahymena pyriformis* were also reported by Pyne, in 1968, on whole embedded and sectioned cells analyzed at the ultrastructural level [143]. These included a careful quantitative comparison of label over basal bodies and mitochondria. Based on his calculations, Pyne concluded that the amount of DNA that might be present in basal bodies but not detectable by his methods would be too little to code for a protein of average length. While this study would seem to put the idea of basal body DNA genes to rest, it did not entirely rule out the presence of DNA in basal bodies, perhaps a short DNA molecule with an atypical function. Based on the amount of DNA in a *Tetrahymena* mitochondrion reported earlier by Suyama and Preer [144], Hufnagel [145] estimated that even if there were 100 times less DNA in a basal body than in a mitochondrion, it still could code for about 1000 amino acids.

In contrast with these early negative results, in 1965, Randall and Disbrey [146] obtained positive evidence for DNA associated with basal bodies in isolated pellicles of *Tetrahymena*, using staining with acridine orange and labeling with radioactive isotopes specific for DNA. Analogous evidence from *Paramecium*, using similar methods, was obtained by Smith-Sonneborn and Plaut, in 1967 [147]. However, acridine orange binds to double-stranded nucleic acids, whether DNA or RNA [148, 149], and furthermore, the preparative methods did not rule out contamination of the basal body-enriched cell fractions from nucleic acids originating elsewhere in the cell, an issue that has arisen for many of the studies in which positive results were later obtained. Following Seaman's [142] earlier lead, in 1965, Hoffman [150] and Argetsinger [151] independently attempted to extract and quantitate the DNA associated with isolated basal bodies of *Tetrahymena*, using electron microscopy to show more clearly the presence and degree of abundance of basal bodies in their pellicle isolates. While both detected tiny amounts of DNA by chemical analysis, they concluded that their studies could not rule out contamination by DNA of nuclear origin.

For this reason, in studies reported in 1966 and 1969, Hufnagel [115, 145] further characterized DNA extracted from basal body-rich pellicle fractions of *Paramecium aurelia*, using density gradient centrifugation in cesium chloride. The expectation was that a basal body-specific DNA would have a unique density, different from that of DNA of nuclear or mitochondrial origin, and would therefore appear as a unique satellite peak in the density gradient analyses. *Paramecium* pellicles were prepared from cells homogenized in buffered raffinose and were characterized by negative staining electron microscopy to assure that basal bodies were present [115, 116]. The majority of the DNA extracted from basal body-enriched pellicle fractions resembled nuclear DNA in density, even when pellicles were washed extensively or prepared from strains having different nuclear DNA densities [145]; this was interpreted to demonstrate that nuclear DNA tends to stick to the pellicles during their preparation and then is not readily removed by washing. While satellite peaks were observed, they occurred sporadically, and their densities resembled those of bacteria known to contaminate the pellicle fractions, namely, *Aerobacter aerogenes*, the bacterium on which *Paramecium aurelia* was grown, and *Bacillus subtilis*, a bacterium being used for research in a neighboring laboratory! However, a small amount of a unique basal body DNA might not have been sufficient to form a visible peak, and could have hidden under the broad peak of nuclear origin. To circumvent the issue of contamination of pellicles by bacterial DNA, Hufnagel and Gall (unpublished) conducted additional CsCl density gradient experiments on isolated pellicles of axenically grown *Tetrahymena*. Again, no apparent satellite peaks that could be identified readily as unique and of possible basal body origin were observed. The results seemed to add to the growing evidence against the presence of basal body DNA.

However, a provocative idea does not die easily! In 1972, Younger et al. [152] reported a study that took advantage of the fact that the oral membranellar bands (MBs) of *Stentor coeruleus* have 15–20,000 closely spaced basal bodies that are replaced almost synchronously when the MBs are regenerated after being experimentally shed. They conducted autoradiographic studies on regenerated MBs that were shed a second time following treatment of the initially shed cells with H^3-thymidine. The MBs from the second shedding were collected and mounted on slides and fixed with a combination of glutaraldehyde, formaldehyde, and picric acid, and subjected to autoradiographic analysis. As controls, the effects of four different DNA synthesis inhibitors on incorporation of the radioactive label into the MBs were monitored. In addition, whole cells were fixed with Carnoy's (an ethanol/chloroform/glacial acetic acid solution), labeled with H^3-actinomycin D (H^3-AD) and processed for autoradiography. Appropriate controls for specificity of DNA labeling were used. Additional experiments were carried out to test the effects of DNA synthesis inhibitors on basal body regeneration itself. Their results led Younger et al. to conclude that basal body replication is not associated with, nor is it dependent on, DNA synthesis. Although they did detect the presence of DNA associated with basal body sites by autoradiography, they suspected that, because mitochondria were present in the shed MBs, the DNA was of mitochondrial origin. To test this idea, further experiments were carried out to distinguish between a mitochondrial and basal body origin of the DNA. Since mitochondrial DNA was known to be circular, and ethidium bromide was known to complex with closed, circular DNA and prevent mitochondrial DNA synthesis, Younger et al. tested its effect on incorporation of H^3-thymidine into MB DNA. Suitable concentrations of ethidium bromide had no effect on formation of basal bodies during MB regeneration, even after four successive regenerations, as evaluated by light microscopy. However, it did appear to affect the incorporation of H^3-thymidine into MB DNA, although the number of mitochondria in regenerated MBs did not appear to be affected, based on staining with acid aniline fuchsin. As the authors stated it, "incorporation of ^{3}H-T into DNA of isolated MBs could not be clearly detected in the presence of ethidium bromide"; in other words, no specific quantitative data were reported. Although Younger et al. concluded that their results provided no evidence for basal body DNA, their study left open two possibilities: (1) that basal body DNA of *Stentor* is circular and therefore affected by ethidium bromide, or (2) that the methods were not sensitive enough to detect a small amount of non-mitochondrial, noncircular DNA in the basal body-rich MBs. A quantitative analysis of the autoradiograms was not reported, even though in the same study, such an analysis was carried out during autoradiographic experiments that looked at the incorporation into the MBs of H^3-orotic acid as an RNA precursor. As with most, if not all, of the preceding studies, contamination of the isolated MBs with DNA from other cellular sources also remained a possibility, and the studies were conducted at the light microscopic level.

As described above, Younger et al. [152] found in *Stentor* that basal body replication is not dependent on DNA synthesis. A similar conclusion was reached by Rattner and Phillips [153] in 1973, for procentriole formation in mitotic mouse L929 cells, evaluated in serial sections of Epon-embedded cells examined by transmission electron microscopy. DNA synthesis was inhibited by pretreating the cells, during growth on Petri dishes, with arabinosyl cytosine, at a concentration shown to be effective for inhibiting DNA synthesis in L929 cells. The authors considered whether their observations could speak to whether or not centrioles are semiautonomous structures containing their own DNA. They concluded that their data did not demonstrate better than 99% inhibition of DNA synthesis, and thus left open the possibility that a small amount of DNA synthesis, such as might take place even in mitochondria, may have occurred. Therefore, they concluded that their experiments could not address the question of centriolar DNA.

More recent efforts to address whether BB/C DNA exists have come from studies on the biflagellate alga, *Chlamydomonas reinhardtii*, which has the advantages of a long history as a model system for flagellogenesis and a well-studied genome. In 1989, Hall, Ramanis, and Luck [154] reported a new type of positive evidence for the presence of DNA in *Chlamydomonas* basal bodies. *Chlamydomonas* has two basal bodies that generate its flagella and change location to become centrioles for the mitotic spindle. During the meiotic life cycle, centrioles disappear during gametic fusion and reappear during the first meiotic division. In a mutant strain of *C. reinhardtii*, one of the two flagella fails to form; the mutation affects the development of the distal portion of its basal body. This so-called *uni* mutation had been found to segregate 2:2, as for nuclear markers (i.e., exhibiting Mendelian inheritance). However, the *uni* mutation was not linked to any of the 18 known nuclear linkage groups in *Chlamydomonas* [155]. Nevertheless, it was linked to several mutations that influence flagellar function or assembly, via their effects on the basal body, and this "*uni*" linkage group (ULG; aka "*uni* chromosome") was found to be circular [155]. These findings caused Hall et al. [154] to investigate further the properties of the ULG, and examine whether it constituted a basal body-located genome. A restriction fragment length polymorphism (RFLP) was constructed and was demonstrated to be linked to the ULG. Further cloning steps were undertaken, to produce a ULG-specific probe that could be used in pulsed-field gel electrophoresis to identify the *uni* chromosome and characterize its size; this was found to be between 5.7 and 9 megabases. In gametic cells, the *uni* chromosome appeared to be linear.

Using their ULG-specific probe, Hall et al. [154] utilized in situ hybridization to localize the *uni* chromosome in vegetative cells of a cell wall-less mutant. The cells were attached to glass slides, fixed briefly with paraformaldehyde, and treated with detergent (Triton-X100) and pronase. Phase contrast microscopy revealed highly disrupted cell structures, consisting mainly of nuclei, basal bodies, and flagella, with some remnants of chloroplasts (including their nucleoids) and pyrenoids. ULG-specific probes, labeled with biotinylated dUTP by nick translation, were hybridized to the detergent- and pronase-treated cells. Controls were run using probes specific for chloroplast and nuclear rDNA. Positive hybridization of all probes was detected by subsequent application of fluorescein-avidin. The slides were also stained for DNA with the fluorescent label DAPI. The authors presented results showing that DAPI positively stained nuclei, chloroplast nucleoids, and basal bodies. The control in situ hybridization probes gave results consistent with their specific target organelles, nuclear labeling, or chloroplast nucleoid labeling, but no basal body labeling. The authors reported that the ULG-specific probe uniquely labeled basal bodies, usually giving "a cluster of signals" on each pair of basal bodies, in every case corresponding to positive DAPI staining of the basal bodies. Most frequently, the ULG-specific labeling (as well as DAPI staining) of basal bodies was seen in images where proteolysis appeared most extensive and chloroplast DAPI staining was absent (indicating a loss of DNA from chloroplasts during the preparative procedure); however, in some cases, the labeling of basal bodies occurred even when DAPI staining of chloroplasts also occurred, indicating that the cells were less disrupted by the detergent and pronase treatment. Apparently, basal bodies were positively labeled with probes representing two different regions of the *uni* chromosome. To determine whether the *uni*-specific signals could be due to RNA rather than DNA, controls were run in which the preparations were treated with pancreatic RNase either before or after hybridization. In both cases, there was no effect on ULG-specific labeling. However, no positive controls were reported, to demonstrate that the enzyme was active under the conditions used.

The fluorescent signal representing ULG-specific labeling occurred as one or two spots per basal body, with two, three, or four signals per basal body pair. Further experiments led Hall et al. to conclude that each basal body is associated with one *uni* chromosome, and that the variable number of signals per basal body had to do with the accessibility of the probe to the *uni* chromosome. Thus, they concluded, each cell has two *uni* chromosomes. Other possibly significant conclusions reached by Hall et al. are as follows: (1) there was no evidence for *uni*-specific DNA in the nucleus, although according to them the experiments were not sensitive enough to rule out this possibility; (2) the *uni* chromosome would most likely be located in the lumen of the basal body, since pronase treatment was required for its detection; and (3) *uni* chromosome DNA would be in a form more dense than for nuclear DNA [156] but less dense than the T4 phage head DNA. Hall et al. also recognized and discussed various issues inherent during meiosis in *Chlamydomonas*, when centrioles disappear and reappear, and recombination between parental *uni* chromosomes is known to occur [155].

While there are many questions one can raise about the methods used in these in situ experiments on *Chlamydomonas*, the most apparent issue has to do with the pronase treatment. Was it needed to expose a tightly packed DNA molecule present in basal bodies, or did it allow contaminating DNA to be released from the nuclei or chloroplasts and contaminate the basal bodies? The specificity of the results speaks against the latter possibility, unless the binding of nuclear DNA is somehow selective for a contaminating *uni* chromosome of nuclear origin, a conclusion that is hard to accept without further evidence; however, see results of Hall and Luck [157], described below. Another concern, addressed by Hall et al., and by others, including Goodenough [158], Johnson and Dutcher [159], and Johnson and Rosenbaum [160], is whether the amount of DNA that constitutes the *uni* chromosome can be packed into the space available in a basal body. Even if it could, there would be spatial constraints, so Hall et al. envisioned a situation in which the DNA chromosome would be inactive during interphase but transcriptionally and synthetically active during basal body duplication, and they raised the possibility that the *uni* chromosome might migrate from the lumen to the surface of the basal body, perhaps also cycling between the nucleus and basal bodies, and they proposed that the chromosome might alternate between a linear form and a circular form in concert with meiotic events.

The ULG is also known as linkage group XIX. In 1991, Johnson and Dutcher [159] reported studies on linkage group XIX of *C. reinhardtii*, which appeared to provide evidence against the localization of the ULG in basal bodies. Johnson and Dutcher hypothesized that if linkage group XIX is found in the basal bodies, then haploid cells (which have two basal bodies) will have at least two copies of ULG DNA, whereas if linkage group XIX is located in the nucleus rather than the basal bodies, then the number of copies of ULG DNA should be the same as for known nuclear linkage groups, whether the strain examined is haploid or diploid. To test this idea, they sought to develop methods that would be sensitive enough to distinguish a twofold difference in copy number between different chromosomes, and to do this made use of so-called Gulliver elements that are present on linkage group XIX and on other linkage groups as well. After a series of experiments was conducted to develop and validate their methods, evidence was obtained, through quantitative analysis of autoradiograms of DNA separated on agarose gels, that linkage group XIX and other nuclear linkage groups are present in the same number per cell, whether the cells are haploid or diploid. They also measured relative copy number (hybridization intensity) of linkage groups in mutant strains with altered basal body number, including a strain in which over 98% of cells have no recognizable basal bodies (having the so-called *bald2* mutation). Relative hybridization intensity was the same as for wild-type strains, even though basal bodies were absent in >98% of the cells. In addition, they also looked at relative copy number of linkage group XIX in strains with a mutation (*vft2*) causing cells to have a variable number of basal bodies, and found no effect on relative copy number of linkage groups; however, since most cells have between zero and two basal bodies per cell, they concluded that their results with the vft2 mutant were not sensitive enough to distinguish between a nuclear and basal body location for linkage group XIX. In their discussion of results, Johnson and Dutcher touched on several alternatives and issues. One issue is that linkage group XIX is present in cells lacking detectable basal bodies; this would seem to argue strongly against a basal body location of the *uni* chromosome. Another is a packing issue, having to do with how two copies of the group XIX chromosome could be packed inside one basal body, in cells having only one basal body but two XIX copies; this was especially concerning, since questions had already arisen about how even one *uni* chromosome could fit inside a basal body (*see* above; also Hall et al., 1989 [154]; Goodenough,

1989 [158]; Johnson and Rosenbaum, 1990 [160] for discussion of this issue). A third issue was with cells having only one detectable linkage group XIX copy but two basal bodies. Johnson and Dutcher concluded that their measurements of copy number were more accurate than those of Hall et al.; their detailed arguments for this seem reasonable. Furthermore, they pointed out several possible weaknesses in the in situ methods and results described by Hall et al. They even suggested that the Hall et al. results could be due to the presence of RNA, even though RNase controls had been run, since evidence for RNA in basal bodies had been previously reported (*see* more on this important concern in Subheading 9). Based on these issues and their own observations, Johnson and Dutcher concluded that linkage group XIX must reside in the nucleus and not in the basal bodies. Their extremely detailed discussion of the known and unknown properties of linkage group XIX leads one to conclude that at the time of their analysis, much remained to be learned about the *uni* chromosome, as well as about the *Chlamydomonas* basal body itself and its composition and assembly.

Around the same time as the positive and negative papers of Hall et al. [154] and Johnson and Dutcher [159], two additional papers appeared, both of which presented strong evidence against the presence of DNA in basal bodies of *Chlamydomonas* [160, 161]. In the first of these, reported in 1990, Johnson and Rosenbaum [160] expressed concern about the disruptive preparative methods that had been used in previous cytological studies that had purported to demonstrate the presence of DNA in basal bodies, and suggested that these methods might allow DNA from other cellular sources to contaminate C/BBs. To circumvent cell disruption, they fixed *Chlamydomonas* cells with glutaraldehyde, embedded them in LR White resin, cut ultrathin sections, immunogold labeled them using an anti-DNA monoclonal antibody (MAb030; from Chemicon), and viewed the sections by transmission electron microscopy. Labeling was evidenced by the presence of gold particles on the lightly stained thin sections. This procedure was thought to prevent the migration of DNA to the basal bodies from other locations in the cells, and it also provided better resolution than in previous LM studies. The monoclonal antibody they used has high activity for double-stranded DNA, has moderate activity for single-stranded DNA, and does not cross-react with RNA, or with synthetic DNA base homopolymers. Controls included pre-application of DNase 1 prior to treatment with the primary antibody. Accessibility of primary and secondary antibodies was demonstrated through control experiments in which an antiserum specific for alpha-tubulin was used in place of the anti-DNA monoclonals.

In examining their results from vegetative cells, Johnson and Rosenbaum [160] found moderate DNase-removable labeling with gold particles over nuclei, similar to that reported by others using the same methods. Most of the label was over chromatin masses arranged around the periphery of the nuclei or around nucleoli; however, while they sometimes saw label over the nucleoli themselves, this apparently was sporadic. Label was also seen over the chloroplast nucleoids. Only occasionally, gold particles were seen over mitochondria. No gold labeling above background levels was seen over sectioned basal bodies. Some sections of the Epon-embedded cells were treated with proteinase K in an effort to expose more of the available DNA to labeling. While this increased somewhat the amount of labeling over nuclei and chloroplast nucleoids, no significant labeling was revealed over basal bodies in their sections. Similar experiments, using gametic cells and a different anti-DNA monoclonal antibody (MAb031), gave similar results, with label over nuclei and chloroplast nucleoids and none over basal bodies. DNase 1 applied to proteinase K-treated sections prior to labeling "almost completely" abolished the nuclear and chloroplast labeling, while treatment with gold-labeled secondary antibody alone gave "essentially no signal," supporting the specificity of the primary antibody and the procedure for DNA. Antiserum against alpha-tubulin, followed by gold-labeled secondary antibody, gave positive labeling of basal body microtubules, demonstrating accessibility of the antibodies and reagents to the basal bodies in sections. To test whether DNA in a compacted form would label with their procedure, they tested the procedure on the very compact, transcriptionally inactive DNA of sea urchin sperm nuclei. Label was seen over the nuclei, but none over the nearby basal bodies.

In reporting experiments with negative results, it is important to apply quantitative analysis to the data; Johnson and Rosenbaum attempted to do this. They compared the amount of DNA expected to be in a basal body, 6–9 Mb (Mb = million base pairs or Mbp) (based on the size of the *uni* chromosome [154]), with the amount estimated to be in a chloroplast nucleoid (~2 Mb). It was concluded that the amount of label over a basal body should be several times greater than over a nucleoid, when in fact no detectable label was observed. Similar comparisons were made with the estimated DNA in a mitochondrion of *Chlamydomonas*. Assuming that each mitochondrion contained one to ten copies of the 16 kb mitochondrial genome, they estimated that their method could detect less than 200 kb of DNA (1 Mb = 10^3 kb). Since 200 kb is a significantly smaller number than 6–9 Mb, they expected that the DNA in a basal body should have been detectable. However, this conclusion was based on the assumption that an entire *uni* chromosome occurs in each basal body. This might not be the case. Also, they only occasionally observed label over the mitochondria, thus calling into question their conclusions about the resolution of their methods. Because there was no detectable label over basal bodies, obviously no quantitative measurements of number of gold particles in a unit area over basal bodies versus over DNA-containing organelles could be done, to provide a way to estimate the upper limit on the amount of DNA that might be present but not detectable by their methods. However, they could have done an analysis of the number of gold particles over the mitochondria alone, as had been done previously by Pyne [143], to get a more quantitative idea of the limit of resolution of their experiments. Furthermore, an interesting additional control might have been to trap known concentrations of DNA in protein gels, fix, embed, and section the gels and determine the minimum concentration of DNA that could be detected by their methods.

In light of the Hall et al. [154] experiments, which appear to show the need for both detergent and pronase treatment to make basal body DNA accessible to in situ labeling, the experiment of Johnson and Rosenbaum, showing that the compacted DNA of sea urchin sperm nuclei can be labeled, is of interest. However, sperm nuclei contain a very large concentration of DNA, yet the concentration of gold particles over the nucleus shown by Johnson and Rosenbaum is not overwhelming. Furthermore, compacted nuclear DNA, as in mammalian sperm, has been shown to be altered, for example, by the formation of disulfide bonds [162], and to exhibit a progressive diminution in binding of H^3-actinomycin D during compaction [163], and this could also be the case for compacted basal body DNA. Therefore, the results obtained by Johnson and Rosenbaum, on labeling of sea urchin sperm DNA, may not demonstrate that their methods are suitable for detecting basal body DNA if it were compacted. A small amount of compacted DNA, perhaps less than would code for a typical protein, might have been undetectable by their methods.

Additional evidence against DNA in *Chlamydomonas* basal bodies was reported by Kuroiwa et al., also in 1990 [161]. These investigators recognized the need to resolve a growing conflict over whether basal bodies might contain trace amounts of DNA. They determined that the volume of the *C. reinhardtii* basal body core is less than about $0.0098 \mu m^3$, or about half the volume of the nucleoid of *Escherichia coli*. Also, they determined that the size of the *uni* chromosome in *Chlamydomonas* (6–9 Mb) is similar to that of the *E. coli* genome (9 Mb; [164]). Kuroiwa and his associates [165] had developed an improved, high-resolution epifluorescence system to measure amounts of DNA as little as 50×10^3 base pairs (50 kb). With this system, they had found that bacterial DNA can be seen easily after staining the cells with DAPI or ethidium bromide [166, 167]. Assuming that the *uni* chromosome is located in the basal body core, Kuroiwa et al. [161] surmised that the *uni* DNA in the basal body should be readily visible too. They prepared slides of whole vegetative cells, gametes, and isolated cell structures, called FBC complexes, that consisted of the cell nucleus, a pair of basal bodies, and their flagella; FB complexes, prepared from FBC complexes by vigorous pipetting to remove the nuclei, were also attached to slides. The slides were immunolabeled for alpha-tubulin using a rat monoclonal antibody followed by an FITC-labeled secondary antibody. They were post-stained for DNA with DAPI, and viewed by epifluorescence microscopy. The slides were viewed and photographed; in addition, fluorimetric measurements were made using VIMPICS (Hamamatsu Photonics, Ltd.). In summary, they found no evidence of staining of DNA in basal bodies in vegetative cells, gametes, and zygotes and calculated that if small amounts of DNA were present, the quantity would be less than 1.5×10^6 base pairs (1.5 Mb) per basal body. The same held true for basal bodies of isolated FB complexes, while added *E. coli* and nuclei did show staining for DNA. However, when cells were treated with pronase, DNA was detected in association with the basal bodies, suggesting to them that pronase treatment released nuclear DNA that then became associated with the basal bodies. These studies of Kuroiwa et al. appear to further call into question the positive findings of Hall et al. [154], and support the nuclear source of basal body-associated DNA in cells treated with pronase. However, they do not eliminate the possibility that a functionally significant amount of DNA (less than 1.5 Mb) might be present and not be detectable by their methods.

In 1991, Johnson and Rosenbaum reviewed the accumulated positive and negative evidence regarding the presence of DNA in basal bodies or centrioles, particularly that obtained from studies on *Chlamydomonas* [168]. They concluded that "DNA does not exist in centrioles." Perhaps in response to concerns about their methods expressed by Johnson and Rosenbaum and others, but also to extend their earlier experiments showing evidence for the presence of the *uni* chromosome in basal bodies [154], Hall and Luck [157], in a study reported in 1995, sought to improve on their methods for in situ hybridization labeling of the *uni* chromosome in *Chlamydomonas reinhardtii*. One goal was to determine whether the *uni* chromosome could be found in nuclei, which required methods that allowed better access of the labeling reagents to nuclear DNA. The methods selected are those commonly used to map genes on mouse or human metaphase chromosomes, at the resolution of standard fluorescence light microscopy. Synchronized and vegetative cells of a cell wall-less strain of *C. reinhardtii* were fixed with 3:1 methanol/acetic acid, either before or after attachment to glass slides. Three different *uni*-linked YACs (yeast artificial chromosomes, used for cloning DNA) were labeled with biotin or digoxigenin and used to detect the *uni* chromosome; biotin-tagged sequences were detected with streptavidin-fluorescein, and digoxigenin-tagged sequences were detected with a rhodamine-labeled anti-digoxigenin IgG. A mouse monoclonal antibody was used to detect acetylated alpha-tubulin, using a secondary antibody tagged with either fluorescein or Texas Red. To develop and validate their methods, *Chlamydomonas* metaphase spreads from synchronous cultures were used, along with a probe for a "dispersed moderately repetitive element called TOC1," which is about 6 kb long and occurs 22–24 times in the strain used in the studies. Using these new, validated in situ methods, and the *uni*-linked YAC probes, Hall and Luck found that in mitotic cells, the *uni* chromosome appeared to be found among other chromosomes at the metaphase plate, as expected for a typical nuclear chromosome. However, in interphase cells, reproducible results were obtained that localized the *uni*-linked probes, and therefore the *uni* chromosome, to the anterior pole of the nucleus, opposite to the chloroplast nucleoid and pyrenoid, and away from probes identifying other nuclear chromosomes. This location of the *uni* chromosome, at the anterior edge of the nucleus, was also shown to be close to the location of the flagellar basal bodies, which were localized using immunofluorescence and anti-tubulin antibodies. Greater than 80% of 62 cells examined gave results confirming this location of the *uni* chromosome in interphase cells. Contrary to their earlier evidence, Hall and Luck also found that the *uni* chromosome appeared to be present in a normal copy number per cell. Without any ultrastructural evidence to the contrary, it seems evident from their results that the *uni* chromosome has a nuclear location in mitotic cells, but within the interphase nucleus, it assumes a position, close to the basal bodies, that would facilitate access of at least some of its gene products to their places of action at or near the basal body.

In spite of a variety of types of evidence to the contrary, by 2000 it appeared convincing to many, including Marshall and Rosenbaum [169], as well as Pederson [170], that basal bodies and centrioles do not contain DNA genomes. However, the possible presence of genetic material, either DNA or RNA, within these organelles, continued to remain a topic of discussion among cell biologists [169, 170, 62]. In fact, as recently as 2008, another interested researcher, Mark Alliegro, still considered the question of DNA in centrosomes (and thus centrioles?) to be "unresolved" [171]. The observations on *Chlamydomonas* by Hall and Luck [157], that the *uni* chromosome is nuclear, but may come to reside in a region of the nucleus that is very close to the basal bodies in interphase cells, suggests the possibility that some or all of the *uni* chromosome might be able to shuttle into and out of the basal bodies, an idea that had been suggested earlier by Goodenough [158]. Alternatively, the close positioning of the *uni* chromosome near the basal bodies may facilitate the transfer of *uni*-derived messenger RNA to the environs of the basal body. In either case, further studies in *Chlamydomonas*, and in ciliated protists as well, using cryoelectron microscopy and newer methods of super-resolution light microscopy, in combination with single molecule labeling, seem called for.

So, do C/BBs contain DNA? The evidence seems less and less convincing. But living cells have ways of surprising us, so let us remain open to this possibility. Perhaps there are DNA sequences that have evaded our detection in most studies, due to size or how they are packaged; perhaps DNA can shuttle into and out of C/BBs, at certain times and for certain purposes that have so far been unrecognized. Perhaps C/BB DNA is only found in some organisms. What does seem likely is that basal bodies probably do not have independent DNA genomes; if coding sequences are present, they probably are also found in the nucleus. Or perhaps the DNA molecules serve in noncanonical ways. At any rate, the advent of new imaging and labeling technologies, mentioned above and also below, may provide novel ways to look into C/BBs without disturbing their molecular architecture or introducing contaminating molecules, and with the possibility of observing molecular behavior in living cells. These new methods may yet reveal evidence for DNA in basal bodies.

9 Nucleic Acids and Basal Bodies/Centrioles: RNA

Early evidence to suggest the presence of RNA in association with microtubule-containing organelles, in particular MTOCs such as C/BBs, has been summarized by others [60, 61, 169, 171, 172]. Interestingly, although the spindle pole body of yeast is considered to be functionally equivalent to the centrosome [55, 173], and was early reported by Unger to contain RNA [174], there do not appear to be any follow-up studies to validate Unger's claim in this easily studied organism. Of particular note here are early reports by Seaman [142], Argetsinger [151], and Hartman et al. [175] of the presence of RNA in isolated basal body fractions of the ciliate *Tetrahymena pyriformis.*

In 1976, Zackroff et al. [61] followed up earlier reports that suggested a role for RNA in MTOC-directed microtubule assembly, by conducting experiments that showed that RNAse A and other polycations, as well as anionic RNA homopolymers, such as polyguanylic acid, polyuridylic acid, and polyadenylic acid (but not polycytidylic acid), can affect astral fiber (microtubule) length during aster reassembly from crude "microtubule organizing center"--containing homogenates of oocytes of the surf clam, *Spisula solidissima.* RNase A is a polycation, and might have exerted its effects by sequestering tubulin, which is anionic. Other polycations without RNAse activity also caused structural alterations in asters that were similar to those caused by RNAse A. Nevertheless, Zackroff et al. also found that the enzymatically active form of RNAse caused reduction in astral fiber length at concentrations lower than the enzymatically inactive (oxidized) form. Polyguanylic acid, polyuridylic acid, and polyadenylic acid, on the other hand, caused augmentation of astral fiber length. Polyguanylic acid also enhanced pole-to-pole fiber assembly in isolated spindles. Polyguanylic acid (PGA) is an inhibitor of some RNases [176], and Zackroff et al. suggested that PGA might enhance astral fiber length by interfering with endogenous oocyte RNase activity, present in the crude homogenates. Indeed, PGA also reversed the inhibitory effect on astral fiber length by exogenously added RNase A. Since this effect was specific for certain RNA homopolymers, Zackroff et al. postulated that one possible explanation for these results is that *Spisula* oocytes might contain a centrosome-associated RNA that might have a role in controlling microtubule assembly during aster formation. Furthermore, specific RNA sequence(s) appeared to be required, as some, but not all, RNA homopolymers promoted astral fiber assembly, while unfractionated yeast RNA had no apparent effect on fiber length. It was therefore suggested that specific RNA homopolymers might mimic or replace an endogenous *Spisula* centrosomal RNA in the centrosomal microtubule organizing center. If this were the case, then RNA in centrosomes might function either as a scaffold or in regulating centrosome-nucleated microtubule assembly. The results of these experiments therefore suggested that *Spisula* oocytes might contain a species-specific centrosome-associated RNA and that this RNA might have a role in controlling microtubule assembly during aster formation.

Similar evidence, that basal bodies contain RNA capable of nucleating aster formation, was reported in 1977, by Heidemann et al. [177]. In contrast to the studies on *Spisula* homogenates, their experiments were conducted in vivo. The experiments were based on their earlier finding that when purified basal bodies from *Chlamydomonas reinhardtii* and *Tetrahymena pyriformis* were injected into unfertilized oocytes of *Xenopus laevis*, asters were induced to form, whereas microtubular structures, such as flagellar axonemes and brain microtubules, did not induce aster formation [178]. In 1977, Heidemann et al. [177] isolated basal bodies from *Chlamydomonas* and *Tetrahymena* and injected them into *Xenopus laevis* oocytes, with or without pretreatment of the basal body fractions with RNase A, RNase T1, DNase, or proteolytic enzymes. Untreated or DNase-treated basal bodies induced the formation of asters; however, treatment with RNases or some proteases knocked out the aster-forming activity. But unlike the *Spisula* experiments, in their studies RNase treatment did not interfere with microtubule elongation.

In 1979, ultrastructural evidence for RNA associated with centrioles of cultured newt (*Taricha granulosa*) lung epithelial cells was reported by Rieder [179]. Selected metaphase or early anaphase cells were fixed in buffered glutaraldehyde, embedded in Epon, serially sectioned to produce thick sections, stained for ribonucleoprotein (RNP) by a modification of Bernard's uranyl-

EDTA-lead method [180], and viewed and photographed by high-voltage electron microcopy. To detect staining due to RNP, after fixation some cells were treated with pancreatic ribonuclease followed by perchloric acid (PCA) extraction, before being embedded in Epon. The RNase had been pretreated by boiling to remove protease activity. In metaphase and early anaphase cells, the centrioles were the most intensely stained parts of the spindle. The RNase-removable, electron-opaque staining was closely associated with the external surface of the centrioles, extending along the lengths of the microtubule triplets. Pericentriolar material (PCM) did not appear to stain. While staining associated with the centriolar triplets was removed by RNase treatment, it was not affected by cold PCA extraction itself or when other positively charged proteins were substituted for the RNase. Similar RNase-sensitive staining was present on the kinetochores, structures known to contain ribonucleoprotein. Rieder discussed his findings in relation to previous proposals for how centriolar RNA could function. The possibilities included (1) initiation of aster formation (as discussed earlier by Heidemann et al. [177]), (2) a role in procentriole formation [181, 182], and (3) serving as a centriolar genome that replicates by way of a DNA intermediate, using a reverse transcriptase [183]. It now appears that a fourth possibility should also be considered: direct RNA-templated RNA polymerization, as described by others, including Jain et al., in 2020 [184]. As will be seen later, a new type of evidence, supporting more than one of these ideas, took over a decade to appear.

Meanwhile, in 1980, Pepper and Brinkley [185] provided additional microscopical evidence supporting the presence of RNA in centrosomes, and its participation in microtubule assembly. Their studies utilized transmission electron microscopy on rat kangaroo cells (strain PTK_2), preincubated with colcemid to disrupt spindle microtubules, and then lysed with Triton-X100 and incubated with tubulin, with or without RNase. The presence of centrosomal RNA, with a potential role in mitosis and assembly of microtubules, was further supported in 1978, by experimental studies of Peterson and Berns [186] on the effect of nucleic acid-binding psoralens, in combination with focused 365 nm laser light, on mitosis and centriole-associated microtubules, in PTK_3 cells.

In 2000, Marshall and Rosenbaum [169] reviewed the existing evidence for centrosomal RNA (cnRNA). After accepting its probable existence, they considered several possible roles: (1) cnRNA might act as a scaffold for microtubule-nucleating proteins, as suggested by Zackroff et al. [61] and others; (2) cnRNA might be needed to assemble a functional centrosome surrounding a centriole; and (3) the centrosome might serve as a way station for the movement of RNA within the cell. Marshall and Rosenbaum further suggested a possible centrosomal role in the intracellular trafficking of messenger RNAs. Additional support for this last idea took some time to emerge. It came through extensive studies to characterize the RNA associated with *Spisula* oocyte centrosomes [62, 63, 171, 172, 187].

Spisula oocyte centrosomes are composed of a pair of centrioles within a pericentriolar matrix. In their initial studies, in 2006, Alliegro et al. [187] isolated a unique set of five RNAs from purified centrosomes of *Spisula* oocytes; one from this set (cnRNA11) was further analyzed by biochemical, in situ hybridization, and sequencing methods, and shown to be more or less associated with centrosomes and to contain an open reading frame (ORF) encoding a predicted 54,000 kDa polypeptide that included "a highly conserved 200-aa reverse transcriptase domain." This led to their proposal that RNA-templated RNA synthesis (RNA replication) might occur within the centrosome, as had already been proposed in 1977, by Went [183]. The results reported by Alliegro et al. also suggested that 18S ribosomal RNA (rRNA) might be present in the centrosomal fraction. Based on controls used to reduce the likelihood of contamination from RNA originating elsewhere in the cell, the results seemed convincing that purified *Spisula* oocyte centrosomes contain more than one type of RNA, but the significance of this finding remained unresolved (*see* Pederson's 2006 analysis of their evidence and conclusions [170]).

Soon after these findings from *Spisula* were reported, in situ hybridization evidence for specific centrosomal RNA sequences was also reported for the invertebrate *Ilyanassa* by Kingsley et al. [188], and for the vertebrate *Xenopus* by Blower et al. [189], as well as in *Drosophila*. Using their newly refined high-resolution fluorescence in situ hybridization (HR-FISH) methods [190] on *Drosophila* embryos and tissues, Le'cuyer et al. [191], in 2007, found that of the 3370 mRNA sequences screened, 71% were localized to specific locations in the cell, including centrosomes, and they concluded that "mRNA localization is a major mechanism for controlling cellular architecture and function." They also demonstrated the usefulness of co-localizing mRNAs and proteins in the same preparations, and established an online accessible database of their HR-FISH results for *Drosophila*. Such methods have continued to be improved [192, 193] and promise to be important in future studies to dissect the mechanisms behind local control of pattern in the cilioprotists as well.

In 2008, Alliegro and Alliegro [62] also focused on two other *Spisula* cnRNAs, cnRNA15 and cnRNA194. Their findings strengthened their conclusion that presumed cnRNAs are indeed localized to *Spisula* centrosomes and revealed: (1) a possible correlation of cnRNA sequences with those of intron-poor nuclear genes; (2) a possible association of the centrosomes with a nuclear structure, the nucleolinus (NLi) (*see* below), following the nuclear membrane breakdown that is brought about by oocyte activation; and (3) a similarity of the cnRNA15 antisense sequence to the sequence of various presenilin genes. Presenilins, proteins implicated in Alzheimer's disease, previously had been reported to be associated with centrosomes, based on both light and electron microscopic immunolabeling experiments that showed co-localization with γ-tubulin and pericentrin [194, 195], suggesting a role of centrosomes in chromosome segregation [194], and implying a role for presenilins in spindle formation, and in aging. It should be mentioned that a role for centrioles in cellular ageing has been proposed and discussed at length by Chichinadze and Tkemaladze in a series of articles dating from 2005 to 2013 [36, 196–198].

Following their initial discovery of cnRNA in surf clam oocytes, Alliegro and Alliegro [63] obtained evidence for trafficking of ribosomal RNA (rRNA)-transcribed spacer domains to maternal procentrosomes of *Spisula* oocytes, as well as to a nuclear sub-compartment known as the "nucleolinus" (NLi). The NLi is a long-ignored intra-nucleolar or nucleolus-associated body distinguishable by light microscopy using specialized staining methods [199, 200]; it has been identified in a variety of invertebrate and vertebrate cells, including *Spisula* oocytes [201, 202]. Evidence was obtained, using phase and differential interference microscopy (DIC) on live oocytes, as well as in situ hybridization to localize specific RNAs and immunofluorescence to localize proteins and riboprobes, that (after their formation) specific transcribed spacer domains are first trafficked to the NLi in the nuclei of unfertilized oocytes, and that (after fertilization) the NLi, along with the transcribed spacer domains, come to be associated with the developing centrosomes and spindle of the fertilized egg [63, 201].

In 2008, based on some of the discoveries described above and those reported by many others, Mark Alliegro [171] reconsidered the possible functions and activities of cnRNA. The candidates included the following: (1) centrosomes might serve as a more or less temporary "way station" for RNAs moving about within the cell or "targeted to the centrosome for distribution to selected blastomeres in order to localize developmental information," as hypothesized earlier for *Ilyanassa* embryos by Kingsley et al. [188]; (2) specific messenger RNAs might be targeted to the centrosome for localized translation into centrosome and/or spindle proteins; (3) cnRNAs might serve "as a scaffold upon which organelle-specific proteins can assemble"; and (4) cnRNAs might be "remnants or derivatives of a once independent genome," presumably derived from a proposed symbiotic progenitor of the centriole (*see* Subheading 10). This last idea had been supported by earlier evidence reported by Alliegro et al. for reverse transcriptase activity in *Spisula* centrosomes [187].

In support of Alliegro's first two hypotheses, and as discussed by him in some detail in 2008 [171], there has been growing evidence for mRNA translation, as well as RNA polyadenylation at spindle poles [203, 189]. There has also been evidence for the localization of specific mRNAs, like *cen* and *ik2 cis*-natural antisense mRNAs, to the centrosome in *Drosophila* [204]. Even as early as 1997, there was evidence that the formation of basal bodies, flagella, and other microtubular structures in *Naegleria* is accompanied by the trafficking of relevant mRNAs to the basal region of growing flagella [205]. However, it was also becoming clear, at least for *Spisula* and *Ilyanassa*, that a large percentage of cnRNAs had sequences that did not correlate with known DNA sequences in available genome databases [188, 171]. This suggested that at least some cnRNAs might not be mRNAs, in which case they might have scaffolding functions in keeping with Alliegro's third hypothesis. Through further investigations, reviewed in 2020, by Ito et al. [60], these noncoding RNAs (ncRNAs) associated with the centrosome have come to be recognized as members of a large class of ncRNAs that include long ncRNAs (lncRNAs; >200 nt) and small ncRNAs (<200 nt). NcRNAs include transfer RNA (tRNA), small nuclear RNA (snRNA), small nucleolar RNA (snoRNA), PIWI-interacting RNA (piRNA), and microRNA (miRNA) [206].

Of particular interest to Ito et al. [60], and to us as well, is that some lncRNAs appear to have scaffolding functions related to their interaction with RNA-binding proteins, and thus may provide a link between centrosomes and assembly of the mitotic apparatus. In at least one case, ncRNAs with a role in assembly of cell structures have been referred to as "architectural RNAs" (arcRNAs) [207]. Also of possible significance is the fact that noncoding RNAs are known to be abundant in the macronucleus of cilioprotists, where they have an important role in guiding genome editing and rearrangements during macronuclear development and appear to facilitate epigenetic inheritance (*see* Allen and Nowacki, 2020, for a useful summary and discussion of macronuclear ncRNAs in cilioprotists [208]). This raises the possibility that ncRNAs associated with cilioprotist basal bodies may mediate an influence of the basal bodies on gene expression in macronuclei. At a more theoretical level, Chapman and Alliegro [209] have called attention to the coevolution of the eukaryotic nucleus and the C/BB with its associated cytoskeletal complex (together called the "karyomastigont") and have proposed "that the proteomes and transcriptomes of karyomastigont components should be assembled in their entirety and analyzed using a shotgun approach."

The evidence seems clear that centrosomes from a variety of invertebrate and vertebrate cells contain a panoply of RNA molecules with several likely functions, including an architectural role in spindle assembly. However, a variety of questions arise when considering the relevance for basal body-containing ciliates. Firstly, where within the centrosome are each of these cnRNAs localized, and by extension, where in the cortex of ciliates? Secondly, are some cnRNAs closely associated with or incorporated into the structure of the centriole or basal body? What about the peri-basal body region? There seems to be a lack of cytological studies at sufficient resolution to answer these questions. In an early study, Dippell [210] examined this question for basal bodies of *Paramecium*, using RNase treatment and ultrastructural analysis. She observed that RNase completely removed a complex of fibers and granules located within the core of the basal body (the "luminal complex"), but the interpretation of this finding remains unresolved, since her RNase may have been contaminated with proteases (*see* Alliegro, 2011 [172], for further discussion of this issue). Clearly, further studies along these lines are badly needed, using the most purified enzyme reagents, advanced fluorescent probes, and high-resolution cryoelectron microscopy.

If RNAs do exist within the basal bodies of ciliates, what functional role(s) might they have in relation to cell architecture? Are there architectural RNAs that guide or otherwise affect the assembly of the microtubular structures associated with their basal bodies? Do specific RNAs exist that help to specify the dimensions of cortical units in different parts of the cell cortex? Do RNA molecules play a part in helping surgically altered cells to gradually return to their species-specific architectures? It seems surprising that studies to answer these questions have not already been conducted in the ciliates, especially those whose genomes have been sequenced, such as *Tetrahymena*. Perhaps of great significance here are experiments conducted years ago by Frankel [211, 212], as well as other, earlier, studies discussed by him. Frankel tested the effects of high temperatures [211] (1964) and nucleic acid (NA) antagonists [212] (1965) on the development of oral structures in *Tetrahymena pyriformis*. Elevated temperatures caused abnormal oral patterns to develop, while NA antagonists (5-fluorodeoxyuridine, 5-fluorounidine, actinomycin D, and p-fluorophenylalanine) were found to inhibit cell division and oral morphogenesis at higher concentrations, but did not affect oral patterning (size or completeness). Frankel suggested that heat might selectively affect the assembly of specific cytoskeletal components, thus having more localized effects, whereas NA antagonists might "reduce the supply of correct oral precursor macromolecules" (presumably proteins), and thus might have a more global effect on cytoskeletal events. Frankel's experiments appear to speak against the idea of RNA having localized control over cellular pattern. However, it is possible that when all RNA synthesis was inhibited, any local activity of architectural RNA molecules in cytoskeletal assembly would have been masked by general effects of NA antagonists on the regulatory mechanisms underlying cell division and oral morphogenesis.

10 Origin of Centrioles and Basal Bodies

The increasingly positive evidence for RNA in centrosomes, and possibly within C/BBs themselves, has had an effect on long-ongoing discussions about the origin of cilia, flagella, and C/BBs. As mentioned earlier, one idea was that C/BBs arose gradually, by direct filiation from simpler cellular structures [43]. A version of this idea of an endogenous origin was argued in detail by Rizzotti [213]. However, the early appearance of C/BBs and C/BB-like organelles in the eukaryotes seems to argue against this idea. Also, early C/BB-like structures appeared at the outset to have the ability to self-reproduce. In part for these reasons, in 1967, Sagan [214] proposed that a spirochaete symbiont was the origin of the basal body/cilium complex. Margulis later argued in favor of this idea [215, 216]. However, with the discovery of RNA in centrioles and other supportive evidence, in 2007, Satir et al. [104] drew attention to weaknesses in the bacterial endosymbiont theory and proposed an origin from an enveloped RNA virus. A key principle underlying these endosymbiont proposals was the property of self-replication exhibited by the C/BB, which Satir et al. argued could not readily be explained by proposals for de novo origin. The idea of a symbiotic prokaryotic or viral origin was reviewed in 2009, by Marshall [217], who had come to believe that the C/BB does not carry its own genome. He felt that it had been proven beyond doubt that C/BBs do not contain their own DNA and that the RNA associated with them is encoded by nuclear genes. Furthermore, there had been many reports of de novo formation of C/BBs. Therefore, Marshall concluded that a symbiotic origin is not required, and C/BBs probably did originate through evolutionary stages from simpler structures, although the stages of their evolution were still unknown.

Because of their elaborate organization, the cilioprotists appear to represent an advanced stage in the evolution of the C/BB, and therefore not good candidates for uncovering C/BB origin. However, as members of the Alveolata, the cilioprotists do have deep roots in the tree of life [218], and therefore must have had a long time to develop complexity. Nevertheless, if we can better understand how cilioprotist architectural diversity has been evolving, perhaps we may uncover clues regarding how C/BBs originated and the steps in their evolution.

11 Unanswered Questions and Unused Methods

11.1 First of All, Why Use Cilioprotists as Models?

In reviewing the vast literature about C/BBs and the cytoskeleton, it seems clear that enormous advances have been made within the past 20 years in our quest to understand the assembly and functioning of C/BBs and their interactions with the cell cytoskeleton. A great many of these new discoveries have been made using model systems other than cilioprotists, including *Chlamydomonas*, *Drosophila*, *Spisula*, mammals, and even human cells. Many new technical advances have made research on cells of higher organisms much easier than in the past. However, many recent advances have been based on earlier discoveries about cilioprotist cells. As we have seen, even recently, cilioprotists have contributed to our understanding of eukaryotic cell architecture and its molecular composition, assembly, and disassembly. But, what about the future? Will cilioprotists still have advantages over mammalian cells?

Most clearly, they will! Cilioprotists have the advantage of numerical supremacy, having abundant C/BBs instead of just a few. Also, they are generally easy to culture and to expose to experimental conditions. In addition, their cytoskeletons contain many conserved proteins, and they have highly patterned, globally repetitive but locally varied cytoskeletal architectures that have already been shown to be easy to analyze using high-resolution fluorescence immunocytochemistry [64, 125, 130, 136, 137] and in situ hybridization. The behaviors and interactions of their cytoskeletal macromolecules will be easy to quantitate using advanced microscopic technologies in combination with automated image analysis. Furthermore, their unique cytoskeletal diversification seems to have evolved hand in hand with the development of complexity in their chromosomal organization, raising questions whose answers may have broad implications for how all eukaryotic cells function. With the availability of genetic toolboxes for several different species, it may be possible to reveal basic molecular mechanisms that would be much harder to discover in other organisms. Besides, as Aubusson-Fleury et al. [125, 130] and others have demonstrated so elegantly, cilioprotists, like butterflies, have a certain cellular beauty that is difficult to resist!

11.2 Using Cilioprotists as Model Systems, What Do We Still Need to Learn?

11.2.1 Nucleic Acids in C/BBs

DNA

If DNA is eventually found in C/BBs, is it a general property of C/BBs or only specific to certain cells, such as *Chlamydomonas*? In other words, is it also found in ciliate BBs and if so where in the BB is it located? In the lumen, closely associated with the microtubule triplets, on the outside of the triplets, or in the cartwheel structure? Is it part of the procentriole or other BB precursor complex? What proteins are associated with it? Can it migrate to and from the nucleus, or elsewhere in the cell? If so, what are the specialized molecular systems responsible for this movement? What is its relationship to nuclear genes? What is its relationship to RNA species that may be present in the basal bodies or nearby cortical regions of ciliates?

RNA

It seems likely that RNA is present in association with C/BBs in other organisms, and probably cilioprotists as well. What more can we learn from cilioprotists? Where is RNA located in relation to basal body structure? Is it closely attached to the outside of the basal body triplets, as in the case of the centrioles of newt lung epithelial cells? Are the RNAs that have been found in pellicle fractions of ciliates as diversified as those found in *Spisula*? Do they also include mRNAs, rRNAs, and small noncoding RNAs, and if so, how does each of these function in relation to basal body assembly and assembly of the complex cytoskeletal architecture typical of the ciliates? Do pellicle-associated RNAs include species with reverse transcriptase activity, as in *Spisula*? Do mRNAs that find their way to different regions of the cell surface code for or bind to different proteins? What is their relationship to the coding and noncoding sequences of nuclear DNAs? Do any participate directly in the assembly of the microtubular cortical structures of ciliates, such as the transverse and postciliary ribbons? Is there self-reproducing RNA in ciliate basal bodies? Do ciliate basal bodies serve as storage sites, way stations, or processing stations for noncoding RNAs used to edit chromosomes during macronuclear development?

11.2.2 Proteins

Much of the attention regarding cell cytoskeleton assembly has focused on control of the time and location of assembly of microtubules. However, as described above, there are many other conserved structural proteins in cilioprotists that form highly organized and visible components of the cytoskeleton. All are associated in some way with the ciliary basal bodies. Understanding the properties of these proteins and how their assembly is controlled and coordinated in cilioprotists will have the potential of providing guidance for future research on other model systems, including humans.

11.2.3 Nucleolini

Do ciliates have nucleolini? In *Spisula*, NLi appear to be nuclear staging locations for RNA destined for centrosomes. In their experiments, the Alliegros developed a specific probe for the *Spisula* NLi (*see* Subheading 9). Perhaps this probe could be used to determine whether ciliate nuclei have NLi.

11.3 Newly Emerging Methods

In 1975, while reviewing what was known then about the assembly and disassembly of microtubular organelles, such as basal bodies in *Tetrahymena*, Norman E. Williams [219] pointed out the need for in vivo analysis for the "total solution of many -----problems!" The present author [220] called attention to the drawback of studies on cells embedded in resins and the opportunities afforded by newly developing microscopical methods to analyze the structure and behavior of individual molecules in their watery environment. Unfortunately, it has taken many years for these dreams to be fulfilled. However, as demonstrated above, in recent years, many new *super-resolution fluorescence microscopy methods* have been developed that are starting to make these dreams accessible [221]. Some of these super-resolution methods, which defy the resolution limits of conventional light microscopy, emerged in the early 2000s to win for some of their developers the Nobel Prize in Chemistry in 2014 [222]. For the most part, the new technologies have been applied to proteins (their structure, movements, and interactions), but they have also been adapted to nucleic acids, and promise to finally answer the old questions concerning whether basal bodies contain nucleic acids, and if so what roles they play in the theater of the cytoskeleton. Some of these newer methods have been referenced above, and are included among the following: *cryoelectron tomography*, which eliminates the need to expose cell components to fixatives, embedding media, and other reagents [44–46]; *expansion microscopy* (ExM) which improves the resolution of light microscopy [92–95]; *single-molecule localization microscopy* (SMLM), which improves the localization of fluorescent labels at the molecular level [96, 97]; *proximity-based labeling methods* (BioID) which are used to analyze protein interactions [101]; *cross-reacting antibodies and in situ probes*, which provide new tools for use in a species of interest; *automated 3D image analysis* [137], which provides the means to obtain quantitative data quickly; refined methods for *high-resolution fluorescence in situ hybridization* (HR-FISH), used to localize specific nucleic acid sequences [190]; and *super-resolution time-lapse microscopy* which extends the effective resolution of fluorescence imaging beyond the diffraction limit [223]. Some of these technologies make use of preserved cells, either chemically fixed or cryofixed, but many are being applied to live cells, with the goal of tracing macromolecular interactions in real time. As an example, the reader is referred to a short paper by Rita Stack that summarizes efforts, now several years old, to observe the translation of single messenger RNAs in living cells [224].

In the future, we can expect other emerging imaging technologies to be adapted to studies on the cytoskeleton of ciliprotists. Examples include (1) *microfluidic chips* and advanced *compression chambers* [225–229] that can be used with the newest super-resolution technologies to follow interactions of fluorescently labeled proteins and nucleic acids in immobilized or semi-immobilized live cells; (2) *improved fluorescent probes* that cause less interference in cellular processes and help improve resolution of differentially labeled macromolecules [230]; (3) *improved data processing and storage* for live cell imaging and quantitative analysis [231]; and (4) new imaging technologies, such as *light sheet microscopy* [232], used to minimize fluorophore bleaching and phototoxic effects while improving resolution of light microscopy.

12 Final Thoughts

12.1 Future Evolution of the Cytoskeleton in the Ciliated Protists

It is intriguing to speculate that many eons from now, when a future catastrophic event wipes out all nonmicrobial life, diverse ciliate species may still survive and their cytoskeletons may evolve further to produce a true, complex, Lilliputian society composed of unimaginably capable, creative, and intelligent microorganisms! At present, to consider this idea further, one can only look at the evidence for past evolution within the cilioprotists and for their present molecular and genetic diversity. These topics were addressed in great detail, in 1981, by Lynn [1], who documented the evidence for evolution of architectural complexity and for the existence of macromolecular diversity in ciliates. However, Lynn also stressed the possibility of conservation of structure through time (and thus function as well?) and that this conservation would be somehow "inversely related to the level of biological organization," thus implying that the evolutionary capability of a single-celled organism could not be as great as that of a multicellular one. Perhaps this is true. However, since Lynn's 1981 treatise, we have learned more and more about the biochemical and functional complexity of living cells, and especially the cilioprotists, and might want to consider that this complexity itself is not at a standstill but is continuing to diversify, and could give rise to unimagined new subcellular capabilities in the future.

Acknowledgement

The author wishes to thank Robert V. Zackroff for calling attention to certain historical events related to our understanding of the role of RNA in centrosome functioning and for his critical reading of this manuscript.

References

1. Lynn DH (1981) The organization and evolution of microtubular organelles in ciliated protozoa. Biol Rev 56:243–292
2. Soares H, Carmona B, Nolasco S, Melo LV (2019) Polarity in ciliate models: from cilia to architecture. Front Cell Dev Biol 7:1–23
3. Bloodgood RA (2010) Sensory reception is an attribute of both primary cilia and motile cilia. J Cell Sci 123(Pt 4):505–509
4. Nanney DL (1968) Cortical patterns in cellular morphogenesis. Science 160:496–502
5. Lynn DH (1991) The implications of recent descriptions of kinetid structure to the systematics of the ciliated protists. Protoplasma 164:123–142
6. Nanney DL (1977) Molecules and morphologies: the perpetuation of pattern in the ciliated protozoa. J Protozool 24:27–35
7. Sonneborn TM (1963) Does preformed cell structure play an essential role in cell heredity? In: Allen JM (ed) The nature of biological diversity. McGraw-Hill, pp 165–221
8. King RC, Stansfield WD, Mulligan PK (2007) A dictionary of genetics, 7th edn. Oxford University Press, p 78
9. Iftode F, Fleury-Aubusson A (2003) Structural inheritance in *Paramecium*: ultrastructural evidence for basal body and associated rootlets polarity transmission through binary fission. Biol Cell 95:39–51

10. Simon M, Plattner H (2014) Unicellular eukaryotes as models in cell and molecular biology: critical appraisal of their past and future value. Int Rev Cell Mol Biol 309:141–198
11. Beisson J, Sonneborn TM (1965) Cytoplasmic inheritance of the organization of the cell cortex in *Paramecium aurelia*. Proc Natl Acad Sci U S A 53:275–282
12. Bornens M (2008) Organelle positioning and cell polarity. Nat Rev Mol Cell Biol 9:874–886
13. Russell JJ, Theriot JA, Sood P, Marshall WF, Landweber LF, Fritz-Laylin L, Polka JK, Oliferenko S, Gerbich T, Gladfelter A, Umen J, Bezanillan M, Lancaster MA, He S, Gibson MC, Goldstein B, Tanaka EM, Hu C-K, Brunet A (2017) Non-model model organisms. BMC Biol 15:55–86
14. Bayless BA, Navarro FM, Winey M (2019) Motile cilia: innovation and insight from ciliate model organisms. Front Cell Dev Biol 7:1–8. article 265
15. Orias E (2012) *Tetrahymena thermophila* genetics: concepts and applications. In: Collins K (ed) Methods in cell biology: *Tetrahymena thermophila*, vol 109. Academic, pp 302–325
16. Ruehle MD, Orias E, Pearson CG (2016) Tetrahymena as a unicellular model eukaryote: genetic and genomic tools. Genet 203:649–665
17. Arnaiz O, Cain S, Cohen J, Sperling L (2007) ParameciumDB: a community resource that integrates the *Paramecium tetraurelia* genome sequence with genetic data. Nucleic Acids Res 35:D439–D444
18. Arnaiz O, Meyer E, Sperling L (2020) ParameciumDB 2019: integrating genomic data across the genus for functional and evolutionary biology. Nucleic Acids Res 48: D599–D605
19. Slabodnick MM, Ruby JG, Reiff SB, Swart EC, Gosai S, Prabakaran S, Witkowska E, Larue GE, Fisher S, Freeman RM Jr, Gunawardena J, Chu W, Stover NA, Gregory BD, Nowacki M, Derisi J, Roy SW, Marshall WF, Sood P (2017) The macronuclear genome of *Stentor coeruleus* reveals tiny introns in a giant cell. Curr Biol 27:569–575
20. Landweber LF (2017) Life with 16,000 chromosomes: *Oxytricha* as a model system to study genome biology, epigenetic inheritance, and somatic differentiation. In: Russell JJ, Theriot JA, Sood P, Marshall WF, Landweber LF, Fritz-Laylin L, Polka JK, Oliferenko S, Gerbich T, Gladfelter A, Umen J, Bezanillan M, Lancaster MA, He S, Gibson MC, Goldstein B, Tanaka EM, Hu C-K, Brunet A (eds) Non-model model organisms, BMC Biol, vol 15, pp 55–86
21. Gall JG (1986) The molecular biology of ciliated protozoa. Academic
22. Collins K (ed) (2012) Methods in cell biology: *Tetrahymena thermophila*, vol 109. Academic. 442 pp
23. Wloga D, Frankel J (2012) From molecules to morphology: cellular organization of *Tetrahymena thermophila*. In: Collins K (ed) Methods in cell biology: *Tetrahymena thermophila*, vol 9. Academic, pp 84–140
24. Bayless BA, Galati DF, Pearson CG (2016) Tetrahymena basal bodies. Cilia 5:1
25. Ehret CF, McArdle EW (1974) The structure of Paramecium as viewed from its constituent levels of organization. In: Van Wagtendonk WJ (ed) Paramecium: a current survey. Elsevier, pp 263–338
26. Tassin A-M, Lemullois M, Aubusson-Fleury A (2016) *Paramecium tetraurelia* basal body structure. Cilia 5:6
27. Dutcher SK, O'Toole ET (2016) The basal bodies of *Chlamydomonas reinhardtii*. Cilia 5:18
28. Bornens M (2012) The centrosome in cells and organisms. Science 335:422–426
29. Carvalho-Santos Z, Azimzadeh J, Pereira-Leal JB, Bettencourt-Dias M (2011) Tracing the origins of centrioles, cilia, and flagella. J Cell Biol 194:165–175
30. Stearns T (2015) Centrioles in absentia. Science 348:1091–1092
31. Avidor-Reiss T, Mazur M, Fishman EL, Sindwani P (2019) The role of sperm centrioles in human reproduction—the known and the unknown. Front Cell Dev Biol 7:article 188
32. Chapman MJ, Dolan MF, Margulis L (2000) Centrioles and kinetosomes: form, function and evolution. Q Rev Biol 75:409–429
33. Zhou F, Roy S (2015) Snapshot: motile cilia. Cell 162:224
34. Jana SC, Marteil G, Bettencourt-Dias M (2014) Mapping molecules to structure: unveiling secrets of centriole and cilia assembly with near-atomic resolution. Curr Opin Cell Biol 26:96–106
35. Sathanathan AH, Ratnam SS, Ng SC, Tarin JJ, Gianaroli L, Trounson A (1996) The sperm centriole: its inheritance, replication and perpetuation in early human embryos. Hum Reprod 11:345–356
36. Chichinadze K, Tkemaladze J, Lazarashvili A (2012) Discovery of centrosomal RNA and

centrosomal hypothesis of cellular ageing and differentiation. Nucleosides Nucleotides Nucleic Acids 31(3):172–183

37. Afzelius BA (1976) A human syndrome caused by immotile cilia. Science 193:317–319
38. Marteil G, Guerrero A, Vieira AF, de Almeida BP, Machado P, Mendonca S, Mesquita M, Villarreal B, Fonseca I, Francia ME, Dores K, Martins NP, Jana SC, Tranfield EM, Barbosa-Morais NL, Paredes J, Pellman D, Godinho SA, Bettencourt-Dias M (2018) Over-elongation of centrioles in cancer promotes centriole amplification and chromosome missegregation. Nat Commun 9:1258
39. Davis EE, Katsanis N (2012) The ciliopathies: a transitional model into systems biology of human genetic disease. Curr Opin Genet Dev 22:290–303
40. Oh EC, Katsanis N (2012) Cilia in vertebrate development and disease. Development 139:443–448
41. Wheatley DN (2018) The primary cilium—once a "rudimentary" organelle that is now a ubiquitous sensory cellular structure involved in many pathological disorders. J Cell Comm Sign 12:211–216
42. Fulton C (1971) Centrioles. In: Reinert J, Ursprung H (eds) Results and problems in cell differentiation, vol 2: origin and continuity of cell organelles. Springer, pp 170–221
43. Bornens M, Azimzadeh J (2007) Origin and evolution of the centrosome. Adv Exp Med Biol 607:119–129
44. Li S, Fernandez J-J, Marshall WF, Agard DA (2019) Electron cryo-tomography provides insight into procentriole architecture and assembly mechanism. elife 8:e3434
45. Goldman E (2002) A new window on the cell's inner workings. Science 298:1155–1157
46. Medalia O, Weber I, Frangakis AS, Nicastro D, Gerisch G, Baumeister W (2002) Macromolecular architecture in eukaryotic cells visualized by cryoelectron tomography. Science 298:1209–1213
47. Fritz-Laylin LK, Fulton C (2016) Naegleria: a classic model for de novo basal body assembly. Cilia 5:10
48. Goshima G, Wollman R, Goodwin SS, Zhang N, Scholey JM, Vale RD, Stuurman N (2007) Genes required for mitotic spindle assembly in *Drosophila* S2 cells. Science 316:417–421
49. Dupuis-Williams P, Fleury-Aubusson A, Garreau de Loubresse N, Geoffroy H, Vayssie L, Galvani A, Espigat A, Rossier J (2002) Functional role of e-tubulin in the assembly of the centriolar microtubule scaffold. J Cell Biol 158:1183–1193
50. Marshall WF, Rosenbaum JL (2003) Tubulin superfamily: giving birth to triplets. Curr Biol 13:55–56
51. Ruiz F, Beisson J, Rossier J, Dupuis-Williams P (1999) Basal body duplication in Paramecium requires gamma-tubulin. Curr Biol 9:43–46
52. Garreau de Loubresse N, Ruiz F, Beisson J, Klotz C (2001) Role of delta-tubulin and the C-tubule in assembly of Paramecium basal bodies. BMC Cell Biol 2:4
53. Shang Y, Li B, Gorovsky MA (2002) *Tetrahymena thermophila* contains a conventional gamma-tubulin that is differentially required for the maintenance of different microtubule-organizing centers. J Cell Biol 158:1195–1206
54. Pearson CG, Osborn DPS, Giddings TH Jr, Beales PL, Winey M (2009) Basal body stability and ciliogenesis requires the conserved component Poc1. J Cell Biol 187:905–920
55. Graf R (2018) Comparative biology of centrosomal structures in eukaryotes. Cell 7:202
56. Jana SC, Bettencourt-Dias M, Durand B, Megraw TL (2016) *Drosophila melanogaster* as a model for basal body research. Cilia 5:22
57. Lattao R, Kovacs L, Glover DM (2017) The centrioles, centrosomes, basal bodies and cilia of *Drosophila melanogaster*. Genet 206:33–53
58. Lee J, Kand S, Choi YS, Kim H-K, Yeo C-Y, Lee Y, Roth J, Lee J (2015) Identification of a cell cycle-dependent duplicating complex that assembles basal bodies *de novo* in *Naegleria*. Protist 166:1–13
59. Marshall WF, Rosenbaum JL (2000) How centrioles work: lessons from green yeast. Curr Opin Cell Biol 12:119–125
60. Ito KK, Watanabe K, Kitagawa D (2020) The emerging role of ncRNAs and RNA-binding proteins in mitotic apparatus formation. Non-Coding RNA 6:13–27
61. Zackroff RV, Rosenfeld AC, Weisenberg RC (1976) Effects of RNase and RNA on *in vitro* aster assembly. J Supramol Struct 5:577–589
62. Alliegro MC, Alliegro MA (2008) Centrosomal RNA correlates with intron-poor nuclear genes in *Spisula* oocytes. Proc Natl Acad Sci U S A 105:6993–6997
63. Alliegro MC, Alliegro MA (2013) Localization of rRNA transcribed spacer domains in the nucleolinus and maternal procentrosomes of surf clam (*Spisula*) oocytes. RNA Biol 10:391–396

64. Fleury A, Le Guyader H, Iftode F, Laurent M, Bornens M (1993) A scaffold for basal body patterning revealed by a monoclonal antibody in the hypotrich ciliate *Paraurostyla weissei*. Dev Biol 157:285–302
65. Frankel J (2000) Cell biology of *Tetrahymena thermophila*. In: Asai DJ, Forney JD (eds) Methods in cell biology, vol 62. Academic, pp 27–125
66. Peck RK (1977) The ultrastructure of the somatic cortex of *Pseudomicrothorax dubius*: structure and function of the epiplasm in ciliated protozoa. J Cell Sci 25:367–385
67. Pitelka DR (1963) Electron microscopic structure of protozoa. Pergamon Press
68. Beisson J, Clerot J-C, Fleury-Aubusson A, Garreau de Loubresse N, Ruiz F, Klotz C (2001) Basal body-associated nucleation center for the centrin-based cortical cytoskeletal network in *Paramecium*. Protist 152:339–354
69. Frankel J (1989) Pattern formation: ciliate studies and models. Oxford University Press
70. Frankel J (1991) The patterning of ciliates. J Protozool 38:519–525
71. Nanney DL (1980) Experimental ciliatology. Wiley, 304 pp
72. Jurand A, Selman GG (1969) The Anatomy of *Paramecium aurelia*. Macmillan
73. Vivier E (1974) Morphology, taxonomy and general biology of the genus *Paramecium*. In: Van Wagtendonk WJ (ed) Paramecium: a current survey. Elsevier, pp 1–89
74. Hufnagel LA, Torch R (1967) Intraclonal dimorphism of caudal cirri in *Euplotes vannus*: cortical determination. J Protozool 14:429–439
75. Tartar V (1961) The Biology of Stentor. Pergamon Press. 413 pp
76. Heckmann K, Frankel J (1968) Genic control of cortical pattern in Euplotes. J Exp Zool 168:11–38
77. Frankel J (1973) Dimensions of control of cortical patterns in Euplotes—role of preexisting structure, clonal life-cycle, and genotype. J Exp Zool 183:71–94
78. Chen-Shan L (1969) Cortical morphogenesis in Paramecium aurelia following amputation of the posterior region. J Exp Zool 170:205–228
79. Beisson J (2008) Preformed cell structure and cell heredity. Prion 2:1–8
80. Frankel J (1979) Propagation of cortical differences in Tetrahymena. Genetics 94:607–623
81. Bayless BA, Galati DF, Junker AD, Backer CB, Gaertig J, Pearson CG (2016) Asymmetrically localized proteins stabilize basal bodies against ciliary beating forces. J Cell Biol 215:457–466
82. Junker AD, Soh AWJ, O'Toole ET, Meehl JB, Guha M, Winey M, Honts JE, Gaertig J, Pearson CG (2019) Microtubule glycylation promotes attachment of basal bodies to the cell cortex. J Cell Sci 132:jcs233726
83. Dippell RV (1968) The development of basal bodies in Paramecium. Proc Natl Acad Sci U S A 61:461–468
84. Chatton E, Lwoff A (1936) Techniques pour l'etude des protozoaires, specialment de leurs structures superficielles (cinetome et l'argyrome). Bull Soc Franc Microscop 5:25–39
85. Chatton E, Seguela J (1940) La continuite genetique de formations ciliares chez les cilies hypotriches. Le cinetome et l'argyrome au cours de la division. Bull Biol France Belg 74:350–442
86. Lwoff A (1950) Problems of morphogenesis in ciliates. Wiley
87. Nanney DL (1975) Patterns of basal body addition in ciliary rows in *Tetrahymena*. J Cell Biol 65:503–512
88. Ng SF, Frankel J (1977) 180° rotation of ciliary rows and its morphogenetic implications in *Tetrahymena pyriformis*. Proc Natl Acad Sci U S A 74:1115–1119
89. Johnson KA, Rosenbaum JL (1992) Replication of basal bodies and centrioles. Curr Biol 4:80–85
90. Azimzadeh J, Marshall WF (2010) Building the centriole. Curr Biol 20(18):R816–R825
91. Guichard P, Hamel V, Gonczy P (2018) The rise of the cartwheel: seeding the centriole organelle. BioEssays 40:e1700241
92. Chen F, Tillberg PW, Boyden ES (2015) Expansion microscopy. Science 347:543–548
93. Dodt H-U (2015) The superresolved brain. Science 347:474–475
94. Chozinski TJ, Halpern AR, Okawa H, Kim H-J, Tremel GJ, Wong ROL, Vaughan JC (2016) Expansion microscopy with conventional antibodies and fluorescent proteins. Nat Methods 13:485–488
95. Geertsema H, Ewers H (2016) Expansion microscopy passes its first test. Nat Methods 13:481–482
96. Winterflood CM, Ewers H (2014) Single-molecule localization microscopy using mCherry. ChemPhysChem 15:3447–3451

97. Franke C, Repnik U, Segeletz S, Brouilly N, Kalaidzidis Y, Verbavatz J-M, Zerial M (2019) Correlative single-molecule localization microscopy and electron tomography reveals endosome nanoscale domains. Traffic 20:601–617
98. Zwettler FU, Reinhard S, Gambarotto D, Bell TDM, Harnel V, Guichard P, Sauer M (2020) Molecular resolution imaging by post-labeling expansion single-molecule localization microscopy (Ex-SMLM). Nat Commun 11:3388
99. Loncarek J, Bettencourt-Dias M (2018) Building the right centriole for each cell type. J Cell Biol 217:823–835
100. Gonczy P, Hatsopoulos GN (2019) Centriole assembly at a glance. J Cell Sci 132:jcs228833
101. Arzlanhan MD, Gulensoy D, Nur Firat-Karalar E (2020) A proximity mapping journey into the biology of the mammalian centrosome/cilium complex. Cell 9:1390
102. Sullenberger C, Vasquez-Limeta A, Kong D, Loncarek J (2020) With age comes maturity: biochemical and structural transformation of a human centriole in the making. Cell 9:1429
103. Wong YL, Anzola JV, Davis RL, Yoon M, Motamedi A, Kroll A, Seo CP, Hsia JE, Kim SK, Mitchell JW, Mitchell BJ, Desai A, Gahman TC, Shiau AK, Oegema K (2015) Reversible centriole depletion with an inhibitor of Polo-like kinase 4. Science 348:1155–1160
104. Satir P, Guerra C, Bell AJ (2007) Evolution and persistence of the cilium. Cell Motil Cytoskel 64:906–913
105. Kaneda M, Hanson D (1974) Growth patterns and morphogenetic events in the cell cycle of Paramecium aurelia. In: Van Wagtendonk WJ (ed) Paramecium: a current survey. Elsevier, pp 219–262
106. Fulton C, Dingle AD (1967) Appearance of the flagellate phenotype in populations of Naegleria amebae. Dev Biol 15:165–191
107. Fulton C, Dingle AD (1971) Basal bodies, but not centrioles, in *Naegleria*. J Cell Biol 51:826–836
108. Mizukami I, Gall J (1966) Centriole replication II. Sperm formation in the fern, Marsilea, and the cycad, Zamia. J Cell Biol 29:97–111
109. Lee J (2010) *De novo* formation of basal bodies during cellular differentiation of *Naegleria gruberi*: progress and hypotheses. Sem Cell Dev Biol 21:156–162
110. Meunier A, Spassky N (2016) Centriole continuity: out with the new, in with the old. Curr Opin Cell Biol 38:60–67
111. Rieder CL, Faruki S, Khodjakova A (2001) The centrosome in vertebrates: more than a microtubule-organizing center. Trends Cell Biol 11:413–419
112. Fong CS, Kim M, Yang TT, Liao J-C, Tsou M-FB (2014) SAS-6 assembly templated by the lumen of cartwheel-less centrioles precedes centriole duplication. Dev Cell 30:238–245
113. Wang W-J, Acehan D, Kao C-H, Jane W-N, Uryu K, Tsou M-FB (2015) De novo centriole formation in human cells is error-prone and does not require SAS-6 self-assembly. elife 6:1–13
114. Al Jord A, Lemaitre A-I, Delgehyr N, Faucourt M, Spassky N, Meunier A (2014) Centriole amplification by mother and daughter centrioles differs in multiciliated cells. Nature 516:104–107
115. Hufnagel LA (1966) Fine structure and DNA of pellicles isolated from Paramecium aurelia. Proc Sixth Int Congr Electr Microsc Kyoto, 239–240
116. Hufnagel LA (1969) Cortical ultrastructure of Paramecium aurelia. J Cell Biol 40:779–801
117. Frankel J, Buhse HE (2017) In memorium: Norman E. Williams (1928-2016): pioneer of ciliate architecture. J Euk Microbiol 64:413–418
118. Gavin RH (1980) The oral apparatus of *Tetrahymena*. V. Oral apparatus polypeptides and their distribution. J Cell Sci 44:317–333
119. Gaertig J (2000) Molecular mechanisms of microtubular organelle assembly in *Tetrahymena*. J Euk Microbiol 47:185–190
120. Hoey JG, Gavin RH (1992) Localization of actin in the Tetrahymena basal body-cage complex. J Cell Sci 103:629–641
121. Garces JA, Hoey JG, Gavin RH (1995) Putative myosin heavy and light chains in *Tetrahymena*: co-localization to the basal body-cage complex and association of the heavy chain with skeletal muscle actin filaments in vitro. J Cell Sci 108:869–881
122. Honts JE, Williams NE (1990) Tetrins: polypeptides that form bundled filaments in Tetrahymena. J Cell Sci 96:293–302
123. Nahon P, Coffe G, Le Guyader H, Darmanaden-Delorme J, Jeanmaire-Wolf R, Clerot J-C, Adoutte A (1993) Identification of the epiplasmins, a new set of cortical proteins of the membrane cytoskeleton in *Paramecium*. J Cell Sci 104:975–990
124. Damaj R, Pomel S, Bricheux G, Coffe G, Vigues B, Ravet V, Bouchard P (2009)

Cross-study analysis of genomic data defines the ciliate multigene epiplasmin family: strategies for functional analysis in *Paramecium tetraurelia*. BMC Evol Biol 9:125

125. Aubusson-Fleury A, Bricheux G, Damaj R, Lemullois M, Coffe G, Donnadieu F, Koll F, Vigues B, Bouchard P (2013) Epiplasmins and epiplasm in *Paramecium*: the building of a submembranous cytoskeleton. Protist 164:451–469
126. Vaudaux P (1976) Isolation and identification of specific cortical proteins in *Tetrahymena pyriformis* strain GL. J Protozool 23:458–464
127. Huttenlauch I, Geisler N, Plessmann U, Peck RK, Weber K, Stick R (1995) Major epiplasmic proteins of ciliates are articulins: cloning, recombinant expression and structural characterization. J Cell Biol 130:1401–1412
128. Huttenlauch I, Peck RK, Plessmann U, Weber K, Stick R (1998) Characterization of two articulins, the major epiplasmic proteins comprising the membrane skeleton of the ciliate *Pseudomicrothorax*. J Cell Sci 111:1909–1919
129. Huttenlauch I, Stick R (2003) Occurrence of articulins and epiplasmins in protists. J Euk Microbiol 50:15–18
130. Aubusson-Fleury A, Balavoine G, Lemullois M, Bouhouche K, Beisson J, Koll F (2017) Centrin diversity and basal body patterning across evolution: new insights from Paramecium. Biol Open 6:765–776
131. Gould SB, Tham W-H, Cowman AF, McFadden GI, Waller RF (2008) Alveolins, a new family of cortical proteins that define the protist infrakingdom Alveolata. Mol Biol Evol 25 (6):1219–1230
132. Kloetzel JA, Baroin-Tourancheau A, Miceli C, Barchetta S, Farmer J, Banerjee D, Fleury-Aubusson A (2003) Plateins: a novel family of signal peptide-containing articulins in euplotid ciliates. J Euk Microbiol 50:19–33
133. Epting D, Slanchev K, Boehike C, Hoff S, Loges NT, Yasunaga T, Indorf L, Nestel S, Lienkamp SS, Omran H, Wolfgang-Kuehn E, Ronneberger O, Walz G, Kramer-Zucker A (2015) The Rac1 regulator ELMO controls basal body migration and docking in multiciliated cells through interaction with ezrin. Development 142:174–184
134. Hegan PS, Ostertag E, Geurts AM, Mooseker MS (2015) Myosin 1d is required for planar cell polarity in ciliated tracheal and ependymal epithelial cells. Cytoskeleton 72:503–516
135. Iftode F, Fleury A, Adoutte A (1997) Development of surface pattern during division in Paramecium III. Study of stomatogenesis in the wild type using antitubulin antibodies and confocal microscopy. Eur J Protistol 33:145–167
136. Bengueddach H, Lemullois M, Aubusson-Fleury A, Koll F (2017) Basal body positioning and anchoring in the multiciliated cell Paramecium tetraurelia: roles of OFD1 an VFL3. Cilia 6:6
137. Galati DF, Abulin DS, Tauber GA, Pham AT, Pearson CG (2016) Automated image analysis reveals the dynamic 3-dimensional organization of multi-ciliated arrays. Biol Open 5:20–31
138. Joukov V, De Nicolo A (2019) The centrosome and the primary cilium: the yin and yang of a hybrid organelle. Cell 8:701–747
139. Schnackenberg BJ, Palazzo RE (1999) Identification and function of the centrosome centromatrix. Biol Cell 91:429–438
140. Lwoff A (1962) Biological order. The MIT Press. 101 pp
141. Rampton VW (1962) Kinetosomes of *Tetrahymena*. Nature 195:195
142. Seaman GR (1959) Large scale isolation of kinetosomes from the ciliated protozoon, *Tetrahymena pyriformis*. Exp Cell Res 21:292–302
143. Pyne MCK (1968) Sur l'absence d'incorporation de la thymidine tritiee dans les cinetosomes de *Tetrahymena pyriformis* (cilies Holotriches). C R Hebd Seanc Acad Sci Paris 267:755–757
144. Suyama Y, Preer JR (1965) Mitochondrial DNA from protozoa. Genetics 52:1051–1058
145. Hufnagel LA (1969) Properties of DNA associated with raffinose-isolated pellicles of *Paramecium aurelia*. J Cell Sci 5:561–573
146. Randall Sir J, Disbrey C (1965) Evidence for the presence of DNA at basal body sites in *Tetrahymena pyriformis*. Proc Roy Soc B 162:473–491
147. Smith-Sonneborn J, Plaut W (1967) Evidence for the presence of DNA in the pellicle of Paramecium. J Cell Sci 2:225–234
148. Blake A, Peacocke AR (1968) The interaction of aminoacridines with nucleic acids. Biopolymers 6:1225–1253
149. Zama M, Ichimura S (1976) Induced circular dichroism of acridine orange bound to double-stranded RNA and transfer RNA. Biopolymers 15:1693–1699
150. Hoffman EJ (1965) The nucleic acids of basal bodies isolated from *Tetrahymena pyriformis*. J Cell Biol 25:217–228

151. Argetsinger J (1965) The isolation of ciliary basal bodies (kinetosomes) from *Tetrahymena pyriformis.* J Cell Biol 24:154–157
152. Younger KB, Banerjee S, Kelleher JK, Winston M, Margulis L (1972) Evidence that the synchronized production of new basal bodies is not associated with DNA synthesis in *Stentor coeruleus.* J Cell Sci 11:621–637
153. Rattner JB, Phillips SG (1973) Independence of centriole formation and DNA synthesis. J Cell Biol 57:359–372
154. Hall JL, Ramanis Z, Luck JL (1989) Basal body/centriolar DNA: molecular genetic studies in *Chlamydomonas.* Cell 59:121–132
155. Ramanis Z, Luck DJL (1986) Loci affecting flagellar assembly and function map to an unusual linkage group in *Chlamydomonas reinhardtii.* Proc Natl Acad Sci U S A 83:423–426
156. Maul GG, Deaven L (1977) Quantitative determination of nuclear pore complexes in cycling cells with differing DNA content. J Cell Biol 73:748–780
157. Hall JL, Luck JL (1995) Basal-body associated DNA: *in situ* studies in *Chlamydomonas reinhardtii.* Proc Natl Acad Sci U S A 92:5129–5133
158. Goodenough UW (1989) Basal body chromosomes? Cell 59:1–3
159. Johnson KA, Dutcher SK (1991) Molecular studies of linkage group XIX of Chlamydomonas reinhardtii: evidence against a basal body location. J Cell Biol 113:339–346
160. Johnson KA, Rosenbaum JL (1990) The basal bodies of *Chlamydomonas reinhardtii* do not contain immunologically detectable DNA. Cell 62:615–619
161. Kuroiwa T, Yorihuzi T, Yabe N, Ohta T, Uchida H (1990) Absence of DNA in the basal body of *Chlamydomonas reinhardtii* by fluorimetry using a video-intensified microscope photon-counting system. Protoplasma 158:155–164
162. Calvin HI, Bedford JM (1971) Formation of disulfide binds in the nucleus and accessory structure of mammalian spermatozoa during maturation in the epididymis. J Reprod Fertil Suppl 13:65–75
163. Darzynkiewicz Z, Gledhill BL, Ringertz NR (1969) Changes in deoxyribonucleoprotein during spermiogenesis in the bull. Exp Cell Res 58:435–438
164. Lima de Faria A (1986) Molecular evolution and organization of the chromosomes. Elsevier
165. Kuroiwa T, Miyamura S, Kawano S, Hizume M, Toh-e A, Miyakawa I, Sando N (1986) Cytological characterization of NOR in the bivalents of *Saccharomyces cerevisiae.* Exp Cell Res 165:199–206
166. Kuroiwa T, Nishibayashi S, Kawano S (1981) Visualization of DNA in plant mitochondria, various phages and single DNA molecules by epi-fluorescence microscopy. Proc 13th Int Bot Congr Sydney 13: 250a
167. Kuroiwa T, Suzuki Z (1981) Visualization of DNA in various phages (T4, x, T7, 029) by ethidium bromide epifluorescent microscopy. Exp Dermatol 37:967–970
168. Johnson KA, Rosenbaum JL (1991) Basal bodies and DNA. Trends Cell Biol 1:145–149
169. Marshall WF, Rosenbaum JL (2000) Are there nucleic acids in the centrosome? Curr Topics Dev Biol 49:187–205
170. Pederson T (2006) The centrosome: built on an mRNA? Nat Cell Biol 8:652–654
171. Alliegro MC (2008) The implications of centrosomal RNA. RNA Biol 5:198–200
172. Alliegro MC (2011) The centrosome and spindle as a ribonucleoprotein complex. Chromosom Res 19:367–376
173. Kilmartin JV (2014) Lessons from yeast: the spindle pole body and the centrosome. Phil Trans R Soc B 369:20130456
174. Unger E (1977) RNS im spindelpolkorper von Hefen. Zeitsch Fur Allg Mikrobiol 17:487–490
175. Hartman H, Puma JP, Gurney T (1974) Evidence for the association of RNA with the ciliary basal bodies of *Tetrahymena*. J Cell Sci 16:241–259
176. Schmukler M, Friedling SP, Levy CC (1972) Polyguanylic acid-inhibited ribonuclease of *Klebsiella*: II. Studies with synthetic polyribonucleotides. Biochim Biophys Acta 268:403–410
177. Heidemann SR, Sander G, Kirschner MW (1977) Evidence for a functional role of RNA in centrioles. Cell 10:337–350
178. Heidemann SR, Kirschner MW (1975) Aster formation in eggs of *Xenopus laevis.* J Cell Biol 67:105–117
179. Rieder CL (1979) Ribonucleoprotein staining of centrioles and kinetochores in newt lung cell spindles. J Cell Biol 80:l–9
180. Bernhard W (1969) A new staining procedure for electron microscopical cytology. J Ultrastruct Res 27:250–265

181. Stubblefield E, Defoor P (1972) Effects of inhibitors of nucleic acid synthesis on centriole replication in Chinese hamster fibroblasts. J Cell Biol 55(2, Pt 2):254a
182. Hartman H (1975) The centriole and the cell. J Theor Biol 51:501–509
183. Went HA (1977) Can a reverse transcriptase be involved in centriole duplication? J Theor Biol 68:95–100
184. Jain N, Blauch LR, Szymanski MR, Das R, Tang SKY (2020) Transcription polymerase-catalyzed emergence of novel RNA replicons. Science 368:6487
185. Pepper DA, Brinkley BR (1980) Evidence for nucleic acids in kinetochores and centrosomes. Cell Motil 1:l–15
186. Peterson SP, Berns MW (1978) Evidence for centriolar region RNA functioning in spindle formation in dividing PtK2 cells. J Cell Sci 34:289–301
187. Alliegro MC, Alliegro MA, Palazzo RE (2006) Centrosome-associated RNA in surf clam oocytes. Proc Natl Acad Sci U S A 103:9034–9038
188. Kingsley EP, Chan XY, Duan Y, Lambert J (2007) Widespread RNA localization in the spiralian embryo. Evol Dev 9:527–539
189. Blower M, Feric E, Weis K, Heald R (2007) Genome-wide analysis demonstrates conserved localization of messenger RNAs to mitotic microtubules. J Cell Biol 179:1365–1373
190. Le'cuyer E, Parthasarathy N, Krause HM (2007) Fluorescent *in situ* hybridization protocols in Drosophila embryos and tissues. Methods Mol Biol 420:289–302
191. Le'cuyer E, Yoshida H, Parthasarathy N, Alm C, Babak T, Cerovina T, Hughes TR, Tomancak P, Krause HM (2007) Global analysis of mRNA localization reveals a prominent role in organizing cellular architecture and function. Cell 131:174–187
192. Diot C, Chin A, Le'cuyer E (2017) Optimized FISH methods for visualizing RNA localization properties in *Drosophila* and human tissues and cultured cells. Methods 126:156–165
193. Jandura A, Hu J, Wilk R, Krause HM (2017) High resolution fluorescent in situ hybridization in Drosophila embryos and tissues using tyramide signal amplification. J Vis Exp 128: e56281
194. Li J, Xu M, Zhou H, Ma J, Potter H (1997) Alzheimer presenilins in the nuclear membrane, interphase kinetochores and centrosomes suggest a role in chromosome segregation. Cell 90:917–927
195. Nizzari M, Venezia V, Repetto E, Caorsi V, Magrassi R, Gagliani MC, Carlo P, Florio T, Schettini G, Tacchetti C, Russo T, Diaspro A, Russo C (2007) Amyloid precursor protein and presenilin1 interact with the adaptor GRB2 and modulate ERK1,2 signaling. J Biol Chem 282:13833–13844
196. Tkemaladze JV, Chichinadze KN (2005) Centriolar mechanisms of differentiation and replicative ageing of higher animal cells. Biochemistry 70:1288–1303
197. Chichinadze K, Tkemaladze DV (2008) Centrosomal hypothesis of cellular ageing and differentiation. Adv Geront 21:367–371
198. Chichinadze K, Lazarashvili A, Tkemaladze J (2013) RNA in centrosomes: structure and possible functions. Protoplasma 250:397–405
199. Carleton HM (1920) Observations on an intra-nucleolar body in columnar epithelium cells of the intestine. Q J Microsc Sci 64:329–341
200. Love R, Liles RH (1959) Differentiation of nucleoproteins by inactivation of protein-bound amino groups and staining with toluidine blue and ammonium molybdate. J Histochem Cytochem 7:164–181
201. Alliegro MA, Henry JJ, Alliegro MC (2011) Rediscovery of the nucleolinus, a dynamic RNA-rich organelle associated with the nucleolus, spindle, and centrosomes. Proc Natl Acad Sci U S A 107(13):13718–13723
202. Alliegro MC, Hartson S, Alliegro MA (2012) Composition and dynamics of the nucleolinus, a link between the nucleolus and cell division apparatus in surf clam (*Spisula*) oocytes. J Biol Chem 287:6702–6713
203. Groisman I, Huang Y-S, Mendez R, Cao Q, Theurkauf W, Richter JD (2000) CPEB, maskin and B1 mRNA at the mitotic apparatus: implications for local translational control of cell division. Cell 103:435–447
204. Bergalet J, Patel D, Legendre F, Lapointe C, Bouvrette LPB, Chin A, Blanchette M, Kwon E, Lecuyer E (2020) Inter-dependent centrosomal co-localization of the *cen* and *ik2* *cis*-natural antisense mRNAs in Drosophila. Cell Rep 30:3339–3352
205. Han JW, Park JH, Kim M, Lee J (1997) mRNAs for microtubule proteins are specifically colocalized during the sequential formation of basal body, flagella, and cytoskeletal microtubules in the differentiation of *Naegleria gruberi*. J Cell Biol 137:871–879
206. Cech TR, Steitz JA (2014) The noncoding RNA revolution—trashing old rules to forge new ones. Cell 157:77–94

207. Chujo T, Yamazaki T, Hirose T (2016) Architectural RNAs (ArcRNAs): a class of long noncoding RNAs that function as the scaffold of nuclear bodies. Biochim Biophys Acta 1859:139–146
208. Allen SE, Nowacki M (2020) Roles of noncoding RNAs in ciliate genome architecture. J Mol Biol 432(15):4186–4198
209. Chapman M, Alliegro MC (2012) The karyomastigont as an evolutionary seme. Q Rev Biol 87:315–324
210. Dippell RV (1976) Effects of nuclease and protease digestion on the ultrastructure of Paramecium basal bodies. J Cell Biol 69:622–637
211. Frankel J (1964) The effects of high temperature on the pattern of oral development in *Tetrahymena pyriformis* GL. J Exp Zool 155:403–436
212. Frankel J (1965) The effect of nucleic acid antagonists on cell division and oral organelle development in *Tetrahymena pyriformis.* J Exp Zool 159:113–148
213. Rizzotti M (1995) Cilium: origin and 9-fold symmetry. Acta Biotheor 43:227–240
214. Sagan L (1967) On the origin of mitosing cells. J Theoret Biol 14:225–274
215. Bermudes D, Margulis L, Tzertzinis G (1987) Prokaryotic origin of undulopodia. Application of the panda principle to the centriole enigma. Ann N Y Acad Sci 503:187–197
216. Margulis L, Chapman M, Guerrero R, Hall J (2006) The last eukaryotic common ancestor (LECA): acquisition of cytoskeletal motility from aerotolerant spirochetes in the Proterozoic Eon. Proc Natl Acad Sci U S A 103:13080–13085
217. Marshall WF (2009) Centriole evolution. Curr Opin Cell Biol 21:14–19
218. Azimzadeh J (2014) Exploring the evolutionary history of centrosomes. Phil Trans Roy Soc B 369:20130453
219. Williams NE (1975) Regulation of microtubules in Tetrahymena. Int Rev Cytol 41:59–86
220. Hufnagel LA (1993) Correlative, cross-disciplinary microscopical analysis of cilioprotists in their natural environment: water. J Protozool 40:39A
221. Chi KR (2009) Super-resolution microscopy: breaking the limits. Nat Methods 6:15–18
222. Van Noorden R (2014) Nobel for microscopy that reveals inner world of cells. Nature 514:286
223. Chitnis A, Nogare DD (2018) Time-lapse imaging beyond the diffraction limit. Methods 150:32–41
224. Stack R (2016) Watching translation of single mRNAs in cells. Nat Methods 13:543
225. Wheeler AR, Throndset WR, Whelan RJ, Leach AM, Zare RN, Liao YH, Farrell K, Manger ID, Daridon A (2003) Microfluidic device for single-cell analysis. Anal Chem 75:3581–3586
226. Lockery S (2007) Channeling the worm: microfluidic devices for nematode neurobiology. Nat Methods 4:691–692
227. Yan Y, Jiang L, Aufderheide K, Wright G, Terekhov A, Costa L, Qin K, McCleery WT, Fellenstein JJ, Ustione A, Robertson JB, Johnson CH, Piston DW, Hutson MS, Wikswo JP, Hofmeister W, Janetopoulos C (2014) A microfluidic-enabled mechanical microcompressor for the immobilization of live single- and multi-cellular specimens. Microsc Microanal 20:141–151
228. Zhou Y, Basu S, Wohlfahrt KJ, Lee SF, Kleinerman D, Laue ED, Seshia AA (2016) A microfluidic platform for trapping, releasing and super-resolution imaging of single cells. Sens Activators B 232:680–691
229. Preyadharishini S, Fernandez JG (2020) Characterization, manipulation and isolation of *Paramecium aurelia* using a micro-electromigration chip and computer vision. Adv Mater Technol 5:2000152
230. Viswanathan S, Williams ME, Bloss EB, Stasevich TJ, Speer CM, Nern A, Pfeiffer BD, Hooks BM, Li W-P, English BP, Tian T, Henry GL, Macklin JJ, Patel R, Gerfen CR, Zhuang X, Wang Y, Rubin GM, Looger LL (2015) High-performance probes for light and electron microscopy. Nat Methods 12:568–576
231. Chittajallu DR, Florian S, Kohler RH, Iwamoto Y, Orth JD, Weissleder R, Danuser G, Mitchison TJ (2015) *In vivo* cell-cycle profiling in xenograft tumors by quantitative intravital microscopy. Nat Methods 12:577–585
232. Stelzer EHK (2015) Light-sheet fluorescence microscopy for quantitative biology. Nat Methods 12:23–26

Part III

Cell and Organelle Motility

Chapter 14

3D and 4D Tumorigenesis Model for the Quantitative Analysis of Cancer Cell Behavior and Screening for Anticancer Drugs

Deborah Wessels, Daniel F. Lusche, Edward Voss, and David R. Soll

Abstract

Cancer cells from cell lines and tumor biopsy tissue undergo aggregation and aggregate coalescence when dispersed in a 3D Matrigel™ matrix. Coalescence is a dynamic process mediated by a subset of cells within the population of cancer cells. In contrast, non-tumorigenic cells from normal cell lines and normal tissues do not aggregate or coalesce, nor do they possess the motile cell types that orchestrate coalescence of cancer cells. Therefore, coalescence is a cancer cell-specific phenotype that may drive tumor growth in vivo, especially in cases of field cancerization. Here, we describe a simple 3D tumorigenesis model that takes advantage of the coalescence capabilities of cancer cells and uses this feature as the basis for a screen for treatments that inhibit tumorigenesis. The screen is especially useful in testing monoclonal antibodies that target cell-cell interactions, cell-matrix interactions, cell adhesion molecules, cell surface receptors, and general cell surface markers. The model can also be used for 2D imaging in a 96-well plate for rapid screening and is adaptable for 3D high-resolution assessment. In the latter case, we show how the 3D model can be optically sectioned with differential interference contrast (DIC) optics, then reconstructed in 4D and quantitatively analyzed by computer-assisted methods, or, alternatively, imaged with confocal microscopy for 4D quantitative analysis of cancer cell interactions with normal cells within the tumor microenvironment. We demonstrate reconstructions and quantitative analyses using the advanced image analysis software J3D-DIAS 4.2, in order to illustrate the types of detailed phenotypic characterizations that have proven useful. Other software packages may be able to perform similar types of analyses.

Key words 3D model, Tumorigenesis, Monoclonal antibody, Cancer therapy drug screen, Cell migration, Coalescence

1 Introduction

We describe a straightforward, in vitro 3D model to study tumorigenesis, cell-cell interactions within the tumor microenvironment (TME), and metastasis in real time [1–3]. Importantly, the model is useful for testing the effects of drugs with therapeutic potential, including function-blocking antibodies, on these disease processes

Ray H. Gavin (ed.), *Cytoskeleton: Methods and Protocols*, Methods in Molecular Biology, vol. 2364,
https://doi.org/10.1007/978-1-0716-1661-1_14,

[4]. The model is based on the observation that tumorigenic cell lines and cancer cells isolated from fresh tumor biopsies form aggregates that undergo coalescence when dispersed in Matrigel™, a transparent, naturally derived, reconstituted 3D basement membrane matrix [5]. Coalescence is a dynamic event, mediated by a subset of cells within the aggregates that can move through the matrix, form bridges between aggregates, and contract to pull the aggregates together [1, 2]. In vivo evidence for coalescence suggests it may contribute to tumor growth in tissue where multiple early stage neoplasms are present, i.e., in cases of field cancerization [6, 7].

As a rapid screen for potential anticancer drugs, the coalescence assay can be performed in a 96-well plate [4]. For a higher resolution study, the assay can be optically sectioned using DIC optics and reconstructed in 4D in the presence or absence of anticancer drugs. DIC optics have the advantage of low phototoxicity, permitting long-term recording. The model can also be expanded to include other cell types present in the TME, such as fibroblasts [8] and endothelial cells. Using cells expressing a fluorophore such as GFP and/or vitally dyed cells, the TME model can be imaged with laser scanning confocal microscopy (LSCM), and these methods are included as well. We illustrate how use of the model has revealed a novel role for cancer-associated fibroblasts (CAFs) in promoting coalescence through reciprocal signaling and physical interactions [8]. We present 4D reconstructions and motion analyses performed with the advanced 3D Dynamic Image Analysis Systems, J3D-DIAS 4.2 [9–12], which can be accessed by a visit to the W.M. Keck Dynamic Image Analysis Facility at the University of Iowa. However, other software packages can be employed for some of the analyses described here. Our emphasis here will be on quantitative methods to analyze coalescence, cell crawling, and cell shapes during tumorigenesis and metastasis. The goal of this approach is to generate accurate cell phenotypes and behaviors so that the effect of a given treatment can be precisely defined.

2 Materials

2.1 Cell Culture

1. Fresh biopsy tissue (*see* **Note 1**).
2. MCF medium: Dulbecco's Modified Eagle's Medium (DMEM)/F12, 5% horse serum, 20 ng/mL EGF, 0.5μg/mL hydrocortisone, 10μg/mL insulin, 0.1 μg/mL cholera toxin, and penicillin-streptomycin.
3. Fibroblast growth medium (available commercially).
4. Endothelial cell growth medium (available commercially).

5. RPMI (Roswell Park Memorial Institute) medium (available commercially).
6. Sterile scalpels, forceps, and scissors.
7. 6-well tissue culture plate.

2.2 2D Coalescence Assay and mAb Screen

1. Matrigel matrix in growth factor medium with phenol red aliquoted into 500μL quantities and stored at −20 °C (*see* **Note 2**).
2. Chilled pipette tips.
3. 96-well flat-bottomed plates.
4. Cancer cell lines such as MDA-MB-231 or tumor biopsy material and non-tumorigenic or normal cell lines at 70–80% confluency and appropriate medium (*see* Subheading 2.1, **item 1**).
5. Well-characterized or novel function-blocking monoclonal (mAb) or polyclonal antibodies, or test reagents without preservatives such as azide or proclin (*see* **Note 3**).
6. Inverted microscope equipped with 10× phase contrast objective and connected to a camera (*see* **Note 4**).
7. Computer for image acquisition.

2.3 3D Coalescence Assay for DIC Imaging

1. Matrigel (*see* Subheading 2.2, **item 1**).
2. 35 mm Petri dish with glass insert in the bottom and in the lid such as those from CellVis or iBidi (*see* **Note 5**).
3. Cancer cell lines or biopsy material and normal cell lines or tissue at 70–80% confluency and appropriate medium (*see* Subheading 2.1) (*see* **Note 6**).
4. Microscope with DIC optics and a motorized *z*-axis stage housed in an incubator at 37 °C and 5% CO_2 or equipped with an environmental chamber.
5. Camera controlled by video acquisition software and synchronized with the light source and the motorized stage (*see* **Note 7**).

2.4 3D Coalescence Overlay Assay with Cancer Cells

1. Matrigel (*see* Subheading 2.2, **item 1**).
2. 35 mm Petri dish with glass insert in the bottom and in the top such as those from CellVis or iBidi.
3. Cancer cell lines or biopsy material (*see* Subheading 2.1) and tissue-matched fibroblasts at 70–80% confluency (*see* **Note 8**).
4. Fibroblast growth medium (*see* **Note 9**).
5. Microscope with DIC optics and a motorized *z*-axis stage housed in an incubator at 37 °C and 5% CO_2 or equipped with an environmental chamber.
6. Type 1 human collagen: make a suitable stock to achieve a 1:30 aqueous working solution.

2.5 3D Assay for Imaging with Laser Scanning Confocal Microscopy

1. Matrigel matrix (*see* Subheading 2.2, **item 1**).
2. RPMI media.
3. Deep Red vital dye: dissolve 15μg of powder in 50μL DMSO, and store at −20 °C.
4. GFP-tagged cancer cells or cell dye (*see* **Note 10**).
5. 35 mm Petri dish with glass coverslip inserts in the bottom and lid.
6. LSCM equipped with an environmental chamber.
7. Human umbilical vein endothelial cells (HUVECs).
8. Endothelial cell medium (*see* **Note 11**).
9. Non-tumorigenic MCF-10A or appropriate other non-tumorigenic cells.
10. Computer capable of storing movies greater than 100 GB.

3 Methods

3.1 Culturing Cells from Biopsy Tissue

1. Prepare MCF medium (*see* Subheading 2.1, **item 2**).
2. Place tissue in sterile Petri dish containing MCF medium.
3. Mechanically dissociate tissue in the medium using scalpel, forceps, and scissors.
4. Transfer medium containing cells into the wells of a 6-well plate.
5. Incubate at 37 °C, 5% CO_2, changing MCF medium every 3 days.
6. Cells can be harvested and transferred into tissue culture flasks containing MCF medium when density has reached 70–80% confluency.
7. Presence of desired cell type should be confirmed through immunostaining or FACs sorting with the appropriate cell type-specific marker.

3.2 Preparation of 2D Coalescence Assay

1. Using the pre-chilled pipette tips, coat the bottom of the wells of the 96-well plate with 50μL of Matrigel™ to form a cushion (Fig. 1a) (*see* **Note 12**).
2. Place the plate in a 37 °C incubator for 20 min to allow the Matrigel™ to gel.
3. 3.Harvest cells at 70–80% confluency by trypsinization and determine cell density.
4. Suspend cells in the appropriate cell growth medium at a density of 3.0×10^5 cells per mL.

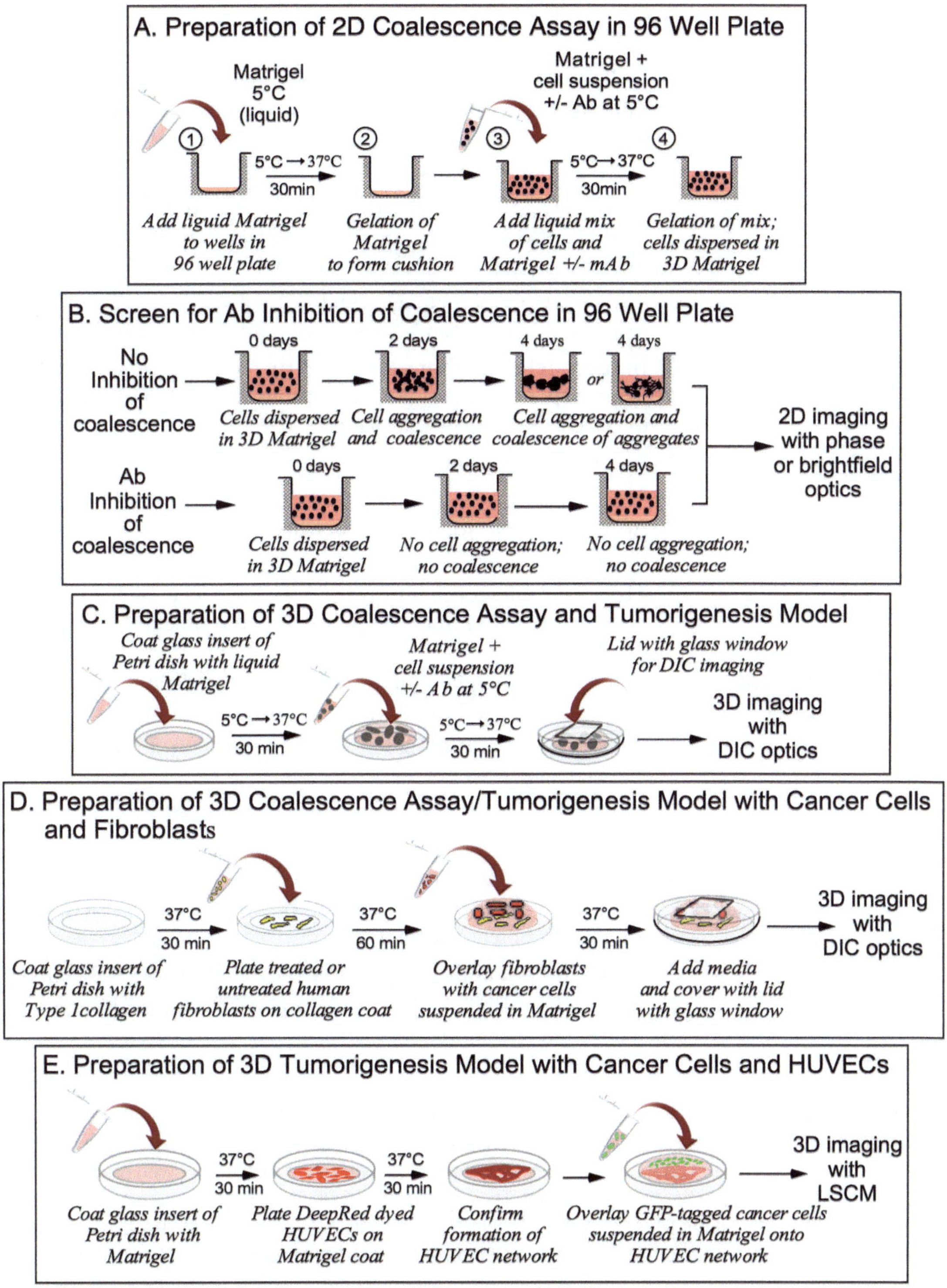

Fig. 1 Preparation of tumorigenesis and coalescence assays for 2D and 3D analyses. (**a**) For the 2D analysis of coalescence, a thin Matrigel™ cushion is applied to the wells in a 96-well plate before adding the cells

5. Strain cell suspension through a 70μm sterile strainer in order to eliminate clumps of cells.
6. Label duplicate tubes for each mAb to be tested.
7. Add purified mAb to 60 mL of appropriate cell growth medium to achieve an antibody concentration between 0.3 and 250μg/mL (Fig. 1b, *see* **Note 13**).
8. Add 80μL of the cell suspension from **step 5** to the mAb medium made in **step 6**. As a control, add 80μL of the cell suspension to 60μL of medium.
9. Add 30μL of chilled Matrigel™ in growth medium to the suspension from **step 7** to achieve a final volume of 170μL.
10. Immediately distribute 80 μL of the cell/Matrigel™ mixture from **step 8** onto the Matrigel cushion (Fig. 1b) of each of the two wells in the 96-well plate.
11. Incubate for 30 min at 37 °C, 5% CO_2 to allow gelation of the Matrigel™.
12. Add 100μL of appropriate culture medium to each well in the dish.
13. Monitor coalescence and record results every 24 h for 4 days using an inverted microscope with a 10× objective (Fig. 1b).

3.3 Preparation of 3D Coalescence Assay

1. Using the pre-chilled pipette tips, coat the 20 mm glass insert of a 35 mm dish with 200μL of Matrigel™ (Fig. 1c).
2. Place the coated dish in a 37 °C incubator for 30 min to allow the Matrigel™ to gel.
3. Harvest cells at 70–80% confluency by trypsinization and determine cell density.
4. Pellet 5.0×10^5 cells at low speed, discard supernatant, and resuspend the pellet in 250μL of chilled MCF media and 500μL of chilled Matrigel™.
5. Using the pre-chilled pipette tips, carefully pipet the cell suspension from step 4 over the Matrigel™ cushion (Fig. 1c).

Fig. 1 (continued) suspended in Matrigel™. (**b**) mAbs or drugs can be tested for their ability to inhibit coalescence by mixing the test agent with the cell suspension. (**c**) The 3D coalescence assay is prepared in a Petri dish with a glass insert coated with Matrigel™. The cell suspension in Matrigel™, with or without mAb or drug, is then overlaid onto the Matrigel™ coat. The dish is covered with a lid with a glass insert for optimal imaging with DIC. (**d**) Interactions between fibroblasts and cancer cells are examined by first plating fibroblasts on a collagen-coated glass insert of a Petri dish. Cancer cells suspended in Matrigel™ are then applied over the fibroblasts, media added, and the dish covered with a lid with a glass insert. (**e**) For LSCM imaging, dye-labeled HUVECs are plated on a coat of Matrigel™ and allowed to form a network. GFP-tagged cancer cells are then overlaid onto the HUVECs

6. Incubate for 60 min at 37 °C, 5% CO_2.
7. Cover the dish with a glass insert window for DIC imaging (Fig. 1c).
8. Flood the dish with appropriate culture medium.
9. Proceed with optical sectioning (*see* Subheading 3.5).

3.4 Preparation of 3D Coalescence Assay with Cancer Cells Overlaid onto Fibroblasts

1. Dilute human Type 1 collagen 1:30 in ultrapure water.
2. Coat the 20 mm glass insert of a 35 mm Petri dish with 500μL of the diluted collagen and incubate at room temperature for 60 min (Fig. 1d).
3. Carefully remove excess collagen and wash three times in sterile PBS (*see* **Note 14**).
4. Plate 5 × 10^4 fibroblasts in 500μL of the appropriate medium on the collagen-coated glass insert (Fig. 1d).
5. Incubate at 37 °C, 5% CO_2 for 30 min (Fig. 1d).
6. Prepare cancer cells suspended in Matrigel™ (*see* Subheading 3.2, **steps 3–4**).
7. Remove excess medium from the fibroblast culture and gently overlay with the cells prepared in the previous step (Fig. 1d).
8. Incubate the dish at 37 °C, 5% CO_2 for 30 min (Fig. 1d) to allow Matrigel™ gelation.
9. Add 5 mL of the medium used for the overlayed cells (Fig. 1d).

3.5 Preparation of 3D Assay for LSCM Imaging

3.5.1 Labeling and Plating HUVECs

1. Harvest HUVECs at 70–80% confluency by trypsinization and determine cell density.
2. Pellet 3.0–5.0 × 10^5 HUVEC cells at low speed, discard supernatant, and resuspend the pellet in 1 mL of RPMI without serum.
3. Add 1μL of Deep Red stock solution to the cell suspension and incubate for 30 min at 37 °C, 5% CO_2, occasionally inverting the tube to maintain an even distribution of the cells in the dye.
4. Pellet the dyed HUVECs, decant the supernatant, and resuspend the pellet in 200μL endothelial cell-specific medium.
5. Disperse the dyed HUVECs on a Matrigel™-coated glass insert (*see* Subheading 3.2, **steps 1** and **2**) (Fig. 1e).
6. Incubate for 4 h at 37 °C, 5% CO_2, after which time the characteristic HUVEC network should be formed (Fig. 1e).

3.5.2 Overlaying HUVECs with Cells in Matrigel™

1. Prepare a suspension of 5 × 10^5 GFP-tagged or dyed cells in 250μL of MCF medium.
2. Using the pre-chilled pipette tips, mix with 500μL ice-cold Matrigel™.

3. Carefully remove excess medium and unattached HUVECs from the plate prepared in Subheading 3.4, **step 1** (*see* **Note 15**).
4. Using the pre-chilled pipette tips, plate the 750μL of cells from **step 2** over the HUVEC network (Fig. 1e).
5. Incubate for 1 h at 37 ° C in 5% CO_2 (Fig. 1e).
6. Add appropriate medium to the dish.
7. As a control, use 5×10^5 non-tumorigenic MCF-10A or other non-tumorigenic cells treated as described in **steps 1–6**.

3.6 2D and 3D Imaging and 3D Optical Sectioning

3.6.1 2D Imaging

1. The 96-well plate should be imaged every 24 h with low-magnification phase or brightfield optics (Fig. 2a).
2. Non-tumorigenic cells such as MCF-10A grow as clumps, while cellular bridges between aggregates, along with distinct large aggregates, become apparent in tumorigenic cells by 48 h (Fig. 2a) (*see* **Note 16**).
3. Inhibition of coalescence by mAb or other drug treatment is evident by day 4 of the assay (Fig. 2b).

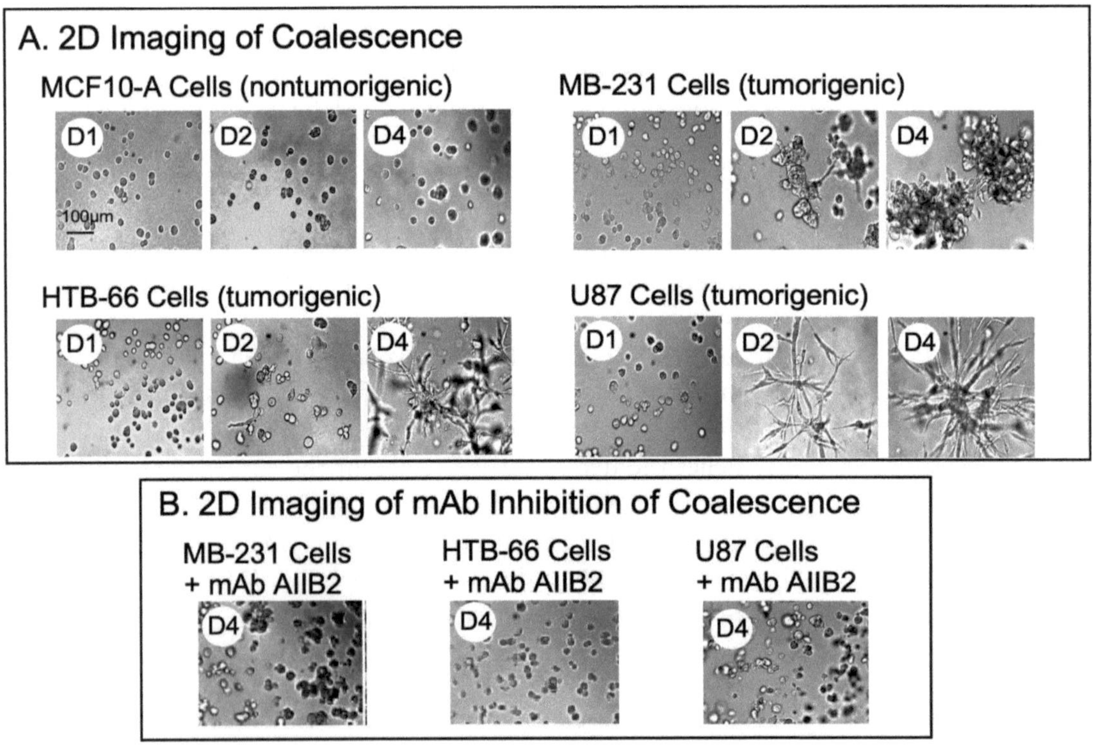

Fig. 2 Imaging the 2D coalescence assay. (**a**) 2D images of cells in the coalescence assay in a 96-well plate show that non-tumorigenic cells such as MCF-10A do not coalesce. In contrast, tumorigenic cells from the breast cancer cell line MB-231, the melanoma cell line HTB-66, and the glioblastoma cell line U87 form aggregates and cell bridges and exhibit coalescence. (**b**) AIIB2, an anti-β1 integrin mAb, inhibits coalescence in all three tumorigenic cell lines shown in **a**. D1, day 1; D2, day 2; D4, day 4

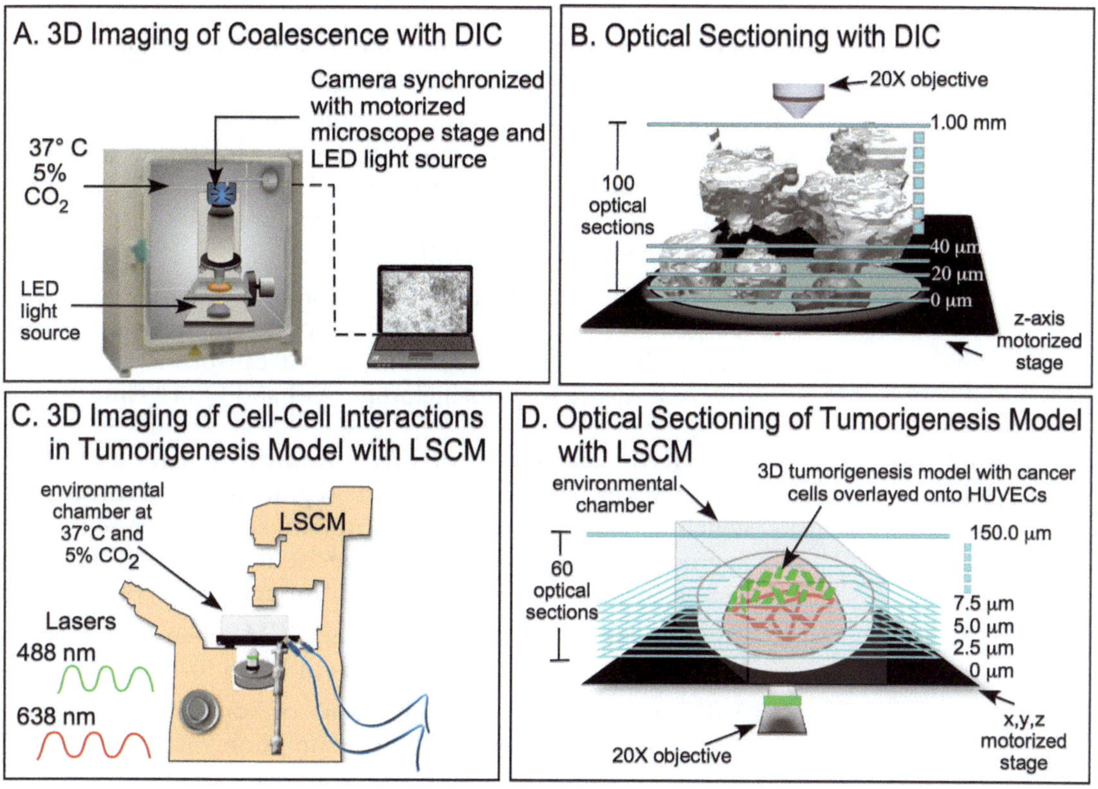

Fig. 3 Optical sectioning of the coalescence and tumorigenesis model with DIC and LSCM. (**a**) The 3D preparation can be maintained for several weeks during optical sectioning by placing a microscope with a motorized *z*-stage in an incubator at 37 °C and 5% CO_2. (**b**) Using the method described here, one can optically section through up to 1.00 mm of sample. (**c**) GFP-tagged cells and/or vitally stained cells placed in an environmental chamber can be optically sectioned with LSCM. (**d**) Optical sections can be obtained at 2.5μm increments with LSCM

3.6.2 3D Imaging and Optical Sectioning with DIC Microscopy

1. Place a microscope with a motorized *z*-stage and long-working distance 20× DIC optics in an incubator maintained at 37 °C and 5% CO_2 or in an environmental chamber if available (Fig. 3a) (*see* **Note 17**).
2. Set the *z*-range, the *z*-height increment, the time interval between consecutive *z*-series, and the total acquisition time in the software controls (Fig. 3b) (*see* **Note 18**).
3. Save images in JPEG format for export into other image analysis software such as J3D-DIAS.

3.6.3 LSCM

1. Place the dish in an environmental chamber with temperature, humidity, and CO_2 controls (Fig. 3c).
2. Select lasers for maximum excitation/emission of fluorophores with a minimum of overlap in their respective spectra.

3. Set the *z*-range, the *z*-height increment, the time interval between consecutive *z*-series, and the total acquisition time in the software controls (Fig. 3d) (*see* **Note 19**).
4. Proprietary file formats can, in some cases, be saved directly in JPEG format. In other cases, it may be necessary to open the files in ImageJ (https://imagej.net/ImageJ) and convert them into JPEG stacks. This step is required for import into J3D-DIAS 4.2 (*see* **Note 20**).
5. Images from each channel can be saved as separate files or as merged files in ImageJ. ImageJ will perform *z*-projections and merge channels into a 2D movie that can be saved in many different movie formats including AVI and QuickTime.

3.7 Segmentation Using J3D-DIAS

1. In order to reconstruct objects and calculate data, J3D-DIAS 4.2 first performs global image segmentation, a process in which an image is converted into similar parts or segments. Segmentation allows for a more precise analysis than that which can be performed on the whole, or unsegmented, object [13].
2. To segment images collected on a LSCM, open JPEG files in J3D-DIAS 4.2 and save as a movie in DIAS format.
3. Automatic bitmap segmentation of optical sections in J3D-DIAS 4.2 is based on the assumption that pixels within an object will exhibit a greater grayscale variation relative to each other than pixels comprising background or out-of-focus objects. Pixels within a matrix containing a high degree of variation are therefore retained, while background is discarded. Bitmap segmentation is preferable for aggregates (Fig. 4a, b) and structures such as a HUVEC network (Fig. 5a) because gaps and spaces are preserved [12].
4. Automatic threshold segmentation is performed similarly to bitmap segmentation except that the interior information is discarded in the former and only an outline is retained [12] (Fig. 5b).
5. The user can manually edit both bitmap and threshold-segmented objects (Fig. 5c).

3.8 Reconstructions with J3D-DIAS

1. Overlapping outlines, obtained from either bitmap or threshold segmentation of DIC images, are stacked in the *z*- axis and smoothed (Fig. 6a).
2. Overlapping outlines, obtained from either bitmap or threshold segmentation of separate LSCM channels, are stacked in the *z*-axis and smoothed and can then be recombined or analyzed separately(Fig. 6b) (*see* **Note 21**).
3. Bitmap pixels are expanded into voxels (3D pixels) and the voxel blocks are wrapped into a continuous surface. Voxel dimensions are determined by the *x*, *y*, and *z* scales.

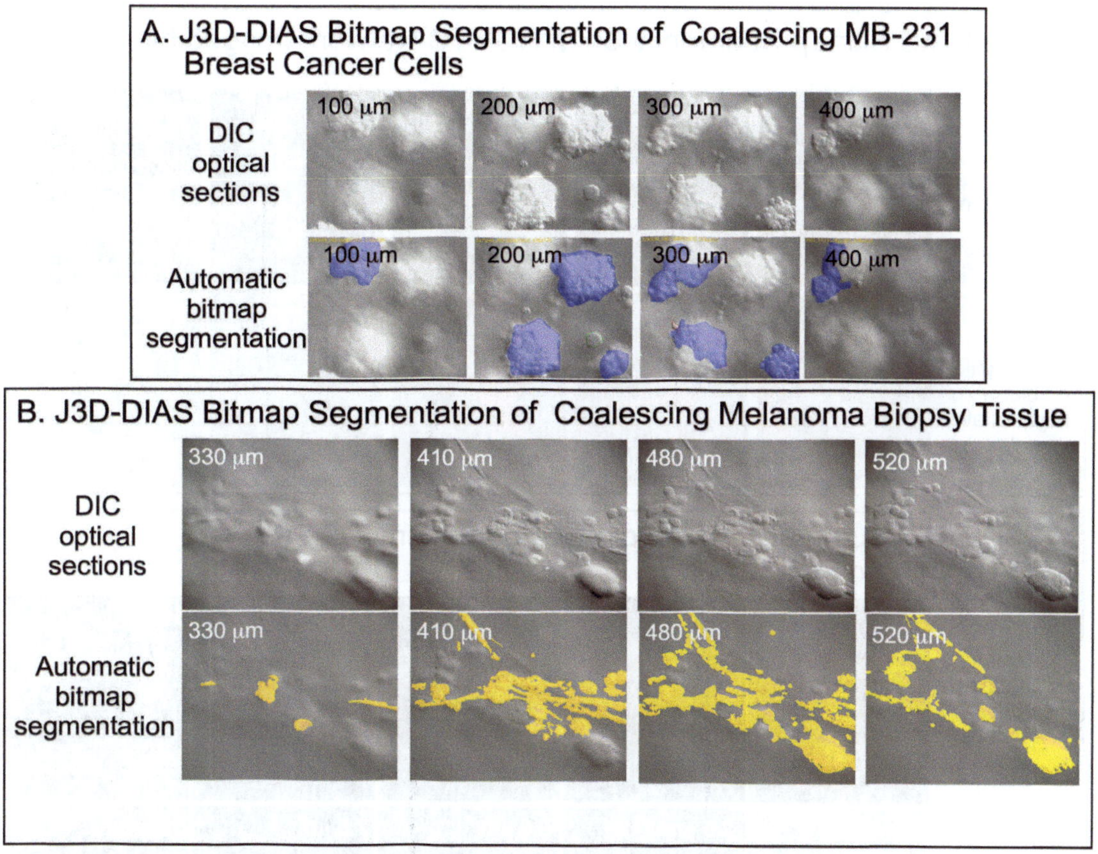

Fig. 4 Automated bitmap segmentation with J3D-DIAS 4.2. (**a**) Representative examples at 100μm increments of optical sections obtained with DIC microscopy of coalescing MB-231 breast cancer cells embedded in Matrigel™ matrix and the corresponding automatic bitmap segmentation (purple). (**b**) Representative examples of optical sectioning with DIC microscopy of coalescing melanoma cells from biopsy tissue embedded in Matrigel™ matrix and corresponding automatic bitmap segmentation (yellow)

4. Pixels detected by the threshold algorithm are replaced using a beta-spline replacement algorithm that approximates the linear outline with curves, in effect smoothing the outline to generate a mathematical model for data calculation.

5. In J3D-DIAS 4.2, Java OpenGL was replaced with the application programming interface (API) known as Metal. Metal provides near direct communication with the graphics chip to accelerate computations and access more recent hardware innovations. A faceted surface from the beta-splines is built using the "marching cubes" algorithm [14].

6. Coalescence of cells from a breast cancer cell line can be observed in 4D reconstructions. In this case, continuous optical sectioning at 10-min intervals was performed with DIC for 1 week (Fig. 7a).

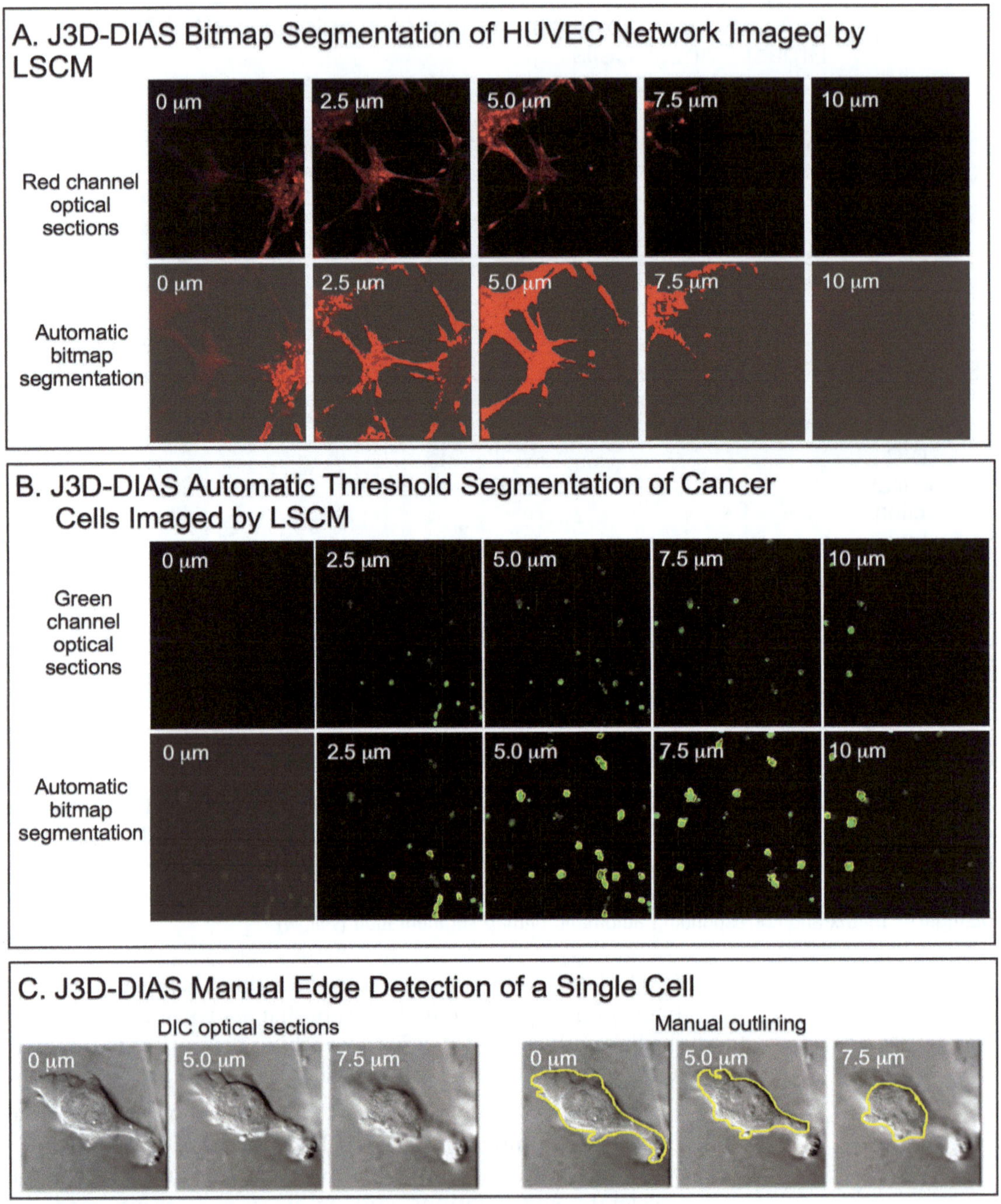

Fig. 5 J3D-DIAS 4.2 performs automatic segmentation of images obtained with a LSCM and allows editing of segmented objects. (**a**) Representative examples of optical sections obtained with LSCM at 2.5μm increments of a Deep Red-dyed HUVEC network imaged with a 638 nm laser line. Images from each channel were converted from the proprietary format to JPEGs, imported into J3D-DIAS 4.2, and segmentation performed by the automatic bitmap method. (**b**) Representative examples of optical sections obtained with a LSCM of GFP-tagged MB-231 breast cancer cells imaged with a 488 nm laser line. The images were converted to JPEGs, imported into J3D-DIAS 4.2, and segmentation performed by the automatic threshold method. The cancer cells were layered over the HUVECs shown in panel A and corresponding levels are shown. (**c**) Manual edge detection by the threshold method in DIC optical sections of a single cell

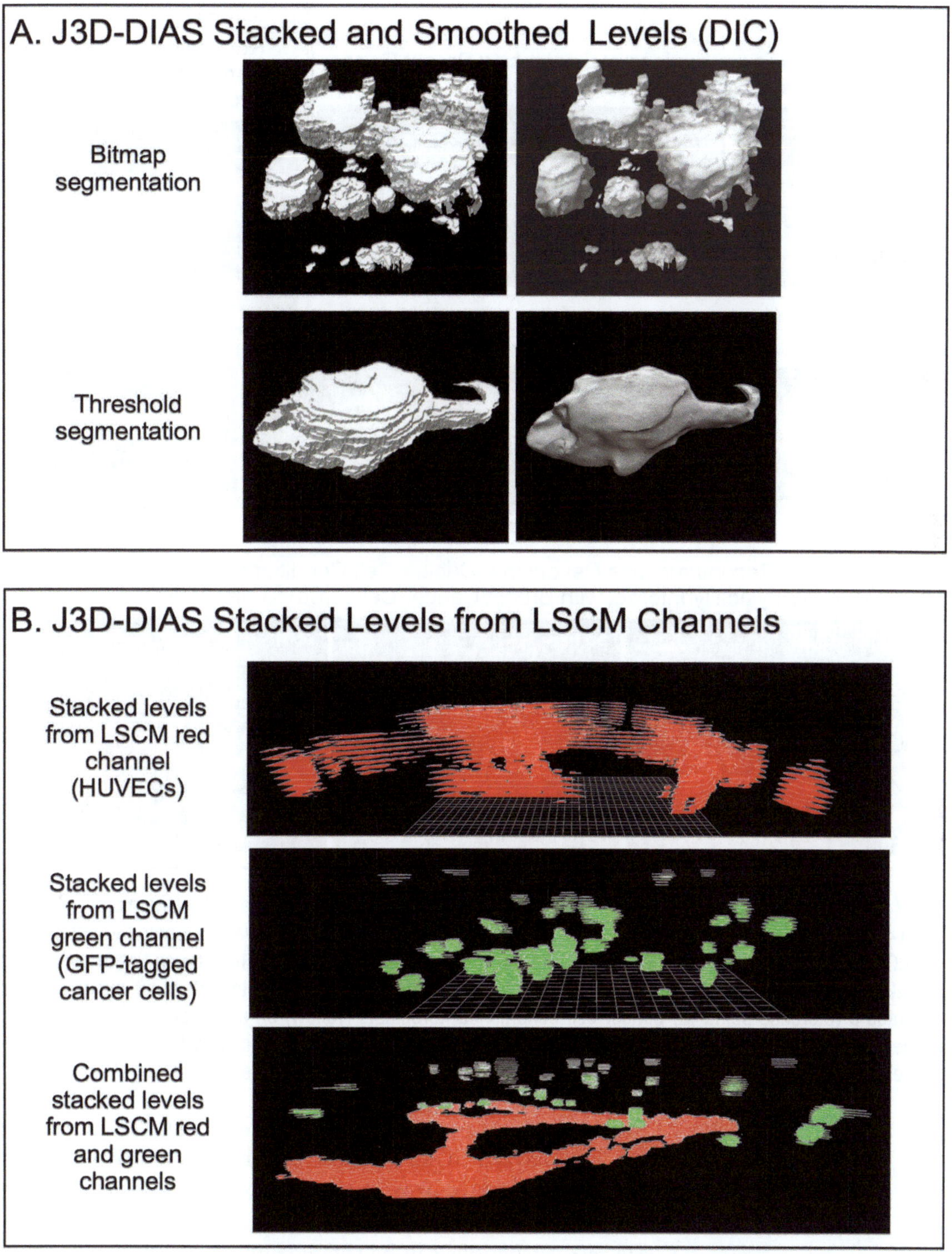

Fig. 6 Stacking and smoothing of segmented images obtained from DIC microscopy and LSCM. (**a**) Stacking and smoothing of segmented DIC optical sections. (**b**) Images from the red and green channels of an LSCM were imported as separate files into J3D-DIAS 4.2. HUVECs (red) and cancer cells (green) were segmented by the bitmap and threshold methods, respectively, and then merged

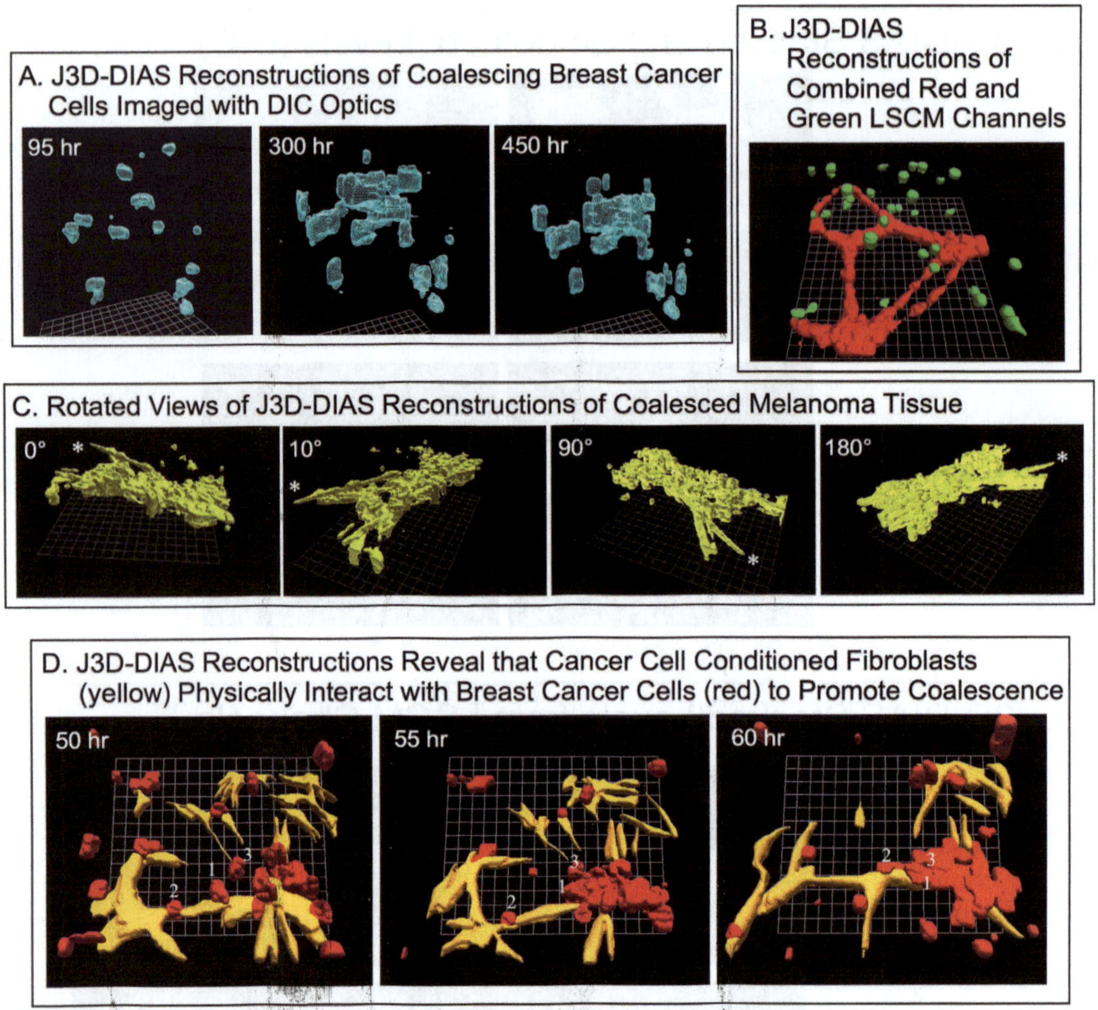

Fig. 7 J3D-DIAS 4.2 4D reconstructions. (**a**) Coalescing breast cancer cells imaged with DIC microscopy from 95 to 450 h. (**b**) GFP-tagged breast cancer cells embedded in Matrigel™ were overlayed onto Deep Red-dyed HUVECs and the preparation imaged with LSCM. (**c**) Rotated views of coalesced melanoma tissue embedded in Matrigel™ and imaged with DIC microscopy. Rotational (swivel) angle is in the top left panel and an asterisk indicates a reference point. (**d**) Cancer cell conditioned fibroblasts (yellow) on a collagen surface attach to cancer cells (red) embedded in Matrigel™ and carry the cancer cells into a coalescing aggregate. Three cancer cells are numbered. Optical sectioning was performed with DIC microscopy

7. Behavior of cancer cells in the presence of a HUVEC network can be analyzed in 4D reconstructions of merged channels. Optical sectioning was performed with LSCM for 72 h (Fig. 7b).
8. 4D reconstructions can be rotated and viewed from any angle (Fig. 7c).
9. 4D reconstructions of optical sections acquired with DIC microscopy reveal that cancer cell conditioned fibroblasts on a

collagen-coated surface physically interact with breast cancer cells to promote coalescence. The breast cancer cells were embedded in Matrigel™ and overlaid onto the fibroblasts, illustrating here the expanded version of the model (Fig. 7d).

10. Comparison of 4D reconstructions in untreated breast cancer cells (Fig. 8a) and breast cancer cells treated with a mAb against β1 integrin (Fig. 8b) reveals mAb inhibition of coalescence. The model was optically sectioned at 10-min intervals using DIC optics for over 17 days.
11. Comparison of 4D reconstructions in untreated melanoma biopsy tissue (Fig. 8c) and melanoma biopsy tissue treated with mAb H4C4 against CD44 (Fig. 8d) reveals mAb inhibition of coalescence. The model was optically sectioned at 10-min intervals using DIC optics for 2 days.

3.9 J3D-DIAS 4.2 Quantitative Analysis of Cell Behavior

1. J3D-DIAS 4.2 calculates 3D cell migration parameters such as velocity, persistence, and direction change from the position of the 3D cell center or centroid. The centroid position at each time point is the average of the *x*-, *y*-, and *z*-coordinates of the interior points of the faceted object. The sequence of centroid positions over time is saved in the 4D path file (Fig. 9a). The central difference method is used to calculate instantaneous velocity [15].
2. J3D-DIAS 4.2 calculates 3D dynamic morphology parameters (volume, roundness, length, and surface area) from the faceted reconstructions [9] (Fig. 9b) and creates a text file to import into other applications such as Excel for the user to perform statistical and graphic operations.

4 Notes

1. Ideally, biopsy material should be obtained from a local hospital and transported to the lab in MCF medium. In our hands, viability is improved if the medium is at room temperature or 37 °C, rather than chilled on ice. Also, it may be necessary to work with the pathologist to improve the chances that tissue will be kept sterile during processing.
2. Matrigel™ is extremely temperature sensitive and will gel rapidly at room temperature. The stock bottle should be stored at −20 °C and thawed on ice at 4 °C overnight prior to aliquoting. Aliquots should be thawed on ice and kept on ice during use. Aliquots should not be thawed and refrozen.
3. Preservatives can be removed from antibody solutions by affinity purification using a protein A, G, or L spin trap column. The specific type of protein column to use depends on the antibody

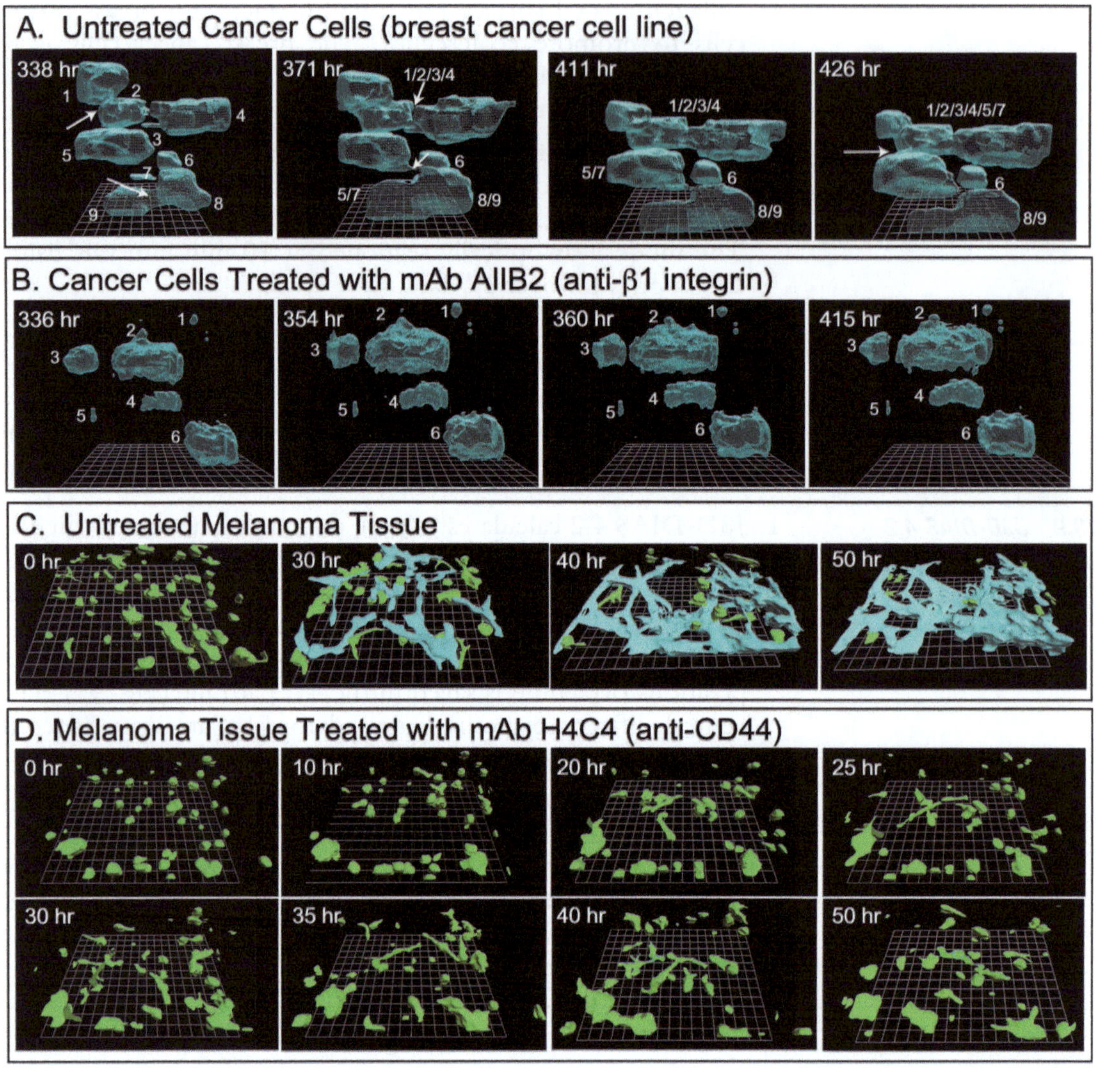

Fig. 8 Monoclonal antibody inhibition of coalescence in the 3D assay. (**a**) Untreated breast cancer cells embedded in Matrigel™ undergo coalescence. Individual aggregates are numbered and numbers are separated by slashes to identify aggregates that coalesced. Time is shown in the upper left of each panel. Optical sectioning was performed with DIC microscopy for 426 h. (**b**) Coalescence is inhibited in breast cancer cells treated with the anti-β1 integrin mAb AIIB2. Aggregates are numbered and time is indicated as described for panel **a**. (**c**) Melanoma tissue embedded in Matrigel™ undergoes rapid coalescence. Single cells are color-coded green and coalesced aggregates are color-coded blue. Optical sectioning was performed with DIC microscopy. (**d**) Melanoma tissue embedded in Matrigel™ and treated with the anti-CD44 mAb H4C4 fails to coalesce. Optical sectioning was performed with DIC microscopy

isotype and the antibody species. This information is available through the supplier and should be determined before purchase.

4. Use a low-magnification objective for a large field of view. Exact magnification depends on magnification of the camera

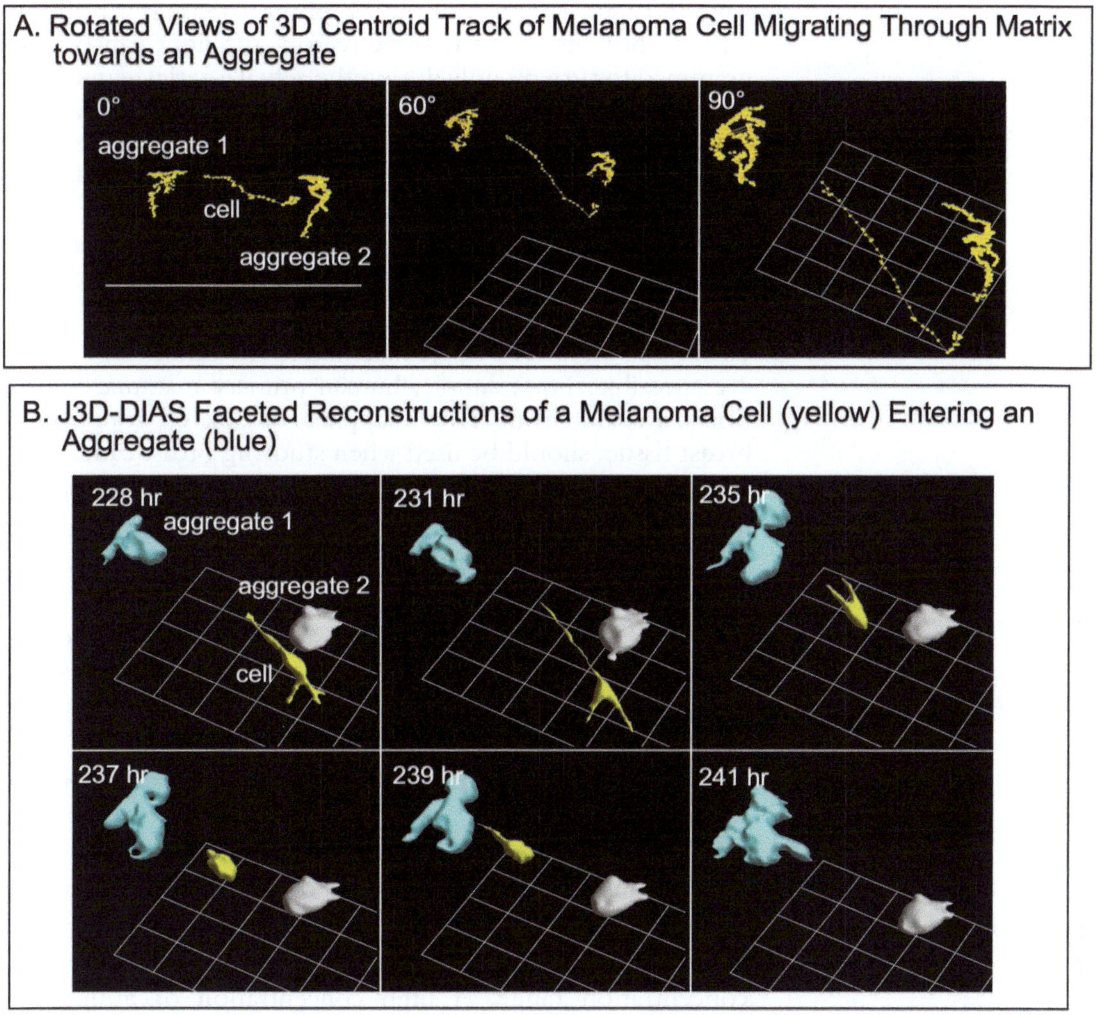

Fig. 9 4D data is calculated from the 4D path file. (**a**) Centroid track (yellow) viewed at different angles (upper left of the panel) of a melanoma cell migrating through the matrix toward an existing aggregate of melanoma cells. (**b**) Color-coded reconstructions of the cell and aggregates shown in **a**

faceplate, so a snapshot of a stage micrometer may be necessary to determine the actual magnification.

5. The working distance of an upright microscope and the z-axis range required for optical sectioning of the sample should be determined prior to setting up the experiment. Low-height dishes, such as those currently available from iBidi, may be necessary in order to optically section within a limited working distance. Lids with glass inserts are necessary for DIC imaging and can also be purchased from iBidi.

6. Controls should be the same cell type or tissue type as the cancer cells or cancer tissue.

7. An LED light source will reduce phototoxicity during long-term recordings. For the same reason, the light can be programmed to turn on only during the actual *z*-series acquisition. An initial test run over a period of several hours is advisable to ensure synchronization of the light, camera, and stage. Synchronization errors accumulate, eventually resulting in blank or black video frames. Also, it may be necessary to allow a brief delay between the time the light is switched on and the onset of image acquisition because the light may take 1 or 2 s to ramp up to full illumination.
8. Fibroblasts should be tissue matched to the cancer type whenever possible. For example, human primary mammary fibroblasts, available from various suppliers or isolated from normal breast tissue, should be used when studying breast cancer.
9. Use only fibroblast-specific medium with the appropriate growth factors.
10. If GFP-tagged cells are not being used, then cancer cells can be dyed with a vital dye. Care must be taken in selection of the dye so that the excitation and emission spectra of the dyes used for the HUVECs and cancer cells are well-separated. It is also imperative to check the laser lines on the LSCM to be sure the appropriate lasers are installed.
11. Use only endothelial cell-specific medium with the appropriate growth factors.
12. A Matrigel™ cushion is necessary to prevent the cells from settling onto the 2D plastic surface rather than remaining embedded in the 3D matrix.
13. Antibodies in the coalescence assay should be tested over a concentration range. A high concentration of 500μg/mL should show blocking activity if present. We typically test a concentration range between 0.3 and 250μg/mL [4].
14. In the coating method described here, the collagen fibers attach to the surface of the coverslip during the 1-h incubation period. Therefore, to avoid scratching the attached fibers when rinsing, it is advisable to remove the excess liquid by tilting the dish and placing the pipette tip on the perimeter of the coverslip.
15. Use a sterile cotton swab to remove media from the rim of the coverslip. This prevents the surface tension from pulling the cell suspension off the coverslip and into the dish.
16. Different cancer cell types exhibit different morphologies during coalescence.
17. Optical sectioning with DIC has the advantage of allowing long-term recording without deleterious phototoxic effects [2, 16]. For example, we have performed continuous optical sectioning using the method described here for 30 days [2].

18. The z-height will vary with the sample, but generally with the optics used here, it is possible to section through up to 1.00–1.5 mm of sample at intervals of 5–10μm. It may be necessary to begin below the first focal plane of the z-range and continue through a few slices above the z-range to accommodate any z-axis drift.
19. If the LSCM has an x, y motorized stage, multiple fields can be acquired. The minimum time interval between consecutive z-series will be a function of the scan rate, image size, and sample height. For most applications, an interval of 10 min is sufficient to get accurate data.
20. ImageJ [17] is freely available, open-source image processing software. FIJI (https://imagej.net/Learn.html#What_is_Fiji.3F) is the version recommended for the types of applications described here.
21. Images from LSCM channels are imported separately into J3D-DIAS 4.2 to permit bitmap segmentation of one channel and threshold segmentation of the other. After the segmentation step, the channels can be recombined.

Acknowledgments

These studies were supported by the Developmental Studies Hybridoma Bank (DSHB), a National Resource created by NIH and housed at the University of Iowa.
The monoclonal antibodies AIIB2 and H4C4 were obtained from the DSHB. AIIB2 was developed by C.H. Damsky. H4C4 was developed by J.T. August and J.E.K. Hildreth.

References

1. Ambrose J, Livitz M, Wessels D, Kuhl S, Lusche DF, Scherer A, Voss E, Soll DR (2015) Mediated coalescence: a possible mechanism for tumor cellular heterogeneity. Am J Cancer Res 5(11):3485–3504
2. Scherer A, Kuhl S, Wessels D, Lusche DF, Hanson B, Ambrose J, Voss E, Fletcher E, Goldman C, Soll DR (2015) A computer-assisted 3D model for analyzing the aggregation of tumorigenic cells reveals specialized behaviors and unique cell types that facilitate aggregate coalescence. PLoS One 10(3): e0118628. https://doi.org/10.1371/journal.pone.0118628
3. Wessels D, Lusche DF, Voss E, Kuhl S, Buchele EC, Klemme MR, Russell KB, Ambrose J, Soll BA, Bossler A, Milhem M, Goldman C, Soll DR (2017) Melanoma cells undergo aggressive coalescence in a 3D Matrigel model that is repressed by anti-CD44. PLoS One 12(3): e0173400. https://doi.org/10.1371/journal.pone.0173400
4. Lusche DF, Klemme MR, Soll BA, Reis RJ, Forrest CC, Nop TS, Wessels DJ, Berger B, Glover R, Soll DR (2019) Integrin α-3 ß-1's central role in breast cancer, melanoma and glioblastoma cell aggregation revealed by antibodies with blocking activity. mAbs:null-null. https://doi.org/10.1080/19420862.2019.1583987
5. Hughes CS, Postovit LM, Lajoie GA (2010) Matrigel: a complex protein mixture required for optimal growth of cell culture. Proteomics 10(9):1886–1890. https://doi.org/10.1002/pmic.200900758

6. Heaphy CM, Griffith JK, Bisoffi M (2009) Mammary field cancerization: molecular evidence and clinical importance. Breast Cancer Res Treat 118(2):229–239. https://doi.org/10.1007/s10549-009-0504-0
7. Slaughter DP, Southwick HW, Smejkal W (1953) Field cancerization in oral stratified squamous epithelium; clinical implications of multicentric origin. Cancer 6(5):963–968
8. Wessels DJ, Pradhan N, Park YN, Klepitsch MA, Lusche DF, Daniels KJ, Conway KD, Voss ER, Hegde SV, Conway TP, Soll DR (2019) Reciprocal signaling and direct physical interactions between fibroblasts and breast cancer cells in a 3D environment. PLoS One 14 (6):e0218854. https://doi.org/10.1371/journal.pone.0218854
9. Kuhl S, Voss E, Scherer A, Lusche DF, Wessels D, Soll DR (2016) 4D Tumorigenesis model for quantitating coalescence, quantitating directed cell motility and chemotaxis, identifying unique cell behaviors and testing anticancer drugs. In: Hereld D, Jin T (eds) Chemotaxis: methods and protocols. Springer
10. Soll DR, Voss E, Johnson O, Wessels D (2000) Three-dimensional reconstruction and motion analysis of living, crawling cells. Scanning 22 (4):249–257
11. Soll DR, Wessels D, Heid PJ, Voss E (2003) Computer-assisted reconstruction and motion analysis of the three-dimensional cell. Scientific World Journal 3(Journal Article):827–841. https://doi.org/10.1100/tsw.2003.70
12. Wessels DJ, Lusche DF, Kuhl S, Scherer A, Voss E, Soll DR (2016) Quantitative motion analysis in two and three dimensions. Methods in molecular biology 1365:265–292. https://doi.org/10.1007/978-1-4939-3124-8_14
13. Kaur DK, V. (2014) Various image segmentation techniques: a review. Int J Comput Sci Mob Comput 3(5):809–814
14. Lorensen WE, Cline HE (1987) Marching cubes: a high resolution 3D surface construction algorithm. ACM SIGGRAPH Comput Graph 21:163–169
15. Maron MJ (1982) Numerical analysis. New York: Macmillan Publishers
16. Dixit R, Cyr R (2003) Cell damage and reactive oxygen species production induced by fluorescence microscopy: effect on mitosis and guidelines for non-invasive fluorescence microscopy. Plant J 36(2):280–290
17. Schneider CA, Rasband WS, Eliceiri KW (2012) NIH Image to ImageJ: 25 years of image analysis. Nat Methods 9(7):671–675

Chapter 15

Melanosome Motility in Fish Retinal Pigment Epithelial (RPE) Cells

Christina King-Smith

Abstract

Several model systems have been developed to investigate mechanisms and regulation of intracellular organelle motility. The fish retinal pigment epithelial (RPE) cell represents an unusual but simple system for the study of actin-dependent organelle motility. Primary cultures of RPE dissociated from the eye are amenable to motility studies using a simple perfusion chamber and conventional phase contrast microscopy. In vivo, melanin-containing pigment granules (melanosomes) within fish RPE migrate distances up to 100 μm in response to light flux. When sheets of RPE are removed from the eye and dissociated, they attach to the substrate with apical projections extending radially from the central cell body. Melanosomes can be chemically triggered to aggregate or disperse throughout the projections. Melanosome migration in RPE apical projections is dependent on actin filaments and thus renders this model system useful for investigations of actin-dependent organelle motility.

Key words RPE, Actin, Pigment granules, Melanosomes, Organelle transport, Microtubules, Motor proteins

1 Introduction

Retinal pigment epithelial (RPE) cells are located at the back of the vertebrate eye. In eyes of fish and other lower vertebrates, RPE melanosomes (pigment granules) undergo light-dependent migration, dispersing out into the cells' long apical projections in the light, and aggregating back into the cell body in darkness. Fish do not have dilatable pupils to control light that enters the retina. Rather, the moveable curtain of RPE melanosomes, together with elongation and contraction of rod and cone photoreceptors, serve to modulate light flux and are collectively called retinomotor movements [1].

RPE cells can be isolated from the eye, dissociated into single cells, and cultured for up to 24 h. In vitro, RPE cells lose their apical-basal polarity, become radially symmetrical, and attach to the

Ray H. Gavin (ed.), *Cytoskeleton: Methods and Protocols*, Methods in Molecular Biology, vol. 2364, https://doi.org/10.1007/978-1-0716-1661-1_15,

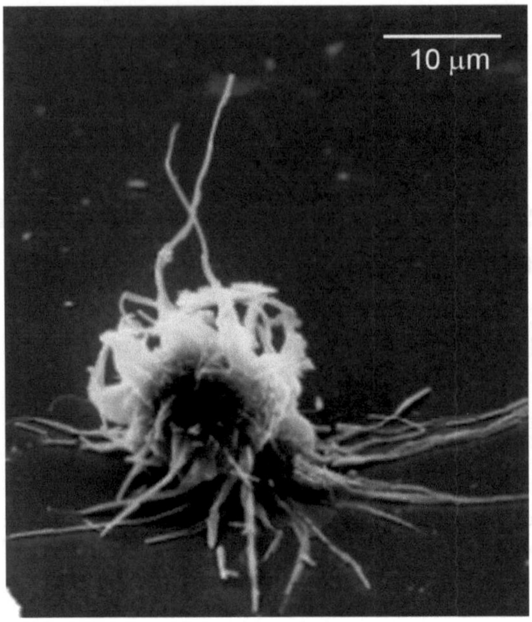

Fig. 1 Scanning electron micrograph of an isolated RPE cell. Formerly apical projections extend from the central cell body of dissociated cells

substrate, with the formerly apical projections extending outward from the central, pigment-filled cell body (Fig. 1). Melanosome aggregation within the (formerly) apical projections can be triggered by application of underivatived cAMP [2]. Dispersion is stimulated by washout of cAMP in medium containing dopamine (Fig. 2).

Time-lapse videomicrography of melanosome motility within apical projections of dissociated RPE cells is easily carried out using a simple perfusion chamber on an upright, phase contrast microscope. Time-lapse videos reveal that patterns of melanosome motility differ during aggregation (retrograde motility) and dispersion (anterograde motility). During dispersion, melanosome motility is bidirectional and saltatory, at rates of 2–3 μm/min in the anterograde direction. In contrast, aggregation is continuous and unidirectional, also at 2–3 μm/min [3]. Studies using cytoskeletal inhibitors, time-lapse and fluorescent microscopy, and quantification of pigment position in fixed cells have demonstrated that while microtubules are plentiful in apical projections, they are not required for melanosome motility. Rather, actin filaments, also plentiful in the RPE cell body and apical projections (Fig. 3), are necessary and sufficient to support melanosome motility in the apical projections of isolated cells, with microtubules supplying structural support [3, 4]. Studies using myosin S1 labeling and platinum replica shadowing have demonstrated that actin filaments within RPE apical projections are uniformly oriented with barbed

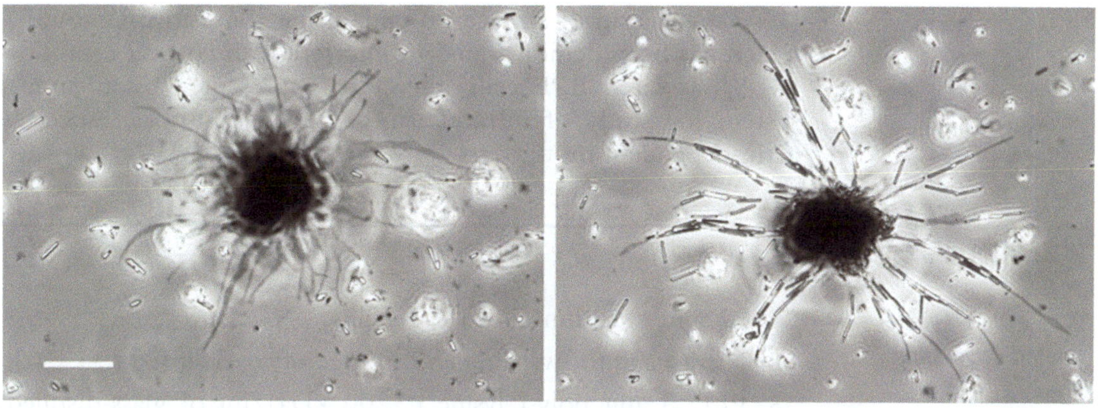

Fig. 2 Isolated RPE cells with aggregated and dispersed melanosomes. The cell on the left was treated with cAMP stimulating migration of melanosomes into the central cell body. Apical projections devoid of melanosomes appear phase-gray. The cell on the right has dispersed melanosomes; these appear phase-bright within the projections. Tips of projections are free of melanosomes because of the bidirectional shuttling of the melanosomes once full dispersion is achieved. Debris surrounding cells are melanosomes from lysed cells. Note melanosomes are mostly cylindrical, not round. Bar: 10 μm

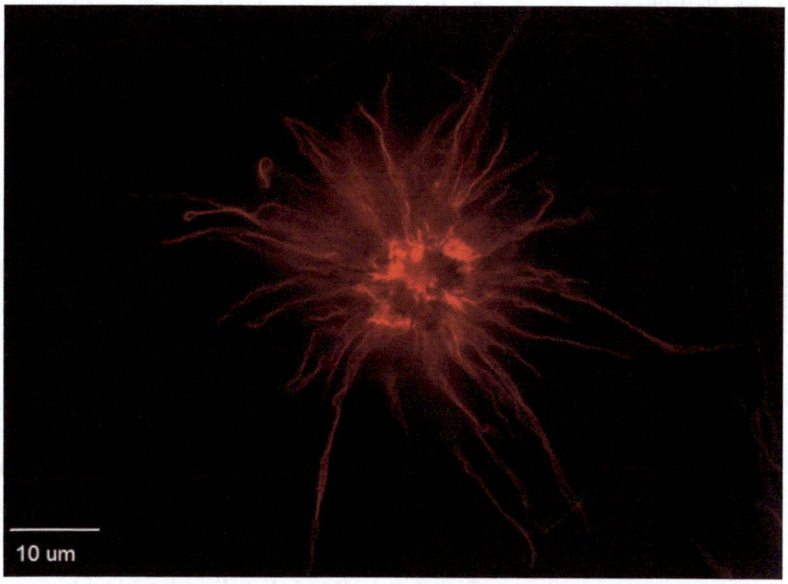

Fig. 3 Isolated RPE cell treated with cAMP to aggregate melanosomes and subsequently fixed and labeled with rhodamine phalloidin revealing F-actin in the cell body and apical projections

ends at the distal tips [5]. Previous work has suggested that melanosome aggregation may rely on retrograde actin flow, based on sensitivity of melanosome aggregation but not dispersion to treatments expected to interfere with actin retrograde flow, including (1) inhibition of myosin II and Rho kinase (ROCK) [6] and

(2) cross-linking of surface proteins using the lectin, ConA [5], which has been demonstrated to interfere with cortical actin flow [7, 8]. Isolated RPE cells, therefore, represent a useful model system for study of intracellular organelle transport, especially actin-dependent motility.

2 Materials

2.1 Poly-L-Lysine-Coated Coverslips

1. 22 mm × 22 mm No. 1 coverslips.
2. Nitric acid: for washing and storage of coverslips.
3. PLL: 1 mg poly-L-lysine (>200 kDa)/ml of glass-distilled, deionized water (ddH$_2$O). Store dry poly-L-lysine at −20 °C desiccated; store PLL solution at −20 °C; it can be reused four times.

2.2 Dissection and Isolation of RPE (Retinal Pigment Epithelial) Sheets

1. Green sunfish (*Lepomis cyanellus*) or bluegill (*Lepomis macrochirus)* 2–4 in. in length, commonly available at fish hatcheries. Maintain on 12 h/12 h light/dark cycle.
2. MS-222 (3-aminobenzoic acid ethyl ester, methanesulfonate salt) anesthetic: 10 g MS-222/L solution in ddH$_2$O. Aliquot in volumes sufficient for dilution to 75 mg/L in fish tank water, and freeze aliquots at −20 °C.
3. Dark adaptation box with aeration.
4. Iris spring scissors (14 mm with curved blades) and forceps (Dumont #3).
5. HEPES-buffered Earles' Ringer (HER): 116.3 mM NaCl, 5.4 mM KCl, 1.8 mM CaCl$_2$, 0.8 mM MgSO$_4$, 1.0 mM Na$_2$HPO$_4$, 25.5 mM glucose, 4.0 mM NaHCO$_3$, 21 mM HEPES, pH 7.2. Prepare using commercially available 10× Earles' balanced salt solution (10× EBSS), supplemented with phenol red in ddH$_2$O, 0.36 g/100 ml glucose, 0.034 g/100 ml NaHCO$_3$, and 0.50 g/100 ml HEPES. Make fresh or less than 24 h before use.
6. 35 mm Petri dish lined with soft dental plastic or parafilm.
7. 90 mm Petri dish lined with white silicon rubber or parafilm.

2.3 Preparation of Dissociated RPE Cells

1. Calcium- and magnesium-free HER (CMFHER): add 1 mM EGTA, 0.36 g/100 ml glucose, 0.034 g/100 ml NaHCO3, and 0.50 g/100 ml HEPES to 10× calcium- and magnesium-free EBSS. Solution pH is 7.4 (*see* **Note 1**).
2. Papain digest solution: 10 U/ml papain in 5 ml CMFHER activated with 0.14 mg/ml DNase and 3 μM L-cysteine. Papain suspension is aliquoted and stored at 4 °C. DNase and cysteine solutions are prepared fresh using 16.6 mg L-cysteine in 500 μl

ddH_2O, and 10 mg DNase in 500 μl ddH_2O, and kept on ice. Five milliliters papain digest solution is prepared per fish. Add sufficient papain and L-cysteine so that the final solution contains 10 U/ml papain, 64 μl L-cysteine, and 36 μl DNase (*see* **Note 2**).

3. 35 mm Petri dish lined with parafilm.
4. 12–15 ml glass conical centrifuge tube.
5. HERB: HER supplemented with 0.4% BSA.
6. HERB supplemented with 0.1 mg/ml DNase.
7. Poly-L-lysine-coated coverslips in 35 mm Petri dishes.
8. 100 cm glass Petri dish with moistened filter paper (Whatman #1) in the cover.
9. 1 ml syringe with 26 5/8 gauge needle.

2.4 Stimulation and Observation of Melanosome Motility

1. 1 mM cAMP in HERB: Made fresh from 10 mM cAMP/HERB stock stored at 20 °C.
2. 10 μM dopamine (3-hydroxytyramine) in HERB made fresh. Dopamine oxidizes easily; solutions should be remade after 1 h.
3. For perfusion chamber: standard microscope slides, double stick tape, wedges of filter paper (Whatman #3).

3 Methods

3.1 Preparation of Poly-L-Lysine-Coated Coverslips

1. Soak coverslips in nitric acid for a few hours to overnight (*see* **Note 3**).
2. Wash coverslips in running tap water for 20–60 min, and rinse 5× in deionized water.
3. Remove coverslips from racks, and place in 10 cm Petri dish with 1 mg/ml poly-L-lysine solution (PLL).
4. Rock or rotate for at least 1 h at room temperature.
5. Remove PLL, and wash coverslips in ddH_2O at least 10× (*see* **Note 4**).
6. Air-dry on filter paper (Whatman # 1); coverslips can be stored for several months.

3.2 Dissection and Isolation of RPE Sheets

1. Thaw MS-222 and add to fish tank water in dark adaptation box (*see* **Note 5**).
2. Add fish and dark adapt 30 min to 2 h.
3. In dim light (0.002 $J/s/m^2$; *see* **Note 6**), enucleate eyes by blunt dissection using closed scissors; cut the optic nerve. Place eyes on a paper towel lens side down.
4. Hemisect the first eye; remove lens.

5. Holding edge of posterior eyecup with forceps (12 o'clock position) invert the eye cup retinal side down onto a paper towel. Gently press the back of the eyecup (sclera side) downward onto the paper towel with the closed scissor blades so that the retina adheres to the paper towel; snip the optic nerve to detach the retina from the eyecup.
6. Transfer the eyecup to the dissection dish; hold the edge of the eyecup with forceps.
7. Gently irrigate the RPE/choroid with HER buffer using a Pasteur pipette to remove sheets of RPE (1–2 mm in diameter). RPE sheets look like black flecks in the medium (*see* **Note 7**).
8. Transfer RPE sheets to approximately 5 ml HER in a wax-lined, 35 mm culture dish using a wide bore pipette or a glass 5 or 10 ml serological pipette (*see* **Note 8**).
9. Place the 35 mm dish containing sheets on a rotator at 30 rpm.
10. Repeat **steps 4–9** with the other eye; pool sheets into one dish.

3.3 Preparation of Dissociated RPE Cells

1. Transfer pooled RPE sheets into the papain solution in minimal volume of medium using a wide-bore or 10 ml serological pipette. Incubate for 30 min while rotating at 30 rpm.
2. After the digest period, transfer sheets to 3 ml HERB + DNase in a 12–15 ml glass conical centrifuge tube.
3. Allow sheets to settle to the bottom of the tube, and then gently resuspend tissue using a Pasteur pipette. Aspirate off debris as tissue is resuspended. Remove solution and repeat in two more changes of 3 ml HERB + DNase (*see* **Note 9**).
4. Using a 5 ml serological pipette, transfer washed sheets to approximately 1 ml of HERB in a 1.5 ml microcentrifuge tube. Thoroughly wet a 1 ml syringe with 26 5/8 gauge needle by passing HERB in and out of the syringe a few times. Slowly dissociate sheets in the microcentrifuge tube by six to eight slow passes through the needle (three to four times in and out). Avoid air bubbles.
5. After dissociation, add 200–400 ml HERB-DNase to dilute the cell suspension. Optimal cell density is about 7.5×10^4 cells/ml (*see* **Note 10**).
6. Plate cells immediately after dissociation by adding 75–100 μl of the cell suspension to completely dry poly-L-lysine-coated coverslips in 35 mm dishes. Dishes should be placed in 100 cm glass Petri plates, which have moistened filter paper in their covers, and left undisturbed for 30 min to 1 h or until cells adhere (*see* **Note 11**).
7. After cells have adhered, change the medium to fresh HERB containing 10 μM dopamine to achieve full melanosome dispersion. Aspirate the old medium from the coverslip using a

Pasteur pipette and immediately replace with 100 μl fresh medium. Repeat twice more for a total of three medium changes.

3.4 Stimulation and Observation of Melanosome Motility

1. To stimulate melanosome aggregation, change medium on coverslips to 1 mM cAMP in HERB using three changes each of 100 μl. Full melanosome aggregation is achieved after 10–25 min. Melanosomes can be redispersed by washout of cAMP using 10 μM dopamine in HERB.
2. For time-lapse video observation, prepare a perfusion chamber using two strips of double-stick tape aligned along both sides of the long edge of a microscope slide. Add a drop of medium to the slide, and invert a coverslip with adherent cells onto the tape. View cells using a 40×–100× phase contrast objective on an upright microscope. Perfuse by pipetting medium on one side of the coverslip and absorbing it on the other side using a wedge of filter paper.

4 Notes

1. pH of CMFHER is adjusted to 7.4 initially; after addition of the cysteine solution, pH will drop to 7.2.
2. Add papain to digest solution 10–15 min prior to dissection to allow 20–30 min activation of the enzyme before adding to RPE sheets. If multiple fish are to be dissected, add papain to separate aliquots of digest solution in parafilm lined, 35 mm plates. Papain should be kept on ice until use.
3. Ceramic coverslip holders are available commercially. Coverslips can be stored in nitric acid indefinitely until use.
4. Complete water rinse is critical, as free PLL is toxic to cells.
5. We use plastic boxes (former rodent shoebox style containers) that hold 3 L of water for up to six fish in each box. Boxes are placed in a custom-made, light- tight wooden box with an aeration hose and bubbler connected to an aquarium pump outside the box. MS-222 will lower the pH of the water; measure pH before and after addition with pH paper and adjust pH up using a few drops of 10 N NaOH. MS-222 is a carcinogen; gloves should be worn.
6. We use a desk lamp with a 25 W incandescent bulb turned against the wall to allow a small amount of light through.
7. After removal of RPE sheets, the back of the eye will remain black because of melanin pigment in the choroid.
8. Wet the inside of the pipette before use so tissue does not adhere to the glass. Do not let the sheets encounter the air-water interface. A wide-bore, fire-polished pipette with a slight bend at the end also works well.

9. The washing step can be carried out in room light. Be sure not to introduce air bubbles into the solution during the washing step; likewise do not remove all medium between washes such that tissue is exposed to the air. Each aliquot of wash solution should be added gently by slanting the tube and slowly trickling the wash solution down the side in order to avoid dissociating cells during washing. By holding the tube up to a black background, melanosomes from lysed RPE cells are visible in solution as the tissue is resuspended. Tissue should be washed in three changes of medium or until melanosomes from lysed cells are no longer visible.
10. Cells may be counted in a hemocytometer to gauge correct cell density; however, cells should be plated immediately after dissociation or they may not adhere to coverslips.
11. Be sure not to move cells after plating or they will not be distributed evenly on the coverslip. The moistened filter paper in the glass Petri dish is important for a humidified environment to discourage evaporation of the medium.

Acknowledgments

Thanks to Stephanie Santioleri for providing the scanning electron micrograph and Jessica Oates for the fluorescence micrograph. This work was supported by an NIH AREA grant # R15 GM066961, a Howard Hughes Medical Institute Science Education Grant, and the Provost's Award for Funded Research from Saint Joseph's University.

References

1. Burnside B, King-Smith C (2010) Chapter 172: Comparative eye: fish retinomotor movements. In: Dartt D, Besharse B, Dana R, Batelle B (eds) The encyclopedia of the eye. Elsevier, Amsterdam, The Netherlands
2. Garcia D, Burnside B (1994) Suppression of cAMP-induced pigment granule aggregation in RPE by organic anion transport inhibitors. Invest Ophthalmol Vis Sci 35:178–188
3. King-Smith C, Paz P, Lee C, Lam W, Burnside B (1997) Bi-directional pigment granule migration in isolated retinal pigment epithelial cells requires actin but not microtubules. Cell Motil Cytoskeleton 38:229–249
4. King-Smith C, Basciano P, Pham N (2001) Effect of the actin stabilizing drug, jasplakinolide, on pigment granule motility in the retinal pigment epithelium (RPE) of green sunfish, *Lepomis cyanellus*. Pigment Cell Res 14:14–22
5. King-Smith C, Vagnozzi RJ, Fischer NE, Gannon P, Gunnam M (2014) Orientation of actin filaments in teleost retinal pigment epithelial cells, and effect of the lectin, Concanavalin A, on melanosome motility. Vis Neurosci 31:1–10
6. Barsoum IB, King-Smith C (2007) Myosin II and rho kinase are required for melanosome aggregation in fish retinal pigment epithelial (RPE) cells. Cell Motil Cytoskeleton 64:868–879
7. Canman JC, Bement WM (1997) Microtubules suppress actomyosin-based cortical flow in Xenopus oocytes. J Cell Sci 110 (Pt 16):1907–1917
8. Rosenblatt J, Cramer LP, Baum B, McGee KM (2004) Myosin II-dependent cortical movement is required for centrosome separation and positioning during mitotic spindle assembly. Cell 117:361–372

Chapter 16

Chemotaxis: Under Agarose Assay

Derrick Brazill and David A. Knecht

Abstract

The unicellular eukaryotic amoeba, *Dictyostelium discoideum*, represents a superb model for examining the molecular mechanism of chemotaxis. Under vegetative conditions, the amoebae are chemotactically responsive to pterins, such as folic acid. Under starved conditions, they lose their sensitivity to pterins and become chemotactically responsive to cAMP. As an NIH model system, *Dictyostelium* offers a variety of advantages in studying chemotaxis, including ease of growth, genetic tractability, and the conservation of mammalian signaling pathways. In this chapter, we describe the use of the under-agarose chemotaxis assay to understand the signaling pathways controlling directional sensing and motility in *Dictyostelium discoideum*. Given the similarities between *Dictyostelium* and mammalian cells, this allows us to dissect conserved pathways involved in eukaryotic chemotaxis.

Key words Chemotaxis, *Dictyostelium*, Amoeba, Folic acid, cAMP

1 Introduction

Efficient chemotaxis requires that cells detect the presence and direction of the source of a diffusing chemoattractant, polarize toward the higher concentration side of the gradient, and move up the gradient. In order to do this, cell surface receptors must transiently bind chemoattractant and detect which part of the cell has the highest percentage of bound receptors. Actin must be polymerized at the front of the cell leading to pseudopod formation [1, 2]. These activities must be coordinated; otherwise, the cell will not be able to respond efficiently to the signal. The molecular mechanisms underpinning these individual activities can be dissected using chemotaxis assays with the unicellular eukaryote *Dictyostelium discoideum* [3, 4].

Dictyostelium and mammalian cells share a number of similarities in how they sense chemoattractants, modify their cytoskeleton, and undergo directional movement [5]. In fact, much of what is known about mammalian chemotaxis was first discovered in

Ray H. Gavin (ed.), *Cytoskeleton: Methods and Protocols*, Methods in Molecular Biology, vol. 2364,
https://doi.org/10.1007/978-1-0716-1661-1_16,

Dictyostelium [6]. *Dictyostelium* are haploid soil amoebae that consume bacteria and divide by binary fission. While in this vegetative state, the amoebae hunt bacteria by chemotaxing toward pteridines such as folic acid, which are released by the bacteria [7, 8]. This process is mediated by G protein-coupled receptors, which have recently been identified [9]. When the amoebae overgrow their bacterial food source and starve, they initiate a developmental pathway, in which amoebae synthesize and secrete cAMP as a chemoattractant to drive aggregation and formation of a multicellular organism consisting of 10^3–10^5 amoebae [10]. To sense the cAMP gradients, the cells downregulate pterin sensitivity and instead express four different cell surface cAMP receptors, all of which are also G protein coupled [11, 12].

Chemotaxis of amoebae toward folic acid and cAMP can be measured and analyzed using under-agarose assays, as well as a variety of other approaches, including micropipette stimulation [13], specialized chemotaxis chambers [14, 15], and commercial systems ECIS/taxis [14, 15]. One major advantage of the under-agarose assay is that it is very simple to set up. It is also very flexible and requires no specialized equipment beyond a microscope for imaging. The under-agarose chemotaxis assay examines the movement of cells as they crawl beneath a layer of agarose toward a chemoattractant. The agarose gel provides a stable environment for diffusion of the chemoattractant to form a gradient. This assay was first used to study migration in leukocytes [16, 17] and then later modified for use with *Dictyostelium* amoebae [18]. The assay is easy to set up. Cell movement is rapid, robust, and highly directional and can be analyzed quantitatively. Mutants with defects in sensing or responding to the signal can be compared to wild-type amoebae. In addition, the cells are required to crawl between the agarose and the plastic Petri dish. Therefore, there is a mechanical aspect to the movement such that defects in cortical structure or force generation can also be identified. For instance, mutants lacking myosin heavy or light chain are unable to move underneath the agarose sheet [19, 20] even though they can sense gradients and move in less restrictive environments. Slight defects, which normally would not be detected, can be exaggerated by using different concentrations of agarose. Because of its simplicity and flexibility, the under-agarose chemotaxis assay is incredibly useful for measuring chemotaxis in a variety of different eukaryotic cell types.

In this chapter, we explain how the motility of wild-type *and* mutant *Dictyostelium cells* can be examined and quantified using the under-agarose assay. The analysis of defective mutants can determine how and where the process of chemotaxis is disrupted. Examination of chemotaxis toward both folic acid and cAMP can uncover whether the defect is general to chemotaxis, or specific to one of the chemoattractants. Briefly, the assay consists of three steps: preparing the agarose plate, preparing and loading the cells and chemoattractant into wells in the agarose sheet, and time-lapse video microscopy of cell movement. The videos are then used to quantify various parameters of chemotaxis: such as chemotactic index, persistence, and speed. Chemotactic index is a measure of how accurately the cell moves up the concentration gradient for each step of movement. It is reported as cos θ, where θ is the angle of deviation between a direct line up the chemical gradient and the actual path of a cell. Persistence is a measure of how frequently a cell changes direction as it proceeds up the gradient. This is defined as the net cellular displacement divided by the total path length of a cell. As a cell makes more turns, the total path length increases relative to the net path length. This parameter can also be examined as turning rate, measured as the angular change from the direction of the last measured step to the current movement step. Persistence is not a parameter directly linked to chemotactic movement because a cell could move persistently but in the wrong direction or alternatively, turn frequently but generally move in the correct direction. The third parameter usually examined is cell speed. Since most chemotactic agents are also chemokinetic [21], the speed of movement of chemotaxing cells is usually higher than that of unstimulated cells. These parameters can be examined as an average over the time of data collection or examined as a function of time. For instance, is cell speed constant or varying significantly over time? These measurements and others provide a quantitative description of cell behavior during chemotactic movement.

2 Materials

2.1 Agarose Plates

1. 10× SorMC buffer: 20.36 g KH_2PO_4 (FW 136), 2.8 g NaH_2PO_4 (FW 142), 0.1 g $MgCl_2$ (FW 203 hexahydrate), 0.055 g $CaCl_2$ (FW 110), water to 1 L (unnecessary to pH or sterilize).
2. SM3 medium: 10 g Bacto or Oxoid peptone, 10 g glucose, 100 mL 10× SorMC, and water to 900 mL. Adjust pH to 6.5–6.7, add water to 1 L, autoclave, and store at 4 °C (*see* **Note 1**).
3. PBM buffer: 2.72 g anhydrous KH_2PO_4, 100μL 0.1 M $CaCl_2$, 1 mL 1 M $MgCl_2$, water to 1 L, adjust to pH 6.1, and autoclave. Store at room temperature.
4. 1% agarose in SM3 medium, made immediately prior to use in Subheading 3.1 (*see* **Note 2**).
5. 0.75% agarose in PBM buffer, made immediately prior to use in Subheading 3.3.
6. Blocking solution:10 g bovine serum albumin (BSA) in 100 ml distilled water. Store frozen.

7. 100 mm × 15 mm or 60 mm × 15 mm polystyrene Petri dishes. Glass bottom Petri dishes can also be used for DIC and fluorescence microscopy (*see* **Note 3**).

2.2 Chemo-attractants

1. 50 mM folic acid stock: dissolve 0.22 g folic acid in 10 mL 0.1 N NaOH. This is stable for months if kept in the dark at 4 °C.
2. Folic acid working solution (100μM to 1 mM): dilute 50 mM folic acid stock solution in SM3 media to the desired working concentration. This is stable for a week if kept in the dark at 4 °C.
3. 10 mM cAMP in PBM. Store at −20 °C.

2.3 Developing Cells for cAMP Chemotaxis

1. Whatman #3125 mm filter paper.
2. Millipore nitrocellulose membrane 47 mm AAWP (white) or AABP (black), 0.8μm pore).
3. 150 mm × 20 mm polystyrene Petri dish.
4. 50 mL centrifuge tube.

2.4 Time-Lapse Imaging

1. Inverted microscope with video camera.
2. Image acquisition software.

3 Methods

3.1 Folic Acid Chemotaxis: Preparation of the Agarose Plate

1. Blocking the surface of the Petri dishes: It is important to prevent adhesion of the agarose to the plastic, which will prevent cells from moving between the agarose sheet and the plastic surface. This can be done by rotating the agarose sheet with a spatula after it solidifies, but it is easier to pre-block the binding sites on the plate with BSA. Add sufficient blocking solution to the Petri dishes to cover the bottom (about 10 mL for 100 mm dishes and 5 mL for 60 mm dishes). Incubate for 5–10 min, and then pipette the blocking solution back into the storage tube (it can be reused many times), and refreeze. Rinse the plate with tap distilled water several times.
2. Make the 1% agarose solution in a 125 mL flask with the screw cap loosely attached (*see* **Note 4**).
3. Heat the flask in a microwave oven to the point where the solution first begins to boil. While wearing a heat protective glove, carefully swirl the flask to gently mix the solution. Place the flask back in the microwave, and heat to a boil again. Remove the flask, and swirl as before. Repeat cycles of boiling and swirling four to five times until the agarose is completely melted.

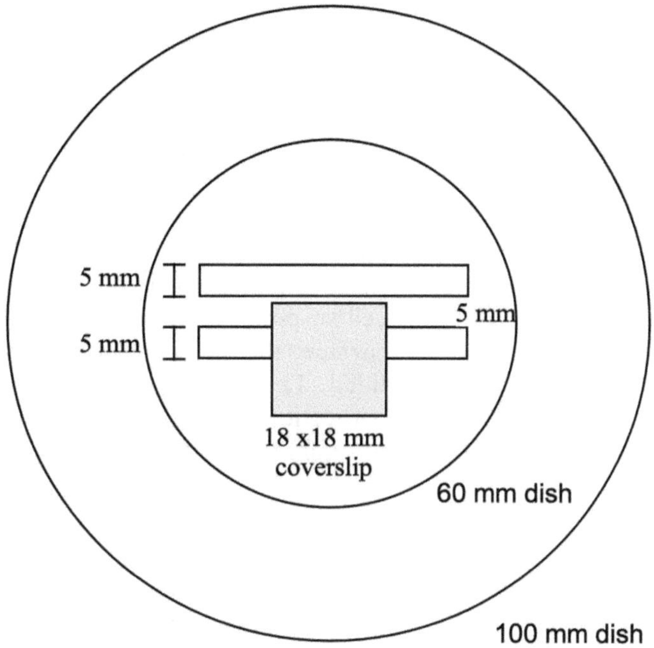

Fig. 1 Template for cutting wells. Place the agarose dish on top of the template and use it as a guide to cut the wells. The wells should be 2–5 mm wide and 5 mm apart

4. Once the flask is cool enough to handle by hand, add 20 mL of agarose solution to a standard 100 mm Petri dish or 8 mL to a 60 mm dish, and allow agarose to cool and harden for at least an hour. Plates can be poured up to 1 day before the experiment.
5. Place the dish on a template copied from Fig. 1. With the template as a guide, cut two wells in the agarose approximately 5 mm wide, 5 mm apart and the length of a standard single-edged razor blade (39 mm long). Shorter blades can be made by repeatedly bending the blade with two pliers until it breaks. The cuts should be as vertical as possible, straight down to the plastic dish without a sawing motion. It is important to avoid cutting the bottom of the Petri dish, as this will impede motility of the amoebae.
6. Cut both ends of the wells with a pointed spatula, and then carefully pry the agarose out with the spatula and discard. Avoid disturbing the surrounding agarose.
7. If you plan to image near the well edge, place a 18 × 18 mm glass coverslip over the cell well, centered so there is a gap on either side of the coverslip. This will aid in imaging cells at the edge of the well (*see* **Note 5**).

3.2 Folic Acid Chemotaxis: Preparation and Loading of Amoebae and Chemoattractant

1. While the agarose is hardening, prepare the cells. For optimal results, use axenically or bacterially growing cells in log phase at a density of less than about 2×10^6 cells/mL. Count the cells, and collect them by centrifugation at $500 \times g$ for 5 min. Remove the supernatant, and resuspend the cells in SM3 to a concentration of $2–3 \times 10^6$ cells/mL.
2. Place the dish on the microscope stage. Immediately before placing the cells in the wells, mix the suspension thoroughly by putting it through a vortex. Fill the cell well to the top with the cell suspension. If you have placed a coverslip on the agarose surface, the meniscus will contact the coverslip as the well is filled. The volume needed to fill the well will depend on how wide the well was cut, so rather than pipetting a specific volume, fill the well to the top (*see* **Note 6**).
3. To create the folic acid gradient, fill the second well with diluted folic acid (100μM–1 M) (*see* **Note 7**). Add enough folic acid to almost fill the well but not enough to cause it to overflow. Filling this well second eliminates the possibility of pressure-driven flow of folic acid under the agarose when one well is full and the other empty.
4. Within a few hours, the amoebae will begin to move out of the cell well and underneath the agarose toward the folic acid well. The amoebae will continue chemotaxis across the bridge toward the folic acid over a period of 8–12 h. Amoebae that have exited the well and are migrating under the agarose will appear flattened (phase dark). This is usually seen once the cells have moved about 50–100μm from the well. Set the field of view to image the cells in the moving wave front (*see* **Note 8**).

3.3 cAMP Chemotaxis: Preparation of the Agarose Plate

Chemotaxis to cAMP is performed similarly to chemotaxis to folic acid with only a few differences.

1. Block the surface of the plates with blocking solution as indicated in Subheading 3.1.
2. Make the 0.75% agarose solution, and pour into Petri dishes. Cut wells as indicated in Subheading 3.1.

3.4 cAMP Chemotaxis: Preparation of cAMP-Responsive (Developing) Amoebae

The following steps must be completed at least 6 h before performing the assay:

1. Pour 5 mL of PBM into a 150 mm × 20 mm Petri dish.
2. Wet and then submerge a Whatman #3125 mm diameter filter into the PBM in the dish.
3. Starting from one edge, use forceps to slowly lower a second Whatman #3125 mm diameter filter directly on top of the first one. It is imperative that there be no bubbles between the two filters.

4. Discard excess PBM leaving approximately 1 mL.
5. Starting from one edge, slowly lower a 47 mm Millipore AAWP or AABP filter onto the Whatman filters.
6. Collect 1×10^7 log-phase amoebae at least 6 h before preparing the assay. Wash the amoebae twice in PBM by centrifugation at $500 \times g$ for 5 min to remove the growth medium and resuspend them in 1 ml PBM.
7. Spread the amoebae evenly over the Millipore filter, avoiding the edges. This is usually done with approximately 250μL of cells at a time. Allow the liquid to wick into the filter before adding more cells. Repeat until all the cells have been added to the filter.
8. After 15 min, use forceps to slide the filter containing the amoebae to another area of the Whatman paper. This ensures that the amoebae develop in a completely fresh environment.
9. Allow the amoebae to develop for 6–8 h at room temperature, in order to express genes required for cAMP chemotaxis.

3.5 cAMP Chemotaxis: Loading the Cells and Chemoattractant into the Wells

1. To collect the starved amoebae, remove the 47 mm filter, and insert it into a 50 mL centrifuge tube. Ensure that the cell-laden side of the filter is positioned upward. Wash the cells off of the filter pad into the PBM buffer with 0.5 mL PBM by pipetting the PBM across the filter. Repeatedly flush the filter for 30 s. Repeat with a fresh 0.5 mL PBM to remove the remainder of the amoebae.
2. Transfer amoebae to a 15 mL Falcon tube and centrifuge at $500 \times g$ for 5 min. Remove the supernatant, and resuspend the amoebae in PBM to a concentration of 5×10^6–1×10^7 cells/mL.
3. Fill one well of the chemotaxis plate to just below the top with the suspension of starving amoebae.
4. To create the cAMP gradient, fill the second well with 5 mM cAMP in PBM buffer.

3.6 Imaging and Analysis

1. Images are best collected with an inverted microscope with phase contrast optics. Cells could be tagged to express a fluorescent protein and imaged by fluorescence microscopy. Depending on the experimental question (many cells over a larger field of view vs. a higher resolution examination of a small number of cells), a 4×, 10×, or 20× objective might be used for imaging. Images should be captured every 20–30 s for as long as necessary to capture the movement of the cells. We use the Open Source ImageJ [22, 23] plug-in Micro-manager [24, 25] for microscope and camera control, but many commercial packages exist as well.

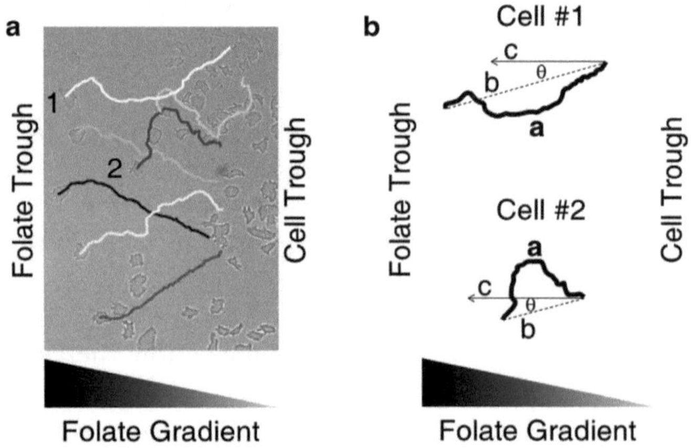

Fig. 2 An image of amoebae taken 3 h into a folic acid chemotaxis assay. (**a**) Several cells have been tracked and their paths determined using the ImageJ Manual Tracking plug-in. (**b**) The paths of two selected cells. Persistence = b/a. Chemotactic index = cos θ

2. Use the Manual Tracking or mTrackJ plug-ins in Fiji [22] to analyze cell movement in the acquired images (Fig. 2a). These plug-ins are similar, but mTrackJ calculates more parameters related to cell movements. The specific instructions for both can be found online with the plug-in (*see* **Note 9**).

4 Notes

1. This simple medium has several functions. Cells will not grow in it, but the nutrients prevent them from starving and initiating development. The glucose blocks adhesion sites present on vegetative cells so that cells move as individuals rather than large islands of adhered cells. The amount of folic acid in the medium is unknown but should be very low. For experiments where the actual concentration of folic acid is critical, a defined medium lacking folic acid can be made (http://dictybase.org/techniques/media/fm-medium.html) or purchased (https://www.formedium.com/us/product-category/products/dictyostelium-discoideum/fm-defined-minimal-media/).
2. Molecular biology-grade agarose used for DNA gel electrophoresis works well.
3. The assay can also be run with a glass bottom dish so that higher resolution DIC optics can be used and epifluorescence imaging of intracellular probes examined in parallel [26]. Various 30 mm glass bottom dishes are commercially available (https://willcowells.com and others), and the assay can be scaled to the smaller dish size. Single-edged razor blades can

be broken by bending with pliers to create whatever sized will is needed. For DIC microscopy, we place a rectangular piece of glass over the dish to prevent drying.

4. The cells move between the non-deformable plastic surface and the agarose sheet. As the cells move, they locally deform the agarose gel, so that it returns to its former position after the cell passes [18]. The stiffer the agarose, the harder it is for the cells to deform it. Different types of agarose will have slightly different stiffnesses at the same concentration. Therefore, it may be necessary to optimize the agarose concentration. For folic acid chemotaxis, 1% agarose is a good starting point. However, we have used between 0.5% and 2% agarose and have found that different concentrations can have an impact on the efficacy of chemotaxis. Mutants that have a slight chemotactic defect may not be detected under the more permissive 0.5% agarose but become quite apparent at the less permissive 2% agarose.
5. Cells first move underneath the agarose in the area adjacent to the cell well resulting in a very bright transition zone, which poses difficulties for observing cells by phase contrast microscopy. This is due to the curvature of the meniscus of the solution at the top of the well creating nonuniform light paths through the agarose. Overlaying with a coverslip creates a flat surface in this region and eliminates this problem so that cells can be imaged as they exit the well and then move up the gradient.
6. Optimally, a near monolayer of cells is desired in the well. This allows for a large number of cells to crawl under the agarose. If the cell density is low after the cells have settled, some buffer can be removed from the well and more of the cell suspension added.
7. Folic acid concentrations can be varied between 0.1 and 5 mM. The concentration is not critical. Although it would seem that the cells are responding to a gradient of chemoattractant diffusing through the agarose, we have recently demonstrated that the situation is actually quite different. The gradient that would be found 5 mm from the chemoattractant well by diffusion would be quite flat [27]. However, cells produce both cell surface and secreted enzymes that degrade the chemoattractants, folate deaminase for folic acid [28] and phosphodiesterase for cAMP [29]. When cells are concentrated in the cell well, local degradation of chemoattractant is highest near the well and less farther from the well; thus, degradation creates a "self-generated" gradient [30]. As the cells move, they continuously degrade chemoattractant and thus continue to generate the gradient as the wave front of cells move toward the source. The self-generated gradient has been confirmed by direct

measurement of folate concentration in the agarose. As a consequence of this phenomenon, the assay can also be done by adding chemoattractant to the melted agarose and cutting a single well for cells. The same chemotactic movement of cells away from the well will be observed.

8. It is important to image cells at the leading edge of the wave of moving cells. These cells tend to be farther spread out so that they do not bump into each other, complicating tracking. In addition, cells immediately behind the front have poorer chemotaxis due to degradation of the gradient by the preceding cells and cells farther back move randomly [30].
9. Use mTrackJ (https://imagescience.org/meijering/software/mtrackj/), or Manual Tracking (https://imagej.nih.gov/ij/plugins/track/track.html), to identify and mark individual cells in each frame of the image sequence (Fig. 2). Given the time between frames, and distance traveled between each frame, the program then calculates the distance moved and speed of the cells between each frame. The change in angle of movement from the previous movement step to the current can also be calculated. The software allows one to designate a "correct" direction; therefore, the angle and portion of movement toward the chemoattractant source can also be measured. Thus, the overall speed of the cell, the total path length that the cell traveled, and the net path length of the cell from its origin can be calculated. This allows the calculation of three basic chemotactic parameters. The first is the average speed of the cell, the speed of the cell over time, or the average speed of all the cells measured. The second measurement is persistence (Fig. 2b). This is a measure of how directly the cell moved and is defined as the ratio of the net path length to the total path length of the cell. A value of 1 represents a direct path with no deviations, where a value of 0 represents a meandering path where the cell ends up back at its original position. In Fig. 2b, Cell #1 has a high persistence, close to 0.8. Cell #2 has a lower persistence, close to 0.5. The third measurement is the chemotactic index (Fig. 2b). This is a measure of how well the cell is able to sense and orient to the chemotactic gradient. It is defined as cos θ, the co*sine* of the angle (θ) of movement relative to a direct line up the diffusional gradient (cos θ) . A value of 1 (co*sine of 0*) represents movement directly toward the source of the chemoattractant, while a value of 0 represents movement directly perpendicular to the source of the gradient (cos 90) and −1 (cos 180) movement opposite to the gradient. In Fig. 2b, both cells have similar chemotactic indices.

References

1. Diez S, Gerisch G, Anderson K, Müller-Taubenberger A, Bretschneider T (2005) Subsecond reorganization of the actin network in cell motility and chemotaxis. Proc Natl Acad Sci U S A 102:7601–7606
2. van Haastert PJM, Devreotes P (2004) Chemotaxis: signaling the way forward. Nat Rev Mol Cell Biol 5:626–634
3. Nichols JM, Veltman D, Kay RR (2015) Chemotaxis of a model organism: progress with Dictyostelium. Curr Opin Cell Biol 36:7–12
4. King JS, Insall RH (2009) Chemotaxis: finding the way forward with Dictyostelium. Trends Cell Biol 19:523–530
5. Artemenko Y, Lampert TJ, Devreotes PN (2014) Moving towards a paradigm: common mechanisms of chemotactic signaling in Dictyostelium and mammalian leukocytes. Cell Mol Life Sci 71:3711–3747
6. Jin T, Xu X, Fang J, Isik N, Yan J, Brzostowski JA, Hereld D (2009) How human leukocytes track down and destroy pathogens: lessons learned from the model organism Dictyostelium discoideum. Immunol Res 43:118–127
7. Pan P, Hall EM, Bonner JT (1975) Determination of the active portion of the folic acid molecule in cellular slime mold chemotaxis. J Bacteriol 122:185–191
8. Van Haastert PJ, De Wit RJ, Grijpma Y, Konijn TM (1982) Identification of a pterin as the acrasin of the cellular slime mold Dictyostelium lacteum. Proc Natl Acad Sci U S A 79:6270–6274
9. Pan M, Xu X, Chen Y, Jin T (2016) Identification of a chemoattractant G-protein-coupled receptor for folic acid that controls both chemotaxis and phagocytosis. Dev Cell 36:428–439
10. Mahadeo DC, Parent CA (2006) Signal relay during the life cycle of Dictyostelium. Curr Top Dev Biol 73:115–140
11. Varnum B, Soll DR (1981) Chemoresponsiveness to cAMP and folic acid during growth, development, and dedifferentiation in Dictyostelium discoideum. Differentiation 18:151–160
12. Parent CA, Devreotes PN (1996) Molecular genetics of signal transduction in Dictyostelium. Annu Rev Biochem 65:411–440
13. Gerisch G, Keller HU (1981) Chemotactic reorientation of granulocytes stimulated with micropipettes containing fMet-Leu-Phe. J Cell Sci 52:1–10
14. Zicha D, Dunn GA, Brown AF (1991) A new direct-viewing chemotaxis chamber. J Cell Sci 99:769–775
15. Muinonen-Martin AJ, Veltman DM, Kalna G, Insall RH (2010) An improved chamber for direct visualisation of chemotaxis. PLoS One 5:e15309
16. Cutler JE, Munoz JJ (1974) A simple in vitro method for studies on chemotaxis. Proc Soc Exp Biol Med 147:471–474
17. Lauffenburger D, Rothman C, Zigmond SH (1983) Measurement of leukocyte motility and chemotaxis parameters with a linear under-agarose migration assay. J Immunol 131:940–947
18. Laevsky G, Knecht DA (2001) Under-agarose folate chemotaxis of Dictyostelium discoideum amoebae in permissive and mechanically inhibited conditions. BioTechniques 31:1140–1149
19. Laevsky G, Knecht DA (2003) Cross-linking of actin filaments by myosin II is a major contributor to cortical integrity and cell motility in restrictive environments. J Cell Sci 116:3761–3770
20. Xu XS, Lee E, Chen T, Kuczmarski E, Chisholm RL, Knecht DA (2001) During multicellular migration, myosin ii serves a structural role independent of its motor function. Dev Biol 232:255–264
21. Devreotes PN, Zigmond SH (1988) Chemotaxis in eukaryotic cells: a focus on leukocytes and Dictyostelium. Annu Rev Cell Biol 4:649–686
22. Schindelin J, Arganda-Carreras I, Frise E, Kaynig V, Longair M, Pietzsch T, Preibisch S, Rueden C, Saalfeld S, Schmid B, Tinevez JY, White DJ, Hartenstein V, Eliceiri K, Tomancak P, Cardona A (2012) Fiji: an open-source platform for biological-image analysis. Nat Methods 9:676–682
23. Schneider CA, Rasband WS, Eliceiri KW (2012) NIH image to ImageJ: 25 years of image analysis. Nat Methods 9:671–675
24. Edelstein A, Amodaj N, Hoover K, Vale R, Stuurman N (2010) Computer control of microscopes using µManager. Curr Protoc Mol Biol Chapter 14: Unit14.20
25. Edelstein AD, Tsuchida MA, Amodaj N, Pinkard H, Vale RD, Stuurman N (2014) Advanced methods of microscope control using µManager software. J Biol Methods 1
26. Lemieux MG, Janzen D, Hwang R, Roldan J, Jarchum I, Knecht DA (2014) Visualization of

the actin cytoskeleton: different F-actin-binding probes tell different stories. Cytoskeleton (Hoboken) 71:157–169
27. Lauffenburger DA, Tranquillo RT, Zigmond SH (1988) Concentration gradients of chemotactic factors in chemotaxis assays. Methods Enzymol 162:85–101
28. Pan P, Wurster B (1978) Inactivation of the chemoattractant folic acid by cellular slime molds and identification of the reaction product. J Bacteriol 136:955–959
29. Aubry L, Firtel R (1999) Integration of signaling networks that regulate Dictyostelium differentiation. Annu Rev Cell Dev Biol 15:469–517
30. Tweedy L, Knecht DA, Mackay GM, Insall RH (2016) Self-generated chemoattractant gradients: attractant depletion extends the range and robustness of chemotaxis. PLoS Biol 14: e1002404

Part IV

Genetic and Proteomic Protocols

Chapter 17

Functional Analysis of Actin-Binding Proteins in the Central Nervous System of *Drosophila*

Qi He and Christopher Roblodowski

Abstract

Using *Drosophila* actin-binding protein Dunc-115 as an example, this chapter describes a MARCM (mosaic analysis with a repressible cell marker)-based method for analyzing cytoskeletal components for their functions in the nervous system. Following a concise description about the principle, a step-by-step protocol is provided for generating the needed stocks and for histological analysis. Additional details and explanations have been given in the accompanying notes. Together, this should form a practical and sufficient recipe for performing at the single cell level loss-of-function and gain-of-function analyses of proteins associated with the cytoskeleton.

Key words *Drosophila*, Dunc-115, Actin-binding protein, MARCM, Confocal microscopy

1 Introduction

It has been well established that the growth cone, a highly sensitive and motile structure at the leading edge of a growing axon, receives and transduces guidance signals through its surface receptors to regulate actin, microtubules, and other components of the cytoskeleton [1, 2]. Consequently, actin-binding proteins form a critical linkage to the cytoskeleton during the process of relaying guidance signals, and studying the functional roles of those proteins holds promises of revealing the molecular underpinning of axon pathfinding. More importantly, accumulated evidence has shown that the cytoskeletal proteins play critical roles in human neurological disorders [3], further highlighting the clinical significance of those proteins including actin-binding proteins. This chapter describes methods for generating genotypes, dissecting and staining the visual system, and analyzing the function of a *Drosophila* actin-binding protein Dunc-115, a protein that has been implicated in regulating axon pathfinding in our earlier analyses [4, 5]. Other cytoskeletal components can be analyzed similarly.

Ray H. Gavin (ed.), *Cytoskeleton: Methods and Protocols*, Methods in Molecular Biology, vol. 2364,
https://doi.org/10.1007/978-1-0716-1661-1_17,

A major step is to analyze axon pathfinding using a powerful technique, MARCM (mosaic analysis with a repressible cell marker), developed in *Drosophila* [6]. As shown in Fig. 1, MARCM utilizes recombination mediated by the FLP/FRT system (*see* Fig. 1 legend) to generate homozygous mutant clones for a gene of interest such as Dunc-115. Defects of Dunc-115 mutation in these clones during axon projection can then be analyzed by confocal microscopy. By creating the UAS-Dunc115 construct, ectopic Dunc-115 expression can be achieved in homozygous clones (Fig. 1) and thus gain-of-function of Dunc-115 can be analyzed. This approach also allows the analysis of genes that are homozygous lethal and provides a practical method for studying axon pathfinding functions of both viable and lethal genes.

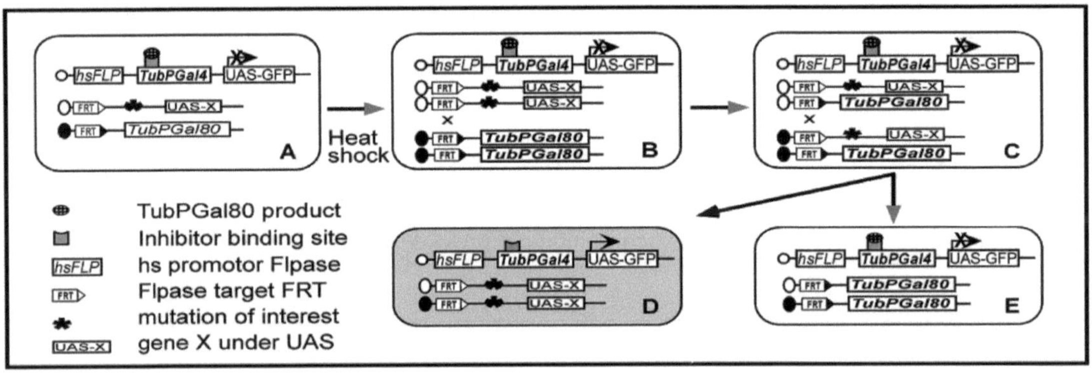

Fig. 1 The MARCM method. Rectangles represent individual cells. (**a**) Flippase (FLP) recombinase is under the control of a heat-shock promoter (hs), and both Gal4 and Gal80 are controlled by a universal tubulin 1 promoter (TubP). Transcription factor Gal4 binds to its unique upstream activator sequence (UAS) to drive the expression of GFP or gene X. Gal80 functions as a repressor of Gal4 activity, and thus when Gal80 and Gal4 are coexpressed in the same cell, Gal80 (hatched oval) binds to a site on Gal4 to repress its activity. As a result, both GFP and gene X cannot be expressed in such a cell (white cross over black arrow, shown only for GFP). (**b**) A heat-shock activates FLP that catalyzes the exchange of two FRT targets during mitotic cell division after DNA duplication (black X), resulting in the exchange of chromosome arms distal to the FRT sites shown in (**c**). Following the completion of cell division, one daughter cell (**d**) did not inherit Gal80 gene and is also homozygous for the mutant (black star, Dunc-115 mutation in this chapter) (or transgene X such as Dunc-115 under UAS though for the purpose of expressing gene X, UAS-X does not need to be on the same chromosome arm carrying the FRT site). So in daughter cell (**d**) (and its offspring), Gal4 is active and drives the expression of GFP (and UAS-X). The other daughter cell inherited two copies of Gal80 and thus has no Gal4 activity or GFP expression, while the background will be cells that did not undergo recombination as in (**b**) and thus they will be like the cell in panel (**a**), also without Gal4 activity or GFP expression. Thus, the only cells expressing GFP will be from panel (**d**). For simplicity, gene X (UAS-X) is depicted as on the same chromosome arm as the mutant (black star). In reality, the scheme can be used with a wild-type arm or with a mutant carrying arm, and the UAS-X transgene can be on any chromosome arm except the one carrying Gal80. Heat-shock is carried out at 37 °C for a desired period of time, and by controlling heat-shock time, it is possible to generate single neuron clones. UAS-GFP is UAS-mCD8-GFP fusion protein (where mCD8 is mouse T cell surface protein CD8) that will be anchored in the plasma membrane. TubP is fused with Gal4 or Gal80 coding sequence. Schematics were inspired by and based on Lee and Luo [6] and Wu and Luo [7]

2 Materials

2.1 Equipment

1. Standard incubator for culturing *Drosophila*.
2. Orbital shaker platform.
3. Dissecting microscope.
4. Confocal microscope.

2.2 Reagents

1. Standard cornmeal medium: mix 850 mL of water with 7.9 g of agar, 27.5 g of torula yeast, 52 g of cornmeal, and 110 g of dextrose. Heat to a boil and cool to 80 °C. Add 2.4 g of tegosept dissolved in 92 mL ethanol and pour into culture vials and bottles (*see* **Note 1**).
2. G418 medium: prepare standard cornmeal medium and cool to 50 °C. Add G418 (Geneticin antibiotic) to 1 mg/mL final concentration.
3. PBS: 37 mM NaCl, 2.7 mM KCl, 8 mM Na_2HPO_4, and 2 mM KH_2PO_4 pH 7.4.
4. PBT: add 0.3% Triton X-100 to 1× PBS.
5. PBTN: add 5% normal goat serum to PBT.
6. 4% formaldehyde: to make 1 mL stock, mix 0.792 mL H_2O, 0.1 mL 10× PBS, and 0.108 mL of 37% formaldehyde stock. The solution should be made fresh daily, and unused portion should be discarded at the end of the day.
7. Antibodies: mouse anti-24B10, rat anti-N cadherin (both from Developmental Studies Hybridoma Bank), rabbit anti-GFP (Molecular Probes), FITC-conjugated goat anti-mouse, Cy5-conjugated goat anti-rat (both from Jackson Lab), and Alexa647-conjugated donkey anti-rabbit (Molecular Probes).
8. 25% glycerol in PBS.
9. 50% glycerol in PBS.
10. Anti-quenching agent: 75% glycerol, 0.1% p-phenylenediamine in PBS.

2.3 *Drosophila* Strains and Culture (Available at the Bloomington Stock Center)

1. A Dunc-115 mutant *dunc-115*^{KG03651} (85E4) has been isolated and analyzed [4].
2. The following strains are used and they are available at the Bloomington Stock Center:

   yw/Y; P[ry, FRT]82B.

   yw/Y; TM2/TM6B.

   yw hsFlp UAS-mCD8-GFP; Sp/CyO; TM2 Ubx/TM6B Hu.

   yw/Y; TubPGal4; FRT82B TubPGal80.

3 Methods

3.1 MARCM Analysis

3.1.1 Preparation of FRT Dunc-115^{KG03651}, Gal4, and Gal80 Strains

1. *FRT dunc-115* strain (*see* **Note 2**):
   Set up the following cross:
   yw; dunc-115^{KG03651} X *yw/Υ; P[ry, FRT]82B*.
   Collect females and mate with *yw/Υ; TM2/TM6B*:
   yw; dunc-115^{KG03651}/P[ry, FRT]82B X *yw/Υ; TM2/TM6B*.
   Culture this cross in *neomycin* (G418) media and select w^+ and cross to *yw/Υ; TM2/TM6B* (single animal cross) (*see* **Note 3**):
   yw; P[ry, FRT]82B dunc-115^{KG03651}/TM6B X *yw/Υ; TM2/TM6B*.
   Collect both males and females to establish the following stock:
   yw; P[ry, FRT]82B dunc-115^{KG03651} or *yw; FRT82B dunc-115^{KG03651}*.
2. Gal4 and Gal80 strains (*see* **Note 4**):
   Set up the following cross:
   yw hsFlp UAS-mCD8-GFP; Sp/CyO; TM2 Ubx/TM6B Hu X *yw/Υ; TubPGal4; FRT82B TubPGal80*.
   Collect males of *Sp* and *Hu* and set up:
   yw hsFlp UAS-mCD8-GFP/Υ; TubPGal4/Sp; FRT82B TubP-Gal80/TM6B Hu X *yw hsFlp UAS-mCD8-GFP; Sp/CyO; TM2 Ubx/TM6B Hu*.
   Select both males and females of *Cy* and Hu^+ and set the following cross:
   yw hsFlp UAS-mCD8-GFP/Υ; TubPGal4/CyO; FRT82B TubP-Gal80/TM2 Ubx X *yw hsFlp UAS-mCD8-GFP/yw hsFlp UAS-mCD8-GFP; TubPGal4/CyO; FRT82B TubPGal80/TM2 Ubx*.
   Select both males and females of Cy^+ and Ubx^+ to establish the following stock:
   yw hsFlp UAS-mCD8-GFP; TubPGal4; FRT82B TubPGal80.

3.1.2 Generation of MARCM Clones (See **Note 5**)

To generate MARCM clones, strains created in Subheading 3.1.1 will be crossed to generate the needed genotype subject to heat-shock to induce mitotic recombination.

1. Set up the following cross:
   yw/Υ; FRT82B dunc-115^{KG03651} X *yw hsFlp UAS-mCD8-GFP; TubPGal4; FRT82B TubPGal80*.
2. Collect eggs during a window of every 2 h, i.e., switch the animals to a new vial every 2 h. Animals with the following genotype will be the target of analysis:

yw hsFlp UAS-mCD8-GFP/Y or *yw; TubPGal4/+; FRT82B dunc-115^{KG03651}/FRT82B TubPGal80.*

3. Leave the collected eggs at 25 °C for 96 h.
4. Heat-shock at 37 °C for 1 h.
5. Dissect third instar larvae and stain with antibodies as described below.

3.2 Histology

3.2.1 Larval Eye Disc/Brain Specimen Preparation

Dissect the visual system of the third instar larvae in PBS [4, 8] and fix in 4% formaldehyde for 40 min at room temperature (RT) (*see* **Note 6**).

3.2.2 Antibody Staining

1. Following fixation, wash specimens three times for 5 min each time with PBT. Make certain all specimens have sunk to the bottom of the tube.
2. Block nonspecific-binding sites on the specimens by incubating with PBTN for 30 min at RT.
3. Following the blocking step, primary antibody is added and the incubation is at 4 °C overnight (O/N). Typically, three primary antibodies are used in this analysis: (1) anti-24B10 labels photoreceptor neurons, anti-N cadherin stains neuropils in the optic lobes, and anti-GFP labels GFP expressed in the MARCM clones (*see* Fig. 1). These antibodies need to be diluted with PBTN, and the folds of dilution depend on the nature of the antibody and will have to be determined empirically.
4. Wash three times for 15 min each with PBT.
5. Add second antibodies. The dilution will be decided experimentally but typically around 200–500 times and is done with PBTN. The three secondary antibodies are tagged with green (FITC-conjugated goat anti-mouse), red (Cy5-conjugated goat anti-rat), and blue (Alexa647-conjugated donkey anti-rabbit) fluorescent tags and will be detected separated by the confocal microscope according to the wavelength. The incubation is as **step 3**.
6. Wash as in **step 4**.
7. Go through 5-min serial washes with 25%, 50%, and 75% glycerol in PBS with the 75% glycerol containing 0.1% p-phenylenediamine (anti-quenching agent to prolong the fluorescent intensity).
8. Mount the specimens on glass slides (*see* **Note 7**) and examine under a confocal microscope.

4 Notes

1. Pre-made fly food mix such as Jazzmix from Applied Scientific can be used by following the manufacture's instruction, while 10X PBS can be purchased from many vendors such as Fisher Scientific.
2. This procedure is for loss-of-function analysis (*dunc-115* mutant) using MARCM. For gain-of-function analysis (ectopic expression), a UAS construct (UAS-Dunc-115 fusion) will need to be built first using standard molecular biology methods and transgenic animals produced by embryo injection. The resulted UAS-Dunc-115 transgenic flies will be prepared for MARCM the same way as with the FRT strain. Similar analysis will reveal the effect of overexpressing Dunc-115 in a MARCM clone.
3. The structure P[ry FRT] carries two copies of the *neomycin* gene controlled by the heat-shock promoter (*hs-neo*), thus conferring the *neomycin* resistance [9], while the *dunc-115*^{KG03651} strain carries the *mini-white* gene and thus has the w^+ phenotype. By selecting *neomycin* resistance and w^+, the only possible genotype is FRT *dunc-115*^{KG03651}.

   The chromosomal location of *dunc-115*^{KG03651} is 85E4 and the FRT used is at 82B. Given that the distance between the two is not long, it is necessary to set up several vials to increase the chances of recombination. For genes that have greater distances from the FRT site, the number of vials can be reduced. It would be informative to set up a pilot test to see the frequency of a given recombination.
4. The general principle behind MARCM is to bring all elements into the same animal after a final cross. Thus, it is often necessary to prepare the two stocks needed for the final cross so that the needed constructs will be in place. For the FRT, it is required to have the Gal80 construct to be on the same chromosomal arm distal to the FRT site (Fig. 1), while the other elements such as Gal4 or UAS-mCD8-GFP can be on any chromosomes. Traditionally, heat-shock-controlled flippase (hs Flp in Fig. 1) is linked with UAS-mCD8-GFP and located on the first chromosome. For a more detailed discussion, *see* [7].
5. It is preferable to use females from the UAS-mCD8-GFP strain in the cross since both male and female offspring will have one copy of the GFP fusion protein and thus will be usable. Heat-shock time may vary depending on the actual setup. It is thus highly recommended to test the time first to make certain a desired result. The clone size in MARCM depends on at which

developmental stage the heat-shock is given and for how long. A condition can be determined where single cell clones can be generated (*see*, e.g., [3]).

6. It is important to have all specimens submerged in the fixative to ensure proper fixation. Fixation and other incubation steps should be carried out on an orbital shaker platform.
7. To examine along the dorsal-ventral axis, the visual system needs to be cleaned so that mouth hooks and other accessories are removed, and the whole specimen sits on its ventral base. For a lateral view, the two optic lobes need to be separated with the use of a surgical knife, and each lobe should rest on the cut-phase. Finally, one piece of coverslip should be placed on each side of the prepared specimen, and a third coverslip should be placed at the top of the first two so that the specimen will not be squashed by the coverslip to preserve its morphology.

Acknowledgments

We would like to thank the Bloomington Stock Center for providing fly strains, Junaid Syed for assistance, and Ray Gavin for reviewing the manuscript. This work was supported in part by a PSC-CUNY grant.

References

1. Huber AB, Kolodkin AL, Ginty DD, Cloutier JF (2003) Signaling at the growth cone: ligand-receptor complexes and the control of axon growth and guidance. Annu Rev Neurosci 26:509–563
2. Stoeckli E (2018) Understanding axon guidance: are we nearly there yet? Development 145:1–10
3. Muñoz-Lasso DC, Romá-Mateo C, Pallardó FV, Gonzalez-Cabo P (2020) Much more than a scaffold: cytoskeletal proteins in neurological disorders. Cell 9:358–399
4. Garcia MC, Abbasi M, Singh S, He Q (2007) Role of *Drosophila* gene *dunc-115* in nervous system. Invertebr Neurosci 7:119–128
5. Roblodowski C, He Q (2017) *Drosophila* Dunc-115 mediates axon projection through actin binding. Invertebr Neurosci 17:1–8
6. Lee T, Luo L (2001) Mosaic analysis with a repressible cell marker (MARCM) for Drosophila neural development. Trends Neurosci 24:251–254
7. Wu JS, Luo L (2006) A protocol for mosaic analysis with a repressible cell marker (MARCM) in *Drosophila*. Nat Protoc 1:2583–2589
8. Kunes S, Wilson C, Steller H (1993) Independent guidance of retinal axons in the developing visual system of Drosophila J. Neuroscience 13:752–767
9. Xu T, Rubin GM (1993) Analysis of genetic mosaics in developing and adult *Drosophila* tissues. Development 117:1223–1237

Chapter 18

Using a Hand-Held Gene Gun for Genetic Transformation of *Tetrahymena thermophila*

Michael Gotesman, Roland E. Hosein, and Selwyn A. Williams

Abstract

Biolistic bombardment is widely used as a means of delivering vector-coated microparticles into microorganisms, cultured cells, and tissues. The first particle delivery system contained a helium propulsion unit (the gun) mounted in a vacuum-controlled chamber. In contrast, the hand-held gene gun does not operate within a chamber. It is completely hand-held, easy, and efficient to use, and it requires minimal space on the laboratory bench top. This chapter describes protocols for using a hand-held gene gun to deliver transformation vectors for overexpression of genes or gene replacement into the macronucleus of *Tetrahymena thermophila*. The protocols provide helpful information for preparing *Tetrahymena* for biolistic bombardment, preparation of vector-coated microcarriers, and basic gene gun operating procedures.

Key words *Tetrahymena*, Biolistic bombardment, Gene gun, Overexpression, Gene replacement, Microcarriers

1 Introduction

Tetrahymena thermophila exhibits nuclear dimorphism [1], in which separate nuclei are used for vegetative and sexual processes. A macronucleus (MAC) provides for vegetative functions, while a micronucleus (MIC), which provides for the sexual process of conjugation, is transcriptionally silent during vegetative growth. The MAC contains approximately 225 chromosomes [2] with approximately 45 copies of each gene and divides amitotically at cell division. The MIC contains five chromosomes and divides mitotically. *Tetrahymena* genome databases can be assessed at cili ate.org or at the National Center for Biotechnology Information (NCBI) website.

Biolistic bombardment is the preferred method for inserting overexpression constructs into the macronucleus and for homologous recombination in somatic and germline transformations in *Tetrahymena* [3]. Overexpression transformations target the

Ray H. Gavin (ed.), *Cytoskeleton: Methods and Protocols*, Methods in Molecular Biology, vol. 2364,
https://doi.org/10.1007/978-1-0716-1661-1_18,

extrachromosomal rDNA at the point in conjugation when the new MAC is under formation. Somatic and germline gene replacements target a respective gene of interest in the MAC or MIC. DNA from a transformation vector engages in homologous recombination with the gene of interest. After many cycles of vegetative growth, a process known as phenotypic assortment [4] ensures that the MAC is homozygous for all alleles, and therefore all copies of the gene of interest have been replaced. However, if the replaced gene is essential, phenotypic assortment may not be completed as the MAC retains a few copies of the endogenous allele, a condition referred to as a knockdown. At conjugation, the MAC is degraded, and a new MAC is formed from a zygotic MIC, effectively destroying the transformations that were achieved with the vector. After conjugation of a somatic transformation, the vegetative cell and its newly formed MAC reflect the genetic constitution of the zygotic MIC. In contrast, germline transformations target the MIC and are expressed in the newly formed MAC after conjugation [5].

Plasmid vectors for transformation of *Tetrahymena* contain a cassette consisting of a gene for antibiotic resistance, transcriptional start and stop sequences, and flanking regions that allow for homologous recombination with the endogenous gene of interest. Some of these vectors have been designed for either disruption or replacement of endogenous alleles [6–9]. The pVGF-1 vector [10] and its pIGF-1 derivatives are frequently used for overexpression of genes in *Tetrahymena*. This vector targets ribosomal genes located in an extrachromosomal segment that replicates autonomously. In current use, the pIGF-1 vector contains genes for antibiotic resistance, a metallothionein promoter that is inducible by ionic cadmium [11], a GFP gene, and a transcriptional stop sequence (Fig. 1). Genes of interest are cloned 3′ to the GFP site in pIGF-1 and therefore are expressed with an N-terminus GFP tag (Fig. 1). Transformation vectors have been used for diverse studies involving overexpression or replacement of *Tetrahymena* genes [12–16]. A list of plasmid vectors is available from the National *Tetrahymena* Stock Center (*see* **Note 1**).

The advent of particle bombardment via the hand-held gene gun led to a more efficient method for transfection of plants, animals, and plant and animal cells in culture and for therapeutic delivery of nucleic acids into cells [17–23]. The basic operation of the gene gun involves adhering microcarriers coated with the biological material of interest (DNA, RNA, or other substances) to the inner surfaces of small plastic tubes. A positively charged organic polymer such as polyethylenimine can be used to neutralize the negative charge on DNA. Polycaprolactone, a biodegradable polyester, is another effective charge neutralizer [24]. High-pressure helium is used to dislodge the microcarriers ("bullets") and force them into the cell. Gold or tungsten microcarriers are widely used for particle bombardment protocols with diverse

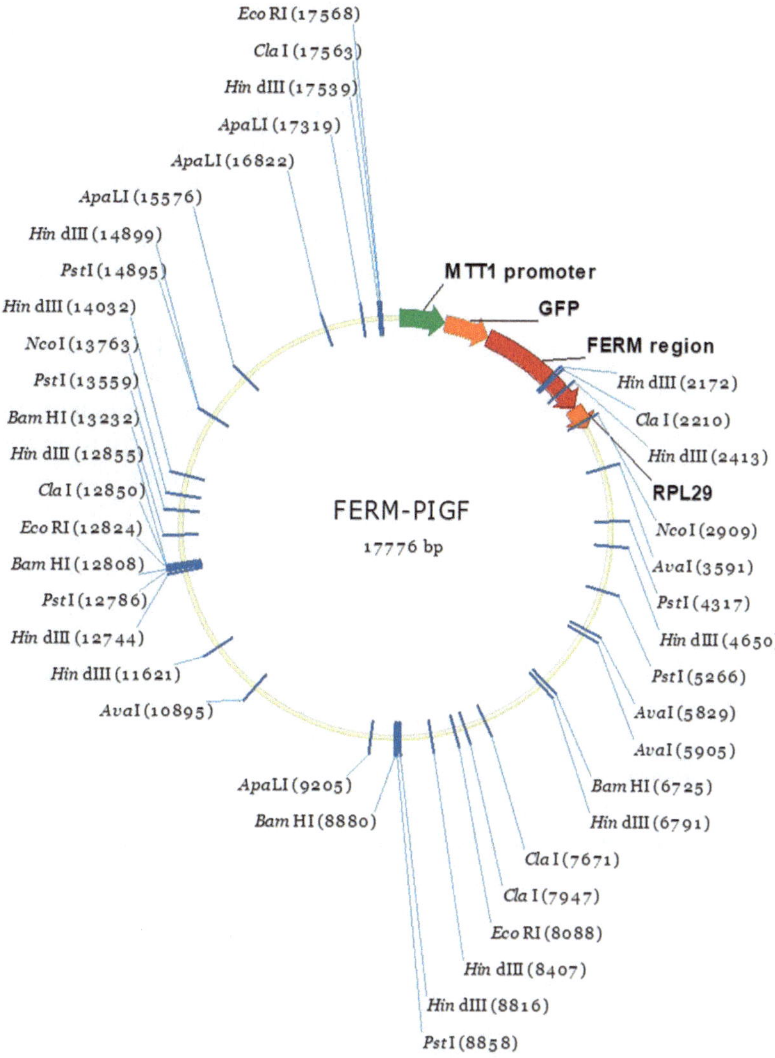

Fig. 1 A pIGF-1 vector was used to prepare a construct for transforming *Tetrahymena thermophila* for overexpression of the FERM domain in a myosin gene [4]. Unique restriction endonuclease cloning sites MTT1 (green) GFP (orange) FERM insert (red) and the transcriptional termination site RPL (darker orange) are pictured from 5′ → 3′

experimental systems [17–23, 25–28]. However, some nanoparticles such as silver can be cytotoxic [29]; therefore, great care must be taken in choosing the correct DNA carrier. Bacterial spores and other degradable compounds have recently been used as microcarriers for both in vitro and in vivo transfection of mice [30] and delivery of therapeutic proteins [31]. Interestingly, *Tetrahymena* is known to ingest microcarriers [32] via phagocytosis. Phagosomes can travel to the nucleus [33], and this raises the possibility that ingested microcarriers could be targeted to the nucleus via phagocytosis.

The Bio-Rad chamber model PDS-1000/He Particle Delivery System (Cat#: 1652257) and the Helios® Gene Gun System (Cat #: 1652431) were recently quoted at a similar price for academic use. However, the chamber requires the purchase of a hepta adaptor (Cat#: 1652225) which increases the overall cost of the chamber by more than 20%. The significant price difference for supplementary components to the chamber model vs hand-held gun model may urge burgeoning *Tetrahymena* researchers to choose the latter model, which is easy to use and requires minimal space on the bench top.

This chapter describes protocols for using a hand-held gene gun to deliver transformation vectors for overexpression of genes or gene replacement in the macronucleus of *Tetrahymena thermophila*. Protocols provide helpful information for preparing cells for biolistic bombardment, preparation of DNA-coated gold particle "bullets" as adapted from a JoVE video article by Woods and Zito [34], and basic gene gun operating procedures for the Bio-Rad Helios Hand-Held Gene Gun based on the Bio-Rad Handbook for the Helios Gene Gun. We are not aware of any other maker of a hand-held gene gun. However, the protocols described in this chapter should be adaptable to other versions of the gene gun.

2 Materials

2.1 Growth and Starvation Media

1. Modified Neff's growth medium [35]: 0.25% proteose peptone, 0.25% yeast extract, 0.5% glucose, 333 μM $FeCl_3$. Add 500 mL of dH_2O to an autoclavable bottle. Add 2.50 g proteose peptone, 2.50 g yeast extract, 5.0 g glucose, and 1 mL of 1000× 33.3 μM $FeCl_3$. Bring the solution up to 1 L with dH_2O, and mix vigorously. Autoclave and store media at 4 °C after sufficient cooling has taken place.
2. Dryl's starvation medium [36]: 2 mM $C_6H_5Na_3O_7$, 1 mM NaH_2PO_4, 1 mM Na_2HPO_4, and 15 mM $CaCl_2$. Add 50 mL of dH_2O to an autoclavable bottle. Add 590 mg Na citrate•$2H_2O$, 140 mg NaH_2PO_4•H_2O, 140 mg Na_2HPO_4. Bring the solution up to 100 mL with dH_2O, and mix vigorously. Add 50 mL of dH_2O to a separate autoclavable bottle. Add 130 mg $CaCl_2$. Bring the solution up to 100 mL with dH_2O, and mix vigorously (*see* **Note 2**). Autoclave each solution for 20 min, and allow it to cool before using. Mix the two solutions under aseptic conditions by allowing each solution to pass through the same filter sterilization column, and adjust the volume to 1 L with dH_2O.

2.2 DNA Coating of Microcarriers

1. Vector DNA: ~1 μg/μL in dH_2O (*see* **Note 3**).
2. 100% ethanol.

3. Spermidine: Add 1.45 g of spermidine to 100 mL pure ethanol to make a 0.10 M solution.
4. 1.0 M calcium chloride.
5. Polyvinylpyrrolidone (PVP) stock: Dissolve 200 mg PVP into 10 mL of dH_2O to make a 20 mg/mL stock solution. Prepare a working solution of PVP: Add 175 μL of 20 mg/mL PVP stock solution to a sterile screw cap container, and bring the volume up to 3.5 mL with 100% ethanol to yield a concentration of 0.01 mg/mL PVP (*see* **Note 4**).
6. Gold (Au): 0.6–1.0 μm gold particles (*see* **Note 5**).

2.3 Gene Gun

1. Helios Gene Gun System (Bio-Rad 165-2431) includes the helium hose assembly with regulator tubing, prep station, syringe kit, Tefzel tube cutter, and optimization kit.
2. Tank of compressed helium.
3. Tank of compressed nitrogen.

2.4 Reagents for Selecting Transformants

1. 100× stock paromomycin: Dissolve 120 mg of paromomycin in 10 mL of sterile dH_2O.
2. 100× stock cycloheximide: Dissolve 30 mg of cycloheximide in 10 mL of sterile dH_2O.
3. 100× antibiotic/antimycotic mix: Use a commercially available product.
4. 1000× stock cadmium chloride: Dissolve 10 mg of $CdCl_2$ into 10 mL of sterile dH_2O.

3 Methods

3.1 Coating Vector DNA to Gold (AU) Microcarriers

1. Add 50 μg (0.6–1.0 μm) gold particles to a sterile microfuge tube containing 50 μL of ethanol and 50 μL of 0.10 M spermidine. Sonicate for 1–2 min, and put the mixture through a vortex for 2–3 s. Next, add between 50 and 100 μg of 1 μg/μL vector DNA to the solution, and gently sonicate (*see* **Note 6**).
2. Dropwise, add 100 μL of 1.0 M calcium chloride to the previously prepared solution (**step 1**), and sonicate for 2–3 min. Allow the mixture to remain at room temperature, with the microfuge tube cap open, for 10 min.
3. After the 10-min incubation, briefly sonicate, and then pulse-centrifuge the mixture for 15 s. Carefully remove and discard supernatant (*see* **Note 7**).
4. Repeat three washes by adding 1 mL of fresh ethanol, and pulse-centrifuge the mixture for 15 s. Carefully remove and discard supernatant.

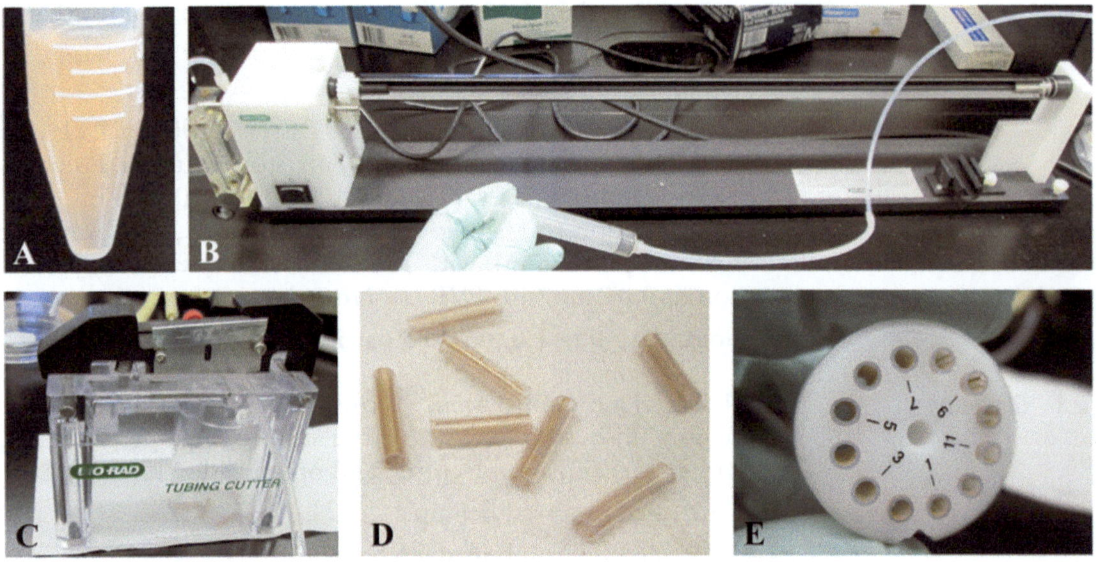

Fig. 2 Preparation of "bullets" for use in the gene gun. (**a**) DNA-gold mixture suspended in solution. (**b**) Loading the mixture in the drying chamber. (**c**): Cutting "bullets" in the tube cutter. (**d**) Cut bullets. (**e**) Loading the bullets into the barrel of the gene gun

5. After the final wash, use 3.5 mL PVP ethanol working solution to transfer the DNA-coated gold particles to a sterile 10 mL screw cap container by repeatedly adding 200 μL of PVP (0.01 mg/mL) aliquots.
6. Complete the transfer of the DNA-coated gold particles, and add the remaining PVP working solution to the container, cap, and invert the container several times to ensure an even suspension (Fig. 2a).

3.2 Coating the Plastic Tubing with Gold Microcarriers (See Note 8)

1. Insert Tefzel tube into prep station rotator (Fig. 2b) until the free end of the tube reaches the "O" ring inside the tube prep station (*see* **Note 9**).
2. Cut the overhanging exposed end of the Tefzel tube ~2 in. (~5 cm) past the external hole of the prep station rotor (right side of Fig. 2b) to form a ~30 in. (~80 cm) Tefzel tube.
3. Dry the inner surface of the tube by adjusting the flow nob (immediately above "B" in Fig. 2b) to allow nitrogen flow between 0.35 and 0.4 L/min for a minimum of 15 min.
4. Draw up the DNA-coated gold particles (prepared in Subheading 3.1, **step 6**) into a 5 mL syringe.
5. Use connector tubing to attach the filled syringe to the dried Tefzel tube (Fig. 2b).
6. Quickly transfer the DNA-coated gold particles from the syringe to the dried tubing situated in the tubing prep station (*see* **Note 10**).

7. Mark off the distal region of the tube where there are no microcarriers. Exclude this region when preparing bullets in a subsequent step.
8. Allow the microcarriers to settle for 3–5 min. Then remove the mixture at a rate of 0.5–1.0 in./s. The tube should be emptied within 30–45 s.
9. Rotate the tube 180° by holding the tubing prep station on/off switch to position II, and allow the gold to coat the inside of the tubing for 3–4 s.
10. Turn the tubing prep station switch to position I for 20–30 s to rotate the tube and smear it with gold.
11. Allow nitrogen to flow through the tube at a rate of 0.35–0.4 L/min for 3–5 min.
12. Turn off the tubing prep station switch to position 0 (off), and close the nitrogen valve.
13. Remove the section of tubing that does not contain DNA-coated gold particles (*see* Subheading 3.1, **step** 7).

3.3 Tube Cutting and Barrel Preparation

1. Transfer the gold smeared tube from the tubing prep station to the tube cutter.
2. Ensure that the end of the tubing contacts the rear end of the prep station. Cut the tubing into sections approximately 1 in. (2.5 cm) in length (Fig. 2c).
3. Load precut tubes (Fig. 2d) into the barrel of the gun (Fig. 2e) or store the tubes at −20 °C until cells have been prepared.

3.4 Preparing Cells for Transformation Using Overexpression Vectors

1. The pIGF-1 vector and its derivatives target developing macronuclei at conjugation. Therefore, two different mating-competent strains are required for this type of transformation. Mating-competent cells can be purchased from the National *Tetrahymena* Stock Center (*see* **Note 11**).
2. Separately grow wild-type *Tetrahymena* cells (e.g., strains 427 and 428) in bottles containing 30 mL of modified Neff's medium at 30 °C overnight, usually 12–18 h dependent upon the concentration of the inoculum.
3. Initiate a starvation regime while cells are still in exponential phase of growth. Gently centrifuge cells at 300 × *g* for 5 min in a sterile glass centrifuge tube. Carefully decant the Neff's medium without disturbing the soft pellet of cells. Resuspend the pellet in Dryl's starvation medium and wash 3× using the gentle centrifugation regime with fresh Dryl's starvation medium for each wash.
4. After the third wash, resuspend each strain in 30 mL of Dryl's starvation medium to yield an optical density of 0.2 as measured at 540 nm. Incubate the cells overnight (approximately 18 h) at 30 °C (*see* **Note 12**).

5. The following day, add ~2.5 mL of each starved *Tetrahymena* strain to a sterile 100 × 15 mm plastic Petri dish, and incubate at 30 °C. Expect conjugal pairing in 2–3 h after the initial cell mixing.

3.5 Preparing Conjugal Pairs for Biolistic Bombardment

1. Concentrate ~30 mL of conjugal *Tetrahymena* pairs (8–10 h after mixing) by gentle centrifugation at 300 × *g* for 5 min in a sterile glass centrifuge tube.
2. Gently decant the Dryl's medium, and resuspend the conjugal pairs in the small volume (approximately 2–3 mL) of Dryl's that remains in the centrifuge tube.
3. Decant the cell suspension onto the surface of sterile 24 cm GF/A Whatman filter paper placed in 100 × 15 mm plastic Petri dishes.

3.6 Using the Helios Hand-Held Gene Gun for Particle Bombardment

1. Load the stored bullets (Fig. 2d) into the barrel of the gun (Fig. 2e); load the barrel into the gun, and reassemble the remaining components (Fig. 3a).
2. Pressurize the gun to 200 psi with helium.

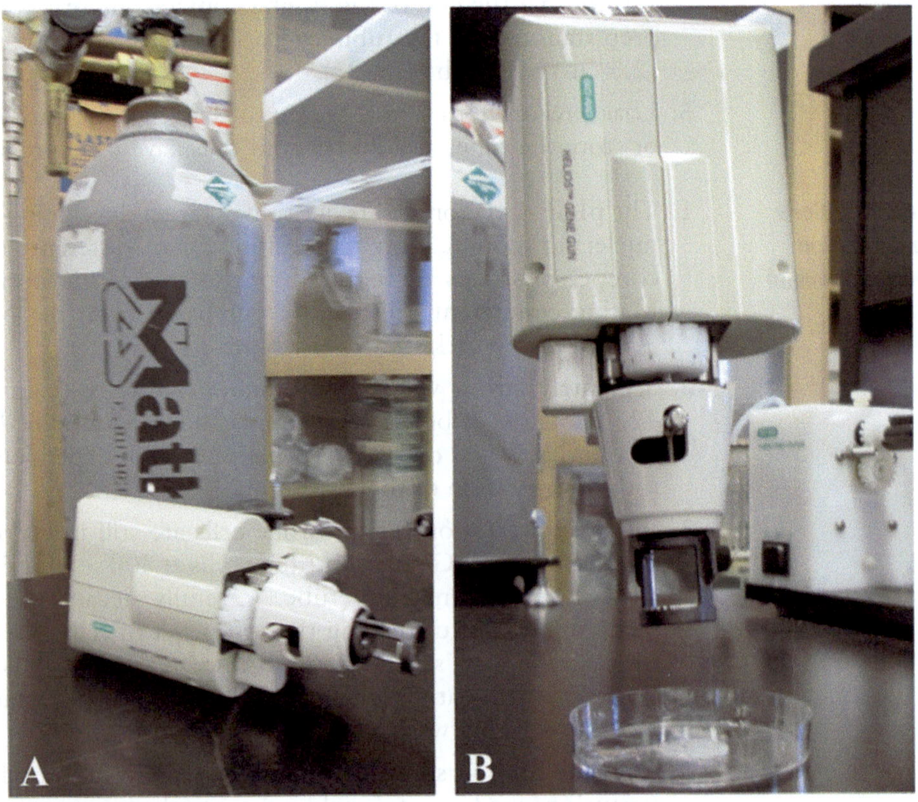

Fig. 3 Assembled gene gun. (**a**) Gene gun attached to a helium tank. (**b**) Firing position of gun

3. Hold the gene gun approximately 20 cm above the Petri dish containing the conjugal pairs (Fig. 3b), and discharge the gene gun with a pressure of 120–180 psi (*see* **Note 13**).

3.7 Screening for Overexpression Transformants

1. Wash the bombarded cells from the Petri dish into Neff's medium containing antibiotic/antimycotic mix, and incubate at 30 °C overnight.
2. Add the appropriate selection antibiotic either paromomycin (120–300 μg/mL) or cycloheximide (30–120 μg/mL) depending on the vector of choice (*see* **Note 14**). Vectors that contain a metallothionein promoter can be activated for expression of antibiotic resistance by addition of 1.0 μg/mL $CdCl_2$ (*see* **Note 15**).
3. Next dispense 250 μL aliquots of cell culture to each well of a 96-well plate, and incubate at 30 °C.
4. Examine each well for antibiotic-resistant clones after 3–5 days. Antibiotic-sensitive cells (non-transformers) will die within 3–5 days. Transformants will display rapid movement, and over time the population density will increase dramatically (*see* **Note 15**).
5. Remove 100 μL of culture from each well containing transformants, and put each aliquot into fresh 3 mL tubes. Add the appropriate concentration of selection marker either 120 μg/mL paromomycin or 30 μg/mL cycloheximide.
6. Repeat **step 4**, except using higher concentration of the appropriate selection antibiotic (either 180–240 μg/mL paromomycin or 45–60 μg/mL cycloheximide) (*see* **Note 15**).

3.8 Somatic Transformations

Somatic transformations are achieved by using plasmid vectors that target genes in the macronucleus of vegetative cells. Therefore, conjugal pairs are not used for this technique.

1. Grow wild-type *Tetrahymena* cells (e.g., strain 427 or 428) in bottles containing 30 mL of modified Neff's medium at 30 °C overnight, usually 12–18 h dependent upon the concentration of the inoculum.
2. Initiate a starvation regime while cells are still in exponential phase of growth. Gently centrifuge cells at 300 × *g* for 5 min in a sterile glass centrifuge tube. Carefully decant the Neff's medium without disturbing the soft pellet of cells. Resuspend the pellet in Dryl's starvation medium and wash 3× using the gentle centrifugation regime with fresh Dryl's starvation medium for each wash.
3. After the third wash, resuspend each strain in 30 mL of Dryl's starvation medium to yield an optical density of 0.2 as measured at 540 nm. Incubate the cells overnight (approximately 18 h) at 30 °C.

4. Transfer concentrated cells onto the surface of sterile 24 cm GF/A Whatman filter paper placed in 100 × 15 mm plastic Petri dishes.
5. Prepare the gene gun and use it as described in Subheading 3.6, **steps 1–3**.

3.9 Screening for Somatic Transformants

Screen for transformants using the protocols described in Subheading 3.7, **steps 1–4**.

4 Notes

1. National *Tetrahymena* Stock Center Plasmid list can be found at (last accessed 10.25.2020) https://tetrahymena.vet.cornell.edu/plasmids.php. The pVGF-1 plasmid can be purchased under Plasmid ID: PID00008.
2. Alternatively Dryl's can be prepared as 10× aliquots.
3. Prepare a suitable transformation vector for either overexpression or replacement of genes in the macronucleus. A list of plasmid vectors is available from the National *Tetrahymena* Stock Center (*see* **Note 1**).
4. Polyvinylpyrrolidone (PVP) serves as an adhesive during the cartridge preparation process. The working concentration of PVP should be between 0.01 and 0.1 mg/mL. We found that 0.01 mg/mL of PVP worked well for us.
5. We have experienced variability in the effectiveness of some commercially available gold particles. However, we achieved good results with gold microcarriers from Seashell Tech: (http://www.seashelltech.com/dnadel.shtml).
6. We used the Branson Model B200 Ultrasonic Cleaner. For gentle sonication, 20–30 s of sonication is sufficient.
7. We used a standard benchtop microcentrifuge for pulse centrifugations. Do not disturb the pellet when decanting the supernatant; it is acceptable to leave a small fraction of the supernatant behind.
8. Tefzel tubing is supplied with the Bio-Rad Gene Gun. It is advisable to begin drying the tube shortly before coating the gold with DNA.
9. Insert the Tefzel tube at the front end of the tubing prep station (right side of Fig. 2b), and gently guide the Tefzel tube until it reaches the internal contact point at the rear end of the tubing prep station (left side of Fig. 2b).
10. Work rapidly and carefully when coating the plastic with gold. A vacuum pump can be used to fill and to empty the Tefzel

tube with the gold-DNA mixture. Alternatively, with a bit of practice, this step can be performed manually with a syringe.

11. *Tetrahymena* mating strains can be purchased from the National *Tetrahymena* Stock Center (last accessed 10.25.2020): https://tetrahymena.vet.cornell.edu/strains.php. We achieved successful transformations with mating pairs CU427.4 (Stock ID: SD00715) and CU428.2 (Stock ID: SD00178).
12. In order to prevent low mating efficiency, the Neff's medium must be completely washed from the culture and replaced with Dryl's starvation medium in order to ensure adequate starvation of cells. Incompletely starved cells will continue to feed and will not mate. Excessive centrifugal force will lyse some of the cells, and the cell debris becomes nutrient material for the cells and consequently, they are not starved. In our hands, optimal starvation time is 18 hours assuming no nutrient medium is present in the culture.
13. *Tetrahymena* cells are very sensitive to biolistic force and will splatter out of the Petri dish when subjected to bombardment at pressures greater than 200 psi. We observed very good results at psi range between 120 and 180. For higher discharge pressures, preparing cartridges with higher PVP concentrations can increase the total number of particles delivered but may increase the number of cells that splatter out of the Petri dish.
14. Begin screening with low concentration of antibiotic (120 μg/mL paromomycin or 30 μg/mL cycloheximide). Increase antibiotic concentration until cells can no longer tolerate a further increase in concentration. It is advisable to replenish the 96-well plates with fresh Neff's medium containing antibiotic/antimycotic mix and the appropriate selection marker after 2–3 days. $CdCl_2$ should not be added to the replenishment media.
15. We observed basal expression of GFP or GFP-tagged epitopes with the pIGF-1 vector in the absence of $CdCl_2$ consistent with another observation [37]. It is possible that the $FeCl_3$ in the Neff's media mildly activates the MTT1 promoter.

Acknowledgments

Protocols in this chapter are based on research performed at Brooklyn College-CUNY and supported by Grants MCB 0517083 and DBI 0619460 from the National Science Foundation awarded to Professor Emeritus, Dr. Ray H. Gavin. The authors would also like to thank Professor Emeritus, Professor Ray H. Gavin for organizing this *Methods in Molecular Biology: Cytoskeleton* series and his constant friendship and professional support throughout the years.

References

1. Karrer KM (2000) *Tetrahymena* genetics: two nuclei are better than one. In: Asai DJ, Forney JD (eds) Methods in cell biology: *Tetrahymena thermophila*, vol 62. Academic, pp 127–186
2. Eisen JA, Coyne RS, Wu M et al (2006) Macronuclear genome sequence of the ciliate *Tetrahymena thermophila* a model eukaryote. PLoS Biol 4:e286. https://doi.org/10.1371/journal.pbio.0040286
3. Merriam EV, Bruns PJ (1988) Phenotypic assortment in *Tetrahymena thermophila*: assortment kinetics of antibiotic-resistance markers tsA death and the highly amplified rDNA locus. Genetics 120:389–395
4. Chalker DL (2012) Transformation and strain engineering of Tetrahymena. In: Collins K (ed) Methods in cell biology: *Tetrahymena thermophila*, vol 109. Academic, pp 327–345
5. Karrer KM (2012) Nuclear dualism. In: Collins K (ed) Methods in cell biology: *Tetrahymena thermophila*, vol 109. Academic, pp 29–52
6. Gaertig J, Gorovsky MA (1992) Efficient mass transformation of *Tetrahymena thermophila* by electroporation of conjugants. Proc Natl Acad Sci 89(19):9196–9200
7. Gaertig J, Gu L, Hai B, Gorovsky MA (1994) High frequency vector-mediated transformation and gene replacement in *Tetrahymena*. Nucleic Acids Res 22:5391–5398
8. Cassidy-Hanley D, Bowen J, Lee JH, Cole E, VerPlank LA, Gaertig J, Gorovsky M, Bruns PJ (1997) Germline and somatic transformation of mating *Tetrahymena thermophila* by particle bombardment. Genetics 146:135–147
9. Bruns PJ, Cassidy-Hanley D (2000) Biolistic transformation of macro- and micronuclei. In: Asai DJ, Forney JD (eds) Methods in cell biology: *Tetrahymena thermophila*, vol 62. Academic, pp 501–502
10. Yao MC, Yao CH (1991) Transformation of *Tetrahymena* to cycloheximide resistance with a ribosomal protein gene through sequence replacement. Proc Natl Acad Sci 88:9493–9497
11. Shang Y, Song X, Bowen J et al (2002) A robust inducible-repressible promoter greatly facilitates gene knockouts conditional expression and overexpression of homologous and heterologous genes in *Tetrahymena thermophila*. Proc Natl Acad Sci U S A 99 (6):3734–3739. https://doi.org/10.1073/pnas.052016199
12. Gotesman M, Hosein RE, Gavin RH (2010) A FERM domain in a class XIV myosin interacts with actin and tubulin and localizes to the cytoskeleton phagosomes and nucleus in *Tetrahymena thermophila*. Cytoskeleton 67:90–101
13. Gotesman M, Hosein RE, Gavin RH (2011) MyTH4 independent of its companion FERM domain affects the organization of an intramacronuclear microtubule array and is involved in elongation of the macronucleus in *Tetrahymena thermophila*. Cytoskeleton 68:220–236
14. Cole ES, Cassidy-Hanley D, Pinello JF, Zeng H, Hsueh M, Kolbin D, Clark TG (2014) Function of the male gamete-specific fusion protein HAP2 in a seven-sexed ciliate. Curr Biol 24:2168–2173. https://doi.org/10.1016/j.cub.2014.07.064
15. Horrell SA, Chalker DL (2014) LIA4 encodes a chromoshadow domain protein required for genomewide DNA rearrangements in *Tetrahymena thermophila*. Eukaryot Cell 13:1300–1311
16. Turkewitz AP, Bright LJ (2011) A Rab-based view of membrane traffic in the ciliate *Tetrahymena thermophila*. Small GTPases 2:222–226. https://doi.org/10.4161/sgtp.2.4.16706
17. Klein TM, Wolf ED, Wu R, Sanford JC (1987) High-velocity microprojectiles for delivering nucleic acids into living cells. Nature 327:70
18. Christou P, McCabe DE, Swain WF (1988) Stable transformation of soybean callus by DNA-coated gold particles. Plant Physiol 87:671–674
19. Yang NS, Burkholder J, Roberts B, Martinell B, McCabe D (1999) In vivo and in vitro gene transfer to mammalian somatic cells by particle bombardment. Proc Natl Acad Sci 87:9568–9572
20. Fynan EF, Webster RG, Fuller DH, Haynes JR, Santoro JC, Robinson HL (1993) DNA vaccines: protective immunizations by parenteral, mucosal, and gene-gun inoculations. Proc Natl Acad Sci 90:11478–11482
21. Johnston SA, Tang DC (1994) Gene gun transfection of animal cells and genetic immunization. In: Roth MG (ed) Methods in cell biology: protein expression in animal cells, vol 43. Academic, pp 353–365
22. O'Brien JA, SCR L (2006) Biolistic transfection of neuronal cultures using a hand-held gene gun. Nat Protoc 1:977–981
23. Belyantseva IA (2009) Helios Gene Gun-mediated transfection of the inner ear sensory epithelium. In: Sokolowski B (ed) Auditory and vestibular research. Methods and protocols, Methods Mol Biol, vol 493. Humana Press, pp 103–124

24. Reytblat I, Lipovsky A, Gedanken A (2016) DNA microspheres coated with bioavailable polymer as an efficient gene expression agent in yeasts. J Nanomater 2016:1–8
25. Acanda Y, Wang C, Levy A (2019) Gene expression in citrus plant cells using Helios® Gene Gun System for particle bombardment. In: CataraMoshe AF, Licciardello B-JG (eds) Citrus tristeza virus. Humana Press, pp 219–228
26. Bednenko J, Harriman R, Mariën L, Nguyen HM, Agrawal A, Papoyan A, Bisharyan Y, Cardarelli J, Cassidy-Hanley D, Clark T, Pedersen D (2018) A multiplatform strategy for the discovery of conventional monoclonal antibodies that inhibit the voltage-gated potassium channel Kv1. 3. MAbs 10:636–650
27. El-Haddad H, Przyborski JM, Kraft LG, McFadden GI, Waller RF, Gould SB (2013) Characterization of TtALV2, an essential charged repeat motif protein of the *Tetrahymena thermophila* membrane skeleton. Eukaryot Cell 12:932–940
28. O'Brien JA, Lummis SCR (2007) Diolistics: incorporating fluorescent dyes into biological samples using a gene gun. Trends Biotechnol 25(11-2):530–534. https://doi.org/10.1016/j.tibtech.2007.07.014
29. Fuentes-Valencia MA, Fajer-Ávila EJ, Chávez-Sánchez MC, Martínez-Palacios CA, Martínez-Chávez CC, Junqueira-Machado G, Lara HH, Raggi L, Gómez-Gil B, Pestryakov AA, Bogdanchikova N (2020) Silver nanoparticles are lethal to the ciliate model *Tetrahymena* and safe to the pike silverside Chirostoma estor. Exp Parasitol 209:107825
30. Aps LR, Tavares MB, Rozenfeld JH, Lamy MT, Ferreira LC, Diniz MO (2016) Bacterial spores as particulate carriers for gene gun delivery of plasmid DNA. J Biotechnol 228:58–66
31. Zilony-Hanin N, Rosenberg M, Richman M, Yehuda R, Schori H, Motiei M, Rahimipour S, Groisman A, Segal E, Shefi O (2019) Neuroprotective effect of nerve growth factor loaded in porous silicon nanostructures in an Alzheimer's disease model and potential delivery to the brain. Small 15:1904203
32. Dürichen H, Siegmund L, Burmester A, Fischer MS, Wöstemeyer J (2016) Ingestion and digestion studies in *Tetrahymena pyriformis* based on chemically modified microparticles. Eur J Protistol 52:45–57
33. Hosein RE, Gavin RH (2007) Myo1 localizes to phagosomes, some of which traffic to the nucleus in a Myo1-dependent manner in *Tetrahymena thermophila*. Cell Motil Cytoskeleton 64:926–935
34. Woods G, Zito K (2008) Preparation of gene gun bullets and biolistic transfection of neurons in slice culture J Vis Exp (12):e675. https://doi.org/10.3791/675
35. Orias E, Hamilton EP, Orias J (2000) *Tetrahymena* as a laboratory organism: Useful strains cell culture and cell line maintenance. In: Asai DJ, Forney JD (eds) Methods in cell biology: *Tetrahymena thermophila*, vol 62. Academic, pp 189–211
36. Dryl S (1959) Antigenic transformation in Paramecium aurelia after homologous antiserum treatment during autogamy and conjugation. J Protozool 6(25):96
37. Couvillion MT, Collins K (2012) Biochemical approaches including the design and use of strains expressing epitope-tagged proteins. In: Collins K (ed) Methods in cell biology: *Tetrahymena thermophila*, vol 109. Academic, pp 347–355

Chapter 19

Proteomic Tools for the Analysis of Cytoskeleton Proteins

Carlos Barreto, Andriele Silva, Eliza Wiech, Antonio Lopez, Avdar San, and Shaneen Singh

Abstract

Proteomic analyses have become an essential part of the toolkit of the molecular biologist, given the widespread availability of genomic data and open source or freely accessible bioinformatics software. Tools are available for detecting homologous sequences, recognizing functional domains, and modeling the three-dimensional structure for any given protein sequence, as well as for predicting interactions with other proteins or macromolecules. Although a wealth of structural and functional information is available for many cytoskeletal proteins, with representatives spanning all of the major subfamilies, the majority of cytoskeletal proteins remain partially or totally uncharacterized. Moreover, bioinformatics tools provide a means for studying the effects of synthetic mutations or naturally occurring variants of these cytoskeletal proteins. This chapter discusses various freely available proteomic analysis tools, with a focus on in silico prediction of protein structure and function. The selected tools are notable for providing an easily accessible interface for the novice while retaining advanced functionality for more experienced computational biologists.

Key words Proteomics, Homology modeling, Comparative modeling, Threading, Sequence similarity, Multiple sequence alignment, Protein domains, Secondary structure prediction, Structure analysis, Protein-protein interactions, Docking analysis

1 Introduction

The cytoskeleton is made up of a diverse family of structural proteins that operate in conjunction with interaction partners to form a conspicuous and dynamic internal superstructure within a cell. The proteins comprising the cytoskeletal system in any particular cell type represent a bewildering array of distinct molecular architectures: combinations of various modular domains, the function of which may or may not be currently known. Although a wealth of structural and functional information is available for representative members of the various classes of cytoskeletal proteins, many more remain partially or fully uncharacterized. However, this dearth of

Ray H. Gavin (ed.), *Cytoskeleton: Methods and Protocols*, Methods in Molecular Biology, vol. 2364,
https://doi.org/10.1007/978-1-0716-1661-1_19,

experimental data may be supplemented by predictive analyses. Computational tools that are freely and easily accessible to the scientific community currently provide an excellent starting point for prediction of the structural and functional properties of these uncharacterized protein sequences. These predictions can lead to well-designed experiments that can further probe the hypothesized function.

Proteomic tools are available for a variety of tasks, ranging from sequence analysis and characterization to the generation of high-resolution 3D models. For example, tools exist for finding sequentially similar sequences and putative homologs (e.g., BLAST, HMMER), recognizing and characterizing functional protein domains and residues (e.g., ScanProsite, Pfam), generating tertiary structure models for a given sequence (e.g., MODELLER, I-TASSER, HHpred), and predicting interactions of proteins with other proteins or other macromolecules (e.g., AutoDock Vina, LeDock, Haddock, ClusPro) (*see* **Note 1**, Table 1). These tools may be used alone as a platform for studying uncharacterized proteins or to map experimentally observed functionality onto a structural model. Alternatively, proteomic analyses may be carried out in conjunction with traditional wet-lab techniques to design more meaningful experiments. For example, a preliminary bioinformatics analysis may identify targets that are more likely to be pertinent to a given study. This chapter discusses various proteomic analysis tools with a focus on protein structure prediction and functional characterization. These in silico experiments are designed to begin with a raw sequence, which may have no functional annotation associated with it, and in a stepwise manner derive sequence and structural information leading up to a three-dimensional model of the sequence or its constituent domain(s), which can be analyzed for its biophysical properties and probed for interaction with putative binding partners.

2 Materials

2.1 Computer with Internet Access

A computer with any operating system may be used to perform the computational analyses described in these methods. However, some tools may not be available on some operating systems or there may be differences across platforms. Certain techniques utilize command line interfaces and may require knowledge of operating system-specific commands.

For tools requiring Linux, open-source operating systems such as Ubuntu are accessible, beginner-friendly options that either may be installed as a second independent operating system (dual-boot) or may run concurrently with another operating system as a virtual machine on emulators such as VirtualBox. However, as some tools are quite resource intensive, the dual-boot option is recommended to ensure system stability.

Most web servers are accessible via any Internet-connected device. Some require an educational e-mail for registration and verification.

2.2 Visualization Software

2.2.1 Visualization Software: Chimera

To visualize and manipulate three-dimensional protein structures, a molecular visualization system is required. Chimera is a comprehensive software package for rendering and manipulating three-dimensional protein structures [1]. In addition to providing an expansive toolset related to protein structural analyses, Chimera also includes limited functionality with respect to sequence analysis and alignment, protein-ligand docking, and molecular dynamics. While many software packages are available for each of these discrete functions, having the toolset available in a single operating environment may result in a more efficient workflow.

Chimera is available for Windows 7 or later, Mac OS X, and Linux, and all are freely accessible for noncommercial use. Only 64-bit builds are currently supported, but 32-bit builds are still available for download as of publication. The most recent version of Chimera, 1.14, was released on November 13, 2019. To install Chimera, follow these steps:

1. Chimera binaries are available at www.cgl.ucsf.edu/chimera/download.html.
2. Download the binary installer that is appropriate for your operating system of choice.
3. Run the installer (on Windows) to install the program. You will be prompted to select an installation directory. On Mac OS X, load the installer's ".dmg" file and drag the enclosed executable to your applications folder. On Linux, open the terminal and use the command chmod +x FILENAME.bin to make the .bin file executable. Then run the executable using the command ./FILENAME.bin, making sure to change FILENAME to the name of the installer that you downloaded. Advanced users may make changes to the installation or simply hit enter at each prompt to accept the default options (*see* **Note 2**).

2.2.2 Visualization Software: PyMOL

PyMOL is another popular open-source visualization package available at www.pymol.org [2]. PyMOL and Chimera each provide a perfectly adequate graphical user interface (GUI) for molecular visualization and annotation. However, PyMOL is likely more accessible to beginners, who may be more comfortable with PyMOL's GUI and who could be overwhelmed by Chimera's more advanced manipulation toolset.

PyMOL2.4 is the most current version of PyMOL. The software is monetized but made available for free under an academic license. Alternatively, you may compile open-source PyMOL directly from the source code or install it via a pre-compiled version that must be installed by the end user using Python (*see* **Note 3**). PyMOL allows for numerous ways to analyze and compare molecular structures in PDB, MOL, MOL2, and SDF filetypes (among others) straight from the graphical user interface. PyMOL's additional versatility lies in the command line interface, and the ready availability of scripts online makes this feature accessible to those not versed in Python. Visit https://pymol.org/ and find your operating system; these instructions will focus on a Windows installation and interface.

1. Download and run the executable (EXE) file and follow the installation prompts.
2. *Obtaining a license.* To purchase a license, visit https://pymol.org/2/buy.html and follow the instructions there. To obtain an academic license, visit https://pymol.org/edu/?q=educational, and complete the survey.
3. Schrodinger will send an e-mail indicating a username, password, and link to download the educational license. Follow the link and type in the relevant information from the e-mail, select the destination folder, and click "Save."
4. Once the license has been downloaded, open PyMOL and a license prompt should appear. Click "Browse for License File," and search for the license file and open. A confirmation window will appear when this is done successfully.

2.2.3 Alternative Molecular Visualization Tools

The Research Collaboratory for Structural Bioinformatics Protein Data Bank (RCSB PDB) maintains a list of many other molecular graphics programs at https://www.rcsb.org/pages/thirdparty/molecular_graphics.

2.2.4 Modeling Software

MODELLER is free for academic use and provides a full-featured modeling package that rivals commercial modeling software suites [3].

1. MODELLER binaries and installation guides are available at https://salilab.org/modeller/download_installation.html.
2. Download the binary installer that is appropriate for your operating system of choice.
3. Use the installation guide for your specific operating system to install the program.
4. Enter your MODELLER license key when prompted (*see* **Note 4**).

3 Methods

3.1 Protein Sequences Databases

The International Nucleotide Sequence Database Collaboration is an international consortium dedicated to collecting and cataloging sequence data submitted by researchers around the world. The consortium is comprised of the DNA Data Bank of Japan (DDBJ) [4]; the European Bioinformatics Institute, part of the European Molecular Biology Laboratory (EMBL-EBI) [5]; and the National Center for Biotechnology Information (NCBI) [6] which exchange data daily to ensure that each member's respective databases have up-to-date sequence data.

For proteomic studies, the Universal Protein Resource (UniProt) [7] administered by EMBL, the Swiss Institute of Bioinformatics (SIB), and the Protein Information Resource (PIR) focuses on developing and maintaining a well-curated repository of protein sequences and annotations.

While the underlying dataset is shared between participating institutions, there are notable differences in interface and crucially, in annotation. While there have been efforts to implement and improve automatic annotation systems [8], manual annotation is the far more valuable, albeit labor intensive, means of understanding the function and context of any given snippet of sequence data. Consequently, it is useful to cross-reference across databases, and it is important to be aware of the source and quality of annotation.

1. Automatic annotation: Based on sequence similarity and presumed homology to proteins of known function, this is the least reliable source of annotation. While it may be a useful starting point, it should not be taken at face value and should be evaluated by the researcher.
2. Submission annotation: Researchers submitting sequence data may also include annotations related to their research project. Such submissions may include links to published article (s) associated with the submission or useful contextual information. Such annotations may be quite useful but are not independently reviewed before being added to the database, and so they only reflect the judgment of the submitting researchers.
3. Curated annotation: The highest quality of annotation, curation, involves a comprehensive analysis of sequence records, including a critical review of associated experimental and predicted data, numerous computational analyses, and an evaluation of the literature to identify pertinent biological information. Such records can combine many submissions and substantial data into a single nonredundant record that provides a wealth of useful data.

3.2 Retrieval of Protein Sequences

Protein sequences, and their annotations, from both NCBI and UniProt are available in several formats (*see* **Note 5**). Protein sequences may be located directly using a record's accession number, by keywords, or via a BLAST search (*see* Subheading 3.2 for details on performing a BLAST search).

3.2.1 Retrieving Protein Sequences from NCBI's GenPept

The following steps explain how to retrieve a specific protein sequence record from the NCBI's GenPept protein database:

Accession Number Search

If you know the accession number assigned to a particular protein of interest, you may search for the record directly.

1. Navigate to www.ncbi.nlm.nih.gov/ for the main portal entrance and simply enter the accession number into the search bar (*see* **Note 6**).
2. You may click download to save the protein sequence as an .faa file (*see* **Note 7**). Alternatively, click on the FASTA link to open a webpage where you may view the FASTA sequence. You may also open the protein's entry itself for more information and links to related records.

Keyword Search

1. You may search for protein sequences from the main portal by selecting "Protein" from the drop-down box to the left of the search bar. Alternatively, you may navigate to www.ncbi.nlm.nih.gov/guide/proteins/ for NCBI's list of useful proteomic databases and resources, and then click the link to "Protein Database."
2. For a simple search, you can type directly into the search box (*see* **Note 8**). For more complex searches, click on the "Advanced" link below the search box.
3. In the advanced tab, you may build complex queries that search against specific fields in the record, e.g., Author, Organism, Journal, Gene Name, Keywords, etc., and link them with Boolean operators. Select the field you wish to search from the drop-down box on the left and type the search query for that field. Click on the "+", plus sign, to the right of the search box to add more search terms as needed, making sure to specify the correct Boolean relationship in the drop-down box to the left of the field drop-down. If you wish to remove a particular search term, simply click on the "−", minus sign, to the right side of that search term.
4. From the search results, you may read through summaries of the returned records to identify proteins of interest or further filter the results using the filter terms to the left of the results list.

5. Upon identifying a protein of interest, you may click on the title directly to open the record or click on a viewing option that navigates to a specific section of the record. Viewing options include "Identical proteins," a list of the records for all submissions with sequences identical to that of the current protein; "FASTA," a minimally annotated sequence in FASTA format (*see* **Note 5**); and "Graphics," for a schematic illustration of the gene locus that encodes the selected protein.
6. From the FASTA view, the protein's sequence can be copied and pasted or saved to a text file, as follows. From the "Send to" drop-down list, choose "File" and "FASTA" format. A FASTA-formatted text file containing the sequence will be downloaded to your browser's default downloads folder.
7. Similarly, the FASTA for multiple entries may be saved from the search results list by clicking on the checkbox to the left of each result and then using the "Send to" drop-down list as in the previous step.

3.2.2 Retrieving Protein Sequences from UniProt

The following steps explain how to retrieve a specific protein sequence record from the EMBL-EBI's UniProt protein database:

Accession Number Search

1. Navigate to www.uniprot.org/ and enter the accession number into the search bar (*see* **Note 6**).
2. If the accession number is correct, the record will open automatically. To quickly access the FASTA, you can click on the "Format" tab at the top of the record. This will open a menu to change the view format where you can select the FASTA view. To copy and paste: Click the FASTA(canonical) link in the "Format" tab menu and copy the text.

   To save to file: Right click on the FASTA(canonical) link in the "Format" tab menu and click "Save link as." It will be saved as a .fasta file to your default download directory.

Keyword Search

1. You may search for protein sequences from the main portal www.uniprot.org/ by selecting "UniProtKB" from the drop-down box to the left of the search bar.
2. Both simple and advanced searches of the UniProtKB database may be conducted as described above for keyword searches of GenPept (**steps 2** and **3**, respectively).
3. By default, UniProt search results will display basic information about the returned records: Entry ID and name, Protein names, Gene names, Organism, and Length. However, these columns may be customized to tailor the search results to information relevant to your criteria. By clicking the "Columns" tab at the top of the search results, columns may be added to display data found in an entry's record directly in the

search results, allowing you to scan large numbers of records for specific criteria like interactions, function, subcellular location, domains, etc. at a glance.

4. Entries may be viewed by clicking on the accession number listed in the entry column. To retrieve the sequence of a protein of interest from within its record, click on the "Format" tab at the top of record and save its FASTA sequence as described above.
5. To retrieve multiple FASTA entries at once from the search results page, click on the selection box (the first column of each entry) for the entries you wish to save and click on the "Download" tab. Select the "Download selected" radio button and choose FASTA(canonical) from the "Format" drop-down menu. For large numbers of FASTA entries, it is advisable to select the "Compressed" option which will download the file as a compressed .gz file. Otherwise, selecting "Uncompressed" and clicking on GO will link to a page with the requested FASTA. You may either copy and paste the FASTA from this page or right click on the page and click "Save as," allowing you to save the page as a text file.

3.3 Text Formatting

It is highly recommended that a simple plain text reader such as Microsoft's Notepad application be used for sequence collection and manipulation. Programs that allow more advanced document formatting options, such as Microsoft Word, are likely to add invisible formatting characters (e.g., an invisible tab or new line symbols) that may cause errors when introduced into other programs. Such word processors should be avoided entirely, as even simply cutting and pasting into them can cause problems. More advanced users will likely prefer basic text editors such as Notepad++ which retain substantial functionality without introducing problematic formatting.

3.4 Searching for Similarities and Inferring Homologies

On its face, the flood of sequence data let loose by advances in sequencing technology has created an unmanageably large pile of discrete sequences. For these databases to be useful, however, we must be able to identify sequences of interest. Broadly speaking, we approach this problem in two complementary ways: theoretically and computationally. Theoretically, we take advantage of our understanding of structural biology and evolution to apply concepts such as conservative substitution and the principle of homology to organize and structure our databases of individual protein sequences. In doing so, we can greatly reduce the computational problem, making large databases of sequence data searchable and preventing them from becoming so unwieldy that they are reduced to data graveyards.

The BLAST algorithm [9], for example, takes advantage of the fact that conservative substitutions are unlikely to impair protein structure and function to score how similar one sequence is to another, allowing it quickly to find sequences within a database that are sequentially similar to a query sequence (*see* **Note 9**). Thus, the essential function of the BLAST algorithm is to identify the sequences most similar to one's query. The theoretical assumption is that such sequences share a functional and/or evolutionary relationship (i.e., putative homology) to the query. If substantiated, these similarities allow us to make claims about the nature of the evolutionary relationship across species and may allow us to extend information related to the properties and function of homologs to our query.

While BLAST searches most easily identify homologous relationships, it can also be used to identify paralogous relationships, as the correlation between structure and function is also captured by the algorithm. By way of example, proteins of different function may share conserved domains or motifs that would elevate its BLAST score relative to a true homolog, spuriously implying a homologous relationship where one does not exist. The nature of the evolutionary relationship cannot be assumed by reference to the scoring algorithm alone, but the types of inferences necessary will not be addressed here.

General use cases for BLAST searches include (a) tracing the lineage of a given protein across species; (b) comparing homologs to identify the impact of structural differences on protein function and properties such as kinetics, binding, etc.; (c) identifying appropriate templates for homology modeling; and (d) identifying and grouping homologs to build a multiple sequence alignment that illuminates conserved sequences and functionally important motifs.

3.4.1 Conducting a BLASTP (Protein-Specific BLAST) of the Nonredundant NCBI RefSeq Database

1. Navigate to NCBI BLAST page at blast.ncbi.nlm.nih.gov/Blast.cgi.
2. Select "protein blast" by clicking on its hyperlink.
3. Cut and paste your sequence of interest (in FASTA format) into the "Enter Query Sequence" textbox (*see* **Note 10**). You can also browse and upload the sequence from a file (in FASTA format).
4. In the "Databases" field, select the appropriate database from the pull-down list. The nonredundant (nr) database is the default as it is the most inclusive database (*see* **Note 11**).
5. (Optional) In the "Organism" field, you may constrain your search to certain taxonomic groupings, either limiting the search to certain groups or excluding groups from the results by clicking on the exclude checkbox to the right of the entry field. You may add multiple organism criteria by clicking on the

"+", plus sign, on the right side of the organism field. Typing in the field will create a drop-down list of possible matches to the NCBI taxonomic database with the corresponding taxid in parenthesis.

6. In the "Program Selection" field, select the "blastp" algorithm (*see* **Note 12**).
7. Advanced users can change important algorithm parameters by clicking on the "Algorithm parameters" link at the bottom. Beginners should avoid making any changes here except for the "Max target sequences" parameter which will increase the number of presented results from the default of 100 to up to 5000 hits.
8. Click on "BLAST" to initiate the search and wait for the results to load. The output page begins with details such as the RID number (*see* **Note 13**), the version of the program used, the citation for BLAST, query length, etc.
9. To download a file containing the FASTAs of selected results: In the "Descriptions" tab, click on "Download" and select "FASTA(complete sequence)." The selected FASTAs will be saved to a text file.
10. Searches may be filtered by organism, percent identity, E value, and query coverage in the fields on the upper right of the output page. To browse the results, different pages may be accessed via the tabs at the top of the search results window.
    (a) *Description*: The "Description" tab loads by default and lists the algorithm results, with columns for BLAST scores, query coverage, E value, and percent identity (*see* **Note 14**). Individual results may be selected/deselected by clicking on the checkbox next to the result. The links at the top of this section can then change the presentation of the selected results. The "GenPept" link will open a window with the GenPept summary for each result's record (up to 100 entries). The "Graphics" link will open a window with a graphical representation of the alignment that highlights the regions of difference between the query and each database sequence. The "Distance Tree" link computes a phylogenetic tree that groups results by sequence similarity, useful for identifying clusters within your results. Finally, the "Multiple Alignment" link can send out selected results to COBALT.
    (b) *Graphic Summary*: The "Graphic Summary" tab provides a graphical representation of the query coverage with graphic identifying potential conserved domains at the top.
    (c) *Alignments*: The "Alignments" tab displays the full alignment between the query and each result.

(d) *Taxonomy*: Finally, the "Taxonomy" tab allows you to organize the BLAST results by species/taxonomic classification.

3.4.2 Specialized BLAST Variants

Since the time of its inception, the basic BLAST algorithm has spawned multiple specialized spin-offs in response to particular use cases. A particularly useful divergence from pairwise alignment techniques such as BLAST that use position-independent scoring parameters are algorithms such as Position-Specific Iterated BLAST (PSI-BLAST) [10] and hidden Markov model-based approaches (HMMER) [11]. These algorithms use position-specific scores to create profiles keyed less to bulk physiochemical properties and more to conservation patterns/signatures found in a set of protein sequences. In doing so, these algorithms can detect distant homologs with low overall sequence similarity that would otherwise be indistinguishable from the many unrelated proteins with similarly low overall sequence similarity.

The PSI-BLAST method, for example, allows the user to dynamically create a customized position-specific scoring matrix ("PSSM"; *see* **Note 15**) based on an iterative series of BLAST searches. The search series is repeated until convergence (i.e., no change in the "hit" list) is observed. During these iterations, the user can adjust the alignment scoring function for the current and subsequent BLAST rounds by setting a cutoff value that determines whether to include low-scoring sequence hits into the PSSM. The addition of low-scoring hits gradually reduces the stringency of the search but may lead to the detection of distant homologs that would have been ignored by a standard BLAST search. The relevance of data uncovered in a PSI-BLAST is highly dependent on optimized parameters (e.g., selection of an appropriate cutoff value). It is recommended that novice users review the NCBI's PSI-BLAST tutorial, located at http://www.ncbi.nlm.nih.gov/books/NBK2590/.

3.4.3 Conducting a PHMMER (Profile-HMM) Search

The HMMER method [11] is another useful tool for identifying homologous proteins. Like PSI-BLAST, HMMER relies on a position-specific scoring profile. However, HMMER constructs this profile using a hidden Markov model approach: a dynamic Bayesian network to model the likelihood of each of the 20 standard amino acids occurring at each position in a given query based on a multiple sequence alignment of the query and sequentially similar proteins. In its simplest form, an implementation of the HMMER method follows a two-step process: (1) a profile-HMM is constructed based on the multiple sequence alignment of the query and similar sequences; and (2) the profile-HMM is queried against a database of protein sequences, with results limited based on a selected cutoff value. By using the profile-HMM in place of the original sequence as the search query, HMMER can identify sequentially distant homologs that may have escaped detection in a BLAST or PSI-BLAST search.

To conduct a PHMMER:

1. Navigate to http://hmmer.org/ and click on "SEARCH" at the top of the screen.
2. A quick search may be run by pasting the FASTA of your query sequence into the search field, selecting the sequence target database you wish to search (*see* **Note 16**), and submitting. Alternatively, to access more search options, click on "Alternative search options" below the Quick Search box.
3. From this page, you may submit sequences by either (a) pasting the FASTA into the search field, (b) uploading a file, or (c) using the accession number of a supported database (*see* **Note 6**).
4. Select the sequence target database you wish to search.
5. To limit searches to specific taxa, you may either (a) type into the organism field in the "Restrict by Taxonomy" section, which will load a drop-down list of possible matches, or (b) click on the "Pre-defined representatives" tab, which will load a taxonomic tree where you may check off which taxa you wish to include in your search (*see* **Note 17**).
6. More options may be accessed by clicking on the "Advanced" button at the end of the page. The parameters most likely to be useful to all users are the "Customise Results" section, where you may (a) add additional columns to display additional information from each result in the results list and (b) change the number of rows displayed per page. Other options and advanced modifications of the search algorithm are available. You may click on the question mark symbol above each section for a description of each parameter and the available options.
7. Click on the "submit" button below the query textbox to initiate the search and wait for the results page to load.
8. The PHMMER output page displays a menu bar at the top of the screen that allows the user to view the results sorted by (a) "Score," (b) "Taxonomy," or (c) "Domain." The fourth option on the menu bar, "Download," opens a page to download the results in multiple formats. To download, simply click on the "Download" link and chose which data and format you wish to download. The "FASTA" format option, which downloads the significant hits from the search, is recommended as a starting point.
9. The results page provides the following details from the PHMMER run:

(a) *Score*: The "Score" page begins with the "Sequence Matches and Features" section, a graphical display of the sequence features identified from your search query. Queries submitted to the server are also searched using three additional methods to detect regions of disorder, signal peptides, transmembrane proteins, and coiled-coils. If a search does not return results, it is not displayed in the graphic, but a small green checkmark next to "disorder," "coiled-coil," and "tm & signal peptide" will indicate that the search was performed. Hovering over features of the graphic will load a pop-up with details and links. A small histogram depicts the "Distribution of Significant Hits" as a function of the E-value. The range of hits with the least significant E-values are binned on the left-hand side with hits within more significant ranges of E-values found the further right you go. Each bar is also color coded to represent the percentage of the results within a particular classification (e.g., taxonomic domain, viruses, unclassified, other). The exact range of e-values and the number of hits within each category may be seen by highlighting a column. Lastly, "Significant Query Matches" displays a summary of each hit. Columns with additional information can be added by clicking on the "Customise" button on the upper right corner of the search results list, which will open the same options menu as described in **step 6**. Clicking on the ">" arrow in the first column to the left expands the result to show the pairwise alignment for a given hit. Alternatively, you may click on the "(show all) alignments" overlay link on the bottom left of the search results. You can open the database record for a particular result by clicking on the link in the "Target" column to access more information. If you searched a UniProt database, you may download the FASTA sequence for a particular result as described in Subheading 3.1, **step 2**.

(b) *Taxonomy*: The "Taxonomic Distribution of All Search Hits" section displays an interactive tree that organizes the results by taxonomic group. The number of results within a taxonomic classification is displayed in parentheses at the node representing it. Clicking on the right-hand arrow at a node will expand that taxonomic group, whereas double-clicking on a node will collapse it. Clicking on a node will generate a small histogram on the bottom right representing the distribution of significant hits for that node (akin to the one described for the overall results on the "Score" page in **step 9**). Lastly, the "Species Distribution" section provides a table listing the number of hits for each listed species. The table will dynamically update to reflect the results for the node you have clicked on. The last column is a link back to the score page that will provide more details on the hits associated with that species.

(c) *Domain*: The same "Sequence Matches and Features" section as described on the score page is presented here to facilitate comparison against the list of "Domain Architectures" below. A table with details of your query's features may be opened by clicking the "Show details" link below the graphic. Additional details for this table may be displayed by clicking on the "Advanced" button on the upper right corner of the opened table. The "Domain Architectures" section groups search results that share the same domain architecture. Each grouping lists the number of sequences within each group, lists the domain names and ordering of the domain architecture, and displays a graphic with the domain architecture of a representative of that group. Clicking on "Show all" will expand the group to display the graphics depicting the domain architecture of each result. Hovering over features of this graphic will load pop-ups with details and links for that feature. The black line in each graphic represents where the query aligned to the target sequence. Hovering over the black line will load a pop-up detailing the alignment coordinates for that hit. Lastly, clicking on "View Scores" will link back to the "Score" page and filter the results to that particular group.

10. Variations on the basic protein HMMER algorithm (PHMMER) are available. For example, the iterative JackHMMer algorithm may be preferable for detecting distant homologs. A link to these variations may be accessed on Search page at www.ebi.ac.uk/Tools/hmmer/search/phmmer.

3.5 Multiple Sequence Alignment

Multiple sequence alignments (MSAs) assist in identifying evolutionary relationships between a set of sequences descended from a common ancestor. From the resulting MSA, sequence homology can be inferred, and phylogenetic analysis can be conducted to assess the aligned sequences' shared evolutionary origins. MSAs also assist in identifying sequence conservation of protein sequence motifs and patterns, structural and functional domains, and secondary structure elements.

In a basic sense, MSAs are constructed by aligning a given set of sequences in a manner that maximizes a given scoring function. In order to account for residues present in only one (or a subset) of the aligned sequences, gaps may be inserted into the alignment, though gaps normally carry a penalty that reduces the scoring function more so than a residue mismatch. For pairwise alignments, and small MSAs, it may be possible for a program to effectively sample all the potential alignment permutations to identify the highest-scoring MSA. However, as the set of sequences increased, the computational requirements scale up fast enough that this strategy becomes prohibitive for medium and large sets. This problem has spurred the development of MSA algorithms that can efficiently sample the available solution space to generate a top-scoring MSA with a reasonable degree of confidence. Some of the most popular MSA generation tools include CLUSTAL Omega [12], T-Coffee [13], MAFFT [14], MUSCLE [15], and PROBCONS [16]. An explanation of the underlying algorithm employed by these servers is available in the papers cited above. The output from each of these tools is likely to be highly similar, if not identical, for most datasets. However, discrepancies may arise due to biases in the algorithms implemented by these tools. As a result, it is often useful to submit a given query to multiple servers and to evaluate the result manually. The user may also wish to apply corrections to the MSA to account for information available to the user (e.g., knowledge of a particular protein family's phylogeny) that was likely not factored into the algorithm.

3.5.1 Creation of a MSA Using CLUSTAL Omega

To create a MSA from a set of protein sequences:

1. Visit the CLUSTAL Omega site hosted by EMBL at the following address: http://www.ebi.ac.uk/Tools/msa/clustalo/.
2. Paste your set of protein sequences in FASTA format into the text box or upload a file containing your set of sequences (*see* **Note 18**).
3. Click on "Submit" and wait for the results page to load. The MSA is displayed on the results page under the "Alignments" tab. A consensus line is provided below the alignment to indicate residue positions that are invariant, conserved, or semiconserved (*see* **Note 19**). The MSA may be colored coded by selecting the "Show Colors" option.
4. To view the MSA, interactively click on the "Results Summary" tab and then press the "Start Jalview" button (*see* **Note 20**)
5. To download the MSA, select the "Results Summary" tab and click on the hyperlink below the label "Alignment in CLUSTAL format." The resulting page will display the alignment in CLUSTAL format, which should be saved as a file with the ". ali" or ".clustal" extension. Depending on the browser, you may also be able to shift-click on this link to save the file directly to your default downloads folder.

3.5.2 Visualization of the MSA Using ESPript 3

ESPript 3 is an alignment editor that can be used to annotate and create high-resolution images of the MSA generated using any alignment program [17]. The program accepts alignment files in various formats as the input (*see* **Note 21**).

1. Go to http://espript.ibcp.fr/ESPript/ESPript/.
2. Click "Run ESPript" button on the top left side of the page.
3. Provide aligned sequences by uploading a file that contains your alignment.
4. There are other options for visual elements such as coloring and determining the number of residues per alignment lane and the distance between them, etc.
5. You can also determine the file type and quality of your output files at the bottom of the page (*see* **Note 22**).
6. After determining all the settings, click the "submit" button on the top left side of the page (*see* **Note 23**).
7. A new page in a new window will open and you will be able to download your file (*see* **Note 24**).

3.6 Protein Motifs, Patterns, and Domains

The presence of domains, repeats, and motifs in proteins dictates the protein structure and/or function. The motifs, patterns, and domains detection programs use sequence similarities and evolutionary history to identify and validate boundaries of specific domains, repeats, and motifs. The various detection methods are sensitive but not always specific in determining repeats' boundaries. Therefore, to determine the final boundaries of the repeats, the data from the repeat/domain detection methods should be supplemented with the data on the secondary structure elements determined from the protein sequence. An integrated sequence analysis for the presence of motifs, patterns, repeats, and domains can reveal highly divergent sequence signatures that, at first glance, do not fit the consensus sequence for a specific signature perfectly [18].

3.6.1 InterPro

InterPro [19] is a database that groups proteins into families and provides functional information about the proteins based on sequence analysis and on the presence of domains and motifs. InterPro uses predictive models, known as signatures, provided by several different databases that make up the InterPro consortium.

1. Point your browser to InterPro: https://www.ebi.ac.uk/interpro/.
2. Upload a file with your protein sequence or paste your sequence into the text box provided under "Sequence, in FASTA format" (*see* **Note 25**).
3. Click "Search."
4. After submission, a processing page will be displayed (*see* **Note 26**).
5. Once the search is completed, click on the link under "RESULTS."
6. A page with your results will be displayed.

7. Hover your mouse over the colored bars to obtain information about your protein. For example, to see the domain residues for your protein: hover your mouse over the bars under "Domain," and the residues will be shown inside a box.
8. To take a snapshot of the results page, click "Options" and select "Save as Image" under "Snapshot."
9. To color your results based on domain relationship, click "Options" and select "Domain Relationship" under "Colour By."
10. To download your results, click "Export" and chose your preferred format (TSV, JSON, XML, GFF, or SVG).

3.6.2 PROSITE

PROSITE is a database of protein motifs, patterns, and profiles that represent known functional sites, domains, and families, which may be searched using the ScanProsite tool [20]. To identify these regions/sites in a given protein sequence:

1. Point your browser to PROSITE: http://prosite.expasy.org/.
2. Paste your protein sequence in the text box provided under "Quick Scan mode of ScanProsite."
3. Click "Scan" (*see* **Note 27**).
4. The results page provides a visual schematic and detailed summary of protein domains and families identified in your target sequence. For additional options, click the "ScanProsite" hyperlink (under "Quick Scan mode of ScanProsite").
5. The default option of ScanProsite allows scanning the given protein sequence against the PROSITE collection of motifs (Option 1). In addition, the user-defined motif(s) can be scanned against the PROSITE collection of proteins (Option 2), or one or more protein sequences and motifs can be scanned against each other (Option 3). For motif-based searches, one may wish to consider generating a pattern to be searched using PRATT [21] or to submit a particular (known) pattern of interest.
6. To create a cartoon to represent the identified motifs/domains in your query sequence through PROSITE or any other protein domains/motifs detection method, the MyDomains-Image Creator [22] can be used . Alternatively, one may use the Illustrator for Biological Sequences (IBS), a standalone package for visualizing and annotating protein or nucleotide sequences (http://ibs.biocuckoo.org/) [23].

3.6.3 PFAM

PFAM [24] is a hidden Markov model (HMM) profile-based method for the detection of protein domains and families. To analyze your protein sequence for Pfam matches:

1. Point your browser to PFAM database hosted by EMBL-EBI (https://pfam.xfam.org/).
2. Click "Sequence Search."
3. Paste your protein sequence into the text box.
4. To search against the Pfam-A families (a set of manually curated families), click "Go" (*see* **Note 28**).
5. The results page will display the graphical output from the search, showing domains identified in your protein sequence as well as a table with statistically significant matches. This table provides additional details regarding the hits: descriptions of domains/families, the alignment between the target sequence and the HMM profile match, and relevant E-values.

3.6.4 CDD

The Conserved Domain Database (CDD) [25] can be used to search for the query protein conserved domains against the NCBI protein database that incorporates structural information from the 3D protein models.

To identify conserved domains in a sequence of interest, follow these steps:

1. Point your browser to https://www.ncbi.nlm.nih.gov/Structure/cdd/wrpsb.cgi.
2. Select "CD-Search and Batch CD-Search."
3. Enter a query protein or nucleotide sequence in the text box.
4. Click "Submit." The results page, normally generated within a few minutes, displays conserved domains detected within the query sequence.

In addition, the user can modify the search options including the options to search different databases, to modify the expect value (E-value) threshold (the default setting is set to produce the lowest number of false-positive results), to adjust the maximum number of hits, and to select the format of the results (concise, standard, or full).

3.6.5 SMART

The Simple Modular Architecture Research Tool (SMART) is a database of protein domains with a focus on signaling domains, which often have divergent sequences. It allows for the analysis of protein domain architectures by providing identification and annotation tools for the constituent domains in protein sequences [26]. To analyze the domain architecture of your protein sequence:

1. Point your browser to SMART: http://smart.embl-heidelberg.de/ (*see* **Note 29**). Click on the box to the left, which has "Normal Mode" highlighted.
2. In the "Sequence Analysis" section on the following page, paste your protein sequence into the "Protein sequence" text

box or provide the UniProt accession number of your sequence (the "Sequence ID or ACC" textbox).

3. Click on "Sequence SMART." The results page will display the domain architecture of your sequence in cartoon form. Click on any of the identified domain to get the sequence details and functional annotation ascribed to that domain.
4. Click on the "Information," "Interactions," "PTMs" (post-translational modifications), or "Orthology" tabs to reveal predictions associated with each of these topics.

3.6.6 HHrepID

HHrepID is a de novo repeat identification method that uses hidden Markov model (HMM) profiles [27]. To detect repeats in your protein sequence:

1. Point your browser to HHrepID (http://toolkit.tuebingen.mpg.de/tools/hhrepid).
2. Copy and paste a single query sequence or a multiple sequence alignment into the "Input" field, or upload a file containing one or more query sequences in FASTA format.
3. Click "Submit." The search results are presented as a graphical output with all identified repeats mapped onto the query. In addition, this output is provided in tabular format and as a colored alignment. Probability values (*P*-values) are provided for each individual repeat and for the whole group of homologous repeat groups found together.
4. The default parameters may be modified (*see* the "Parameters" tab). For example, the number of MSA generation steps can be changed, and the *P*-value thresholds for repeat family detection and self-alignment may be adjusted to increase or decrease stringency. In addition, secondary structure prediction and domain boundary detection may be enabled.

3.6.7 REPRO

1. REPRO is a de novo repeat identification method [28]. To detect repeats in your protein sequence using this method:
2. Point your browser to REPRO: http://www.ibi.vu.nl/programs/reprowww/.
3. Paste your protein sequence of interest into the text box in FASTA format or upload a file containing a sequence to be analyzed.
4. Click "Run Repro." An initial list of "top-scoring non-overlapping local alignments" will be displayed when the search is complete.
5. Specify the number of alignments that would be clustered to detect the repeat sets.

6. Click "Run Repro." The results are presented under four categories: "Top alignments," "Stacked alignments," "Evaluation," and "Fragments." The final repeat sets are listed under "Fragments."

3.7 Protein Secondary Structure Prediction

Secondary structure refers to the localized, regularly occurring structure of a given stretch of protein sequence in the form of alpha helices, beta strands, and loops. Secondary structure is dominated by hydrogen bonding between nearby residues and so presents a far less challenging task than tertiary structure prediction, which must account for additional parameters such as long-distance interactions. As a result, high confidence secondary structure prediction can be accomplished quickly for most protein sequences.

The canonical secondary structure motifs include alpha helices, beta strands, and loops. Surveys of the available crystal structures indicate that each of the proteogenic amino acids has a particular propensity to occur as one of these three secondary structural motifs. These biases result from the unique physiochemical properties of each amino acid and its surrounding residues

Modern secondary structure prediction algorithms typically operate by assigning a secondary structure to each residue in a given sequence based upon a propensity table as well as knowledge-based potentials extracted from analyses of crystalized structures. A particular algorithm's accuracy may vary based upon the type of protein (globular or fibrous) or based upon the protein family. As a result, it may be worthwhile to submit a sequence to several different algorithms. PSIPRED [29], Raptor-XSS8 [30], PSSpred [31], SSPro [32], and Jpred [33] are examples of commonly used secondary structure prediction programs that generally provide results with high accuracy.

3.7.1 PSIPRED

PSIPRED utilizes a neural network to analyze the results from a PSI-BLAST search of the query sequence. The current version of PSIPRED can be accessed at http://bioinf.cs.ucl.ac.uk/psipred/.

1. Copy and paste your sequence of interest into the "input" field, in FASTA format. Other protein prediction tools may also be enabled for a query including domain prediction and homology-based modeling.
2. In the "Submission Details" section, you must provide a "Short identifier" for the job and may optionally provide an e-mail address to receive a job completion alert.
3. Select "Predict" to begin the analysis. Results will be available on the following page, which will periodically refresh. Results are normally complete within 30–60 min. If you provided an e-mail address in **step 3**, you can navigate away from this page; a link to the results will be sent to that address.

4. The results page displays the query sequence, with alpha helices and beta strands colored pink and yellow, respectively. Download links are provided for the raw output and for a zip file containing all the results.

3.7.2 PSSpred

PSSpred utilizes multiple back-propagation neural networks to analyze the results from a PSI-BLAST search of the query sequence. The current version of PSSpred (v2) is available online at http://zhanglab.ccmb.med.umich.edu/PSSpred/. A downloadable package is also available at this address.

1. Copy and paste your sequence of interest into the "PSSpred on-line" field, in FASTA format.
2. You must provide an e-mail address in the "e-mail" field below the submission box. A job name is optional.
3. After submitting the job, a page will be generated indicating the address where results will be posted upon completion.
4. The results page displays the query sequence, with a second row "SS:" showing the predicted secondary structure at each position (H, alpha helix; E, beta strand; C, coil) and a row "conf:" displaying the confidence score (1–9: low-high).

3.7.3 RaptorX-SS8

RaptorX-SS8 predicts both three-state and eight-state secondary structure using conditional neural fields generated from PSI-BLAST profiles as a part of its suite of structure prediction tools (*see* **Note 30**). This differs from PSIPRED and PSSpred, which predict secondary structure using a three-state model (helix, strand, coil). The current version of RaptorX-SS8 is available online at http://raptorx.uchicago.edu/StructurePrediction/predict/.

1. Copy and paste 1–20 sequences of interest into the "Sequence for Prediction" field, in FASTA format, and then press "submit."
2. The following page will provide a link to the address where results will be stored, as well as a progress indicator. Note that you can create an account to organize and store past and current jobs.
3. The results page will provide additional information besides secondary structure, including a complete three-dimensional model of the full sequence and any individual domains that were detected. Secondary structure predictions are available under the tab, "Section III. Detailed Prediction Results." Click on the "+" sign to expand this tab.
4. Initially, the "SS3" tab will be active, and the secondary structure prediction for each residue will be displayed graphically as

a three-state system (helix, blue; strand, yellow; coil, red). Higher bars represent higher confidence.

5. Click on the "SS8" tab to switch the results to an eight-state system. The eight-state model includes five additional secondary structure motifs beyond the three basic elements. The eight-state system is colored as follows: helix, blue; isolated beta-bridge, green; extended strand in a beta ladder, light yellow; three-helix, dark yellow; five-helix (pi-helix), tan; hydrogen-bonded turn. red; bend, orange; and coil, brown. Again, higher bars represent higher confidence.
6. A link to download the "detailed prediction results" is displayed on the right under the heading "Section II. Summary Prediction Results."

3.7.4 SCRATCH

SCRATCH is a suite of predictors based on machine learning and saved profiles. Like Raptor-X, the suite includes SSPro and SSPro8 for three-state and eight-state secondary structure prediction along with numerous other predictors for solvent accessibility, disulfide bonds, disorder, and tertiary structure, among other things. SCRATCH is accessible via http://scratch.proteomics.ics.uci.edu/index.html.

1. Enter your e-mail, an identifier for the name of the query, and a protein sequence. The plain protein sequence should be entered without headers, spacing, or new lines.
2. Check off which predictors you would like to use, and click "SUBMIT."
3. Your results will be delivered via e-mail.

3.7.5 JPred

The JPred server's current version runs JPred 4, which incorporates the Jnet algorithm based on application of multiple sequence alignment profiles in order to make more accurate predictions. In addition to protein secondary structure, JPred also makes predictions on solvent accessibility and coil-coil regions. JPred can be accessed at http://www.compbio.dundee.ac.uk/jpred4/index_up.html.

1. Copy and paste the single-letter amino acid sequence in plain text.
2. Clicking on "Advanced options..." just below the text field opens additional options, including the ability to upload a file.
3. Click "Make a Prediction."
4. A new prediction page will be generated. From this page, the data secondary structure cartoon can be opened as an SVG file by clicking "View Results Summary in SVG." Additional methods of downloading the data are also possible.

3.8 Prediction of Intrinsically Disordered Regions of Proteins

Disordered proteins or protein segments lack an organized tertiary structure. Even though they are believed to be intrinsically disordered regions lacking a structural fold typically associated with function, they nevertheless are often associated with essential biological functions. Furthermore, intrinsically disordered protein regions are vastly abundant in many proteins related with several human disorders, making these segments attractive drug targets for potential therapeutics. In 1997, Romero et al. proposed the first protein disordered region predictor which applied data mining algorithms to protein sequence data without fixed protein three-dimensional structures [34]. Since then, numerous predictors of protein disordered regions have become available and include several data mining and machine learning algorithms such as nearest neighbor algorithm, support vector machines (SVMs), neural networks (NNs), artificial neural network (ANNs), regression, sliding window, random forest, and Bayesian Markov chain model.

3.8.1 DISOPRED3

DISOPRED3 [35] can predict intrinsically disordered regions and protein-binding sites within them. Predicted IDRs are annotated as protein binding through a novel SVM-based classifier, which uses profile data and additional sequence-derived features.

To identify these regions in a protein sequence:

1. Point your browser to PSIPRED: http://bioinf.cs.ucl.ac.uk/psipred/.
2. Select "Sequence Data" under "Select input data type."
3. Select "DISOPRED3 under "Popular Analyses."
4. Paste your protein sequence in the text box provided under "Submission details."
5. Click "Submit" (*see* **Note 31**).
6. The results page will display the full protein sequence and highlight the disordered residues, by (1) a purple outer box indicating disorder prediction without protein binding and/or (2) a green outer box indicating disorder prediction with protein binding.

3.8.2 MFDp2

MFDp2 [36] combines per-residue disorder probabilities predicted by MFDp with per-sequence disorder content predicted by DisCon and applies novel post-processing filters to provide disorder predictions with improved predictive quality.

1. Point your browser to MFDp2: http://biomine.cs.vcu.edu/servers/MFDp2/.
2. Upload a file with your protein sequence or paste your sequence in FASTA format into the text box provided under "1. Upload a file with protein sequences, or paste them into text area."

3. Insert your e-mail address in the text box provided under "Please provide your e-mail address to be notified when results are ready."
4. Click "Run MFDp2."
5. After submission, a processing page will be displayed (*see* **Note 32**).
6. Ready Results can be found either by clicking the link under "Results page" or the link sent by MFDp2 to your e-mail provided during your protein sequence submission (*see* **Note 33**).
7. The MFDp2 results summary page will display the full protein sequence in which "red letters" correspond to predicted disordered residues and "green letters" represent residues predicted as ordered.
8. To download the results as a .cvs file, select the "MFDp" box and "DisCon" box under "Select Predictions," and then click on "Download Results as a .cvs file." To download the results as a .fasta file, select "MFDp" box and "DisCon" box under "Select Predictions," and then click on "Download Results as a .fasta file."

3.8.3 PrDOS

PrDOS [37] is a hybrid design that combines a machine learning model with a template-based approach. The output by the SVM model is combined with results of a search for homologs in PDB.

1. Point your browser to PrDOS: http://prdos.hgc.jp/cgi-bin/top.cgi.
2. Paste your protein sequence in the text box provided under "Query amino acid sequence."
3. Set the "Prediction false positive rate" to 5.0% (*see* **Note 33**).
4. Keep the "Do not use template-based prediction" box unchecked (*see* **Note 34**).
5. Check the box "Receive prediction results by e-mail" and input your e-mail address (*see* **Note 35**).
6. Click "Predict."
7. PrDOS page will provide you with an estimated calculation time. Click "OK."
8. After submission, a processing page will be displayed.
9. Ready Results can be found either by remaining on the processing page until job is done or by e-mail. The HTML-formatted predicted results will display your full protein sequence, in which "red letters" correspond to predicted disordered residues and "black letters" represent residues predicted as ordered. A plot is also displayed, in which predicted disordered

residues are seen above the threshold (red, horizontal) line and predicted ordered residues are below the threshold line. To download the results in either CASP format or CSV format, click "Format" under "Download prediction results" and select the preferred file format, and then click "Download."

3.9 Retrieval of Three-Dimensional Coordinates of Known Protein Structures

The column-based Protein Data Bank (PDB) format and newer chemical-based formats such as the macromolecular chemical interchange format (mmCIF) and molecular modeling database format (MMDB) are different standards for encoding the atomic coordinates of amino acid chains in three-dimensional space so that this information can be recognized and manipulated by software. Solved coordinates, obtained from X-ray crystallography and NMR spectrometry, can be downloaded from repositories such as the RCSB Protein Data Bank (described below) or built using molecular modeling software (*see* Subheading 3.9). In the RCSB PDB repository [38], each structure file has a unique four-character alphanumeric code, called a PDB ID code, for example, the PH domain of *Drosophila* β-spectrin has a PDB ID code 1DRO. To obtain a coordinate file from the RCSB PDB:

1. Navigate to the RCSB PDB repository (www.rcsb.org).
2. You can perform simple searches by entering search terms into the search field at the top of the main page. If you know the PDB ID code of a crystal structure, entering it will directly open its record (proceed to **step 4**).
3. Search terms other than a PDB ID code will return a page of search results. The search results may be filtered from the results page by selecting from the "Refinements" list on the left side of the results page. Tabs at the top of the results can change the display of the results between three options: (a) "Summary" which provides a picture of the structure along with a summary of details, (b) "Gallery" which only displays images of the structures, and (c) "Compact" which provides a single line description of each structure. Similarly, the "Tabular Report" drop-down allows the search result list to be reorganized to display various features of potential interest such as ligands present in the structure, biological details, or crystallography/NMR details. If these predefined reports do not provide all the details of interest, you may also select "Create custom report" in the drop-down to specify the data fields that will be displayed.
4. Clicking on the PDB ID or title of a record will open the record. The record page displays information about the structure. Tabs at the top of the page switch between the sections of the record.

(a) *Structure Summary*: The "Structure Summary" tab provides a variety of important details about the structure, including links to the literature, related structures, ligands, and details about the experimental data. Of particular importance for structures that will be used as modeling templates is the "Protein Feature View" section. The graphic in this section will highlight regions that could not be resolved or that were only partially resolved. Hovering over such a highlighted region will load a pop-up, specifying the location.

(b) *3D View*: The "3D View" tab opens a model viewer (*see* **Note 14**) that allows you to visualize the structure, any ligands, or other molecules or atoms that are present in the structure's solution. A drop-down menu provides several options for viewers with different visualization interfaces and features.

(c) *Annotations*: Structures with annotations such as domain classifications, protein families, molecular function, cellular component, etc. will list them in this section. Clicking on the annotation will return a search page with PDB structures that share that annotation.

(d) *Experiment*: Details related to the crystallization or NMR experimental procedure will be found here.

(e) *Sequence*: Lastly, the "Sequence" tab will load the Protein Feature Viewer to provide graphical summaries of protein biological and structural features. Each track represents a different data source, allowing information from multiple resources to be integrated within a single graphic.

5. To download the record, click the "Download Files" button on the upper right of the record and select "PDB format" to open a save window.

3.10 Modeling the Three-Dimensional Structure of a Protein or Domain Sequence

3.10.1 Modeling Algorithms and Approaches

Currently, the best route to create a three-dimensional model for a full-length protein or isolated domain is by "homology modeling" (comparative modeling), which involves predicting the tertiary structure of a given protein sequence by inferring constraints from a sequentially similar template structure. These algorithms operate on the assumption that similar protein sequences will result in a similar tertiary structure. As a result, the quality of homology models depends on the closeness of the evolutionary relationship on which they are based [39]. In this sense, homology modeling is distinguishable from ab initio or de novo methods, which generate models based purely on the underlying physical principles. Homology modeling methods are limited in that in order to model a given sequence, a suitable experimentally crystalized template must be available, with the threshold for suitability often considered to be

≥30% pairwise sequence identity across the aligned target-template region [40]. In the case of multi-domain targets, this limitation may be overcome by selecting templates for individual domains [41].

To generate high-quality models for queries that fail to satisfy the sequence identity threshold for homology modeling, threading methods can be used. Threading methods are the second major family of modeling algorithms that rely on fold recognition (threading) to predict the tertiary structure of a target sequence. Given that structure is more conserved than sequence, threading algorithms operate by extracting spatial constraints from a given template structure. Therefore, threading methods differ from homology modeling in that templates are selected based upon a sequence-structure alignment rather than a sequence-sequence alignment. For example, a simple threading algorithm might operate by predicting the secondary structure of a given query sequence, searching the PDB databank for crystal structures with a similar secondary structure architecture, and extracting 3D coordinates from the aligned template region in order to predict where the aligned residues in the query should be placed. That being said, modern threading algorithms typically employ much more sophisticated scoring algorithms that account for additional physiochemical parameters and energy minimization. Given that structure is more conserved than sequence, threading methods may be able to generate high-quality models for queries that fail to satisfy the sequence identity threshold for homology modeling.

Homology modeling involves the use of four basic steps: (1) identification of a suitable structural template, (2) alignment of the template and target sequences, (3) model building, and (4) model evaluation (*see* **Note 36**). Structural templates may be identified using the sequence similarity-based algorithm, PSI-BLAST, to search for matches in the protein databank (i.e., the "nr PDB" database option). With respect to desktop software, MODELLER is perhaps the most widely used homology modeling tool [3]. MODELLER is free for academic use and provides a full-featured modeling package that rivals commercial modeling software suites. However, MODELLER functions as a command line tool, which may be daunting to the novice, though easy to follow tutorials are provided by the Sali lab (http://salilab.org/modeller/tutorial/). Users unfamiliar with command line software may wish to access MODELLER through a GUI provided by third-party tools (e.g., Chimera) and online servers (e.g., HHpred). To summarize, in cases where a target protein shares low sequence identity with previously crystalized proteins, homology modeling may need to be coupled with fold recognition methods, which identify the likely protein fold even in cases where there is no clear sequence homology. To that end, many modern prediction servers employ a mixture of homology modeling and threading algorithms ("multi-servers"). Examples of multi-servers include HHpred, Phyre2, LOMETS, I-TASSER. Raptor-X, pGenTHREADER, FUGUE, and SPARKS-X (*see* **Notes 1** and **37**). Several novice-level methods for model generation and a basic modeling script for modeling using MODELLER are discussed below.

3.10.2 Using HHpred to Generate a Homology Model Structure

1. HHpred [42] is accessible through your web browser at the following address: http://toolkit.tuebingen.mpg.de/hhpred).
2. Paste your protein sequence in the designated "Input" field or upload a file with the protein sequence you are interested in modeling. Input your sequence in one of the supported formats, e.g., FASTA (*see* **Note 38**).
3. Select the PDB_mmCIF70 or PDB_mmCIF30 database.
4. Click "Submit" button and wait for the results.
5. When the results appear in the browser window, from the "Hitlist" (under "Hits"), select the top structure(s) and/or any other listed structure(s) that you wish to use as a template for the target sequence. Select the "Model using selection" option from the top menu bar to initiate model generation.
6. HHpred will then proceed to align your query against the selected template(s).
7. Click "Forward to MODELLER" button and an alignment file in ".pir" format is generated.
8. Enter your MODELLER license key (*see* **Note 4**). Click the "Submit" button and wait for your results.
9. When the results are ready, a cartoon rendition of the model will appear. Click "Download PDB File" button in the "3D Structure" panel to save the model on your local computer.

3.10.3 Using I-TASSER to Generate a Threading-Based Model

1. Visit http://zhanglab.ccmb.med.umich.edu/I-TASSER/registration.html in order to request an academic license. Submit a registration form using a ".edu" address.
2. Access the main I-TASSER [43] submission page at http://zhanglab.ccmb.med.umich.edu/I-TASSER/.
3. Paste your target sequence (in FASTA format) into the input form.
4. There are options in the submission page that allow the user to assign atom-atom restraints, to assign templates (with or without alignment), to exclude some templates, or to specify secondary structure for specific residues.
5. The user can uncheck "Keep my results public" box if the user wants to keep the job private (an assigned key will be needed to access the results).

6. Select "Run I-TASSER" to initiate the model building process. The resulting page will provide a link to the modeling output (record this address). Model building and optimization often requires 1–3 days to complete.
7. Results will be available at the address provided in **step 6**.
8. Results are available for a specified time period and can be downloaded as a single archive file (".bz2") via a link at the top of the results page.
9. The results page includes several useful sections in addition to the structural modeling output: "Predicted secondary structure," "Predicted solvent accessibility," "Predicted normalized B-factor" (a measure of flexibility), "Function prediction using COFACTOR," and "Predicted binding sites." A description is provided for each of these functions and the associated metrics for measuring the confidence of these predictions.
10. The most important output from an I-TASSER run is provided in the section titled "Top 5 Models Predicted by I-TASSER." I-TASSER's modeling process involves the generation of several hundred structures for a given sequence, which are then clustered based on three-dimensional similarity. The top-scoring structure from each of the five largest clusters is selected for further refinement (e.g., energy minimization) and provided as output to the user. These models are ranked according to the clustering frequency (i.e., Model 1 is from the cluster with the most members). Confidence scores (C-Score and TM-Score), an estimated RMSD, and a graph of the estimated RMSD at each residue position are provided for each of the top five models.
11. Evaluating the initial output: a TM-Score of <0.17 indicates a model is likely incorrect, whereas a TM-Score of >0.5 indicates that a model is likely to have the correct topology. The estimated RMSD and target-template alignments should be inspected; a low overall TM-Score may result from poor modeling at the highly variable distal ends of a protein while intervening functional domains are modeled accurately.

3.10.4 Generating a Homology Model Using MODELLER

1. Align the target protein with the template using MODELLER's **align2d()** function. This function considers structural information when creating a pairwise alignment between the target and template, which may significantly improve alignment quality for templates with low similarity to the target. For templates with high similarity, however, pairwise alignments using standard alignment algorithms would produce the same alignment.
2. Create Python script for align2d. Use a basic text editor to write the Python script to be executed by MODELLER.

```
from modeller import *
env = environ()
aln = alignment(env)
mdl = model(env, file='TEMPLATEPDB',
 model_segment=('FIRST:A','LAST:A'))
aln.append_model(mdl, align_codes='UNIQUETEMPLATE-
CODE',
 atom_files='TEMPLATEPDB')
aln.append(file='SEQUENCEFILE', align_codes='UNI-
QUEIDENTIFIER')
aln.align2d()
aln.write(file='OUTPUTFILENAME.ali', alignment_for-
mat='PIR')
aln.write(file='OUTPUTFILENAME.pap', alignment_for-
mat='PAP')
```

Substitute in the appropriate filenames and codes for the following variables. Assume the variables (and the files they refer to) are case sensitive and don't use spaces or unusual characters to avoid errors.

TEMPLATEPDB: The filename of the PDB to be used as a template. The file extension .pdb is implied; so if you are using a template file named ModelABC.pdb, only ModelABC needs to be entered. This variable is entered into two locations. The first in the **model()** function to read the PDB file and the second in the **append_model()** function to write the name of the PDB file into the alignment file. This second entry corresponds to field 2 of the PIR entry for the structure (*see* **Note 39**).

UNIQUETEMPLATECODE : The sequence from the model will be assigned this identifier. To avoid errors and future confusion, assign it a unique code. SEQUENCEFILE: The filename of the sequence for the target protein in PIR database format. For example, if the sequence for your file is named "XYZ.ali," you will enter that as the variable here.

UNIQUEIDENTIFIER: The target protein sequence read in from SEQUENCEFILE will be assigned this identifier.

OUTPUTFILENAME: The name you wish to assign the output alignments. The PAP file format is designed to be easily readable, but it is the PIR-formatted file that will be used by MODELLER. The names for the output files need not be identical (*see* **Note 40**). Finally, **model_segment()** specifies the starting and ending chain of the PDB file. Typically, this will default to "A," but for models with more than one chain or generated models, it could be another letter or left blank. The chain identifier may be found in the PDB file, under the ATOM coordinates section, listed after the three-letter residue code. Save the text file as a Python script (e.g., align2d.py).

3. Collect files. Place all the files referenced above in the same directory. You will need the script file (e.g., align2d.py), the T EMPLATEPDB file (e.g., ModelABC.pdb), and the SEQUEN CEFILE file (e.g., XYZ.ali).
4. Execute the alignment script. From the MODELLER command line interface, navigate to the directory and execute the script. The files will be created in the directory from which you have run the script.
5. Check the output files for errors and if desired, manually check the alignment.
6. Create Python script to build models. Create another text file using a basic text editor for the following script:

```
from modeller import *
from modeller.automodel import *
env = environ()
a = automodel(env, alnfile='ALIGNMENTFILE',
 knowns='TEMPLATEPDB', sequence='UNIQUEIDENTIFIER')
a.starting_model = 1
a.make()
```

7. Substitute in the proper values for the variables— ALIGN MENTFILE : The alignment file (in PIR database format) between the template and the protein target. If using the alignment generated by the prior step, it would be named "O UTPUTFILENAME.ali." TEMPLATEPDB: The same variable used in **step 2**. UNIQUEIDENTIFIER : The same variable used in **step 2**. Save the text file as a Python script (e.g., model.py).
8. Execute the script. From the MODELLER command line interface, navigate to the directory and execute the script. The files will be created in the directory from which you have run the script.

3.11 Model Refinement

Protein modeling programs are getting better and more advanced, but often still need further refinement. There are multiple programs designed to improve the quality of protein models. It is suggested that all these programs are utilized in the refinement process as each program will uniquely improve the protein models.

3.11.1 ModRefiner

ModRefiner [44] flexes side chains and the protein backbones to find an optimal conformation that is derived from the native state of the residues regarding the hydrogen bonding, backbone topology, and side chain positioning.

1. Go to https://zhanglab.ccmb.med.umich.edu/ModRefiner.
2. Enter the primary structure of the sequence you are investigating or upload your PDB file that contains your model.
3. Enter your e-mail address in the appropriate box (results will be sent there).
4. Click "Run ModRefiner."
5. Results will be e-mailed to your e-mail address.

3.11.2 SCWRL4

SCWRL4 [45] is a program of protein model refinement that works through side chain packing and energy minimization. It is a standalone program and must be downloaded from the licensor.

1. Go to http://dunbrack.fccc.edu/SCWRL3.php/.
2. Click the "License for SCWRL4 for non-profit users: Click here" link at the top of the page.
3. Fill out the form, go to link, download the program, install it, and run "Scwrl4.exe."
4. A command screen will pop up. Type in "Scwrl4 -i modelin.pdb -o modelout.pdb" to run the program that accepts your modelin.pdb as the input file and outputs modelout.pdb file as the refined model (*see* **Note 41**).

3.11.3 3Drefine

3DRefine [46] is a protein refinement program that focuses on hydrogen bonding network and energy minimization based on force field information derived from experimental data from existing databases. The program also incorporates post-refinement model analysis by either MolProbity or RWPlus algorithms.

1. Go to http://sysbio.rnet.missouri.edu/3Drefine/.
2. Type in your job name and enter your e-mail address.
3. Either copy/paste your structural content of your PDB file into the box or upload a PDB file that contains the coordinates of your model.
4. Choose the post-refinement analysis method by clicking the boxes.
5. Click the "Submit" button.

3.11.4 Refold

Refold [47] is a 3D protein model refinement program that utilizes rapid iterative refinement and molecular dynamics. Refold accepts a PDB file and the primary sequence as inputs.

1. Go to https://www.reading.ac.uk/bioinf/ReFOLD/ReFOLD_form.html.
2. Fill out the primary sequence of your protein model in the input box.
3. Upload the PDB file of your 3D protein model.
4. Optionally, provide an e-mail address where the results will be sent to and a short name for the project.
5. Click the "Predict" button.

3.11.5 refineD

refineD [48] is another protein refinement program that works through machine learning-based restrained relaxation of every individual residue. These residues are then scored with energy functions. The program also allows the usage of two different modes with different degrees of restraints: conservative mode, which relies on cumulative restraints, and adventurous mode, which relies on noncumulative restraints. The program accepts PDB files as an input. After completion of the run, five refined models with RMSD scores will be provided as the output.

1. Go to http://watson.cse.eng.auburn.edu/refineD/.
2. Type in your job name and enter your e-mail address.
3. Either copy/paste your structural content of your PDB file into the box or upload a PDB file that contains the coordinates of your model.
4. Choose the refinement mode (conservative or adventurous).
5. Click the "run refineD" button.

3.12 Model Evaluation Methods

Several of the model building programs provide options for assessing model validation internally, e.g., TM-Score in I-TASSER. In addition to these internal validation methods, one may also wish to consider submitting any generated models to external servers. There are multiple programs available that can be utilized for assessing the quality of the protein models based on various approaches. These programs are also useful for testing whether any refinements in generated models result in improvements in the structural quality of the protein. To be considered a high-quality model, the assessed model is expected to consistently score well in all the evaluation programs.

3.12.1 Verify 3D

Verify3D [49] determines the compatibility of the evaluated residues in the models by assessing the compatibility of the primary structure (1D) of the model with its 3D structural coordinates. It calculates the scores by comparing the structural environment of the individual residues of the model to those assessed for individual residues of native 3D structures obtained from the Protein Data Bank. The program accepts PDB files as input and visualizes the evaluation by providing a line graph depicting the scores of individual residues in the model provided. To pass the test successfully, at least 80% of the residues tested in the model must meet the passing criteria.

1. Go to https://servicesn.mbi.ucla.edu/Verify3D/,
2. Upload your PDB file (*see* **Note 42**).
3. Click “Run verify3d” button.
4. Click the “check status” button.
5. Results will be displayed along with a detailed analysis.

3.12.2 VoroMQA

VoroMQA [50] is a program that assesses structural quality of protein models by using the statistical potential energy of molecules based on interatomic distances and solvent contact areas. VoroMQA locally scores individual residues in the protein model and provides a final global score for the whole molecule based on a comparison of the results obtained from high-quality experimentally determined protein structures.

The program accepts PDB files as input, and the results are depicted in a graphical format with scores for each individual residue. A global score is also provided for the entire molecule. A model with a score of <0.300 is considered a bad model and a score >0.400 indicates a good model.

1. Go to http://bioinformatics.ibt.lt/wtsam/voromqa.
2. Upload your PDB file as input.
3. Provide an optional job description and an e-mail address to send the results to.
4. Click” Submit” button.
5. A new tab within the same page will appear that contains the results and the score.

3.12.3 ProSA-web

ProSA-web [51] is a program that validates structural quality of protein models by generating knowledge-based energy values based on comparisons with experimentally obtained high-quality structures in the Protein Data Bank. ProSA-web outputs a *z*-score plotted on a graph containing the *z*-scores of experimentally crystalized structures of various lengths. A homology- or threading-based model should ideally display a *z*-score comparable to those of the displayed similar-length experimentally solved structures. The local residue-based plots give an assessment of the individual regions of the model.

1. Go to https://prosa.services.came.sbg.ac.at/prosa.php.
2. Choose your PDB file to upload, or, alternatively, you can test structures found on the PDB database by providing PDB code and chain ID (*see* **Note 42**).

3. Click "Analyse."
4. The page will reload, and a new page will appear that indicates the *z*-score, local quality plot, as well as a visual heatmap showing the quality of the assessed model.

3.12.4 ProQ3

ProQ3 [52] is a program developed to evaluate protein model structures utilizing both machine learning and deep learning. ProQ3 measures structural features including secondary structure and surface area accessibility and calculates potential energy and number and type of interatomic interactions by using energy functions derived from both full atom modeling and centroid modeling. The program accepts PDB files as inputs and provides a global score by the end of the evaluation process. A higher global score indicates a model with higher structural quality.

1. Go to https://proq3.bioinfo.se/.
2. Upload your PDB file and optionally provide the primary sequence of the molecule in your PDB file. If not provided, the primary sequence will be extracted from the uploaded PDB file (*see* **Note 43**).
3. Provide an optional job name and an e-mail address to receive the results.
4. Click the "Submit" button.

When the job is finished, the page will reload with results.

3.13 Tools for Analyzing Structures

3.13.1 Visualization of Protein Structure in Chimera

1. Open Chimera.
2. From Chimera menu bar, click "File," and then "Open" and search for your locally saved protein structure as a PDB file. Select the PDB file and click "Open."
3. To download a protein structure from a Database, click "File" and then "Fetch by ID"—a pop-up menu will appear. You can select the protein database of your choice and input the protein structure ID. Enter the protein structure ID and click "Fetch."
4. In either step (**2** or **3**), the protein structure will appear in the main window. It is depicted in "ribbon" format by default.
5. *Getting started*: from menu bar, click "Tools," then "General Controls," and then "Model Panel." The model panel lists the models in Chimera and enables many operations upon them, such as "activate," "group," "clipping," and "Ramachandran plot."
6. *Sequence display*: from menu bar, click "Tools" > "Sequence" > "Sequence." The "Show Model Sequence" window will be displayed. There, you will see a list of all polypeptide chains identified in the PDB file. Select the chain of interest and click "Show." A Sequence panel will be

generated showing the amino acid sequence of the selected chain. By default, alpha helices are depicted in gold and beta strands are depicted in green.

7. Changing *background color*: from menu bar, click "Actions," then "Color," and then "all options." The "Color Actions" window will open, select "background," select the desired color, and close the window.
8. *Coloring the protein*: (1) Color Secondary Structure: from menu bar, click "Tools," then "Depiction," and then "Color Secondary Structure." The "Color Secondary Structure" will open, select the secondary structure you desire to color and check the corresponding box, choose the desired color for the secondary structure, and click "Apply" to visualize the changes. To confirm the color changes, click "OK." (2) Color as a rainbow: from menu bar, click "Tools," then "Depiction," and then "Rainbow." The "Rainbow" window will open, and select if you want to change the color for every residue, chain, or model. Chose the desired color range and click "Apply" to visualize the changes. To confirm the color changes, click "OK." (3) Coulombic Surface Coloring: from menu bar, click "Tools," then "Surface/Binding Analysis," and then "Coulombic Surface Coloring." The "Coulombic Surface Coloring" window will open showing options for fine-tuning the representation. Click "Apply" to visualize the changes. To confirm the color changes, click "OK." By default, negatively charged regions are colored red and positively charges regions are colored blue.
9. *Labeling protein residues*: from menu bar, click "Actions," then "Labels," and then "residue," and then you may display the label in many different forms and even customize by clicking "custom." To turn off Protein Residues Labeling, click "Actions," then "Labels," then "residue," and then off.
10. Basic visualization techniques: (1) As atoms/bonds: from menu bar, click "Actions" > "Atoms/Bonds" > "Show"—you may further select your preferred display such as sticks, ball and stick, sphere, and wire (*see* **Note 44**). (2) As surface: from menu bar, click "Actions" > "Surface" > "Show"—you may further select your preferred display such as solid, mesh, and dot. (3) As ribbon: from menu bar, click "Actions" > "Ribbon" > "Show"—you may further select your preferred display such as flat, edged, rounded, and super smooth.
11. *Structure comparison*: Open two PDB files. From menu bar, click "Tools" > "Structure Comparison" > "MatchMaker." The "MatchMaker" window will open. Click "Apply" or "OK." If you would like to generate a sequence alignment based on structural superimposition: click "Tools" >

"Structure Comparison" > "MatchMaker"> select the box titled "After superimposition, compute structure-based multiple sequence alignment" > click "Apply" or "OK."

3.13.2 Visualization of Protein Structure in PyMOL

1. *Opening and viewing a local PDB file.* With PyMOL open, select "File" in the menu bar and select "Open" and browse for your PDB file and select "open."
2. Multiple files may be opened by repeating this previous step or by selecting multiple PDB files in the same folder by holding down the Control Key and clicking on the files before selecting "Open."
3. *Opening and viewing a PDB file from the PDB database.* Files may be accessed directly from the PDB database by selecting "File" and then "Get PDB File." Type the PDB ID in the "PDB ID:" field and select "Download."
4. *Rotating the model in space.* Click and hold the left mouse button while dragging the mouse to move the model along its central axis. Right-click and hold and drag the mouse up to zoom out, and move it down to zoom in. The scroll wheel defines the thickness of the viewing plane, allowing you the creation of a laminate. Additional commands can be access by right-clicking in the viewing window.
5. *Using the object menu panel.* The object menu panel is a vertically oriented panel at the bottom right of the window. A list of the objects available is listed, and to the right of each object are the buttons A, S, H, L, and C. These three panels can be used to manipulate the displayed PDB file.
6. *Menu panel "action" button.* The menu panel "A" button (for "action") allows for the zoom, orient, center, and origin movements of the mouse. Objects can also be edited, deleted, and renamed, and hydrogen and water molecules can be added or removed. The "State" menu allows PDB files with different conformations or poses to be analyzed individual or all at once.
7. *Menu panel "show" button.* The "S" button (for "show") changes the way that the protein or particular selections can be represented. These representations include cartoons (for alpha helices and beta sheets), ribbons, spheres, and balls and sticks. Selecting representations turns different representations on without affecting the state of any current representations, thus allowing for multiple representations to be set to on at the same time. Hovering over "as" in the main menu allows for a similar selection menu that enables one representation while turning off any other active representations.
8. *Menu panel "hide" button.* The "H" button (for "hide") has opposite functionality as the "show button."

9. *Menu panel "label" button.* The "L" button (for "label") allows for the labeling of individual residues, atoms, and chains and by other properties.
10. *Menu panel "color" button.* The multicolored "C" button (for "color") allows for the coloration of individual objects and selections.
11. *Measuring the distance between objects.* Another useful tool is the ability to measure the distance between two objects. This feature can be accessed by clicking the "Wizard" in the Menu bar. Drop-down to "Measurement" and click on the two objects you want to measure; the resulting distance is represented in Angstroms.
12. *Superposition of multiple objects.* In the command line bar, type <<super [object 1 name], [object 2 name]>> and press "Enter."
13. *Aligning proteins of higher sequence similarity.* In the command line bar, type <<align [object 1 name], [object 2 name]>> and press "Enter."
14. *Aligning proteins of lower sequence similarity CE algorithm.* In the command line bar, type <<cealign [object 1 name], [object 2 name]>> and press "Enter."
15. *Displaying surface electrostatics.* Click on the "Plugin" in the Menu Bar and Select "APBS Electrostatics," and ensure that the Selection drop-down menu is set to the target PDB file and select "Run." Click "Yes" when prompted to close the APBS dialogue box.
16. *Exporting an image.* PyMOL can be used for exporting high-quality images. Compose the molecules so that they are positioned to how you want the final image to look. Click on "File," "Export image as..." and select "PNG."
17. In the new window, click on the drop-down menu and select "ray trace with opaque background" for a black background or "ray trace with transparent background" for a transparent background.
18. Click "Save PNG as. . ." to select the folder; when the folder is selected, rename as needed, and click "Save."

3.13.3 Manipulation of Protein Structures Using Chimera

Energy Minimization: Chimera has implemented limited energy minimization functionality and can be used to refine small errors in homology models (e.g., steric clashes) by moving the model toward a local energy minimum. To minimize a structure, select "Tools" from the menu bar, followed by "Structure Editing" and "Energy Minimization." A dialog box will appear. By default, Chimera's energy minimization routine applies 100 steps of steepest descent followed by 100 steps of conjugate gradient minimization. The "Fixed atoms" field may be used to define regions of the structure that will be held fixed in place during minimization. This feature may be used, for example, to refine only a loop region within a protein.

Refining Rotamers/Mutating Individual Residues: Chimera can be used to refine the side chains of specific residues or to mutate individual residues in a protein, using the Dunbrack or Richardson rotamer libraries. To access this functionality, select "Tools" from the menu bar, followed by "Structure Editing" and "Rotamers." A dialog box will appear. Select a residue (e.g., by control-clicking on the model). Choose a "Rotamer type" on the dialog box and the "Rotamer library" and press "Apply." If you intend to refine a side chain, the "rotamer type" should be the same as the amino acid being refined, whereas if a mutation is intended, the "rotamer type" should be the replacement amino acid. A second dialog box will appear showing a table of various side chain conformations (chi angles and associated probabilities). Select a side chain conformation from the table. The 3D model will be updated to display the selected conformation. By default, the "Existing side chain(s):" field on the second dialog box is set to "replace." Thus, pressing "Apply" will replace the selected residue with the active amino acid rotamer, refining or mutating the given residue. If a residue has been mutated, it is recommended that the energy minimization routine discussed above be applied to the structure.

3.14 Prediction of Protein-Protein Interactions

Protein-protein interactions (PPIs) play an essential role in several biological processes including those that involve cytoskeletal proteins. To fully understand the pathogenesis and treatments of many disorders, it is indispensable to study the detail of these interactions. Nevertheless, the present experimental methods suffer from a high rate of false-positives as well as false-negatives. While high-throughput experimental techniques have created vast amounts of protein-protein interaction (PPI) data, their coverage is still limited and the PPI data is quite noisy. Computational prediction of protein-protein interaction is gaining popularity as it allows for predictions of potential interactions when there is limited information available from experimental methods.

3.14.1 IntAct

IntAct [53] provides a freely available, open-source database system and analysis tools for molecular interaction data. All interactions are derived from literature curation or direct user submissions.

1. Point your browser to IntAct: https://www.ebi.ac.uk/intact/.
2. Type or paste your protein term in the search box provided under ("Search IntAct"), e.g., BRCA2 or LCK_HUMAN (*see* **Note 45**).
3. Click "Search."

4. A page showing your results should appear within minutes.
5. The full amount of interactions can be seen on top of the page under ("IntAct Search Results").
6. For details on all predicted interactions, select the "Interactions (#)" tab. You may get further details on each interaction by clicking on the magnifying glass icon preceding the molecule name.
7. For details on the particular interactors, select the "Interactors" tab. You may narrow your interactors information down by selecting any of the follow tabs: Proteins, Complexes, Compounds, Nucleic Acids, or Genes.
8. *Visualizing interactions*: select the "Graph" tab. IntAct will create a network visualization graph. You may use the "CytoscapeWeb Controls" to change the graph layout by selecting either "force directed," "radial," or "circle." You may also choose to merge edges by selecting either "on" or "off" (*see* **Note 46**).
9. *Selecting first neighbors*: while looking at the graph created by IntAct, right-click on your protein. A list with options will be displayed; select "Select first neighbours."
10. To download your results, select the "Interactions" and scroll down the list under ("Select format to Download"), chose the format, and click "Download."

3.14.2 STRING

STRING [54] is a database of known and predicted protein-protein interactions. The interactions include direct and indirect associations; they stem from computational prediction, from knowledge-based inference, and from interactions aggregated from other databases.

1. Point your browser to STRING: https://string-db.org/.
2. Type or paste your protein name in the text box provided under ("Protein Name"), e.g., CDC15
3. Select the type of organism under "Organism," e.g., *Saccharomyces cerevisiae*. You may also select "auto-detect."
4. Click "Search."
5. By default, a network of predicted protein interactions will be displayed. Nodes can be moved; pop-ups provide information on nodes and edges (*see* **Note 47**).
6. *To get details on interactors*: while still on network view, select the interactor of interest and click on its node. A window will appear showing detailed information about the interactor.
7. *To visualize more/less interactors in your network*: select the "+ More" tab. Every click will increase the number of interactors shown. Conversely, select the "- Less."

8. *Exporting your current network*: select the "Exports" tab and choose a preferred format (PNG or SVG), and then click "Download."
9. *Displaying clusters*: select the "Clusters" tab, choose your preferred option, and click "APPLY."
10. *Obtaining network stats*: select the "Analysis" tab to get information about total number of nodes, total number of edges, average node degree, and average local clustering coefficient.

3.14.3 BioGRID

The Biological General Repository for Interaction Datasets (BioGRID) [55] is a public database that archives and disseminates genetic and protein interaction data from model organisms and humans. It holds over 1,740,000 interactions curated from both high-throughput datasets and individual-focused studies, as derived from over 70,000+ publications in the primary literature.

1. Point your browser to BioGRID: https://thebiogrid.org/.
2. Select "By Identifier" from the list shown above the search box. Type or paste your protein term in the search box provided under ("Search"), e.g., CDC15 (*see* **Note 48**).
3. Select the type of organism under "Organism," e.g., *Saccharomyces cerevisiae*. You may also select "All Organisms."
4. Click "Submit Identifier Search."
5. The Result Summary page will display a table with all the predicted protein interactions.
6. *To get details on interactors*: From the ("Switch View") tab, select "Interactors." A list with all predicted interactors will be shown. You may click on the interactor to read more info.
7. *To get details on interactions*: From the ("Switch View") tab, select "Interactions." A list with all predicted interactions will be shown. The Role and Experimental Evidence Code are provided.
8. *To download the Published Interactions for This Protein*: Above the ("Switch View") tab, click "Download Published Interactions For This Protein."
9. *Visualizing interactions*: From the ("Switch View") tab, select "Network." A network of predicted protein interactions will appear.
10. *Changing visualization layouts*: While still on "Network" tab, select "LAYOUTS." A list showing options for different layouts will appear. Select your desired layout and your network will change accordingly.
11. *Saving your network image*: while still on "Network" tab, select "FILE," and then select "EXPORT TO PNG."

12. Post-translation modifications information: from the ("Switch View") tab, select "PTM Sites," and then a graph and a list containing information about post-translation modifications will appear. You can also access the publication source directly from the list. Alternatively, you may search by publication. Select "By Publication" from the list shown above the search box. Type or paste the PubMed ID (e.g., 29301708) or keyword term.
13. Click "Submit Publication Search."

3.15 Docking Analysis

3.15.1 AutoDock Vina

AutoDock [56] is a molecular docking tool for receptor-ligand binding. AutoDock Vina uses the PDBQT file format, which can be generated from PDB file formats using AutoDock Tools, as detailed here.

1. The AutoDock Tools software can be obtained as part of the MGL Tools suite at http://mgltools.scripps.edu/downloads.
2. After installing AutoDock Tools, open AutoDock Tools.
3. *Selecting the receptor and creating the PDBQT file.* Select the molecule by mousing over the toolbar, selecting "File," and then "Read Molecule, and selecting the containing folder and file of your target molecule.
4. To add polar hydrogens to your molecule, mouse over the toolbar and select "Edit," "Hydrogen, then "Add," and then "Polar Only." Hit "OK" (*see* **Note 49**).
5. The next step is to save the molecule as a PDBQT. Click on "Grid" and then select "Macromolecules" and click on "Choose." Click on your target molecule and then click "Select Molecule" and then choose an adequate location to save your PDBQT file and hit "Save."
6. *Defining the Grid Box.* The Grid Box is the search area on the macromolecule where Vina will search for potential binding sites. It is important to minimize the size of the search box for more accurate results.
7. Begin by mousing over "Grid" and selecting "Grid Box."
8. A new window appears, which allows you to modify the size and center of the search box.
9. First, change the spacing to 1.000 Å.
10. Then, use the dials to modify the *X*, *Y*, and *Z* dimensions and centers of the search box. Do not close the program yet.
11. *Selecting the ligand and creating a PDBQT file.* Use your cursor to select "Ligand," "Input," and "Open," and then find and select your desired ligand PDB file; once located, select "Open." Add hydrogens as needed per the instructions

above. Then mouse over "Ligand" and "Output," and click "Save as PDBQT..." and select your desired folder and hit "Save."

12. *Creating a configuration file.* While it is possible to operate Vina without a configuration file, creating a configuration file is invaluable when performing multiple docking runs. The configuration file includes the center and size of the search box defined using AutoDock Tools and the names of the ligand and receptor files. Exhaustiveness, the number of parallel runs the software performs during the docking query in its search for a global minimum, can also be manually defined in the file. If not specified, the default exhaustiveness is set to 8; the appropriate exhaustiveness setting is determined by the specifications of the user's computer and the amount of time the user would like to spend on a trial.
13. Open a text editor like Notepad or Notepad ++ to create a TXT file.
14. Copy the information for the Grid Box into the TXT file and specify the name of the ligand, receptor, and exhaustiveness. A sample of the TXT script would look like this:

```
receptor = model.pdbqt
ligand = model.pdbqt
center_x = 36.516
center_y = -7.389
center_z = -31.081
exhaustiveness = 50
size_x = 55
size_y = 40
size_z = 40
```

15. Save the TXT configuration file.
16. *Docking using AutoDock Vina.* AutoDock Vina is run using command prompts. Ensure that your configuration file, ligand PDBQT file, and receptor PDBQT file are all in the same folder. In this instance, we are using C:\Vina Docking as our containing folder
17. Open the command prompt and change the directory to the containing folder by typing "cd" followed by a space and the path of the containing folder. In our sample case, the command line should read <<cd C:\Vina Docking>>.
18. Hit the "Enter" key.

19. To run Vina, type the file path leading to where Vina.exe is saved in—this should be entered in quotations. We will use the default file path to which Vina is saved; therefore, the command line should read <<C:\Vina Docking “C:\Program Files (x86)\The Scripps Research Institute\Vina\vina.exe”>>.
20. Press “Enter.” The text should appear if this is done correctly; the text includes common commands for Vina, but we will be relying on our configuration file.
21. Next, we run the configuration file, which in this example has been named “conf.txt.” The command line should read <<C:\Vina Docking> "C:\Program Files (x86)\The Scripps Research Institute\Vina\vina.exe" --config conf6.txt>> (*see* **Note 50**).
22. Hit “Enter.” Vina will begin calculating and a progress bar will appear.
23. You will know that Vina is done calculating when a chart is populated. This chart lists the poses and affinity values.
24. When Vina finishes, a new ligand PDBQT file will be made. The file will be automatically named by appending “_out” to the name of the ligand PDBQT file. In this instance, since our ligand file was named “ligand.pdbqt,” our output file is “ligand_out.pdbqt.”
25. *Viewing the results.* Open a visualization software like PyMOL to open the output file (ligand_out.pdbqt) and the receptor file (receptor.pdbqt). Use the software to cycle through the poses.

3.15.2 LeDock

LeDock [57] is a streamlined software system for flexible docking of small ligands. LeDock uses complementarity along with hydrogen bonding and electrostatics to calculate the docking scenarios.

1. *Obtaining the software.* LeDock can be downloaded from http://www.lephar.com/download.htm.
2. *Creating docking files in LePro.* Before docking with LeDock, it is necessary to create an input file and a PDB file in LePro. Begin by opening LeDock, and select the LePro tab.
3. In the “Protein input” line, select “Choose” and select the PDB file of your protein of interest. The protein output file path and filename will be automatically populated into the same folder from which you retrieved the protein; the default name is pro.pdb. Likewise, the docking parameters file path will be automatically populated into the same folder and be named dock.in by default.
4. Click “Add Hydrogen.” This makes the text field below editable in order to specify the search parameters of the dock.in file.
5. In the text field, specify the search box coordinates under “Binding Pocket” by replacing “xmin,” “xmax,” “ymin,” etc.

with their respective values. Given its lightweight profile, LeDock lacks a native tool for designating an enclosure box for the search; therefore, an alternate program for designating the enclosure box must be used (*see* **Note 51**).

6. Edit the number of desired binding poses; the default setting is 20.
7. Select "Save."
8. *Docking*. Begin by selecting the LeDock tab.
9. The protein input and docking input file paths should be automatically filled if they were just made with LePro. If not, specify the file paths.
10. Input the ligand file path. LeDock requires that the ligand be in MOL2 file format, to convert a PDB ligand file to MOL2 (*see* **Note 52**).
11. Select "Start Docking."
12. LeDock will generate a DOK output file into the same folder as your target ligand.
13. *Visualizing the results*. Rename the *.dok file to *.pdb. Open the new *.pdb output file along with pro.pdb in PyMOL to view the results.

3.15.3 ClusPro

ClusPro [58] is a protein-docking algorithm designed to model protein complexes from the protein subunits in their unbound state. ClusPro server is a rigid-body docking algorithm and directs the transformations relying on searching the entire three-dimensional space of one of the subunits using fast Fourier transform (FFT) to simulate binding.

1. Point your browser to ClusPro: https://cluspro.bu.edu/login.php (*see* **Note 53**).
2. Login or sign up for an account (use a ".edu" e-mail address).
3. Click on the "Dock" tab above the banner.
4. Enter your job name.
5. For the receptor and ligand, enter PDB IDs or upload a PDB file. There is an option to specify chains to be used in docking.
6. There are advanced options that allow the user to specify attraction and repulsion of residues ("Attraction and Repulsion" option), to remove unstructured terminal residues ("Structure Modification" option), to use the uploaded receptor as units of a dimer or a trimer ("Multimer Docking" option), to set receptor as antibody and ligand as antigen ("Antibody Mode" option), to use heparin as a ligand ("Heparin Ligand" option), to submit small-angle X-ray scattering

(SAXS) data ("Saxs Profile" option), and to use restraints to filter results if the interacting residues are known ("Restraints" option).

7. Submit the job by clicking on the "Dock" tab.
8. To access the results, click on the "Results" tab above the banner. You can view the results by clicking on the job ID in "My Finished Jobs" list.
9. The results can be downloaded as ".pdb" files. (balanced, electrostatic-favored, hydrophobic-favored, and VdW+Elec coefficient sets can be selected).
10. The user can click on the "Display Models" drop-down menu to specify the number of models to be displayed for each coefficient set.
11. To view and to initially analyze the models, the user can use PyMOL [2].
12. To specifically identify residues involved in the selected protein-protein interfaces, the protein complexes can be analyzed with the PDBSum Generate server (http://www.ebi.ac.uk/thornton-srv/databases/pdbsum/Generate.html) [59].

4 Notes

1. All analyses described can be carried out using more than one program. Listed below are some alternative software choices for each type of analysis described earlier. This list is not exhaustive, but refers the reader to some of the more popular tools available

Function	Program or database	URL	Primary references
Sequence and structure databases	DDBJ	http://www.ddbj.nig.ac.jp/	[4]
	EMBL-EBI	http://www.ebi.ac.uk/	[5]
	GenBank (proteins)	http://www.ncbi.nlm.nih.gov/protein	[6]
	UniProt	http://www.uniprot.org/	[7]
	RefSeq	http://www.ncbi.nlm.nih.gov/refseq/	[60]
	RCSB PDB	http://www.rcsb.org/	[38, 61]
	neXtProt	http://www.nextprot.org/	[62]
	PIR	http://pir.georgetown.edu/	[63]

(continued)

Function	Program or database	URL	Primary references
Sequence search	BLAST	http://blast.ncbi.nlm.nih.gov/Blast.cgi *Note*: PSI-BLAST, DELTA-BLAST, and other variants are available at this address	[9]
	HMMER	http://hmmer.janelia.org/ *Note*: JackHMMer, HMMsearch, and other HMMer variants are available at this address	[11]
	HHblits	http://toolkit.tuebingen.mpg.de/hhblits	[64]
	FASTA Suite	http://www.ebi.ac.uk/Tools/sss/fasta/ *Note*: searches are available with SSEARCH (local), GGSEARCH (global), and GLSEARCH (global query, local database)	[65]
Multiple sequence alignment	CLUSTAL Omega	http://www.ebi.ac.uk/Tools/msa/clustalo/	[12]
	T-Coffee	http://tcoffee.crg.cat/apps/tcoffee/do:regular	[13]
	MAFFT	http://mafft.cbrc.jp/alignment/server/	[14]
	MUSCLE	http://www.ebi.ac.uk/Tools/msa/muscle/	[15]
	PROBCONS	http://probcons.stanford.edu/	[16]
	PRALINE	http://www.ibi.vu.nl/programs/pralinewww/	[66]
	COMPASS	http://prodata.swmed.edu/compass/compass.php	[67]
	PROMALS3D	http://prodata.swmed.edu/promals3d/promals3d.php	[68]
	Probalign	http://probalign.njit.edu/probalign/login	[69]

(continued)

Function	Program or database	URL	Primary references
Secondary structure prediction	PSIPRED	http://bioinf.cs.ucl.ac.uk/psipred/	[29]
	RAPTOR-XSS8	http://raptorx.uchicago.edu/StructurePrediction/predict/	[30]
	PSSpred	http://zhanglab.ccmb.med.umich.edu/PSSpred/	[31]
	SSPro	http://scratch.proteomics.ics.uci.edu/index.html	[32]
	Jpred4	http://www.compbio.dundee.ac.uk/jpred4/index.html	[33]
	NetSurfP	http://www.cbs.dtu.dk/services/NetSurfP/	[70]
	PredictProtein	https://www.predictprotein.org/	[71]
	PORTER	http://distill.ucd.ie/porter/	[72]
	SOPMA	https://npsa-prabi.ibcp.fr/cgi-bin/npsa_automat.pl?page=npsa_sopma.html	[73]
	Yaspin	http://www.ibi.vu.nl/programs/yaspinwww/	[74]
	SABLE	http://sable.cchmc.org/	[75]
Sequence analysis (domain, motif, active site annotation, etc.)	InterPro	http://www.ebi.ac.uk/interpro/	[19]
	ScanProsite	http://prosite.expasy.org/.	[20]
	PRATT	http://www.ebi.ac.uk/Tools/pfa/pratt/	[21]
	Pfam	http://pfam.sanger.ac.uk	[24]
	CDD	http://www.ncbi.nlm.nih.gov/Structure/cdd/cdd.shtml	[25]
	SMART	http://smart.embl-heidelberg.de/	[26]
	HHrepID	http://toolkit.tuebingen.mpg.de/hhrepid	[27]
	REPRO	http://www.ibi.vu.nl/programs/reprowww/	[28]

(continued)

Function	Program or database	URL	Primary references
Model generation (homology modeling, threading, ab initio)	MODELLER	http://salilab.org/modeller/	[3]
	HHpred	http://toolkit.tuebingen.mpg.de/hhpred	[42]
	I-TASSER	http://zhanglab.ccmb.med.umich.edu/I-TASSER/	[43]
	Robetta	http://robetta.bakerlab.org/	[76]
	Quark	http://zhanglab.ccmb.med.umich.edu/QUARK/	[77]
	RAPTOR-X	http://raptorx.uchicago.edu/	[78]
	LOMETS	http://zhanglab.ccmb.med.umich.edu/LOMETS/	[79]
	Phyre2	http://www.sbg.bio.ic.ac.uk/phyre2/	[80]
	pGenThreader	http://bioinf.cs.ucl.ac.uk/psipred/?pgenthreader=1	[81]
	SWISS-MODEL	https://swissmodel.expasy.org/interactive	[82]
	SPARKS-X	http://sparks-lab.org/yueyang/server/SPARKS-X/	[83]
	Geno3D	http://geno3d-pbil.ibcp.fr/cgi-bin/geno3d_automat.pl?page=/GENO3D/geno3d_home.html	[84]
	IntFOLD	https://www.reading.ac.uk/bioinf/IntFOLD/IntFOLD5_form.html	[85]
	3D Jigsaw	https://bmm.crick.ac.uk/~svc-bmm-3djigsaw/help_crick.html	[86]
	Pcons	http://pcons.net/	[87]
Model quality assessment	VERIFY3D	https://genesilico.pl/toolkit/unimod?method=Verify3D	[49]
	VoroMQA	http://services.mbi.ucla.edu/SAVES/	[50]
	ProSA	http://prosa.services.came.sbg.ac.at/prosa.php	[51]

(continued)

Function	Program or database	URL	Primary references
	ProQ	http://www.sbc.su.se/~bjornw/ProQ/ProQ.cgi	[52, 88]
	ModFOLD	http://www.reading.ac.uk/bioinf/ModFOLD/	[89]
	QMEAN	http://swissmodel.expasy.org/qmean/cgi/index.cgi	[90]
	TM-Score	http://zhanglab.ccmb.med.umich.edu/TM-score/	[91]
	PSVS	https://montelionelab.chem.rpi.edu/PSVS/	[92]
	ModEval	http://modbase.compbio.ucsf.edu/evaluation/	[93]
Model refinement tools	ModRefiner	https://zhanglab.ccmb.med.umich.edu/ModRefiner/	[44]
	SCWRL4	http://dunbrack.fccc.edu/SCWRL3.php/	[45]
	3Drefine	http://sysbio.rnet.missouri.edu/3Drefine/	[46]
	Refold	https://www.reading.ac.uk/bioinf/ReFOLD/	[47]
	RefineD	watson.cse.eng.auburn.edu/refineD/	[48]
	MolProbity	http://molprobity.biochem.duke.edu/	[94]
	Refine^{pro}	http://sysbio.rnet.missouri.edu/REFINEpro/	[95]
Miscellaneous visualization tools	ESPript3	http://espript.ibcp.fr/ESPript/ESPript/ *Note*: generates publication quality images from multiple sequence alignment data	[17]
	My domains	http://prosite.expasy.org/mydomains/ *Note*: generates cartoons of domain architecture	[22]

(continued)

Function	Program or database	URL	Primary references
	Illustrator for Biological Sequences (IBS)	http://ibs.biocuckoo.org/online.php *Note*: generates cartoons of domain architecture	[23]
	BoxShade	http://www.ch.embnet.org/software/BOX_form.html *Note*: generates publication quality images from multiple sequence alignment data	n/a
	Jalview	http://www.jalview.org/	[96]
	Various alignment editors	https://en.wikipedia.org/wiki/List_of_alignment_visualization_software *Note*: comprehensive list with capabilities outlined for each program	n/a
	Pro-Origami	http://munk.cis.unimelb.edu.au/pro-origami/ *Note*: generates 2D representations of protein domain architecture from a given PDB	[97]
	WebLogo3	http://weblogo.threeplusone.com/ *Note*: generates publication quality sequence logos from multiple sequence alignment data	[98]
	DISOPRED3	http://bioinf.cs.ucl.ac.uk/psipred/?disopred=1	[35]
Prediction of intrinsically disordered proteins or protein regions	MFDp	http://biomine.cs.vcu.edu/servers/MFDp2/	[36]
	PrDOS	http://prdos.hgc.jp/cgi-bin/top.cgi	[37]
	DisEMBL	http://dis.embl.de/	[99]
	MetaDisorder	http://iimcb.genesilico.pl/metadisorder/	[100]
	ESpritz	http://protein.bio.unipd.it/espritz/	[101]

(continued)

Function	Program or database	URL	Primary references
Protein-protein interaction prediction tools	IntAct	https://www.ebi.ac.uk/intact/	[53]
	String	https://string-db.org/	[54]
	BioGRID	https://thebiogrid.org/	[55]
	GeneMANIA	https://genemania.org/	[102]
Docking software and webservers	AutoDock Vina and AutoDock Tools	http://vina.scripps.edu/ http://mgltools.scripps.edu/	[56]
	LeDock	http://www.lephar.com/software.htm	[57]
	ClusPro	https://cluspro.bu.edu/login.php	[58]
	SwissDock	http://www.swissdock.ch/	[103]
	Haddock	https://wenmr.science.uu.nl/haddock2.4/	[104]
	PatchDock	https://bioinfo3d.cs.tau.ac.il/PatchDock/	[105]
Downloadable modeling and visualization software	Chimera	http://www.cgl.ucsf.edu/chimera/	[1]
	PyMOL	http://www.pymol.org	[2]
	MODELLER	https://salilab.org/modeller/	[3]
	VMD	https://www.ks.uiuc.edu/Research/vmd/	[106]

2. Detailed instruction on installation is available at https://www.cgl.ucsf.edu/chimera/download.html.
3. For instructions, refer to https://pymolwiki.org/.
4. To obtain a MODELLER license key to install and run MODELLER, fill out the following license agreement at https://salilab.org/modeller/registration.html.
5. FASTA is the most popular format for protein sequence data and is widely used in most sequence-based proteomic software tools. It is characterized by a comment line which is a single line description, followed by lines of sequence data. The description line is distinguished from the sequence data by a greater than (">") symbol in the first column. It is recommended that all lines of text be shorter than 80 characters in length. The sequence data is single letter codes in uppercase with no white spaces between them. An example sequence in FASTA format is:

>NP_005564.1 lamin-B1 isoform 1 [Homo sapiens]
MATATPVPPRMGSRAGGPTTPLSPTRLSRLQEKEEL
RELNDRLAVYIDKVRSLETENSALQLQVTEREEVR
GRELTGLKALYETELADARRALDDTARERAKL
QIELGKCKAEHDQLLLNYAKKESDLNGAQIKL
REYEAALNSKDAALATALGDKKSLEGDLEDLKDQIAQ
LEASLAAAKKQLADETLLKVDLENRCQSLTEDLE
FRKSMYEEEINETRRKHETRLVEVDSGRQIEYEYKLA
QALHEMREQHDAQVRLYKEELEQTYHAKLENARLSSE
MNTSTVNSAREELMESRMRIESLSSQLSNLQKESRA
CLERIQELEDLLAKEKDNSRRMLTDKERE
MAEIRDQMQQQLNDYEQLLDVKLALDMEISAYRKLLE
GEEERLKLSPSPSSRVTVSRASSSRSVRTTRGKRKRVD
VEESEASSSVSISHSASATGNVCIEEIDVDGKFIRLKNT
SEQDQPMGGWEMIRKIGDTSVSYKYTSRYVLKAGQTV
TIWAANAGVTASPPTDLIWKNQNSWGTGEDVK
VILKNSQGEEVAQRSTVFKTTIPEEEEEEEEAAGVV
VEEELFHQQGTPRASNRSCAIM

6. While the databases do cross-reference each other, using the accession number from one database may not return a result when searched in another database. This is especially true when searching within curated databases such as UniProtKB, which may weed out redundant submissions in its curation process.
7. The .fasta file format may contain either nucleotide or amino acid codes, whereas .faa is specific to amino acids. Both files are simple text files that can be opened with a basic text editor. If your operating system does not open the file directly, you may open it from within a text editor or install a text editor such as Notepad++ that provides an "Edit with . . ." right-click option.
8. The search engine recognizes the Boolean operators "AND," "OR," and "NOT." These operators are processed in a left-to-right sequence. Terms inside of parentheses are processed first and then integrated into the overall term. You may also place search terms within quotes to restrict matches to records including the exact text but be aware that it may interfere with phrase indices that allow the search engine to recognize queries such as protein kinase c or insulin dependent as complete phrases and other automatic term mapping performed by the search engine.
9. BLAST, a heuristic sequence alignment and search algorithm, is used to query a database and uncover potentially related sequences. Alignments are scored and used to rank the similarity of the "hits" with respect to the query. Scoring is based on the alignment length, gaps/inserts, residue similarities, and statistical significance with respect to the database size. Residue similarities, at each position of the query-hit alignment, are calculated based on biochemical natures and adjusted

according to evolutionary parsimony. BLAST quantifies this biochemical similarity by utilizing a substitution matrix (e.g., BLOSUM or PAM) that assigns a substitution value to each pair of amino acids. Penalties are minor for conservative substitutions, becoming increasingly severe for more radical substitutions. Since this scoring is based on pairwise comparison of residue identity regardless of where in the sequence it occurs, it is referred to as a position-independent scoring parameter. The user can customize the BLAST scoring, but the standard parameters are frequently sufficient. Many variations on the BLAST search method and its scoring have been developed and can, in some cases, uncover additional hits.

10. As an alternative option, you can paste the accession number of your protein sequence or limit the search to a subregion of your provided sequence.

11. The downside of the nonredundant protein sequences (nr) database being the most inclusive is that it includes all sequences without regard to the quality of the submission, potentially overwhelming the search with records of questionable value. While curated options such as the reference proteins database (RefSeq) [5] and the UniProtKB/SwissProt database (UniProt Knowledge Base) can have very useful annotated records, it pays for that depth with breadth. It is important to be aware of these trade-offs in the context of one's project. Of particular note, is the Protein Data Bank (PDB) database, which is a collection of proteins with solved structures, crucial for identifying templates for homology modeling.

12. Customized versions of the BLAST algorithm, such as PSI-BLAST (*see* above), PHI-BLAST, and DELTA-BLAST, can be selected at this point.

13. By clicking on the RID (Reference ID) hyperlink, a new, identical tab will open, but the HTML will now include the RID for your search. You can return to your search results using this HTML for up to 48 h, after which time the RID will expire.

14. Query Coverage: The percentage of the query sequence that was matched to the database sequence. E-value: The E-value gives a statistical measure of the significance of the alignment and describes the number of hits that would be "expected" to occur by chance when searching a sequence database of a particular size; the lower the E-value, the more significant the match. An E-value of 1 means that it would be expected to find one match with a similar score simply by chance in a database of the same size. Exact and closely matched hits will have E-values of 0. If the query sequence is short (less than 100 nucleotides or amino acids long), the top E-values may be larger than

10–50 even if there is an exact match. Therefore, it is a good idea to check the similarities and identities (given by percent) as well as the E-value. Percent Identity: The percentage of residues in the query that are identical to the corresponding residue in the database sequence.

15. A position-specific scoring matrix (PSSM) is a scoring system applied to each position of a paired sequence alignment that corresponds to the likelihood of a substitution from one residue (of the query) to another (of the hit). In contrast to position-independent matrices such as BLOSUM that assign the same substitution penalty to any given pair of amino acids, regardless of where the substitution occurs, PSSMs can account for the importance of the *location* of the substitution. Motifs or domains in a protein may be highly insensitive to substitutions and even gaps/insertions in a certain region without compromising its function unless they occur in a critical location. Locating such homologs could be quite difficult if guided exclusively by global sequence similarity, so attuning a scoring matrix to these characteristics creates a signature, or profile, that can be used to search a database for distant homologs. Of course, the accuracy of such algorithms is contingent on the quality of the data used to develop the signature, but when based on a broad set of true homologs, these algorithms are effective at finding the proverbial needle in a haystack.
16. PHMMER can check query sequences against a variety of sequence databases. The default database is Reference Proteomes, a representative cross-section of proteomes derived from the UniProtKB database. More information on the database options may be found at https://hmmer-web-docs.readthedocs.io/en/latest/databases.html.
17. If your query sequence has many results, the search may fail to load any results. In such cases, it is recommended that you use "Restrict by Taxonomy" to break your search down into smaller components.
18. Make sure that your sequences have different names as the first 30 characters of the name are significant, and if they are not unique, the program will fail. Also, remove any white space or empty lines from the beginning of your input.
19. "*" denotes identity, ":" denotes a conserved residue, and "." denotes a semi-conserved residue in the MSA.
20. For the automatic links to work, you need to install Jalview Desktop on your computer. Free installation programs are available from the Jalview site.
21. Supported formats of alignment files include Clustal, FASTA, MultAlin, NPS@, or ProDom formats.

22. For outputs with highest quality, you may pick the Tiff Image (600 dpi) option.
23. To rearrange the order of MSA, you may enter your own order in the box called "Defining groups." After entering your new order, simply press the submit button again at the top left corner.
24. The results page is a pop-up window and all pop-up blockers in your browser must be disabled.
25. Alternatively, you may click on the "Search by text" tab to search by families, domains, proteins, keywords, or GO terms, e.g., IPR020422, kinase, O00167, PF02932, GO:0007165, 1t2v, and UP000005640.
26. Your InterProScan search results will be listed on the processing page after the job is done. Each search may run for different times. You can navigate to other pages if you prefer. Once the job is finished, you will be notified, and the results will be available for 7 days.
27. The default scan excludes patterns with a high probability of occurrence to reduce the number of false-positive hits. You can uncheck this option if you are interested in doing a more complete but less stringent scan of your sequence. The "ScanProsite" advanced form allows you to further tweak the scan.
28. A set of automatically generated, non-HMM-based entries called Pfam-B is maintained at the Pfam FTP site [Pfam-B.tgz] and may be accessed there. In the future, Pfam-B may be integrated in the main website.
29. When your browser first encounters the SMART website, you will be asked to choose between the "normal" and "genomic" mode, after which the mode setting will be saved for future visits. The models differ in the underlying protein database used, and only the proteomes of completely sequenced genomes are used in the genomic mode. For most routine purposes, "normal" mode is chosen, but the mode can be changed by accessing the "Setup" link on the top of the page at any time.
30. Instead of using three classes (helix, strand, and the rest) to assign the secondary structure of a protein, the eight-state prediction adopts the full DSSP eight-class output classification: H, alpha-helix; G, 3–10-helix; I, pi-helix (extremely rare); E, extended strand; B, beta-bridge; T, turn; S, bend; and C, the rest.
31. Sequence must be in FASTA format. Adding your e-mail is highly recommended.
32. Processing page shows the job is being performed and it will get refreshed every now and then. Closing the processing page

will not affect the results and when the results will be sent to the e-mail you provided when submitting your protein sequence.

33. 2.5% is the default false-positive rate. If you want to recover at least 60% of disordered regions, you should set false-positive threshold at 4%.
34. The template prediction is useless when you want to know just the disorder tendency of local amino acid composition. You can predict disordered regions without template prediction by this option.
35. If "Receive prediction results by e-mail" checkbox is not checked, the page will show a prediction progress report until all prediction processes are finished and finally return HTML-formatted prediction results. Although it depends on the length of query protein and server conditions, PrDOS usually takes from 5 to 10 min to predict a protein sequence. Thus, using "Receive prediction results by e-mail" option is strongly recommended.
36. These steps often need to be repeated and tweaked until a satisfactory model is obtained. The second step is crucial to making a reliable model since the alignment dictates the relationship between your protein sequence and the template sequence.
37. You may want to parse your sequence into its constituent domains and submit each domain sequence individually since structural templates will, in most cases, be identified separately for each domain. Rarely, there will be structural templates that can be used to model the entire multi-domain proteins
38. Sequence alignment formats that are allowed include A3M, CLUSTAL, FASTA, and STOCKHOLM.
39. The first line of a sequence in the PIR database format should begin with a ">P1," indicating a protein code, followed by an identifier for the sequence or structure. This sequence/alignment code will be referenced by the Python scripts used by MODELLER so this field must have a unique identifier for all proteins in the file. For the sake of simplicity and to avoid cross-platform errors, assume it is case sensitive, avoid spaces, and only use alphanumeric characters. The second line is formatted to read ten fields (each field is divided by a ":", colon, resulting in nine colons total on this line). Generally, only the first two fields are required. The first field specifies if a 3D structure is available and, if so, its type: structureX, X-ray; structureN, NMR; structureM, model; sequence, sequence with no structure; and structure, a structure of unspecified source. The second field is the name of the file associated with the sequence. The file extension .pdb is assumed, so entering "R2D2" in this

field will automatically load R2D2.pdb. For the purposes of this procedure, fields 3–10 are not required but details are available on the MODELLER website (https://salilab.org/modeller/9v8/manual/node454.html). The colons demarcating the fields must be present however for the file to be properly read, so ensure that there are nine colons total, one between each anticipated field. The sequence/alignment is then read in until the program detects the termination character "*", an asterisk.

40. *Warning*: MODELLER will overwrite files in the same directory without requesting confirmation or warning. It is advised that you keep track of filenames or move/copy files to different directories between runs.
41. Before proceeding, please make sure your input model files are in PDB format (e.g., modelin.pdb). The refined model output (e.g., modelout.pdb) can be found in the SCWRL4 program folder
42. The coordinate files should be named as one continuous word based on alphanumeric characters only. Else the file will not be recognized by the program.
43. When different multiple protein chains are not identified with separate chain IDs, you may get no score. In order to edit the chain IDs of your structures in your coordinate files, you can visit http://www.canoz.com/sdh/renamepdbchain.pl.
44. Sometimes you may need to "hide" the current display mode of your protein before changing to another one, e.g., your protein is depicted in "ribbon" format by default, so to change it to "stick," you may need to perform the following: (1) click "Actions" > "Ribbon" > "hide"—your protein will be hidden until you select a new format; (2) click "Actions" > "Atoms/Bonds" > "Show"; and (3) click "Actions" > "stick."
45. Search is based on exact word matches. To search for isoforms, add "*" at the end, e.g., use PI2345* to search for PI2345's isoforms. You may also use gene ontology (GO) to find interactions of all proteins associated with a particular process.
46. To be able to visualize your protein interaction network, you need to have Adobe® Flash® Player installed.
47. Colored nodes indicate query proteins and first shell of interactors. White nodes indicate second shell of interactors. Empty nodes indicate proteins of unknown 3D structure, while filled nodes indicate that some 3D structure is known or predicted.
48. Alternatively, you may search by publication. Select "By Publication" from the list shown above the search box. Type or paste the PubMed ID or keyword term, e.g., 29301708. Click "Submit Publication Search."

49. This step is recommended for most PDB files as they usually do not contain hydrogens.

50. If you want to generate a log file, the complete command will be <<C:\Vina Docking> "C:\Program Files (x86)\The Scripps Research Institute\Vina\vina.exe" --config conf6.txt –log log.txt>>. In this instance, the log file will be named "log.txt."

51. One such option is to use the GetBox plugin for PyMOL. Another option is to designate the coordinate box using AutoDock Tools, and converting the AutoDock Tools coordinates into LeDock values. The conversion format is accessible in the LeDock User Guide which can be accessed at http://www.lephar.com/download/LeDock.pdf. The conversions are as follows:

$$xmin = [centerx] - (0.5 \times [sizex])$$
$$xmax = [centerx] + (0.5 \times [sizex])$$
$$ymin = [centery] - (0.5 \times [sizex])$$
$$ymax = [centery] + (0.5 \times [sizey])$$
$$zmax = [centerx] + (0.5 \times [sizex])$$
$$zmax = [centerz] - (0.5 \times [sizez])$$

52. If your ligand is in PDB format, PyMOL may be used to convert the ligand. First open the ligand PDB file in PyMOL. Select "File" and then "Export Molecule," and click "Save." In the "Save Molecule As" navigation pane that opens, click the "Save as type" drop-down to specify the output file type and select "MOL2 (*.mol2)." Name your molecule and select the destination folder.

53. You can sign up for an account or use the server without an account.

References

1. Pettersen EF et al (2004) UCSF Chimera—a visualization system for exploratory research and analysis. J Comput Chem 25 (13):1605–1612
2. The PyMOL Molecular Graphics System, Version 2.4, Schrödinger, LLC
3. Webb B, Sali A (2016) Comparative protein structure modeling using modeller. Curr Protoc Bioinformatics 54:5.6.1–5.6.37
4. Tateno Y et al (2002) DNA Data Bank of Japan (DDBJ) for genome scale research in life science. Nucleic Acids Res 30(1):27–30
5. Kulikova T et al (2007) EMBL nucleotide sequence database in 2006. Nucleic Acids Res 35(Database issue):D16–D20
6. Benson DA et al (2014) GenBank. Nucleic Acids Res 41:D36
7. UniProt C (2014) Activities at the Universal Protein Resource (UniProt). Nucleic Acids Res 42(Database issue):D191–D198
8. MacDougall A et al (2020) UniRule: a unified rule resource for automatic annotation in the UniProt Knowledgebase. Bioinformatics 36 (17):4643–4648. https://doi.org/10.1093/bioinformatics/btaa485
9. Altschul SF et al (1990) Basic local alignment search tool. J Mol Biol 215(3):403–410
10. Altschul SF et al (1997) Gapped BLAST and PSI-BLAST: a new generation of protein

database search programs. Nucleic Acids Res 25(17):3389–3402

11. Finn RD, Clements J, Eddy SR (2011) HMMER web server: interactive sequence similarity searching. Nucleic Acids Res 39 (Web Server issue):W29–W37
12. Sievers F et al (2011) Fast, scalable generation of high-quality protein multiple sequence alignments using Clustal Omega. Mol Syst Biol 7:539
13. Di Tommaso P et al (2011) T-Coffee: a web server for the multiple sequence alignment of protein and RNA sequences using structural information and homology extension. Nucleic Acids Res 39(Web Server issue): W13–W17
14. Katoh K et al (2002) MAFFT: a novel method for rapid multiple sequence alignment based on fast Fourier transform. Nucleic Acids Res 30(14):3059–3066
15. Edgar RC (2004) MUSCLE: multiple sequence alignment with high accuracy and high throughput. Nucleic Acids Res 32 (5):1792–1797
16. Do CB et al (2005) ProbCons: probabilistic consistency-based multiple sequence alignment. Genome Res 15(2):330–340
17. Robert X, Gouet P (2014) Deciphering key features in protein structures with the new ENDscript server. Nucleic Acids Res 42 (W1):W320–W324
18. Wiech EM, Cheng HP, Singh SM (2015) Molecular modeling and computational analyses suggests that the Sinorhizobium meliloti periplasmic regulator protein ExoR adopts a superhelical fold and is controlled by a unique mechanism of proteolysis. Protein Sci 24 (3):319–327
19. Mitchell AL et al (2019) InterPro in 2019: improving coverage, classification and access to protein sequence annotations. Nucleic Acids Res 47(D1):D351–D360
20. de Castro E et al (2006) ScanProsite: detection of PROSITE signature matches and ProRule- associated functional and structural residues in proteins. Nucleic Acids Res 34 (Web Server issue):W362–W365
21. Jonassen I, Collins JF, Higgins DG (1995) Finding flexible patterns in unaligned protein sequences. Protein Sci 4(8):1587–1595
22. Hulo N et al (2008) The 20 years of PROSITE. Nucleic Acids Res 36(Database issue): D245–D249
23. Wenzhong L et al (2015) IBS: an illustrator for the presentation and visualization of biological sequences. Bioinformatics 31 (20):3359–3361
24. Finn RD et al (2014) Pfam: the protein families database. Nucleic Acids Res 42(Database issue):D222–D230
25. Marchler-Bauer A et al (2011) CDD: a Conserved Domain Database for the functional annotation of proteins. Nucleic Acids Res 39 (D):225–229
26. Schultz J et al (2000) SMART: a web-based tool for the study of genetically mobile domains. Nucleic Acids Res 28(1):231–234
27. Biegert A, Soding J (2008) De novo identification of highly diverged protein repeats by probabilistic consistency. Bioinformatics 24 (6):807–814
28. George RA, Heringa J (2000) The REPRO server: finding protein internal sequence repeats through the Web. Trends Biochem Sci 25(10):515–517
29. Buchan DW et al (2013) Scalable web services for the PSIPRED Protein Analysis Workbench. Nucleic Acids Res 41(Web Server issue):W349–W357
30. Wang Z et al (2011) Protein 8-class secondary structure prediction using conditional neural fields. Proteomics 11(19):3786–3792
31. Yan R, Xu D, Yang J, Walker S, Zhang Y (2013) A comparative assessment and analysis of 20 representative sequence alignment methods for protein structure prediction. Sci Report 3:2619
32. Pollastri G et al (2002) Improving the prediction of protein secondary structure in three and eight classes using recurrent neural networks and profiles. Proteins 47:228–235
33. Drozdetskiy A et al (2015) JPred4: a protein secondary structure prediction server. Nucleic Acids Res 4(W1):W389–W394
34. Romero O, Dunker K (1997) Sequence data analysis for long disordered regions prediction in the Calcineurin Family. Genome Inform Ser Workshop Genome Inform 8:110–124
35. Ward JJ et al (2004) Prediction and functional analysis of native disorder in proteins from the three kingdoms of life. J Mol Biol 337:635–645
36. Mizianty MJ et al (2013) MFDp2-accurate predictor of disorder in proteins by fusion of disorder probabilities, content and profiles. Intrinsically Disordered Proteins 1(1):e24428
37. Ishida T, Kinoshita K (2007) PrDOS:prediction of disordered protein regions from amino acid sequence. Nucleic Acids Res 35(Web Server issue):W460–W464
38. Berman HM et al (2000) The Protein Data Bank. Nucleic Acids Res 28(1):235–242

39. Moult J (2005) A decade of CASP: progress, bottlenecks and prognosis in protein structure prediction. Curr Opin Struct Biol 15 (3):285–289
40. John B, Sali A (2003) Comparative protein structure modeling by iterative alignment, model building and model assessment. Nucleic Acids Res 31(14):3982–3992
41. Fernandez-Fuentes N et al (2007) Comparative protein structure modeling by combining multiple templates and optimizing sequence-to-structure alignments. Bioinformatics 23 (19):2558–2565
42. Soding J, Biegert A, Lupas AN (2005) The HHpred interactive server for protein homology detection and structure prediction. Nucleic Acids Res 33(Web Server issue): W244–W248
43. Zhang Y (2008) I-TASSER server for protein 3D structure prediction. BMC Bioinform 9:40
44. Dong X, Yang Z (2011) Improving the physical realism and structural accuracy of protein models by a two-step atomic-level energy minimization. Biophys J 101:2525–2534
45. Krivov GG, Shapovalov MV, Dunbrack RL (2009) Improved prediction of protein side-chain conformations with SCWRL4. Proteins 77(4):778–795
46. Bhattacharya D et al (2016) 3Drefine: an interactive web server for efficient protein structure refinement. Nucleic Acids Res 44 (W1):W406–W409
47. Shuid AN, Kempster R, McGuffin LJ (2017) ReFOLD: a server for the refinement of 3D models of proteins guided by accurate quality estimates. Nucleic Acids Res 45:W422–W428
48. Bhattacharya D (2019) refineD: improved protein structure refinement using machine learning based restrained relaxation. Bioinformatics 35:3320–3328
49. Eisenberg D, Luthy R, Bowie JU (1997) VERIFY3D: assessment of protein models with three-dimensional profi les. Methods Enzymol 277:396–404
50. Olechnovič K, Venclovas Č (2017) VoroMQA: assessment of protein structure quality using interatomic contact areas. Proteins 85(6):1131–1145
51. Wiederstein M, Sippl MJ (2007) ProSA-web: interactive web service for the recognition of errors in three-dimensional structures of proteins. Nucleic Acids Res 35(Web Server issue): W407–W410
52. Uziela K et al (2017) ProQ3D: Improved model quality assessments using Deep Learning. Bioinformatics 33(10):1578–1580
53. Hermjakob H et al (2004) IntAct: an open source molecular interaction database. Nucleic Acids Res 32:D452–D455
54. Jensen LJ et al (2009) STRING8—a global view on proteins and their functional interactions in 630 organisms. Nucleic Acids Res 37: D412–D416
55. Oughtred R et al (2019) The BioGRID interaction database: 2019 update. Nucleic Acids Res 47(D1):D529–D541
56. Morris GM et al (2009) AutoDock4 and AutoDockTools4: automated docking with selective receptor flexiblity. J Comput Chem 16:2785–2791
57. Zhang N et al (2006) Enriching screening libraries with bioactive fragment space. Bioorg Med Chem 26(15):3594–3597
58. Vajda S et al (2017) New additions to the ClusPro server motivated by CAPRI. Proteins 85(3):435–444
59. Laskowski RA et al (2018) PDBsum: structural summaries of PDB entries. Protein 27:129–134
60. O'Leary NA et al (2016) Reference sequence (RefSeq) database at NCBI: current status, taxonomic expansion, and functional annotation. Nucleic Acids Res 44(D1):D733–D745
61. Goodsell DS et al (2020) RCSB Protein Data Bank: enabling biomedical research and drug discovery. Protein Sci 29:52–65
62. Lane L et al (2012) neXtProt: a knowledge platform for human proteins. Nucleic Acids Res 40(Database issue):D76–D83
63. Barker WC et al (2001) Protein Information Resource: a community resource for expert annotation of protein data. Nucleic Acids Res 29(1):29–32
64. Remmert M et al (2012) HHblits: lightning-fast iterative protein sequence searching by HMM-HMM alignment. Nat Methods 9 (2):173–175
65. Madeira F et al (2019) The EMBL-EBI search and sequence analysis tools APIs in 2019. *Nucleic Acids Res* 47(W1):W636–W641
66. Bawono P, Heringa J (2014) PRALINE: a versatile multiple sequence alignment toolkit. Methods Mol Biol 1079:245–262
67. Sadreyev RI et al (2009) COMPASS server for homology detection: improved statistical accuracy, speed and functionality. Nucleic Acids Res 37(Web Server issue):W90–W94
68. Pei J, Grishin NV (2014) PROMALS3D: multiple protein sequence alignment enhanced with evolutionary and three-dimensional structural information. Methods Mol Biol 1079:263–271

69. Chikkagoudar S, Roshan U, Livesay D (2007) eProbalign: generation and manipulation of multiple sequence alignments using partition function posterior probabilities. Nucleic Acids Res 35(Web Server issue):W675–W677
70. Klausen MS et al (2019) NetSurfP-2.0: improved prediction of protein structural features by integrated deep learning. Proteins 87:520–527
71. Yachdav G et al (2014) PredictProtein—an open resource for online prediction of protein structural and functional features. Nucleic Acids Res 42(Web Server issue):W337–W343
72. Pollastri G, McLysaght A (2005) Porter: a new, accurate server for protein secondary structure prediction. Bioinformatics 21 (8):1719–1720
73. Geourjon C, Deleage G (1995) SOPMA: significant improvements in protein secondary structure prediction by consensus prediction from multiple alignments. Comput Appl Biosci 11(6):681–684
74. Lin K et al (2005) A simple and fast secondary structure prediction method using hidden neural networks. *Bioinformatics* 21:152–159
75. Adamczak A, Porollo A, Meller J (2004) Accurate prediction of solvent accessibility using neural networks based regression. Proteins 56:753–767
76. Yang J et al (2020) Improved protein structure prediction using predicted interresidue orientations. Proc Natl Acad Sci 117 (3):1496–1503
77. Xu D, Zhang Y (2013) Toward optimal fragment generations for *ab initio* protein structure assembly. Proteins 81:229–239
78. Ma J et al (2013) Protein threading using context-specific alignment potential. Bioinformatics 29(13):i257–i265
79. Wu S, Zhang Y (2007) LOMETS: a local metathreading-server for protein structure prediction. Nucleic Acids Res 35 (10):3375–3382
80. Bennett-Lovsey RM et al (2008) Exploring the extremes of sequence/structure space with ensemble fold recognition in the program Phyre. Proteins 70(3):611–625
81. Lobley A, Sadowski MI, Jones DT (2009) pGenTHREADER and pDomTHREADER: new methods for improved protein fold recognition and superfamily discrimination. Bioinformatics 25(14):1761–1767
82. Waterhouse A et al (2018) SWISS-MODEL: homology modelling of protein structures and complexes. Nucleic Acids Res 46(W1): W296–W303
83. Yang Y et al (2011) Improving protein fold recognition and template-based modeling by employing probabilistic-based matching between predicted one-dimensional structural properties of query and corresponding native properties of templates. Bioinformatics 27 (15):2076–2082
84. Combet C et al (2002) Geno3D: automatic comparative molecular modelling of protein. Bioinformatics 18:213–214
85. McGuffin LJ et al (2019) IntFOLD: an integrated web resource for high performance protein structure and function prediction. Nucleic Acids Res 47:W408–W413
86. Bates PA et al (2001) Enhancement of protein modelling by human intervention in applying the automatic programs 3D-JIGSAW and 3D-PSSM. Proteins (Suppl 5):39–46
87. Wallner B et al (2003) Automatic consensus based fold recognition using Pcons, ProQ and Pmodeller. Proteins (Suppl 6):534–541
88. Wallner B, Elofsson A (2003) Can correct protein models be identified? Protein Sci 12 (5):1073–1086
89. McGuffin LJ, Buenavista MT, Roche DB (2013) The ModFOLD4 server for the quality assessment of 3D protein models. Nucleic Acids Res 41(Web Server issue):W368–W372
90. Benkert P, Kunzli M, Schwede T (2009) QMEAN server for protein model quality estimation. Nucleic Acids Res 37(Web Server issue):W510–W514
91. Zhang Y, Skolnick J (2004) Scoring function for automated assessment of protein structure template quality. Proteins 57(4):702–710
92. Bhattacharya A, Tejero R, Montelione GT (2007) Evaluating protein structures determined by structural genomics consortia. Proteins 66(4):778–795
93. Shen MY, Sali A (2006) Statistical potential for assessment and prediction of protein structures. Protein Sci 15(11):2507–2524
94. Williams CJ et al (2018) MolProbity: more and better reference data for improved all-atom structure validation. Protein Sci 27:293–315
95. Bhattacharya D, Cheng J (unpublished) REFINEpro: a conformation ensemble approach for protein structure refinement. http://sysbio.rnet.missouri.edu/REFINEpro/faq.html
96. Waterhouse AM et al (2009) Jalview Version 2—a multiple sequence alignment editor and analysis workbench. Bioinformatics 25 (9):1189–1191

97. Stivala A et al (2011) Automatic generation of protein structure cartoons with Pro-origami. Bioinformatics 27(23):3315–3316
98. Crooks GE et al (2004) WebLogo: a sequence logo generator. Genome Res 14 (6):1188–1190
99. Linding R et al (2003) Protein disorder prediction: implications for structural proteomics. Structure 11(11):1453–1459
100. Kozlowski LP, Bujnicki JM (2012) MetaDisorder: a meta-server for the prediction of intrinsic disorder in proteins. BMC Bioinformatics 13:111
101. Walsh AJM, Martin T, Di Domenico T, Tosatto SCE (2012) Espritz: accurate and fast prediction of protein disorder. Bioinformatics 28(4):503–509
102. Warde-Farley D et al (2010) The GeneMANIA prediction server: biological network integration for gene prioritization and predicting gene function. Nucleic Acids Res 38 (Web Server issue):W214–W220
103. Grosdidier A, Zoete V, Michielin O (2011) SwissDock, a protein-small molecule docking web service based on EADock DSS. Nucleic Acids Res 39(Web Server issue):W270–W277
104. Van Zundert GCP et al (2016) The HADDOCK2.2 webserver: user-friendly integrative modeling of biomolecular complexes. *J Mol Biol* 428:720–725
105. Schneidman-Duhovny D et al (2005) PatchDock and SymmDock: servers for rigid and symmetric docking. Nucleic Acids Res 33: W363–W367
106. Humphrey W et al (1996) VMD—visual molecular dynamics. J Mol Graph 14:33–38

Index

Ray H. Gavin (ed.), *Cytoskeleton: Methods and Protocols*, Methods in Molecular Biology, vol. 2364,
https://doi.org/10.1007/978-1-0716-1661-1,

E

F

G

H

I

J

K

L

M

U

V

Z

MIX
Papier aus verantwortungsvollen Quellen
Paper from responsible sources
FSC® C105338

If you have any concerns about our products,
you can contact us on
ProductSafety@springernature.com

In case Publisher is established outside the EU,
the EU authorized representative is:
Springer Nature Customer Service Center GmbH
Europaplatz 3, 69115 Heidelberg, Germany

Printed by Libri Plureos GmbH
in Hamburg, Germany